USED BOOK

URI BOOKSTORE

USED

3⁹⁵

SOIL MECHANICS

Second Edition

SOIL MECHANICS

Tien Hsing Wu

Ohio State University

Allyn and Bacon, Inc.

Boston
London
Sydney

Copyright © 1976, 1966 by Allyn and Bacon, Inc.,
470 Atlantic Avenue, Boston, Massachusetts 02210.
All rights reserved. Printed in the United States of
America. No part of the material protected by this
copyright notice may be reproduced or utilized in any
form or by any means, electronic or mechanical,
including photocopying, recording, or by any information storage and retrieval system, without written
permission from the copyright owner.

Library of Congress Cataloging in Publication Data

Wu, Tien-hsing, date
   Soil mechanics.

   Bibliography: p.
   Includes index.
   1. Soil mechanics. I. Title.
TA710.W8 1976   624'.1513   75-26633
ISBN 0-205-04863-3

Contents

Preface   ix
Notations   xi

Chapter 1   Index Properties and Classification   1

1.1 Introduction, 1 / Soil Particles, 2   *1.2 Particle Size,* 2 · *1.3 Particle-size Distribution,* 3 · *1.4 Particle Shape,* 6 / Soil Aggregates, 7   *1.5 Soil Structure,* 7 · *1.6 Weight-Volume Relationships,* 9 · *1.7 Density of Soils,* 13 · *1.8 Atterberg Limits,* 17 · *1.9 Soil-classification Systems,* 21 / Problems, 27

Chapter 2   Soil Water and Soil Solids   30

Pore Pressure, 30   *2.1 Introduction,* 30 · *2.2 Porewater Pressure in Saturated Soils,* 30 · *2.3 Capillary Action,* 31 / Water Flow through Soils, 34   *2.4 Darcy's Law,* 34 · *2.5 Flow through a Porous Medium,* 35 · *2.6 Coefficient of Permeability,* 39 / Effective-stress Principle, 40   *2.7 Effective Stress in Saturated Soils,* 40 · *2.8 Partially Saturated Soils,* 41 · *2.9 Effective Stresses In Situ,* 42 / Problems, 44

Chapter 3   Seepage Problems   48

Permeability, 48   *3.1 Permeability Tests,* 48 · *3.2 Pumping from Wells,* 50 · *3.3 Permeability of Nonhomogeneous soils,* 51 / Seepage, 52   *3.4 Equation of Continuity,* 52 · *3.5 Solutions of the Continuity Equation,* 54 · *3.6 Potential and Stream Functions,* 55 · *3.7 Flow Net,* 59 · *3.8 Anisotropic Materials,* 64 · *3.9 Nonhomogeneous Materials and Transfer Conditions,* 65 · *3.10 Free Surface Flow and Basic Parabola,* 68 · *3.11 Examples of Flow through Earth Embankments,* 72 · *3.12 Seepage Force,* 74 · *3.13 Limitations of Seepage Theory,* 77 / Problems, 79

Chapter 4 Stresses and Strains   85

4.1 Stresses at a Point, 85 · 4.2 Strains at a Point, 88 · 4.3 Invariants of Stress and Strain, 90 · 4.4 Stress-Strain Relations, 90 · 4.5 Stress-Strain Relationships of Soils, 93 · 4.6 Excess Hydrostatic Porewater Pressure, 94 · 4.7 Measurement of Stress-Strain Properties, 98 / Problems, 103

Chapter 5 Volume Change and Compressibility   107

5.1 Introduction, 107 · 5.2 Measurement of Compressibility, 107 · 5.3 Field Consolidation, 110 · 5.4 Rate of Consolidation, 114 · 5.5 Terzaghi Consolidation Theory, 116 · 5.6 Experimental Time-consolidation Curves, 120 · 5.7 Visco-elastic Models for Consolidation, 125 / Problems, 127

Chapter 6 Elastic Equilibrium   130

6.1 Introduction, 130 · 6.2 Fundamental Relationships of Elasticity, 131 · 6.3 Stress and Displacement Functions, 134 / Loads on the Boundary, 136 6.4 Stresses in a Soil Mass Due to Surface Loads, 136 · 6.5 Influence Charts, 139 · 6.6 Stress Distribution, 142 · 6.7 Deformation, 145 / Displacements on the Boundary, 150   6.8 Beam Subjected to Deflection, 150 · 6.9 Pressure against Yielding Restraints, 153 / Problems, 156

Chapter 7 Settlement Analysis   160

7.1 Introduction, 160 · 7.2 Settlement of Loads on Sand, 161 · 7.3 Settlement of Loads on Clay, 162 · 7.4 Size Effects, 167 · 7.5 Settlement Considerations in Design, 172 · 7.6 Field Observations of Foundations on Clay, 174 · 7.7 Field Observations of Foundations on Cohesionless Soils, 179 / Problems, 183

Chapter 8 Shear Strength   187

8.1 Failure Theories, 187 · 8.2 Mohr-Coulomb Theory of Failure, 187 · 8.3 Direct Shear Test, 189 · 8.4 Triaxial Test, 192 · 8.5 Porewater Pressure and Volume Change, 194 · 8.6 Effective-stress and Total-stress Mohr Envelopes, 196 / Shear-strength Properties of Some Common Soils, 197   8.7 Shear Strength of Cohesionless Soils, 197 · 8.8 Shear Strength of Saturated Clays, 199 · 8.9 Shear Strength of Compacted Unsaturated Clays, 208 / Theoretical Considerations, 211   8.10 The Three-dimensional Yield Surface, 211 · 8.11 The Void-ratio Criterion, 212 · 8.12 Cohesion and Internal Friction, 214 · 8.13 Rate of Loading, 216 / Problems, 217

Chapter 9 Plastic Equilibrium   221

9.1 Introduction, 221 / Rigorous Solutions, 221   9.2 Slip Plane, 221 · 9.3 Rankine's Theory of Earth Pressure, 223 · 9.4 Bearing Capacity, 227 / Approximate Solutions, 232   9.5 Terzaghi Bearing-capacity Theory,

232 · 9.6 Bearing Capacity of Shallow Foundations, 235 · 9.7 Bearing Capacity of Deep Foundations, 237 · 9.8 Modification of Bearing Capacity Equations, 246 · 9.9 Slope Analysis by Circular-arc Method, 247 · 9.10 Slope Analysis by Method of Slices, 254 · 9.11 Effect of Seepage and Porewater Pressure on Stability, 257 / Upper and Lower Bound Solutions, 258 9.12 Statically Admissible and Kinematically Admissible States, 259 · 9.13 Grouser-plate Problem, 260 / Velocity Solution, 263 9.14 Velocity Field, 263 · 9.15 Velocity Field beneath a Strip Foundation, 267 · 9.16 Limitations on Use of Plasticity Theory, 271 / Problems, 272

Chapter 10 Earth-Pressure Problems   276

10.1 Introduction, 276 · 10.2 Effect of Deformation of Earth Pressure, 277 / Walls with Adequate Deflections, 280   10.3 Earth Pressure on Smooth Walls, 280 · 10.4 Influence of Wall Friction, 281 · 10.5 Coulomb's Theory for Cohesionless Soils, 282 · 10.6 Critical Height, 286 · 10.7 Coulomb's Theory for Cohesive Soils, 287 · 10.8 Effect of Surcharge Loads, 289 · 10.9 Effect of Seepage and Porewater Pressure, 292 / Walls with Limited Deflections, 293 · 10.10 Deformation of Braced Excavations, 293 · 10.11 Earth Pressure on Bracing Systems, 294 · 10.12 Field Measurement of Earth Pressure, 297 / Yielding inside a Soil Mass, 299   10.13 Pressure over Yielding Base, 299 · 10.14 Pressure on Tunnels, 303 / Problems, 308

Chapter 11 Problems of Stability   311

11.1 Concepts of Design and Analysis, 311 · 11.2 Conditions of Loading, 312 · 11.3 Conditions of Stability, 314 · 11.4 Stability of Saturated Intact Clays, 315 · 11.5 Stability of Stiff-fissured Clays, 328 · 11.6 Stability of Cohesionless Soils, 328 · 11.7 Stability of Compacted Partially Saturated Clay, 330 · 11.8 Stability of Soils Intermediate between Clay and Sand, 334 · 11.9 Stability of Deep Foundations, 334 · 11.10 Design Considerations, 337 / Problems, 343

Chapter 12 Numerical Solutions   346

12.1 Introduction, 346 · 12.2 Finite-difference Equation, 347 · 12.3 Seepage through Homogeneous Material, 348 · 12.4 Seepage through Nonhomogeneous Material, 350 · 12.5 Consolidation by Vertical Drainage, 352 · 12.6 Consolidation by Radial Drainage, 355 · 12.7 Bearing-capacity Problem, 358 / Problems, 365

Chapter 13 Properties of Natural Soil Deposits   367

13.1 Introduction, 367 · 13.2 Origin of Soils, 369 / Alluvial Deposits, 369   13.3 Braided-stream Deposits, 369 · 13.4 Meander-belt Deposits, 372 / Glacial Deposits, 374   13.5 Glacial Till and Outwash, 375 · 13.6 Glacial-lake Deposits, 378 / Aeolian Deposits, 379   13.7 Wind-blown

Sand, 379 · 13.8 Loess, 380 / Rocks and Residual Soils, 381 *13.9 Residual Soils, 381 · 13.10 Decomposed Rocks, 382* / Exploration of Soil Deposits, 384 *13.11 Borings, 385 · 13.12 Sampling, 385 · 13.13 Soundings, 387 · 13.14 Geophysical Methods, 388 · 13.15 Subsoil Exploration Program, 389*

Chapter 14   Properties of the Clay Fraction   393

14.1 Introduction, 393 · 14.2 Atomic Bonds, 393 · 14.3 Clay Minerals, 395 · 14.4 Repulsive Potential, 398 · 14.5 Attractive Potential, 403 · 14.6 Flocculation and Dispersion, 404 · 14.7 Clay Structure, 405 · 14.8 Adsorbed Water, 407 · 14.9 Ion Exchange, 408 · 14.10 Compression and Swelling of Clay, 409 · 14.11 Shear Strength of Clay, 411 · 14.12 Composition and Properties of Natural Soils, 413

Bibliography   417

Index   431

Preface

The science of soil mechanics began its rapid growth with the pioneering studies of Karl Terzaghi during the early part of this century. Terzaghi developed many of the theories of soil mechanics out of the practical necessity to provide solutions to the many difficult foundation problems introduced by modern construction.

Although they are closely related, soil mechanics and foundation engineering are not synonymous. Foundation engineering as a profession is a subtle combination of soil mechanics and engineering geology with the intuitive art of judgment and innovation where experience exercises an important role. Soil mechanics, on the other hand, is a study of the behavior of a material whose most important characteristic is its particulate composition. Since this characteristic is not unique to soils, the principles and techniques of soil mechanics may find application to a variety of problems in geophysics, materials processes, and most recently, lunar exploration.

This book is about soil mechanics. The major topics are water movement through soils, elastic deformation of soil masses, and failure in soil masses (Chapters 2 through 11). Each section begins with a treatment of the fundamental principles followed by examples of application to some common engineering problems. This is accompanied by a discussion of the reliability and limitations of the solutions as established by field and laboratory observations.

Since the analysis of seepage, deformation, and failure requires an understanding of the principles of hydraulics, elasticity, and plasticity, the fundamentals of mechanics are presented without apology. However, natural soil masses depart appreciably from ideally elastic or plastic materials. Hence I have attempted to emphasize the effect of the material properties on the solutions and to show their significance. The application of theoretical and experimental findings to physical problems often involves inductive arguments and synthesis of our knowledge about the relevant topics. Examples of this approach are given in Chapters 7 and 11 in the analysis of settlement and stability.

Analytical methods encounter serious difficulties when problems include complicated boundary conditions or nonhomogeneous materials, both of which are common in soil mechanics. While the intricate variations of many natural phenomena defy exact analysis, their effects often can be estimated by means of numerical solutions as illustrated in Chapter 12.

In analyzing any problem involving a large soil mass, the nature of the variations in the material properties must be recognized. Hence soil properties on a geologic scale are outlined in Chapter 13. For an understanding of soil behavior, one must examine the microscopic properties of soil particles, and these are presented in Chapter 14.

The revised edition naturally includes some of the important developments that have taken place since the first edition was written. However, a more important motive is to improve the book's effectiveness as a textbook. To this end, Chapter 4 has been added in an attempt to take a more fundamental approach to the topic of stress-strain properties. Settlement analysis has been assigned a separate chapter. I have also tried to separate, as much as possible, the elementary topics from more advanced ones so that the book may be adapted to instruction at different levels. For an introductory course, it is possible to use only Chapters 1 and 2 and the more elementary parts of Chapters 5, 8, 9, and 10. Other sections may be added for a more advanced course.

The first edition, in its various stages, was reviewed by Professors O. B. Andersland, Z. C. Moh, and E. Misiaszek. Professor L. E. Malvern reviewed the chapters on elasticity and plasticity, and Professor J. K. Mitchell reviewed the chapter on clays. Professors K. N. Hendrickson and R. E. Olson reviewed the second edition manuscript. Their suggestions are invaluable.

In writing the second edition I have drawn on the experience gained from teaching soil mechanics at various levels during the past eight years. In these ventures I have benefited from the generous collaboration of Professors H. Gray, J. R. Hopper, and C. A. Moore. I am also grateful to many of my colleagues and students who have contributed significantly and often indirectly to my understanding of soil mechanics.

Authorship of any sort is a fantastic indulgence of the ego. It is well, no doubt, to reflect on how much one owes to others.

—J. K. GALBRAITH, *The Affluent Society*

T. H. Wu
Worthington, Ohio

Notations

A, B, C, D	constants	K_p	coefficient of passive earth pressure
A	area	L	length
A_r	area ratio	M	moment; constrained modulus; a number
B	width		
C	resultant cohesion	N	a number
C_c	compression index	N_s	stability number
C_e	swelling index	N_γ, N_c, N_q	bearing capacity numbers for shallow foundation
C_s	shape factor		
D	diameter	$N_{p\gamma}, N_{pc}$	bearing capacity numbers for deep foundation
E	modulus of elasticity; energy or work		
F	resultant force on slip surface; field intensity	P	force; resultant pressure
		P_A	resultant active earth pressure
F_n	normal force on slip surface		
F_s	factor of safety	P_p	resultant passive earth pressure
F_t	tangential (shear) force on slip surface		
		Q	concentrated load
G	shear modulus	Q_f	load at failure
G_s	specific gravity	R	radius; dimensionless radial distance
H	thickness; height		
I	group index	S	resultant shearing resistance
I_D	density index (relative density)	S_r	degree of saturation
		S_t	sensitivity
I_L	liquidity index	T	surface tension; dimensionless time; temperature
I_P	plasticity index		
K	coefficient of earth pressure	T_0	tortuosity factor
K_a	coefficient of active earth pressure	T_v	time factor
		U	dimensionless porewater pressure
K_0	coefficient of earth pressure at rest		

U_H	average degree of consolidation of a layer	n	porosity; a number
U_z	degree of consolidation at depth z	n_a	ratio, distance from bottom of wall to point of application of resultant earth pressure/total height of wall
V	volume; velocity of displacement		
V_a	volume of air	p	pressure
V_s	volume of solids	p_a	active earth pressure
V_v	volume of void	p_c	preconsolidation pressure
V_w	volume of water	p_o	overburden pressure
W	weight	p_p	passive earth pressure
W_s	weight of solids	q	rate of flow; distributed load per unit area
W_w	weight of water		
X	body force in x direction	r	radius; radial distance
Y	body force in y direction; dimensionless electric potential	s	shear strength
		t	time
		u, v, w	displacements in x, y, and z directions
Z	body force in z direction; dimensionless distance; dimensionless electric potential	u	porewater pressure
		u_i	initial porewater pressure
		v	velocity; valence
		v_a	velocity of flow in soil voids (same as seepage velocity)
		$\dot{v}_a, \dot{v}_b$	velocity along slip lines
a, b, c, d	constants	w	water content
a	a dimension; area	x	horizontal distance
b	a dimension	y, z	vertical distance
c	a dimension; unit cohesion		
c_a	adhesion between foundation and soil		
c_e	Hvorslev's true cohesion	Φ	potential function; stress function
c_v	coefficient of consolidation		
e	void ratio	Ψ	stream function; repulsive potential
f	shearing resistance between soil and foundation		
		α	angle; angle between failure plane and major principal plane
h	hydraulic head; vertical distance		
h_c	capillary rise	β	angle; angle of downstream slope
i	hydraulic gradient		
i_c	critical gradient	γ	unit wt. of soil
k	coefficient of permeability; Boltzmann constant	γ_d	dry density
		γ_s	unit wt. of solids
l	distance; length	γ_w	unit wt. of water
m	a number	γ'	submerged unit weight
m_c	volume compressibility	δ	angle of friction between soil and structure
m_e	volume expansibility		
m_v	volume compressibility in one-dimensional consolidation	ϵ	strain; electrostatic charge unit
		ζ	dimensionless distance

η	angle; coefficient of viscosity	σ	normal stress
θ	angle	$\bar{\sigma}$	effective normal stress
λ	$\Big($Lame's constant $= \dfrac{\mu E}{(1+\mu)(1-2\mu)}\Big)$; dielectric constant	σ'	principal stress difference $= (\sigma_1 - \sigma_3)$
		$\sigma_1, \sigma_2, \sigma_3$	major, intermediate, minor principal stresses
		τ	shear stress
μ	Poisson's ratio	ϕ	angle of internal friction
ρ	settlement; displacement; charge density	ϕ_e	Hvorslev's true angle of internal friction
		ψ	angle

SOIL MECHANICS

1
Index Properties and Classification

1.1 Introduction

Soils are aggregates of mineral particles that cover extensive portions of the earth's surface. In soil engineering, the forces applied to soil masses very frequently produce relative displacements between particles. Hence a study of soil mechanics requires an appreciation of the particulate nature of the material. Another characteristic of soils is that they are three-phase systems; they are composed of solid particles, water, and air. Since air is very compressible, and water may flow into or out of a soil, the relative proportions of the three components change with time and load. Hence these components often form the basis for the quantitative description of soil behavior.

The first part of this chapter is devoted to a description of the physical characteristics of the soil particles and the means that may be used for their definition and measurement.

In the second part, the three components of the soil aggregate are studied and relationships that define the proportions of the component parts are analyzed. These physical properties of the soil particle and soil aggregate constitute the *index properties*, which are used for classification and identification of soils. Many index properties can be correlated with more complex engineering properties such as permeability or compressibility.

Soil classification consists of the division of soils into groups and subgroups, each with certain distinctive properties. Soil classification makes it possible to describe a soil by its index properties and to convey general information regarding the significant engineering properties of the soil. However, what constitutes the "significant properties" depends primarily on the problem under consideration. Hence there exist a number of soil classification systems; most of them are oriented toward certain specialized problems (pavement design, for example). It follows that different classification systems often make use of different index properties.

The third part of this chapter summarizes several common soil-classification systems. The significance of the index properties with respect to important engineering properties is discussed in succeeding chapters.

SOIL PARTICLES

1.2 Particle size

Many soil descriptions and classifications are based on the size of the soil particles. This is the simplest criteria for soil description. Soils are commonly named gravel, sand, silt, and clay, on the basis of the particle size. The dividing line between these categories is arbitrary, and, as is common to arbitrary definitions, there are several systems in current use. Three of these are given in Fig. 1.1. These systems were initiated independently by various agencies that worked with soils.

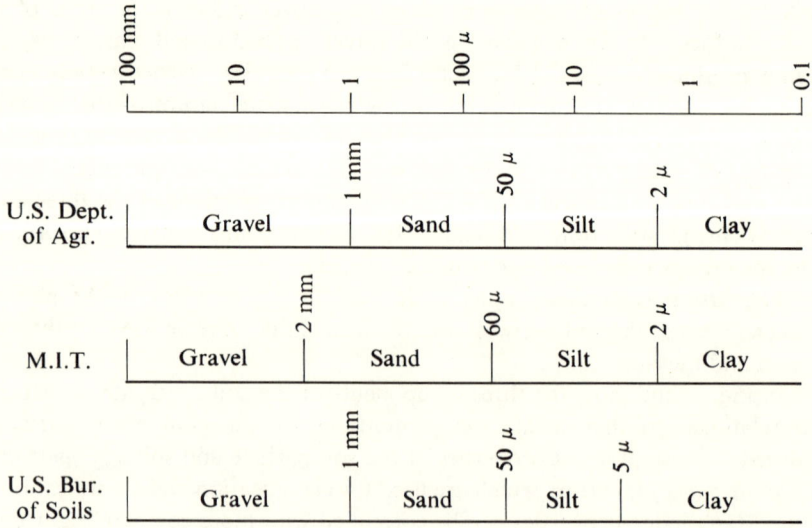

FIGURE 1.1 *Soil classification by size.*

It is obvious that natural soils most frequently consist of particles from more than one size group. In such a case, the soil is named after the principal constituent. For example, a soil that is predominantly clay but also contains some silt is called a silty clay. One convenient method of naming mixed soils is the

Public Roads Administration system, shown in Fig. 1.2. The triangular graph has three coordinate axes representing the percentages of clay, silt, and sand that constitute the soil. Special names are assigned to various combinations of the three components, as designated by the areas within the triangle. Thus if a soil is composed of 40 percent sand, 35 percent silt, and 25 percent clay, it is called a clay loam. This is shown as point A in Fig. 1.2.

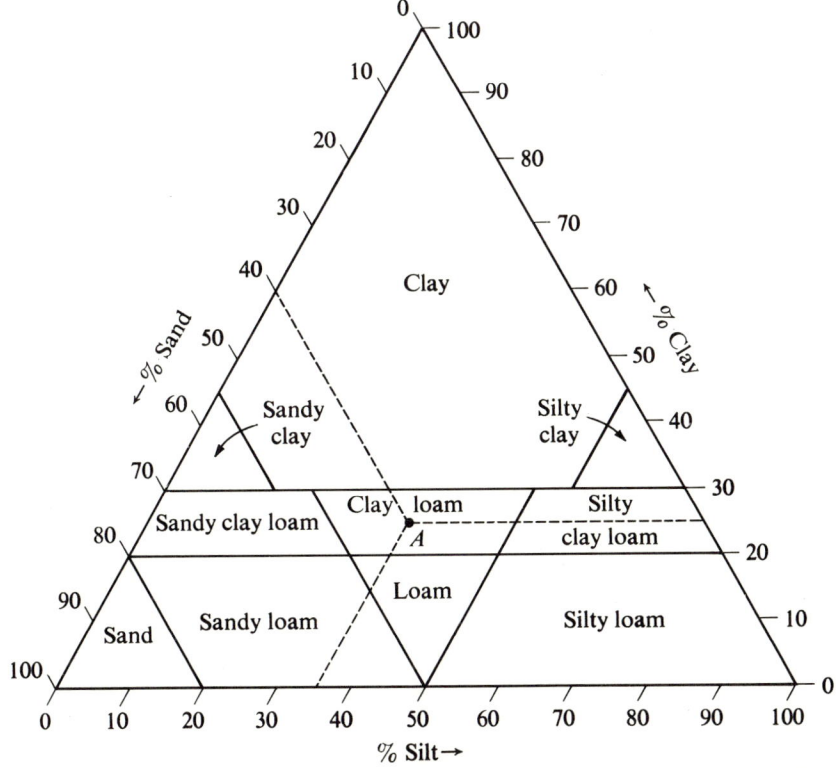

FIGURE 1:2 *Public Roads Administration system of soil classification. Size classification is same as U.S. Bureau of Soils* (Fig. 1.1).

1.3 Particle-size distribution

An adequate description of the particle-size characteristics requires the determination of the percentages of the soil that fall into the different size ranges.

SIEVE ANALYSIS. The particle-size distribution is determined by means of sieve analysis if the particles are sufficiently large. The specimen is shaken through a set of sieves with progressively smaller openings. As an example the results from a sieve analysis are given in Table 1.1. The amount retained on a particular sieve represents the fraction that is larger than the sieve size on which it is retained but smaller than that of the preceding sieve. The result may be presented in the form of a frequency diagram, as shown in Fig. 1.3(a). The vertical axis denotes the percentage of the soil that falls within a particular size-range. In engineering practice, the cumulative percentage [Fig. 1.3(b)] is more commonly used. In this graph the ordinate is the percentage of the soil that is smaller than a given size. Thus the values on the cumulative graph are obtained by summation of the values on the frequency curve. To obtain the percentage finer than a given size, say 0.149 mm (No. 100 sieve), we sum up all the percentages on the frequency curve for all size ranges below 0.149 mm, beginning with the smallest fraction (see last column in Table 1.1). In practice the cumulative graph is usually drawn as a smooth curve. Particle-size-distribution curves of five natural soils are shown in Fig. 13.6.

TABLE 1.1   *Sieve Analysis Data*

Sieve no.	Opening, mm	% retained	% finer
8	2.38	0	100
16	1.19	1	99
30	0.590	11	88
50	0.297	27	61
100	0.149	40	21
200	0.074	16	5
400	0.037	5	0

HYDROMETER ANALYSIS. The finest sieve has an opening of about 0.04 mm. Hence for particles finer than 0.04 mm, it is necessary to use the hydrometer analysis. In the hydrometer analysis, a specimen of soil is dispersed in water and made into a thin suspension. When left standing, the particles settle to the bottom at velocities which are related to their sizes by Stokes' law. The larger particles settle out first. After a given time interval t, only particles finer than a certain size still remain in suspension. According to Stokes' law, the velocity of a spherical particle falling through water (through distance y during time t) is

$$v = \frac{y}{t} = \frac{\gamma_s - \gamma_w}{18\eta} D^2$$

in which η denotes the viscosity of water, D the diameter, γ_s and γ_w the unit weights of the solid particles and water respectively, and y the distance traveled. From this we get

$$D = \sqrt{\frac{18\eta}{\gamma_s - \gamma_w}} \sqrt{\frac{y}{t}} \tag{1.1}$$

Thus above a depth y and after a time t, all particles have a diameter smaller

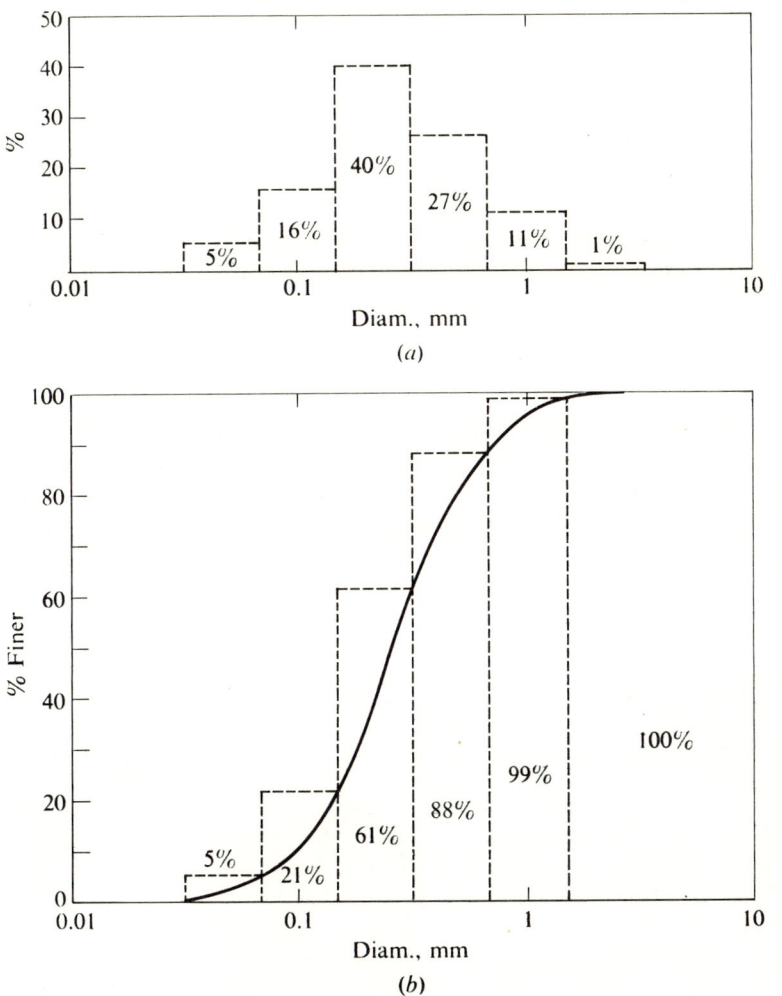

FIGURE 1.3 *Particle-size-distribution curves.*

than D as given by Eq. (1.1). All larger particles would have dropped to depths greater than y.

The amount of soil particles that remain in suspension at depth y is determined by the hydrometer which measures the unit weight of the suspension. Hydrometers are calibrated to read directly the number of grams of soil in suspension. Thus by taking a hydrometer measurement at depth y at time t the amount of soil in suspension is determined, and these particles are all smaller than the size calculated from Eq. (1.1).

From the particle-size-distribution curve, two frequently used constants can be obtained. One is the diameter at 10 percent finer, which means that 10 percent of the soil particles are finer than this size. This is known as the *effective diameter*, D_{10}. The other constant is the *uniformity coefficient*, which is the ratio of the diameter at 60 percent, or D_{60}, to D_{10}. The effective diameter is of considerable practical significance because it may be used to estimate the permeability of the soil (Chapter 2). The uniformity coefficient (as the name implies) is an index of the uniformity of particle size. For a soil in which the particles are of about the same size, D_{60} and D_{10} are close to each other and the uniformity coefficient is close to 1. For a soil with a wide range in particle size, D_{60} is much greater than D_{10}, and the uniformity coefficient is large.

1.4 Particle shape

The larger particles in a soil, namely, silt, sand, and gravel, may be rounded or angular in shape. Figure 1.4 is a micrograph of a sand with angular grains. The shape of the particles reflects the origin and geologic history of the material. Many soil particles produced by weathering and decomposition of rock are initially angular in shape. Subsequent abrasion during transportation by the agencies of air, water, or ice reduces the angularity.

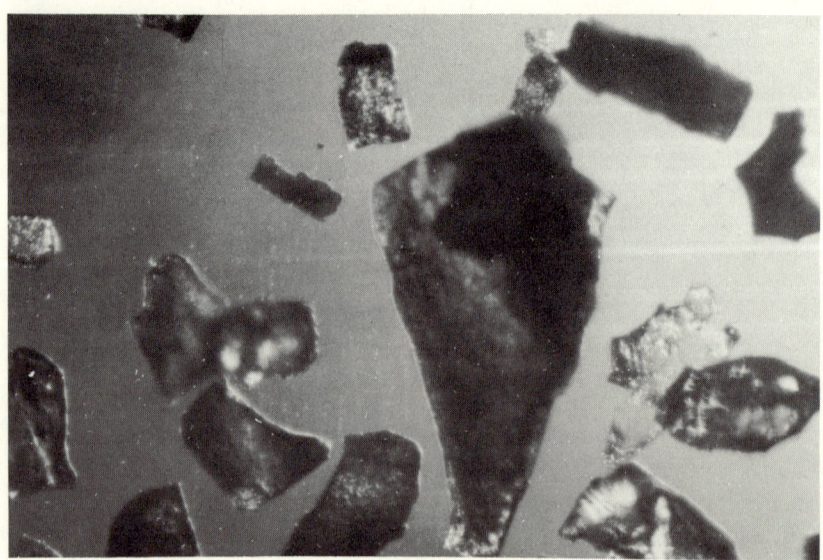

FIGURE 1.4 *Micrograph of an angular sand from Georgia.*

Two indexes are useful in the quantitative description of shape. They are *roundness* and *sphericity*. The roundness of a particle is $\sum_{i=1}^{N} (r_i/R)/N$, where r_i is the radius of a corner, R the radius of the maximum circle inscribed by the particle [Fig. 1.5(*a*)], and N the number of corners in the particle. Thus roundness measures the sharpness of the corners. Sphericity, which is an index of how closely the particle approaches a sphere, is defined as D_d/D_c, where D_d is the diameter of a circle with an area equal to that of the particle projection as it rests on its flat side and D_c is the diameter of the smallest circumscribing circle [Fig. 1.5(b)]. Typical examples are shown in Fig. 1.5(c) and (d). For a more thorough treatment of shape, see Krumbein and Pettijohn (1938).

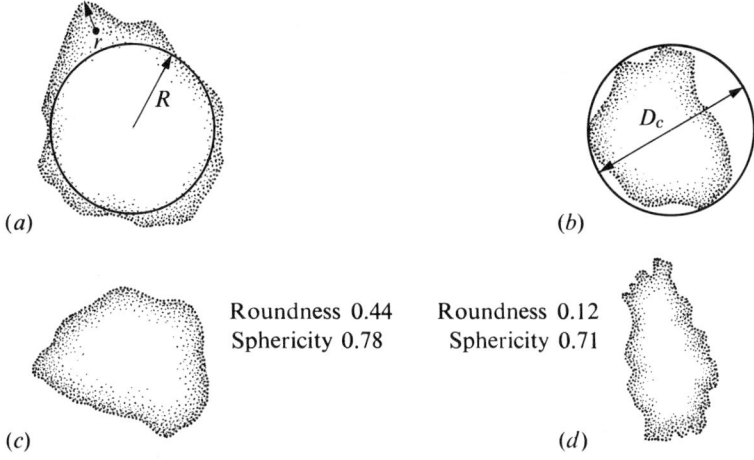

FIGURE 1.5 *Particle shapes.*

The shape of particles of the clay fraction is dependent on their chemical composition and their crystal structure. They may consist of finely ground quartz or other rock minerals. These may be rounded, subangular, or angular depending on the amount of abrasion they have suffered. These particles are often referred to as *rock flour*. There are also the clay minerals, whose crystal structure consists of sheets containing silicon, aluminum, oxygen, and hydrogen (see Sec. 14.3). The most common clay minerals are kaolinite, illite, and montmorillonite. Because of their sheet structure, particles of clay minerals are generally plate-shaped, with few exceptions. For example, the mineral halloysite has a tubular shape. The thickness of the individual clay particle may be no more than 10 Å, but the width is usually many times greater.

SOIL AGGREGATES

1.5 Soil structure

The soil structure refers to the geometric configuration of the particles in a soil aggregate and has a profound effect on the physical properties of the soil. Unfortunately, no satisfactory quantitative measure has yet been devised to describe the structure.

The structure of natural soils is the net product of the interaction between the forces of sedimentation, surface forces of the soil particles, and subsequent geologic forces. If particles of sand are allowed to settle from a suspension in water, the particles tend to take up stable positions to form a single-grained structure [Fig. 1.6(a)]. Very loose sand or silt may have a honeycomb structure as shown in Fig. 1.6(b). If the fine particles consist of clay minerals, the surface forces play an important part. If strong attractive forces exist between the edge or corner and the face of clay plates, a flocculent structure [Fig. 14.10(b)] develops. Otherwise, the clay plates may occupy nearly parallel positions as they settle from suspension. This is called a dispersed structure and is illustrated in Fig. 14.10(a).

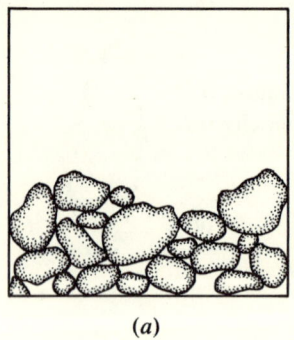
(a)

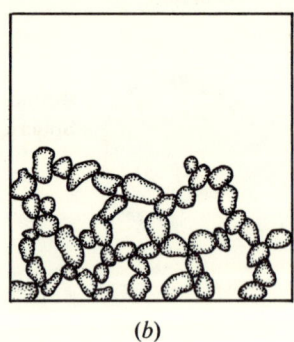
(b)

FIGURE 1.6 *Soil structures: (a) single-grained structure; (b) honeycomb structure.*

Soils with flocculent and honeycomb structures have large voids between solid particles and are held together by surface forces at the contact points. Such structures are generally not very stable. When a load is applied to the soil, the contacts may be broken and part of the structure destroyed, thus compressing the voids to form a more stable structure that can withstand the load. Some soils may be so unstable that the structure collapses with small disturbances. If the void space is filled with water, the soil-water mixture may lose all stability and flow as a viscous liquid. Examples include the spectacular flow slides of very sensitive clays in Scandinavia and the St. Lawrence Valley (Sharpe, 1938). Occasionally very loose deposits of fine sand or silt have been observed to flow after small disturbances such as a seismic tremor, an adjacent slide, or even tidal action (Peck and Kaun, 1948; Terzaghi, 1957). *Sensitivity*, which is the ratio of the strength of an undisturbed soil to that of a soil completely remolded at constant volume, reflects the loss in strength experienced by a soil when its original structure is destroyed by remolding.

1.6 Weight-volume relationships

In this section we consider the relationships that are used to describe the component parts of a soil aggregate. Soils are three-phase systems that consist of air, water, and solids. The components are illustrated schematically in Fig. 1.7. The volumes of air, water, and solids are designated by V_a, V_w, and V_s, respectively, and their weights by W_a, W_w, and W_s, respectively. In addition, the part occupied by air and water is called voids and its volume is denoted by V_v.

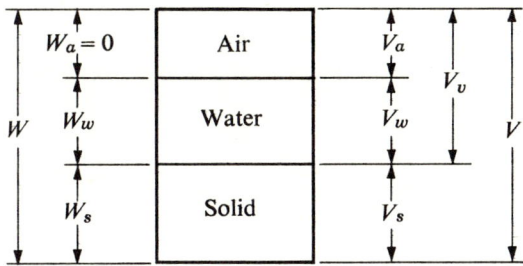

FIGURE 1.7 *Components of soil aggregate.*

The relative amount of voids in a given soil is a most useful quantity. As is demonstrated in succeeding chapters, it is closely related to many aspects of soil behavior. One measure of the relative amount of voids is the *porosity n*, which is the ratio of the voids volume to the total volume of the soil, or

$$n = \frac{V_v}{V} \tag{1.2}$$

The *void ratio e* is the ratio of the volume of voids to the volume of the solids, or

$$e = \frac{V_v}{V_s} \tag{1.3}$$

To express the quantity of water in a soil aggregate, we use the *water content* or *moisture content w*, which is the ratio of the weight of water to that of the solids expressed as a percentage, or

$$w = \frac{W_w}{W_s} \cdot 100 \tag{1.4}$$

The *degree of saturation* S_r is the percentage of voids that is occupied by water, or

$$S_r = \frac{V_w}{V_v} \cdot 100 \tag{1.5}$$

The *unit weight* or density of a material is defined as the weight of a given piece of the material divided by its volume. Since a soil aggregate contains three different phases, it is important to identify clearly the phase or phases to which the density refers. We have the density of the soil γ, which contains all three phases, or

$$\gamma = \frac{W}{V} = \frac{W_s + W_w}{V_s + V_w + V_a} \qquad (W_a = 0) \qquad (1.6)$$

When measuring the degree of compaction of a soil, use is often made of the *dry density*, which is the density of the soil with the weight of water removed while the volume remains constant. This, of course, is not easily accomplished in reality, as soils usually shrink upon drying. The dry density is therefore a fictitious quantity used as a measure of the amount of solids in a unit volume of soil aggregate. This quantity has the same significance as void ratio or porosity, in that it indicates the relative amount of solids in a given soil. The dry density γ_d is

$$\gamma_d = \frac{W_s}{V} \qquad (1.7)$$

Finally, there is the density of the solid particles γ_s, which is

$$\gamma_s = \frac{W_s}{V_s} \qquad (1.8)$$

The density of the solid particles does not vary a great deal, since it represents the density of the minerals. With few exceptions the specific gravity G_s of the solid particles ranges between 2.60 and 2.80. The average value is 2.65 for sand and silt and 2.75 for clay. Table 1.2 gives some examples of void ratios of natural soils.

TABLE 1.2 *Physical Properties of Some Natural Soils*[a]

Soil	Void ratio e	Water content w, %	Liquid limit LL, %	Plastic limit PL, %	Modulus of elasticity E, kg/cm²	Compr. index C_c	Undrained shear strength c_u, kg/cm²
Soft clay, Chicago	0.67	26	32	18	40[b]	0.20	0.20
Stiff clay, London	0.65	24	80	28	700[b]	0.30	5.00
Quick clay, Oslo	1.08	40	28	20	—	—	0.08
Dense sand, Mich.	0.55	—	—	—	600[c]	0.02	—
Very loose sand, Trondheim[d]	0.89	—	—	—	125[c]	—	—

[a] The values given are those measured at specific localities. Properties of these soils can be expected to vary considerably from one site to another.
[b] Initial tangent modulus measured in unconfined compression test.
[c] Initial tangent modulus measured in consolidated-undrained triaxial test with $\sigma_3 = 1.00$ kg/cm².
[d] Disturbed samples.

The quantities defined above can be readily calculated if the weight and volume of the various components are known. However, this is not often the

SEC. 1.6 Weight-volume relationships

case. In practical problems, certain ratios such as water content or unit weight are more easily determined than the volumes of air, water, and solids. It is therefore necessary to calculate ratios such as porosity from other ratios such as water content. This again presents no problems, as long as the unit weights of the components are known. The unit weight of water is known (62.4 pcf, or 1.0 g/cc) and the specific gravity of the solids may be readily determined. The air may be taken as weightless. The unit weight allows us to convert volume to weight and vice versa. To completely determine the quantity of the three components, proportions relating all three must be known. Hence the water content itself is insufficient, because the amount of air remains undetermined. If the degree of saturation is also known, the proportions of all three quantities are established and all other ratios may be calculated from these two.

Example 1.1. A specimen of soil has a wet weight of 167 g and a dried weight of 112 g. The volume of the specimen measured before drying is equal to 102 cc. The specific gravity is 2.65. Find the void ratio, porosity, degree of saturation, and dry density.

Solution. In this problem we are given the several quantities of weight and volume and the specific gravity. This means that all the unknown weights and volumes can be calculated directly:

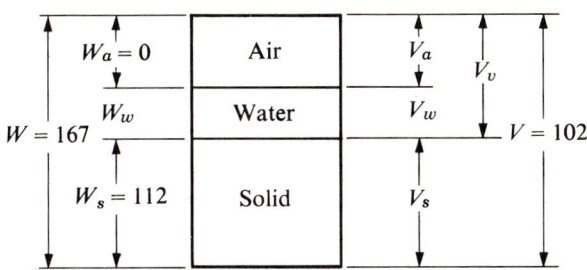

$$W_w = W - W_s = 167.0 - 112.0 = 55.0 \text{ g}$$

$$V_s = \frac{W_s}{\gamma_s} = \frac{112.0}{2.65} = 42.3 \text{ cc}$$

$$V_w = \frac{W_w}{\gamma_w} = \frac{55.0}{1.00} = 55.0 \text{ cc}$$

$$V_a = V - V_s - V_w = 102.0 - 55.0 - 42.3 = 4.7 \text{ cc}$$

Now that the weights and volumes of all the components have been determined, the required ratios are readily found:

$$e = \frac{V_v}{V_s} = \frac{V_a + V_w}{V_s} = \frac{4.7 + 55.0}{42.3} = 1.41$$

$$n = \frac{V_v}{V} = \frac{4.7 + 55.0}{102} = 0.58$$

$$S_r = \frac{V_w}{V_v} 100 = \frac{55.0}{4.7+55.0} 100 = 92\%$$

$$\gamma_d = \frac{W_s}{V} = \frac{112.0}{102.0} = 1.10 \text{ g/cc}$$

Example 1.2. A specimen of soil has a water content of 30%, a unit weight of 112.0 pcf, and a specific gravity of 2.70. Find the void ratio and degree of saturation.

Solution. We are given two ratios and the specific gravity. Since only ratios are required in the answer, the problem is not affected by the weight or volume of the specimen. We can therefore make calculations for a specimen with a certain convenient weight or volume. If we take V to be 1.00 cu. ft. then

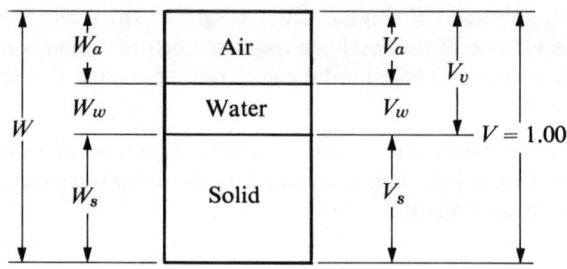

$$W = V\gamma = 112.0 \text{ lb}$$
$$= W_s + W_w = W_s + wW_s = W_s + 0.30W_s = 112.0$$

$$W_s = \frac{112.0}{1.30} = 86.2 \text{ lb}$$

$$W_w = wW_s = 0.30 \times 86.2 = 25.8 \text{ lb}$$

$$V_s = \frac{W_s}{\gamma_s} = \frac{86.2}{2.70 \times 62.5} = 0.51 \text{ ft}^3$$

$$V_w = \frac{W_w}{\gamma_w} = \frac{25.8}{62.5} = 0.41 \text{ ft}^3$$

$$V_a = V - V_s - V_w = 1.00 - 0.51 - 0.41 = 0.08 \text{ ft}^3$$

The ratios are then found to be

$$e = \frac{V_v}{V_s} = \frac{V_w + V_a}{V_s} = \frac{0.41 + 0.08}{0.51} = 0.96$$

$$S_r = \frac{V_w}{V_v} 100 = \frac{0.41}{0.41 + 0.08} 100 = 84\%$$

We note that it does not matter whether we take V as 1.00 ft³ or assume any other convenient quantity. For example, we may also assume W to be 100 lb. However, we can only choose one such quantity and all other weights and volumes must be calculated from it.

Example 1.3. Prove the relationship $\gamma_d = \gamma/(1 + w)$.

Solution. The problem is to define γ_d in terms of γ and w. The solution is carried out in the same way as the previous examples, except that we work with symbols instead of numerals. We put down the various symbols for weight and volume to get

$$\frac{\gamma}{1+w} = \frac{W_s + W_v}{V} \frac{1}{1 + (W_w/W_s)}$$

We then simplify the expression as follows:

$$\frac{W_s + W_v}{V} \frac{1}{1 + (W_w/W_s)} = \frac{W_s + W_v}{V} \frac{W_s}{W_s + W_w} = \frac{W_s}{V} = \gamma_d$$

1.7 Density of soils

When artificially compacted soils are used in the construction of embankments, for example, a very convenient way to control the quality of the construction is to specify that a certain minimum dry density must be attained in the compacted soil. This is equivalent to a limit on the porosity. To determine realistically the dry density to be required, it is necessary to know what is the maximum dry density that can be attained for a given soil under a particular method of compaction. The specification may then require a dry density equal to or greater than, say, 95 percent of the maximum dry density.

MOISTURE DENSITY CURVE. For a given compaction effort, the dry density is governed by the water content. By compacting the soil at various water contents in the laboratory, dry density–moisture curves may be obtained, as shown in Fig. 1.8(*a*). Beginning at a low water content, the moisture serves to render the soil more plastic and workable and facilitates the compaction process.* Therefore, in this range the dry density increases with the water content. The maximum dry density is obtained at a moisture content called the optimum water content, point *a*, curve 1.

The dry density then drops rapidly with further increases in moisture. At this state the soil is near saturation. Compaction is therefore ineffective, since there are little or no air voids left. Near saturation, the amount of voids increases with increasing water volume, and the dry density is reduced. To illustrate this, a "zero-air-void curve" is also drawn on Fig. 1.8(*a*). This is the moisture-density curve for a saturated soil. It can be seen that the different compaction curves all approach this curve at high water contents.

One may well expect the maximum dry density and optimum water content to vary with the type of soil. The value of the maximum dry density is dependent to a large extent on the particle size distribution. Well graded soils can be

---

* This explanation was put forward by Hogentogler (1936). A more recent theory on compaction was suggested by Olson (1962).

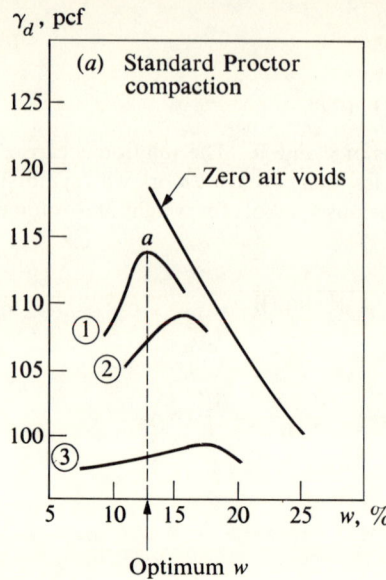
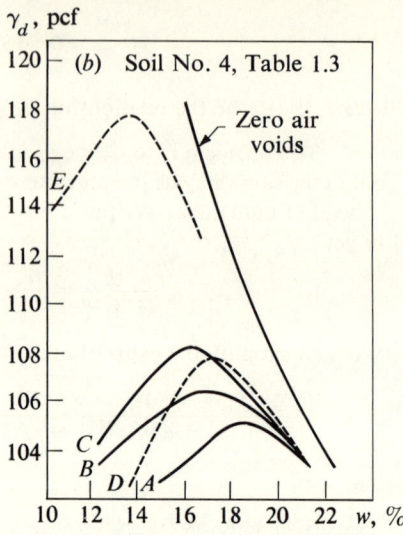
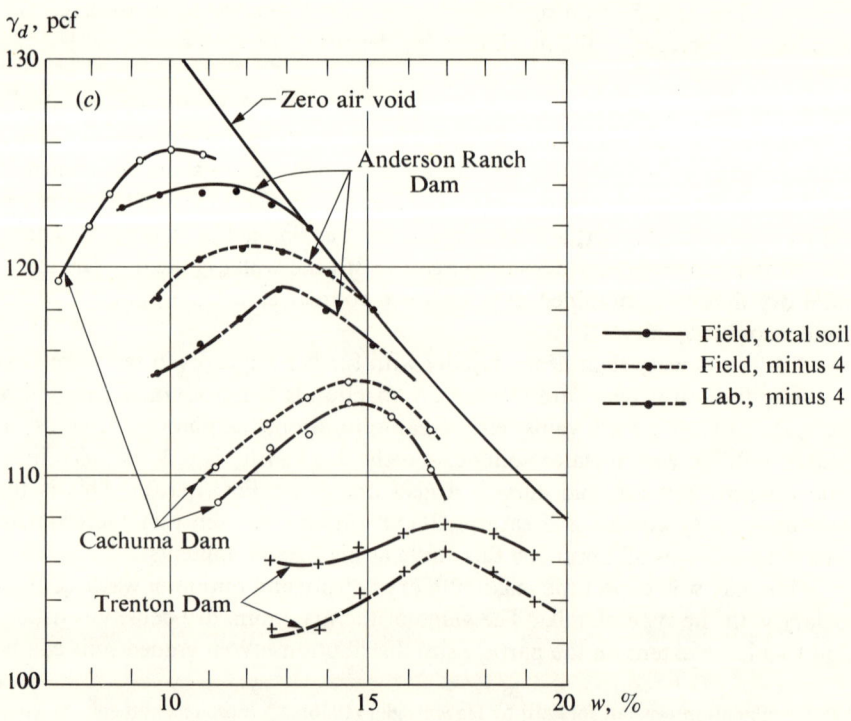

FIGURE 1.8 *Moisture-density curves.* [(*a*) and (*b*) after Johnson and Sallberg (1960); (*c*) after Hilf (1957).]

compacted to higher dry densities than soils with poor gradation. The effect of moisture is rather pronounced in clay soils and relatively unimportant in cohesionless soils, as shown in Fig. 1.8(a). The index properties of the three soils are given in Table 1.3.

TABLE 1.3 *Index Properties of Soils in Fig. 1.8(a) and (b)*

No.	Soil	Sand	Silt	Clay	LL	PL
(1)	Sandy silty clay	32	33	35	28	9
(2)	Silty clay	5	64	31	36	15
(3)	Uniform sand	96	6	0	non-plastic	
(4)	Clayey silt	5	80	15	38	20

COMPACTION EFFORT. The moisture-density relation is naturally dependent on the type of compaction. Hence the justification of a laboratory compaction test is based on its ability to simulate the results obtained with compaction equipment used in the field. Several laboratory compaction methods have been developed. In the Proctor method the soil is compacted into a mold with a volume of $\frac{1}{30}$ ft^3 in three layers by 25 blows from a 5.5-lb hammer falling through a distance of 12 in.† This method was developed by Proctor (1933). Subsequent adoption of heavier field-compaction equipment has led to revisions of laboratory compaction methods. For example, the CBR‡ compaction uses a heavier hammer and larger number of blows per layer. Laboratory compaction may also be accomplished by means of kneading. A tamper with a calibrated spring that delivers a certain static pressure is used instead of the drop hammer. Compaction consists of a certain number of tamps per layer of soil. The principal features of several common laboratory compaction methods are given in Table 1.4.

TABLE 1.4 *Laboratory Compaction Tests*

	Vol. of mold (ft^3)	Hammer wt (lb)	Hammer drop (in.)	Layers	Blows per layer
Proctor	$\frac{1}{30}$	5.5	12	3	25
CBR or modified AASHO	$\frac{1}{30}$	10	18	5	25
Harvard miniature kneading compaction	$\frac{1}{454}$	40-lb force		5	25 tamps

† The AASHO (T99-57) and ASTM (D698-58T) are slight modifications of the standard Proctor test and use the same compactive effort.
‡ California Bearing Ratio.

Figure 1.8(b) compares the moisture-density curves obtained by laboratory compaction and by field compaction with sheepsfoot rollers. Field-compaction effort can be varied by increasing the number of passes of the roller per layer or by increasing the pressure applied by the roller. We see that greater compaction effort produces a higher dry density at a lower optimum water content.

If the soil contains particles larger than the No. 4 sieve (0.187 in.), these large particles are removed and the laboratory compaction is done on the soil passing the No. 4 sieve. For comparison with laboratory compaction results, field density is also measured for the part of the soil finer than the No. 4 sieve. If the amount finer than No. 4 sieve is more than that necessary to fill the voids between the large particles, the density of the entire soil is always greater than that of the minus-4 fraction. Many field studies have been carried out by the U.S. Bureau of Reclamation. Figure 1.8(c) compares the laboratory moisture-density curves of the minus-4 material with the field curves of the minus-4 fraction and of the entire soil obtained from three earth dams. The soil properties are summarized in Table 1.5. Proctor compaction was used in the laboratory tests, and the field compaction consisted of 12 passes of the sheepsfoot roller over 6-in. layers. The rollers had contact pressures of 490 psi.

Figure 1.8(b) and (c) indicate that the compactive effort in the standard Proctor test is still reasonably close to present field-compaction effort.

Example 1.4. The calculation of dry densities may be illustrated by a study of the compaction curves for the Anderson Ranch Dam [Fig. 1.8(c)]. The dry density of the soil is 124 pcf at a water content of 11 percent. What is the dry density of the minus-4 fraction? We note from Table 1.5 that 11 percent of the soil is coarser than the No. 4 sieve and that the specific gravity is 2.64.

Solution. If we consider 1 ft³ of the soil, the weight of the solids is 124 lb. The weight and volume of the coarse particles (> No. 4) are

$$W'_s = 0.11 \times 124 = 13.6 \text{ lb}$$

$$V'_s = 13.6 \times \frac{1}{2.64 \times 62.4} = 0.083 \text{ ft}^3$$

The weight and volume of the remainder are

$$W''_s = 124.0 - 13.6 = 110.4 \text{ lb}$$

$$V''_s = 1.000 - 0.083 = 0.917 \text{ ft}^3$$

The dry density of the minus-4 fraction is

$$\frac{W''_s}{V''_s} = \frac{110.4}{0.917} = 120.5 \text{ pcf}$$

The zero-air-void curve for $w = 11$ percent may be calculated as follows. We take the weight of solids as 1.00 lb and the weight of water as 0.11 lb. For a saturated soil the total volume is the sum of the water volume and solid volume. Hence

$$V = \frac{0.11}{62.4} + \frac{1.00}{2.64 \times 62.4} = 0.00177 + 0.00606 = 0.00783 \text{ ft}^3$$

and

$$\gamma_d = \frac{1.00}{0.00783} = 127.7 \text{ pcf}$$

DENSITY INDEX. Another measure of density that is particularly useful with coarse-grained soils is the *relative density*, defined as

$$I_D = \frac{e_{\max} - e}{e_{\max} - e_{\min}} \tag{1.9}$$

in which e is the void ratio of the soil in situ. The quantity $e_{\max}$ is the void ratio of the soil in the loosest state that can be attained in the laboratory, and the quantity $e_{\min}$ is the void ratio of the soil in the densest state. It follows that $e_{\max}$ and $e_{\min}$ must be obtained by specified procedures if I_D is to remain consistent. The loosest state is usually attained by pouring the soil slowly into a standard container. The maximum density may be accomplished by the compaction procedures described above or by vibration.

1.8 Atterberg limits

It is noted in Sec. 1.4 that the particles of the clay fraction may consist of finely ground rock flour or clay minerals (see also Chapter 14). Because of their crystal structure and shape, clay minerals have strong surface forces that are predominant over the gravity forces. The surface forces attract water molecules to the clay particles. Those closest to the clay particles are tightly held to the clay and their properties are quite different from those of ordinary water. This is called *adsorbed water* and is considered to give clays their cohesive and plastic properties (see, for example, Grim, 1948). Very little water is adsorbed by rock flour, which is nonplastic. Thus we see that the amount of fines present in a soil is a poor indicator of what the soil behavior may be like. Two clay soils have very different properties if one is composed primarily of quartz and the other of a clay mineral such as illite. Furthermore, there exist several kinds of clay minerals whose properties and surface forces differ over a wide range. The Atterberg limits are designed to serve as an index of the property of the clay fraction.

Beginning at a very low water content, a clay soil is first a solid and then becomes plastic as the water content increases. The word plastic refers to the ability of the soil to be molded into various shapes without breaking up. The various states of solid, plastic, and liquid reflect the stiffness, or consistency, of the soil. The Atterberg limits are the water contents at which soil consistency changes from one state to another. They are called the shrinkage, plastic, and liquid limits, respectively (Fig. 1.9).

LIQUID AND PLASTIC LIMITS. The *liquid limit* (LL) is the water content at which the soil on two sides of a groove flows together after the dish which contains the soil has been dropped through a distance of 1 cm 25 times. The dimensions of the soil pat and groove and the dish are shown in Fig. 1.10. According to Casagrande (1932) this test is analogous to a strength test, and at the liquid limit

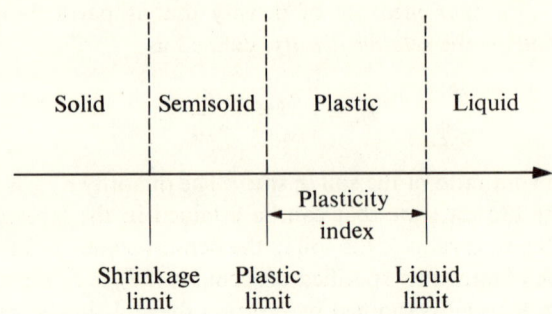

FIGURE 1.9 *Consistency limits.*

the soil has a strength of approximately 1 g/cm². The *plastic limit* (PL) is defined as the water content at which the soil crumbles when it is rolled down to a thread $\frac{1}{8}$ in. in diameter. The difference between the liquid and plastic limits is called the *plasticity index* (I_p).

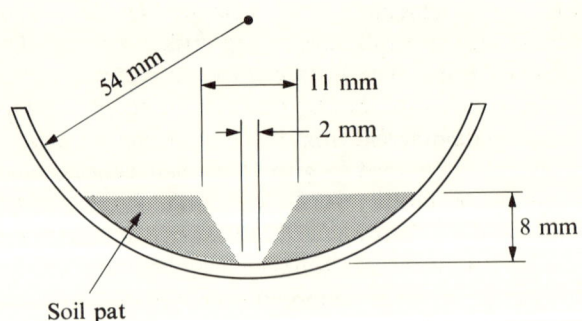

FIGURE 1.10 *Dimensions of soil pat and groove for liquid–limit test.*

Because of differences in the surface forces, soils containing different clay minerals exhibit different consistencies at the same water content. At a given water content, say 60 percent, one soil would be in the plastic state whereas another soil, containing very different clay minerals, would be in the liquid state. The two soils would have very different Atterberg limits. The Atterberg limits of several soils are given in Tables 1.2 and 1.5.

We note that these limits are arbitrary in nature. Furthermore, the consistency changes gradually from one state to another rather than going through an abrupt transition at the limits. However, like any classification system, the Atterberg limits, when used consistently, provide a reliable index to the clay properties.

The practical significance of the liquid and plastic limits lies in their ability to reflect the types and amounts of clay minerals present in the fine fraction.

TABLE 1.5 *Index Properties of Soils from Three Earth Dams*

Dam	> No. 4 sieve, %	< No. 200 sieve, %	I_p, %	LL, %	G_s
Anderson Ranch, Idaho	11	33	13.5	28.5	2.64
Cachuma, California	36	26	6.3	25.3	2.66
Trenton, Nebraska	0	91	6.0	27.0	2.64

This is illustrated by the data in Fig. 1.11. For a given soil such as London clay, the plasticity index increases in proportion to the amount of clay-sized particles present, and the relationship is a straight line that passes through the origin. Data from different soils usually plot as separate lines, as illustrated by the difference between London clay and Horten clay. This is because of the different clay minerals present in the two soils. For example, very different relationships between plasticity index and percent clay fraction are obtained for three clay minerals—kaolinite, illite, and montmorillonite—as plotted in Fig. 1.11. The slope of these lines is called the *activity ratio*. The active clays (large activity ratio) exhibit plastic properties over a wide range in water content. This may be interpreted as the result of the strong interaction between the surface forces and the water molecules. For additional data, see Grim (1962), pp. 205–18.

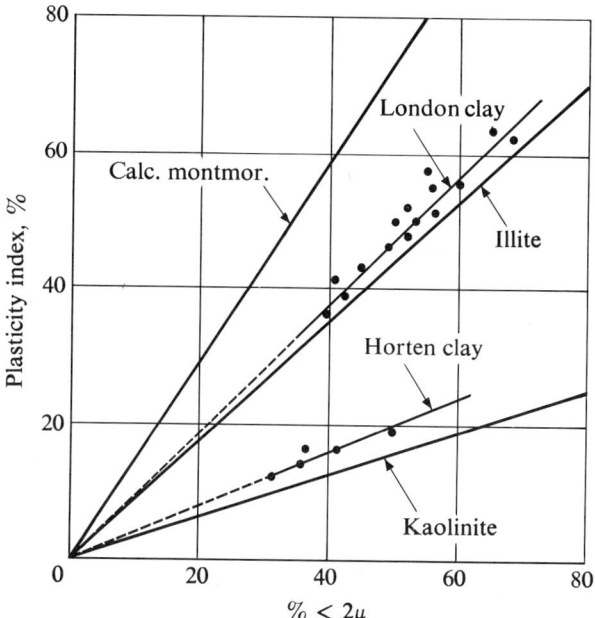

FIGURE 1.11 *Relationship between plasticity index and percent clay fraction.* [After Skempton (1953).]

Casagrande (1936) has shown that the liquid limit may be plotted against the plasticity index on the plasticity chart as shown in Fig. 1.12. Points that represent samples from the same soil stratum or from deposits with very similar mineralogic compositions fall long lines that are approximately parallel to the A-line. The Atterberg limits of inorganic clays usually plot above the A-line while those from organic clays and inorganic silts plot below the A-line. Artificial soils composed of kaolinite, illite, and montmorillonite in various portions have limits that fall within the zone enclosed by the dashed lines [Seed et al. (1964)].

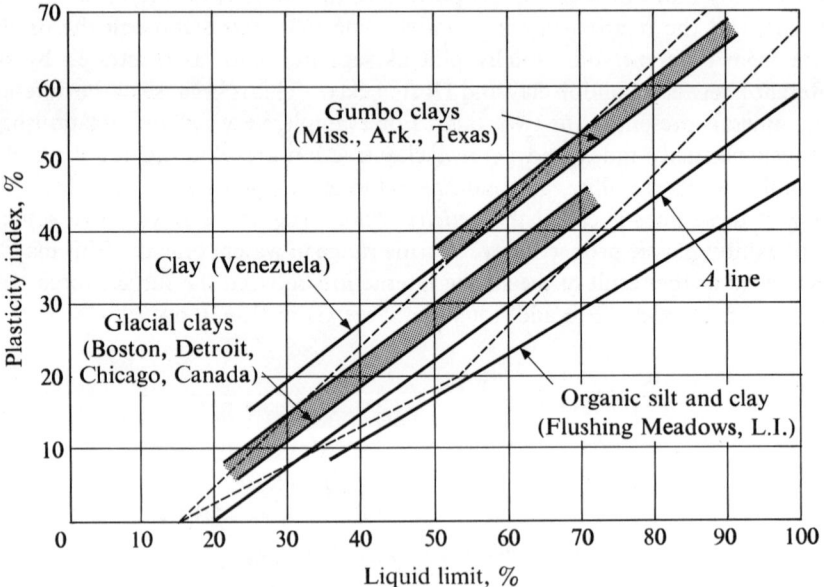

FIGURE 1.12 *Relationship between liquid limit and plasticity index.*

A useful index which reflects the properties of natural soils is the *liquidity index* (I_L), defined as

$$I_L = \frac{w - \text{PL}}{I_p} \tag{1.10}$$

in which w is the in situ water content. For remolded soils, the liquidity index is a measure of consistency. A remolded soil with a liquidity index close to 1 may be expected to have very low strength, and a liquidity index of close to 0 suggests a stiff soil. It must be noted that the liquid and plastic limits are determined only for soils in remolded condition. Hence the strength of an undisturbed soil would be greater than that of the remolded soil at the same liquidity index because of different soil particle structures. Some undisturbed soils have water contents greater than their liquid limits and yet show appreciable strength. However, if the particle structure of the soils is destroyed by some disturbance the soil flows as a viscous liquid. This is characteristic of the very sensitive clays (also called quick clays) mentioned in Sec. 1.5. Soils that have been compressed under heavy loads would have a liquidity index near 0.

SHRINKAGE LIMIT. The *shrinkage limit* (SL) is defined as the water content at which the soil no longer shrinks in volume upon further drying. As a saturated soil is slowly dried, beginning at a high water content (point A, Fig. 1.13), the volume and water content of the soil are represented successively by points B and C. At first, the volume change is equal to the volume of water lost and the soil remains saturated. However, as the water content decreases below the shrinkage limit (point B), air enters the soil. For the determination of the shrinkage limit, the shrinkage line is often replaced by $AB'C$ to simplify the procedure. Then the shrinkage limit is defined by point B'.

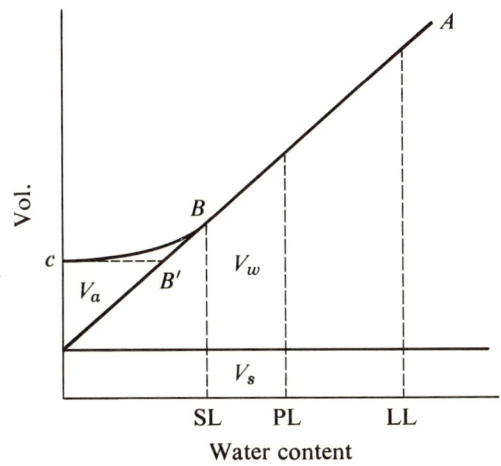

FIGURE 1.13 *Relationship between soil volume and water content during drying of soil.*

1.9 Soil-classification systems

The classification according to particle size is outlined in Secs. 1.2 and 1.3. As the clay fraction may differ widely in its physical properties, classification by size alone is inadequate where the soil contains fines, especially clay minerals. Hence more elaborate soil-classification systems have evolved, taking into account the Atterberg limits as well as the particle-size distribution.

AASHO SOIL CLASSIFICATION SYSTEM. This classification evolved from the older Bureau of Public Roads system. It is designed primarily for the use of highway engineers and classifies soils according to their suitability as highway subgrade material. Soils are classified into seven groups, from A-1 through A-7. The best subgrade material for highways is a well-graded soil consisting mostly of sand and gravel but containing a small amount of fines to serve as a binder. It is rated as A-1 and its characteristics are described in Table 1.6.

Poorly graded soils such as fine sand are difficult to compact to a high density and are less desirable (see soil A-3). Soils containing large amounts of fines are poor subgrade material. These are classified as A-4 through A-7, in decreasing order of their suitability as subgrade material. Clays with high liquid limits and plasticity indexes are subject to large fluctuations in strength during wet and dry cycles, and are very undesirable. When they are present in sufficient quantities, the soil is rated as A-6 or A-7.

This classification system also makes use of a group index, which is defined as

$$I = 0.2a + 0.005ac + 0.01bd \tag{1.11}$$

in which

$a =$ the portion of soil passing No. 200 sieve greater than 35 percent and not exceeding 75 percent, expressed as a number (from 0 to 40)
$b =$ the portion of soil passing No. 200 sieve greater than 15 percent and not exceeding 55 percent, expressed as a number (from 0 to 40)
$c =$ the portion of liquid limit greater than 40 and not exceeding 60, expressed as a number (from 0 to 20)
$d =$ the portion of the plasticity index greater than 10 and not exceeding 30, expressed as a number (from 0 to 20)

UNIFIED CLASSIFICATION SYSTEM. A more comprehensive classification system is the *Unified classification system*. The soils are organized into the following size groups: gravels (G), sands (S), inorganic silts and fine sands (M), inorganic clays (C), and organic silts and clays (O). Each group is then divided into subgroups according to their significant index properties. The gravels and sands with little or no fine materials are subdivided according to their size-distribution properties into well graded (GW and SW) or poorly graded (GP and SP). If the soil contains more than 12 percent fines, their properties must be taken into account. Since the fine fraction in the soils may have substantial influence on soil behavior, the gravels and sands have two other subdivisions. Those with a fine fraction that serves as good binder material (mostly silts) are classed as GM or SM. If the fines contain plastic clays, the soils are classed as GC or SC.

For the fines the important index properties are the Atterberg limits, which are used to subdivide the clays and silts. The liquid limit and plasticity index are plotted on the plasticity chart (Fig. 1.14). The A-line on the plasticity chart is the boundary between the inorganic clays (CL and CH) which are above this line and the inorganic silts and organic clays (ML, MH, OL, and OH) which are below. The clays and silts are further divided into those of high and low compressibility according to the liquid limit. This is based on the empirical observa-

TABLE 1.6 *Classification of Highway Subgrade Materials (with suggested subgroups)*

General classification:	Granular materials (35% or less passing No. 200)							Silt-clay materials (more than 35% passing No. 200)			
	A-1		A-3	A-2				A-4	A-5	A-6	A-7
Group classification:	A-1-a	A-1-b		A-2-4	A-2-5	A-2-6	A-2-7				A-7-5, A-7-6
Sieve analysis, percent passing:											
No. 10	50 max.										
No. 40	30 max.	50 max.	51 min.								
No. 200	15 max.	25 max.	10 max.	35 max.	35 max.	35 max.	35 max.	36 min.	36 min.	36 min.	36 min.
Characteristics of fraction passing No. 40:											
Liquid limit				40 max.	41 min.	40 max.	41 min.	40 max.	41 min.	40 max.	41 min.
Plasticity index	6 max.		N.P.	10 max.	10 max.	11 min.	11 min.	10 max.	10 max.	11 min.	11 min.
Group index	0		0	0		4 max.		8 max.	12 max.	16 max.	20 max.
Usual types of significant constituent materials	Stone fragments, gravel and sand		Fine sand	Silty or clayey gravel and sand				Silty soils		Clayey soils	
General rating as subgrade	Excellent to good							Fair to poor			

TABLE 1.7 Unified Soil Classification System (for Airfield and Highway Subgrade Classification)[a]

Major divisions			Group symbols	Typical names	Field-identification procedures	Laboratory classification criteria
Coarse-grained soils (more than half of material is larger than No. 200 sieve size)	Gravels (more than half of coarse fraction is larger than No. 4 sieve size)	Clean gravels (little or no fines)	(1) GW	Well-graded gravels, gravel-sand mixtures, little or no fines	Wide range in grain sizes and substantial amounts of all intermediate particle sizes	Use grain-size curve in identifying the fractions as given under field identification. Determine percentages of gravel and sand from grain-size curve. Depending on percentage of fines (fraction smaller than No. 200 sieve size), coarse-grained soils are classified as follows: Less than 5%: GW, GP, SW, SP. More than 12%: GM, GC, SM, SC. 5–12%: Borderline cases requiring use of dual symbols. (1) $\frac{D_{60}}{D_{10}}$ (greater than 4); $\frac{(D_{30})^2}{D_{10} \times D_{60}}$ (between 1 and 3)
			(2) GP	Poorly graded gravels, gravel-sand mixtures, little or no fines	Predominantly one size or a range of sizes with some intermediate sizes missing	(2) Not meeting all gradation requirements for GW
		Gravels with fines (appreciable amount of fines)	(3) GM	Silty gravels, gravel-sand-silt mixtures	Nonplastic fines or fines with low plasticity	(3) Atterberg limits below A line or I_p less than 4; (4) Atterberg limit above A line with I_p greater than 7 — Above A line with I_p between 4 and 7 are borderline cases requiring use of dual symbols
			(4) GC	Clayey gravels, gravel-sand-clay mixtures	Plastic fines	
	Sands (more than half of coarse fraction is smaller than No. 4 sieve size)	Clean sands (little or no fines)	(5) SW	Well-graded sands, gravelly sands, little or no fines	Wide range in grain size and substantial amounts of all intermediate particle sizes	(5) $\frac{D_{60}}{D_{10}}$ (greater than 6); $\frac{(D_{30})^2}{D_{10} \times D_{60}}$ (between 1 and 3)
			(6) SP	Poorly graded sands, gravelly sands, little or no fines	Predominantly one size or a range of sizes with some intermediate sizes missing	(6) Not meeting all gradation requirements for SW
		Sands with fines (appreciable amount of fines)	(7) SM	Silty sands, sand-silt mixtures	Nonplastic fines or fines with low plasticity	(7) Atterberg limits below A line or I_p less than 4; (8) Atterberg limit above A line with I_p greater than 7 — Limits plotting in hatched zone (Fig. 1.13) with I_p between 4 and 7 are borderline cases requiring use of dual symbols
			(8) SC	Clayey sands, sand-clay mixtures	Plastic fines	

[a] After Waterways Experiment Station (1953).

SEC. 1.9   Soil-classification system   25

TABLE 1.7 cont.

Major divisions		Group symbols	Typical names	Identification procedures on fraction smaller than No. 40 sieve size		
				Dry strength (crushing characteristics)	Dilatancy (reaction to shaking)	Toughness (consistency near PL)
Fine-grained soils (more than half of material is smaller than No. 200 sieve size)	Silts and clays (liquid limit less than 50)	(9) ML	Inorganic silts and very fine sands, rock flour, silty or clayey fine sands, or clayey silts with slight plasticity	None to slight	Quick to slow	None
		(10) CL	Inorganic clays of low to medium plasticity, gravelly clays, sandy clays, silty clays, lean clays	Medium to high	None to very slow	Medium
		(11) OL	Organic silts and organic silty clays of low plasticity	Slight to medium	Slow	Slight
	Silts and clays (liquid limit greater than 50)	(12) MH	Inorganic silts, micaceous or diatomaceous fine sandy or silty soils, elastic silts	Slight to medium	Slow to none	Slight to medium
		(13) CH	Inorganic clays of high plasticity, fat clays	High to very high	None	High
		(14) OH	Organic clays of medium to high plasticity, organic silts	Medium to high	None to very slow	Slight to medium
Highly organic soils		(15) Pt	Peat and other highly organic soils	Readily identified by color, odor, spongy feel, and frequently by fibrous texture		

tion that the compressibility of a soil increases with the liquid limit (see Sec. 5.2). A soil with a liquid limit in excess of 50 percent is classified as a silt or clay of high compressibility (MH, CH). If the liquid limit is less than 50 percent, it is classified as a silt or clay of low compressibility (ML, CL).

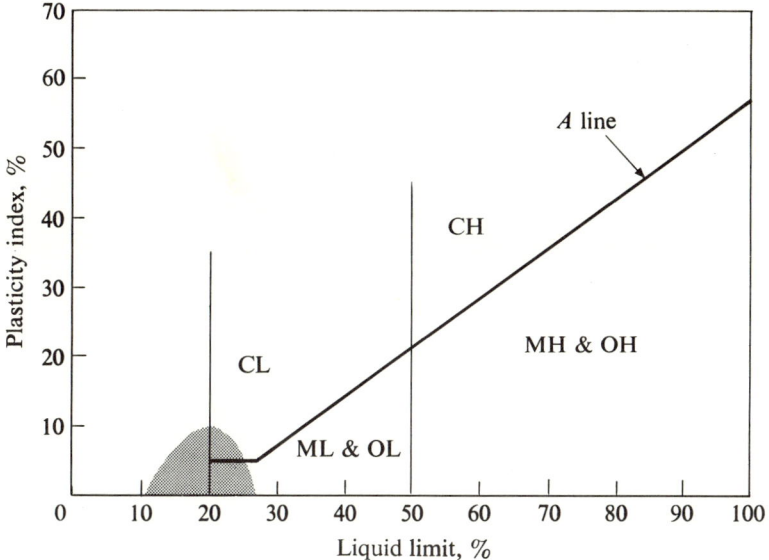

FIGURE 1.14 *Plasticity chart.* [After Casagrande (1948).]

The organic clays may be distinguished from the inorganic silts by their characteristic odor and black color. Also oven-drying greatly alters the Atterberg limits of organic soils whereas its effect on inorganic soils is much smaller. Casagrande (1948) reported that oven drying changed the liquid limit of a Connecticut organic clay from 84 percent to 51 percent and the plastic limit from 50 percent to 42 percent.

The Unified classification system is summarized in Table 1.7. The details of the classification system are explained in the publication by Waterways Experiment Station (1953). A comprehensive summary of the engineering properties of the various soil groups has been made by Casagrande (1948).

PEDOLOGICAL CLASSIFICATION. In addition to the above engineering classification systems there is also the pedological system developed by agronomists. Pedology is a study of soil development near the ground surface. A soil profile generally shows a sequence of layers (called soil horizons) extending 5 to 10 ft below the ground surface. The properties of these horizons reflect the parent material from which the soil is derived and the affect of environmental factors such as climate, ground slope, and vegetation on the process of soil formation. This system classifies soils according to the characteristics of the successive soil horizons. The characteristics used for classification include color, texture, thickness of the horizons, etc. Soils are assigned special names, frequently the names of localities where such soil profiles were first noted. Similar profiles found subsequently at other localities are assigned the same name.

Pedological soil-classification systems are widely used in highway and airfield design. Often they are supplemented by some measurement of soil strength. For examples of empirical design based on soil classification, the reader may refer to Davis and Jones (1954) and McLaughlin and Stokstad (1946).

It should be emphasized that natural soils do not consist of distinct groups but form a broad spectrum. Thus the dividing lines used in classification are necessarily arbitrary. Secondly, all soil classifications are based on their index properties. They can, at best, serve as nomenclature for describing soils and provide some indication of the significant engineering properties. None of the soil-classification systems, however elaborate, should be used as a measure of the soil's engineering properties (such as strength or compressibility) for the very obvious reason that such properties are not measured in the classification systems. Furthermore, the soil properties that are important depend on the type of problem to be solved. It is thus obvious that a rigid classification system cannot hope to include in it all the soil properties that may be needed to solve the many diverse problems in soil mechanics.

Example 1.5. Classify the soil whose particle-size-distribution curve is given as Fig. 13.6, curve one, by the AASHO and Unified classification systems. The Atterberg limits are: PL = 28 percent, LL = 56 percent.

Solution. The glacial lake clay has 56 percent particles finer than 2 microns and a LL of 56 percent. Its group index is

$I = (0.2)(56 - 35) + (0.005)(56 - 35)(56 - 40) + (0.01)(40)(28 - 10) = 13.08$

Hence it is A-7 by the AASHO classification.

Since more than half of the material is finer than 2 microns and the LL exceeds 50 percent, it belongs to the group of clays and silts of high compressibility in the Unified classification system. On the plasticity chart, it plots above the A-line and is therefore a clay (CH).

PROBLEMS

1.1 The particle-size-distribution curves of three soils are shown in Fig. 1.15. Which of the following statements are correct: (a) A has a larger uniformity coefficient than B. (b) A has a larger effective diameter than B. (c) A may be compacted to a higher dry density than B. (d) B may be compacted to a higher dry density than C. (e) C contains the largest percentage of clay-sized particles. *Answer: b, d*

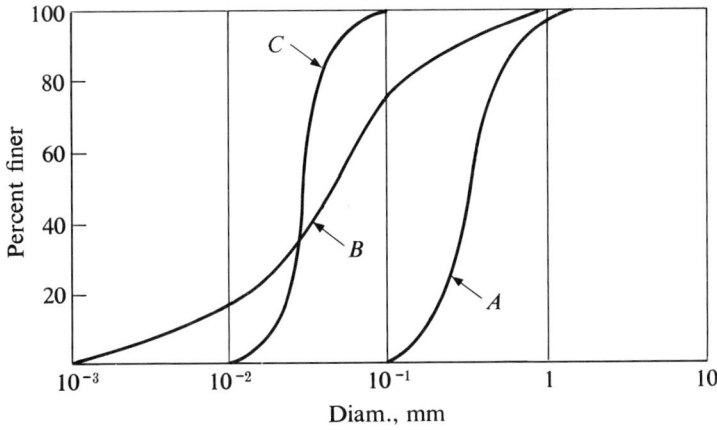

FIGURE 1.15

1.2 An undisturbed specimen of clay was tested in the laboratory and the following results were obtained:

Liquid limit = 25%　　Wet wt. = 210 g
Plastic limit = 13%　　Oven-dried wt. = 125 g
Sp. gr. = 2.70

What is the total volume of the original undisturbed specimen assuming the specimen to be: (a) 100% saturated? (b) 50% saturated?
Answer: 131.0 cc; 216.3 cc

1.3 In an experiment to determine the shrinkage limit, the following data were obtained:

Before shrinkage: vol. of sat'd wet soil = 14.50 cc
wt. of sat'd wet soil = 22.40 g
After shrinkage: vol. of dry soil = 5.52 cc
wt. of dry soil = 12.45 g

Compute the shrinkage limit and the specific gravity of the soil.
Answer: 7.8%; 2.73

1.4 The following index properties were determined for soils A and B:

	A	B
Liquid limit	30%	9%
Plastic limit	12%	6%
Water content	15%	6%
Sp. gr. of solids	2.70	2.68
Deg. of saturation	100%	100%

Which of the following statements are correct? (*a*) Soil A contains more clay than soil B. (*b*) Soil A has a greater wet density than soil B. (*c*) Soil A has a greater dry density than soil B. (*d*) Soil A has a greater void ratio than soil B.
Answer: a, d

1.5 In a compaction test a specimen of compacted soil weighs 4.2 lb and its volume is $\frac{1}{30}$ ft³. The water content of the specimen is 10%. Other index properties of the soil are listed as follows:

$$\text{Liquid limit} = 15\%$$
$$\text{Plastic limit} = 6\%$$
$$\text{Sp. gr. of solids} = 2.68$$

Determine the wet density, dry density, void ratio, and degree of saturation of the sample.
Answer: 126 pcf; 114.5 pcf; 0.453; 60%

1.6 The following properties were obtained from three different soils:

Soil	LL	I_p	% < No. 40	% < No. 200
A	10	5	50%	10%
B	70	20	70%	30%
C	70	50	80%	40%

(*a*) Compare soils A and B as to texture (coarse of fine grained). (*b*) Compare soils A and B as to suitability as a subgrade material. (*c*) Compare soils B and C as to organic content.

1.7 A soil deposit to be used for construction of an earth embankment has an average dry density of 100 lb/ft³. A 100,000 yd³ embankment is to be built and the compacted embankment is to have an average dry density of 115 lb/ft³. The following properties were determined for the soil:

Moisture content = 10%   LL = 18%
Sp. gr. = 2.65   PL = 6%
Effective diam. = 0.003 mm

Determine the volume of soil to be excavated for 100,000 yd³ of embankment the wet weight of the soil to be excavated, and the dry weight of the soil to be excavated.

Answer: 115,000 yd³; 170,500 tons; 155,000 tons

1.8 The soil from a large borrow area has been selected for the construction of an earth embankment. Describe in detail the steps you would take to control compaction of the fill and give the reasons for the procedure you outline.

2

Soil Water and Soil Solids

PORE PRESSURE

2.1 Introduction

Since soil is a three-phase material that is composed of solids, water, and air, it is necessary to consider the contribution of the various components to soil behavior. Consider the soil deposit shown in Fig. 2.1. A microscopic view of the soil element A is shown in Fig. 2.2(a). The soil is located below the water table and the voids are filled with water and the soil is saturated. The soil element B is shown in Fig. 2.2(b). Although this element is located above the water table, it may contain some water in the voids. The water may have entered the ground as surface precipitation. Also, if the voids are small enough, the water can arrive at B via capillary action. In both elements A and B the solid particles form the structural network of the soil.

2.2 Porewater Pressure in saturated soils

In Fig. 2.2(a) the voids in the soil are interconnected. Hence the water in the void space at A has a hydrostatic pressure equal to

$$u = \gamma_w h_w \tag{2.1}$$

where h_w is the vertical distance between point A and the free water surface or water table. This is called the *pressure head* in hydraulics, and the pressure u is called the *porewater pressure* or simply *porepressure*. If a piezometer tube is inserted into the soil so that its tip is at A, water would rise to a distance h_w in the

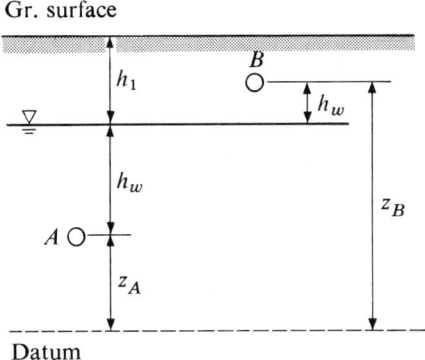

FIGURE 2.1 *Soil deposit and water table.*

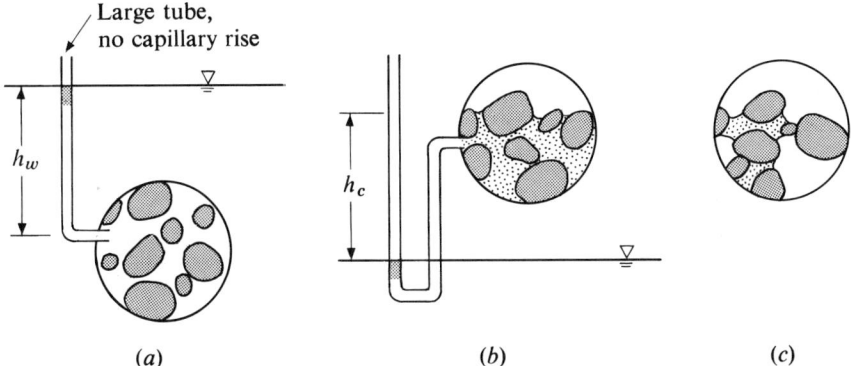

FIGURE 2.2 *Porewater pressure.*

tube. If we choose the datum shown in Fig. 2.1, then the total head in Bernoulli's equation is

$$h = \frac{v^2}{2g} + \frac{u}{\gamma_w} + z = \frac{v^2}{2g} + h_w + z_A \tag{2.2}$$

where v is the velocity of flow, g is the gravitational acceleration, and z is the elevation above datum. In most soil mechanics problems the velocity is small and $v^2/2g$ may be taken as zero.

2.3 Capillary action

Consider the case where the water in element B is drawn from the ground water by capillary action. At the air and water interface, a surface tension exists. The presence of surface tension along the meniscus depresses the pressure in the water immediately below the meniscus.

To study this problem we consider the simplified picture in which the soil voids form a continuous tube. The value of the surface tension T and the angle α are particular for the fluid and the material of the capillary tube. For water and clean glass, T is 0.075 g/cm and α is 0 deg. The pressure u in the water immediately inside the meniscus may be evaluated by considering the equilibrium of the meniscus [Fig. 2.3(a)]. If the atmospheric pressure is taken as the datum, then the air pressure p_a is 0. Summation of the vertical forces gives

$$T 2\pi r \cos \alpha + u \pi r^2 = 0$$

or
$$u = -\frac{2T \cos \alpha}{r} \tag{2.3}$$

Thus we find that u is negative when α is smaller than 90 degrees, as is the case here. The pressure in the water beneath the meniscus is therefore less than atmospheric.

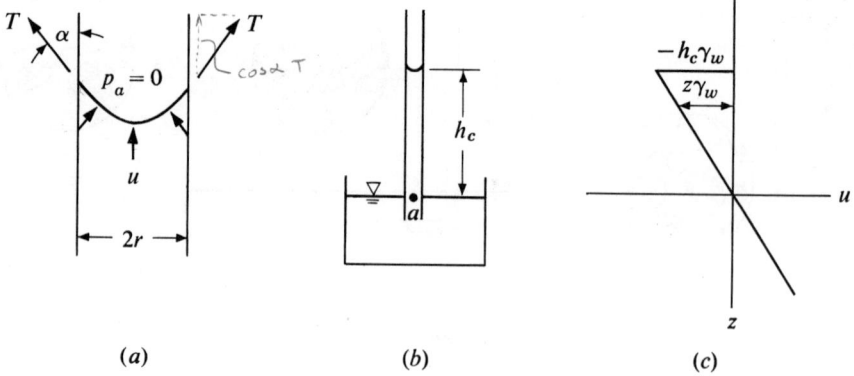

(a) (b) (c)

FIGURE 2.3 *Capillary action.*

To determine the capillary rise h_c, we consider the pressures at point a [Fig. 2.3(b)] located at the same level as the free water surface outside the tube. The downward pressure must be equal to the atmospheric pressure (datum), or water would flow into the tube from the container. We then have

$$h_c \gamma_w - \frac{2T \cos \alpha}{r} = 0$$

$$h_c = \frac{2T \cos \alpha}{r \gamma_w} \tag{2.4}$$

Hence, equilibrium is attained if the water has risen in the tube to height h_c.

We see that the pressure in the capillary water increases hydrostatically from a minimum of $-h_c \gamma_w$ immediately beneath the meniscus to 0 at the elevation of the free water surface [Fig. 2.3(c)], if atmospheric pressure is used as the datum.

For a soil containing capillary water [Fig. 2.2(b)] the porewater pressure is $-h_c \gamma_w$ immediately beneath the meniscus. A piezometer tube would show a water level at a distance h_c below that point.

SOIL MOISTURE ABOVE WATER TABLE. The principles explained in the preceding sections are illustrated in a number of phenomena in soil-water movement. Some of these are described below.

The results of capillary flow are best illustrated by the conditions of soil moisture above the water table. We consider the simplified case in which moisture enters the soil above water table by capillary action only. If the voids in the soil were simply a bundle of tubes of the same diameter, the capillary rise as given by Eq. (2.4) would be the same at every point. This means that within a distance h_c above the water table, the soil would be saturated, while above this the soil would be dry, as shown in Fig. 2.4(a).

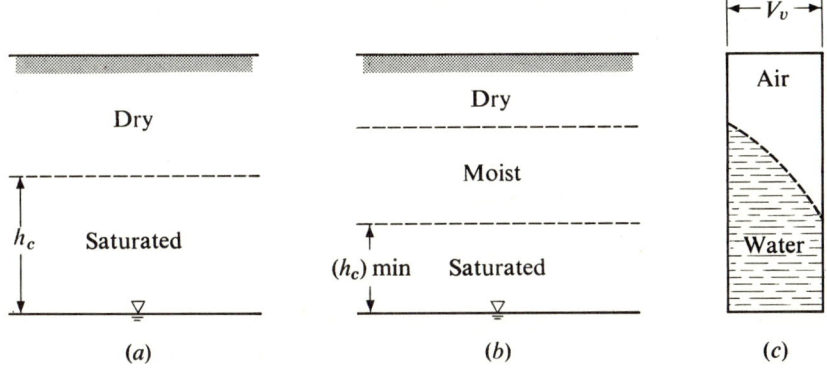

FIGURE 2.4 *Soil-moisture conditions.*

However, the voids in a soil form channels of variable cross section and are interconnected. In such cases, the soil above the water table becomes saturated only up to a distance $(h_c)_{min}$ that corresponds to the largest void size present in the soil. The presence of such a void in a channel prevents the capillary water from rising higher than $(h_c)_{min}$. In a channel where such large voids are absent, the water rises higher. The maximum rise is possible only in the smallest voids present. The result is that above the distance $(h_c)_{min}$, only a fraction of the voids are filled with water. The situation is then represented by Fig. 2.4(b). Above the saturated zone, there is a moist zone. This has its upper limit at a distance $(h_c)_{max}$ above the water table; the value of $(h_c)_{max}$ corresponds to the smallest void. Figure 2.4(c) shows the relative amounts of water and air in the soil voids at different elevations above the water table.

In the saturated zone, the change of porewater pressure with depth is as described in Fig. 2.3(c). Above the saturated zone in Fig. 2.4(b), both air and water are present in the voids, and the situation is described by Fig. 2.2(c). Then the porepressure u is the combined effect of the porewater pressure u_w and pore-air pressure u_a. For the general case, we may write

$$u = \chi u_w + (1 - \chi) u_a \tag{2.5}$$

where χ is a coefficient that indicates the relative importance of air and water pressures. It depends on the relative amounts of air and water present in the voids and also the soil property under consideration.

WATER FLOW THROUGH SOILS

2.4 Darcy's law

When a difference in total head exists between two points as shown in Fig. 2.5, water flows from the point with larger head to the point with smaller head. The difference in total head between A and B is $h_1 - h_2$, and the distance between them is L. The hydraulic gradient i is defined as

$$i = \frac{h_1 - h_2}{L} \tag{2.6}$$

Darcy's law states that for laminar flow, the rate of flow of water across an area A in a given soil is proportional to the hydraulic gradient, or

$$q = ki A = k \frac{h_1 - h_2}{L} A \tag{2.7}$$

where k is the constant of proportionality and has units of velocity. It is called the coefficient of permeability and must be determined experimentally for the particular soil under consideration. The rate q is expressed as volume per unit

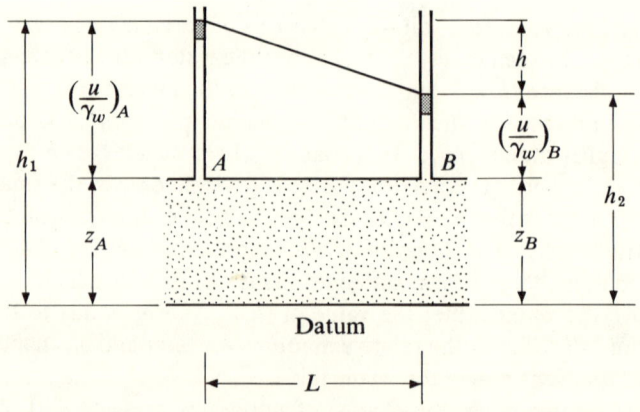

FIGURE 2.5 *Flow under a pressure gradient.*

time. If the rate of flow is expressed in terms of volume per unit time per unit area, then Darcy's law may be written as

$$v = \frac{q}{A} = ki \tag{2.8}$$

where v is the rate of flow across a section of soil with unit cross-section area. Expressed in differential form, it becomes

$$v = k\frac{dh}{dL} \tag{2.9}$$

The term, approach velocity or discharge velocity is commonly used for v. In subsequent sections, it is referred to simply as velocity.

2.5 Flow through a porous medium

In a porous material like soil we consider the pores to be connected together, thereby forming passageways through which water moves. Of course, the passageway is an irregular one, as can be realized by examination of Fig. 2.6, which shows the shape of the intersice between four spheres. Nevertheless, it is useful to analyze the flow as if it moved through a bundle of fine tubes.

FIGURE 2.6 *Interstice between spherical particles.*

We consider first laminar flow between two parallel plates of infinite width, separated by a distance $2H$ [Fig. 2.7(a)]. The forces in the x direction that act on an element of the fluid between the plates are illustrated in Fig. 2.7(b). The water pressures on the faces ac and bd are p and $p + (\partial p/\partial x)\,dx$, and the shear stresses on ab and cd are τ and $\tau + (\partial \tau/\partial y)\,dy$.

The solution for the velocity is*

$$v_a = \frac{H^2 - y^2}{2\eta}\left(-\frac{dp}{dx}\right) \tag{2.10}$$

---

* The derivation for this and the following cases may be found in most books on fluid mechanics, e.g., Streeter and Wylie (1975).

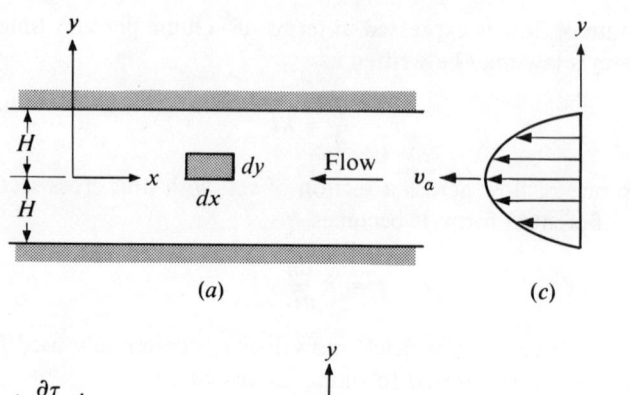
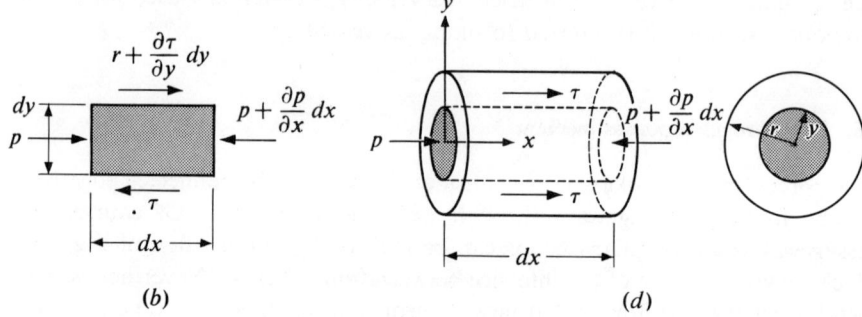

FIGURE 2.7 *Laminar flow.*

Thus the velocity varies across the distance H as a parabola [Fig. 2.7(c)]. The average velocity $\bar{v}_a$ across the section is equal to

$$\bar{v}_a = \int_{-H}^{H} \frac{v_a dy}{2H} = \frac{H^2}{3\eta}\left(-\frac{dp}{dx}\right) \tag{2.11}$$

For flow through a tube of circular cross section with radius r, the average velocity is

$$\bar{v}_a = \frac{r^2}{8\eta}\left(-\frac{dp}{dx}\right) \tag{2.12}$$

in which $\bar{v}_a$ is the average velocity in the tube. This is known as Poiseuille's law.

Since $\quad h = \dfrac{p}{\gamma_w} + z \quad$ and $\quad h_w = \dfrac{p}{\gamma_w}$

we may replace dp/dx by $\gamma(dh_w/dx)$. Thus under a pressure and a positional gradient, Eqs. (2.11) and (2.12) become

$$\bar{v}_a = \frac{H^2 \gamma_w}{3\eta}\left(-\frac{dh}{dx}\right) = \frac{H^2 \gamma_w}{3\eta} i \tag{2.13}$$

$$\bar{v}_a = \frac{r^2 \gamma_w}{8\eta}\left(-\frac{dh}{dx}\right) = \frac{r^2 \gamma_w}{8\eta} i \tag{2.14}$$

SEC. 2.5   Flow through a porous medium

Considering Eqs. (2.13) and (2.14), we see that the flow through a channel of a given cross section may be expressed generally as

$$\bar{v}_a = \frac{r_h^2 \gamma_w}{C_s \eta}\left(-\frac{dh}{dx}\right) = \frac{r_h^2 \gamma_w}{C_s \eta} i \quad (2.15)$$

in which r_h denotes the hydraulic radius, which is the ratio of the area of the channel to the wetted perimeter of the channel. Thus for a circular cross section we have

$$r_h = \frac{\pi r^2}{2\pi r} = \frac{r}{2}$$

and for the space between two plates with width equal to B

$$r_h = \frac{2HB}{2B} = H$$

The term C_s is a shape factor that depends upon the shape of the channel cross section. For the circular cross section, Eq. (2.15) gives

$$\bar{v}_a = \frac{r_h^2 \gamma_w}{C_s \eta} i = \frac{r^2 \gamma_w}{4 C_s \eta} i = \frac{r^2 \gamma_w}{8\eta}$$

Thus C_s equals 2 for a circular cross section. Similarly we obtain C_s equal to 3 for the space between two plates.

Equation (2.15) cannot be used to calculate the average velocity of flow through soils because of the drastic difference between soil voids and a circular tube of uniform cross section. Furthermore, the passageway through the pores is not straight, but has many bends, and the actual length of flow path cannot be calculated. Actually the pores in the soil mass are interconnected, and, for a water particle starting at a given point, many possible flow paths exist. This situation is illustrated in Fig. 2.8(a), which shows the pore space between four soil particles. As a water particle enters this space along route a, it has the choice of continuing along three possible routes, shown as b, c, and d. If we trace the progress of this particle through the soil under a gradient dp/dx, its path will be a zigzag one, as represented by the dashed line in Fig. 2.8(b) [de Josselyn de Jong (1958)]. Thus the flow path through this porous material consists of a succession of channels at various angles to direction x.

We must note here that since the actual flow path for a given channel [Fig. 2.8(b)] is at an angle θ to the direction x, the flow in that particular channel is

$$\bar{v}_a = \frac{r_h^2 \gamma_w}{C_s \eta}\left(-\frac{dh}{dl}\right) = \frac{r_h^2 \gamma_w}{C_s \eta}\left(-\frac{dh}{dx}\right)\cos\theta = \frac{r_h^2 \gamma_w}{C_s \eta}\left(\frac{x}{l}\right) i \quad (2.16)$$

in which l is the actual length of the channel. We may picture the porous medium as composed of a collection of flow channels whose average length is l; also r_h and C_s represent, respectively, the average hydraulic radius and shape factor of the channels. To obtain the rate of flow q across a given cross section perpendicular to x, the velocity $\bar{v}_a$ must be multiplied by the total cross-sectional area of

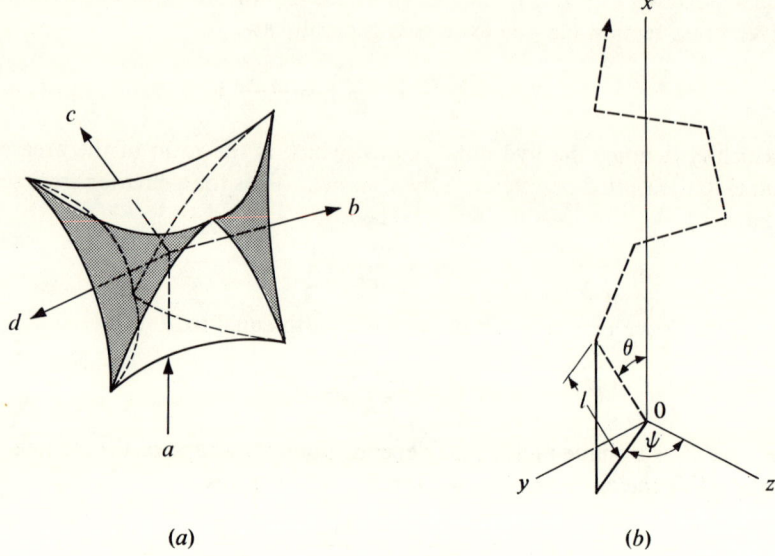

Figure 2.8 *Flow path of a water particle through a porous medium.*

the channels. The velocity $\bar{v}_a$ in Eq. (2.16) is in the direction of the channel at angle θ to x and its component in direction x is $\bar{v}_a(x/l)$. Thus

$$q = \bar{v}_a\left(\frac{x}{l}\right)A_v = \frac{r_h^2 \gamma_w}{C_s \eta}\left(\frac{x}{l}\right)^2 A_v i \qquad (2.17)$$

in which A_v is the area occupied by the voids in the given cross section of the soil. Since r_h^2, C_s, x/l, are geometrical properties of the pores, and the viscosity η and unit weight γ_w are almost constant within most practical temperature ranges, we may simplify Eq. (2.17) to

$$q = k_a A_v i \qquad (2.18)$$

in which k_a is a property of the soil and the liquid and is a constant for a given soil and liquid. It is equal to

$$k_a = \frac{r_h^2 \gamma_w}{C_s \eta}\left(\frac{x}{l}\right)^2 \qquad (2.19)$$

The average velocity of flow through the pores (called seepage velocity), $\bar{v}_a$, is

$$\bar{v}_a = \frac{q}{A_v} = k_a i \qquad (2.20)$$

The relationship between $\bar{v}_a$ and v in Eq. (2.7b) may be obtained by equating the flow rate from Eqs. (2.18) and (2.8)

$$\bar{v}_a A_v = vA$$

and

$$k_a i A_v = kiA$$

2.6 Coefficient of permeability

Thus we obtain

$$v = \bar{v}_a \frac{A_v}{A} \qquad k = k_a \frac{A_v}{A} \qquad (2.21)$$

2.6 Coefficient of permeability

The permeability coefficient as expressed by Eqs. (2.19) and (2.21) can be written as

$$k = k_a \frac{A_v}{A} = \frac{r_h^2}{C_s} \frac{\gamma_w}{\eta} \left(\frac{x}{l}\right)^2 \frac{A_v}{A} \qquad (2.22)$$

The ratio A_v/A is equal to the ratio of void volume to soil volume V_v/V, which is the porosity n. The ratio x/l represents the direction of flow through the pores and is denoted by $1/T_0$. The term T_0 is called tortuosity. Thus Eq. (2.22) may be written as

$$k = \frac{r_h^2 \gamma_w n}{C_s \eta T_0^2} \qquad (2.23)$$

In this equation γ_w and η may be considered to be constants. The hydraulic radius r_h depends on the size of the pores, which in turn depends on the size of the soil particles. The larger the particle size the larger is the value of r_h. The permeability also increases with the porosity n. The shape factor C_s is indeterminate. It is equal to 2.00 for a circular tube and 1.78 for a square tube. Hence, it does not vary over a wide range. The tortuosity cannot be predicted from available theories. To some extent both C_s and T_0 would be affected by the porosity and the shape of the soil particles. More detailed treatments may be found in the papers by Carmen (1956), Scheidegger (1957), Leonards (1962), Michaels and Lin (1956), and Olsen (1960).

Experimental studies of permeability by Hazen (1911) have shown that the permeability of filter sands may be correlated with the effective diameter D_{10}. He gave the following empirical formula:

$$k = C D_{10}^2 \qquad (2.24)$$

in which D_{10} is the effective diameter in centimeters and C is a coefficient whose value lies between 100 and 150. The permeability is in cm/sec. Typical values of the permeability coefficient are given in Table 2.1. For fine or medium clean sands the effect of void ratio on permeability has been found to be

$$k = 1.4 \, k_1 e^2 \qquad (2.25)$$

where k_1 is the permeability at $e = 0.85$.*

---

* Equation (2.25) was attributed to A. Casagrande [Terzaghi and Peck (1967)].

TABLE 2.1 *Typical Values of Permeability Coefficient*

Soil types	Permeability coeff., cm/sec
1. Clean gravel	$1-10^2$
2. Clean sands; sand and gravel	$10^{-3}-1$
3. Very fine sands; silts; mixture of sand, silt, and clay; stratified clay deposits	$10^{-6}-10^{-3}$
4. Homogeneous clays	$<10^{-7}$

EFFECTIVE-STRESS PRINCIPLE

2.7 Effective stress in saturated soils

We consider first the case of a saturated soil. Figure 2.1 shows a section through a soil mass at rest. The voids are filled with water. Since the voids are continuous, the water in the voids, or porewater, is also continuous and communicates freely with the water above the soil. In this case the porewater pressure u at any point A is given by Eq. (2.1).

In addition to the porewater pressure, we see that there also must exist contact stresses between the solid grains of the soil. To study this type of stress, we pass a horizontal section b–b through the soil [Fig. 2.9(a)]. The section does not cut across any solid particle but passes only through points of contact between particles.

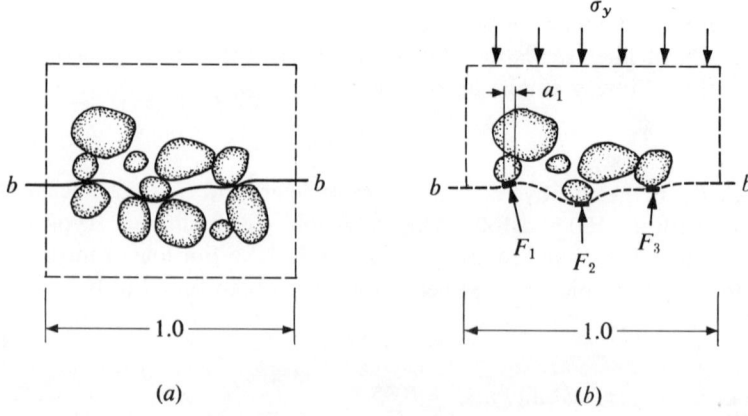

FIGURE 2.9 *Effective stress.*

We now consider the equilibrium of the block in Fig. 2.9(b). The cross-sectional area is 1. The stress σ_y is equal to the weight of the soil and water above the point. This stress is called the total stress. Over the section b–b we have the porewater pressure u (Eq. 2.1) where the section cuts through a void. The forces that act on the particle to particle contacts are shown as $F_1, F_2, F_3, \ldots$ and the contact areas are $a_1, a_2, a_3, \ldots$. We take the sum of the components normal to b–b of $F_1, F_2, F_3, \ldots$. They are located within a unit cross-sectional area of the soil and have the unit of stress. We call this the effective stress $\bar{\sigma}_y$,

or
$$\bar{\sigma}_y = F_{1y} + F_{2y} + F_{3y} + \cdots = \Sigma F_{iy}$$

Equilibrium requires that

$$\Sigma F_{iy} + u(1 - \Sigma a_i) = \sigma_y$$

or
$$\bar{\sigma}_y + u(1 - \Sigma a_i) = \sigma_y \tag{2.26}$$

We note that Σa_i is the area of contact surfaces per unit area of soil. To simplify notations we let $\Sigma a_i = a$. Then

$$\sigma_y = \bar{\sigma}_y + u(1 - a) \tag{2.27}$$

The value of a is not accurately known. However, careful estimates [see Bishop and Eldin (1950)] show that it cannot be greater than a few hundredths of the cross-sectional area of soil. Hence, without serious loss of accuracy, Eq. (2.27) may be replaced by

$$\sigma = \bar{\sigma} + u \tag{2.28}$$

At any point the porewater pressure u acts equally in all directions. It can only cause hydrostatic compression of the solid particles. This amount is usually negligible in soil mechanics problems because the modulus of compression of the solid particles is very large. Deformation or strain in the soil particle skeleton can arise only if there is a change in the effective stress which acts on the skeleton. This fundamental concept is frequently called the principle of effective stress. Stress–strain relationships of soils are described in Chapters 6 and 8.

2.8 Partially saturated soils

In an unsaturated soil, the water is restricted to some of the interstices between the soil particles, as illustrated in Fig. 2.10. Under such conditions, the water in the voids is not continuous but is surrounded by air. In the air voids, the pore pressure is equal to the pore air pressure u_a. In the water, the pore pressure is equal to the porewater pressure u_w. Over a unit area of section b–b,

$$\sigma = \bar{\sigma} + u_w a_w + u_a(1 - a_w - a) - T$$

in which a_w is the area of the section that passes through water and T is the resultant surface tension. Since the value of a is small, it is again taken as zero for simplicity and we have

$$\sigma = \bar{\sigma} + u_w a_w + u_a(1 - a_w) - T \qquad (2.29a)$$

We note that the surface tension is dependent on the curvature of the meniscus (Eq. 2.3), which in turn is dependent on the amount of water present.

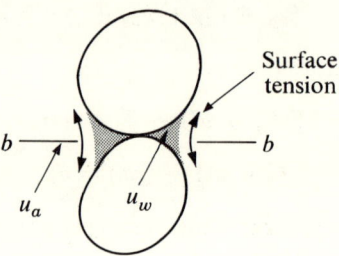

FIGURE 2.10 *Porewater pressure and poreair pressure.*

If we retain our definitions of effective stress (Eq. 2.28) and pore pressure (Eq. 2.5),

$$\sigma = \bar{\sigma} + \chi u_w + (1 - \chi) u_a \qquad (2.29b)$$

Comparison of Eq. (2.29a) with Eq. (2.29b) suggests that χ may be considered to represent the effects of a_w and T. A theoretical derivation of χ has been given by Blight (1967). Equation (2.29b) was suggested by Bishop et al. (1960a) and has also been derived by Croney et al. (1958) and Aitchison (1960).

2.9 Effective stresses in situ

Consider now point A in the soil deposit shown in Fig. 2.1. The value of the total stress σ_y is equal to the weight of the column of soil and water above the point, so

$$\sigma_y = h_1 \gamma + h_w \gamma_{\text{sat}} \qquad (2.30)$$

where γ and γ_{sat} are the unit weights of the soil above and below the water table respectively. From Eq. (2.1) we get

$$u = h_w \gamma_w$$

Then
$$\bar{\sigma}_y = \sigma_y - u = h_1 \gamma + h_w(\gamma_{\text{sat}} - \gamma_w) \qquad (2.31)$$
$$= h_1 \gamma + h_w \gamma'$$

where γ' denotes the submerged unit weight.

In the case of a soil deposit above the water table which is saturated by capillary action (Fig. 2.11), the porewater pressure above the water table is negative if atmospheric pressure is used as the datum (see Sec. 2.3). At the top of the zone of saturation the porewater pressure is $-h_c\gamma_w$. Since the effective stress is the total stress minus the porewater pressure, the negative porewater pressure increases the effective stress.

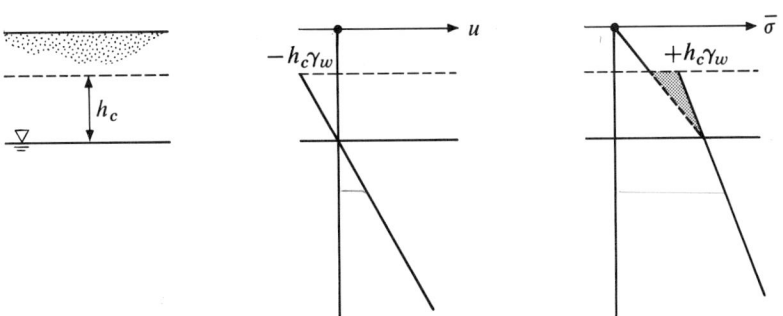

FIGURE 2.11 *Porewater pressure and effective stress in a soil deposit with capillary rise.*

EFFECTIVE STRESS UNDER SEEPAGE. Figure 2.12(a) shows seepage taking place downward through a layer of soil. The water levels A and B are maintained constant. The water flows downward through the layer of silty sand and enters the layer of coarse sand and gravel below it. At elevation a–a the porewater pressure is equal to $h_1\gamma_w$, while at b–b it is $(h_1 + h_2 - h)\gamma_w$. The quantity h is equal to the head loss between a–a and b–b. The gradient and velocity are respectively

$$i = \frac{h}{h_2} \quad \text{and} \quad v = ki = k\frac{h}{h_2}$$

The variation of the porewater pressure with depth is shown graphically in Fig. 2.12(b). The total stresses are shown in Fig. 2.12(c). At b–b the total stress is $(h_1\gamma_w + h_2\gamma_{sat})$. The effective stress is equal to the difference between the total stress and porewater pressure and is $(h_2\gamma' + h\gamma_w)$.

The case of upward seepage is illustrated in Fig. 2.13. The pore pressure and effective stresses at b–b are

$$u = (h_1 + h_2 + h)\gamma_w$$
$$\bar{\sigma}_y = h_2\gamma' - h\gamma_w$$

It should be noted that, in this case, there is an upward limit to the hydraulic head h. If h is steadily increased, a point is reached where the porewater pressure at b–b becomes so large that it is equal to the weight of the soil and water above b–b. This may be written as

$$(h + h_1 + h_2)\gamma_w = h_1\gamma_w + h_2\gamma_{sat} \tag{2.32}$$

or

$$\bar{\sigma}_y = h_2\gamma' - h\gamma_w = 0$$

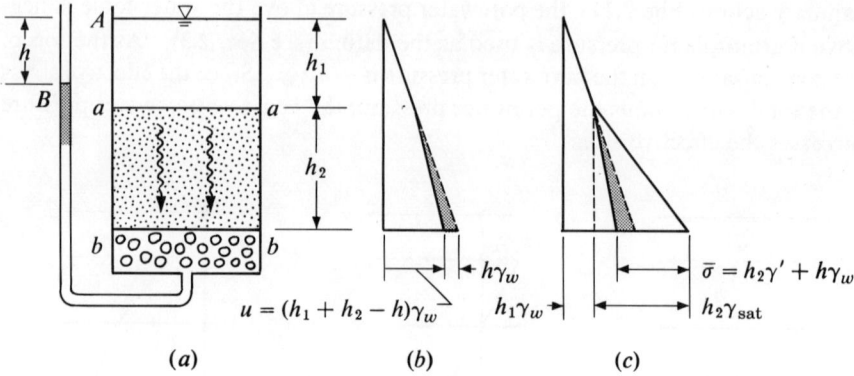

FIGURE 2.12 *Porewater pressure and effective stress in a soil deposit subjected to seepage—downward flow.*

This means the effective stress is zero and there are no contact forces between the solid particles. The soil structure then breaks up. This state is called quicksand, or boiling, or failure by heave.

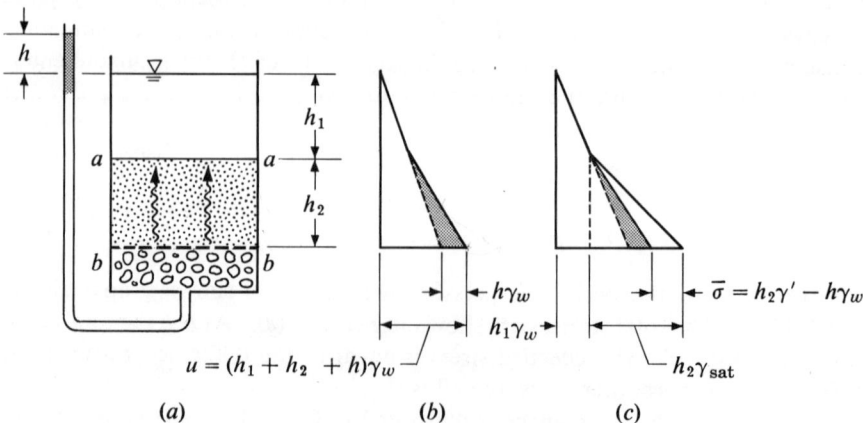

FIGURE 2.13 *Porewater pressure and effective stress in a soil deposit subjected to seepage—upward flow.*

PROBLEMS

2.1 What is the absolute pressure (in psi) in the water just below the meniscus in a capillary tube whose inside diameter is 0.1 mm? The surface tension is equal to 75 dynes/cm and the wetting angle is 10 deg.
Answer: 14.3 psi

2.2 When water at 20°C is added to a fine sand and a silt, a difference in capillary rise of 20 cm is observed between the two soils. The capillary rise in the fine sand is 30 cm. What is the difference in the size of the voids?
Answer: 0.002 cm

2.3 For the case shown in Fig. 2.14, determine the pressure head, the elevation head, and the total head at the entering end, the exit end, and point A in the soil sample.
Answer: 9 ft, 0 ft; 5 ft, 0 ft; 6 ft; 0 ft; elevation of A is taken as datum

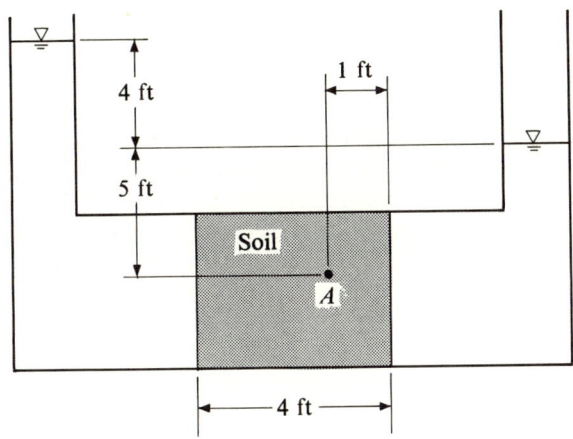

FIGURE 2.14

2.4 The soil deposit shown in Fig. 2.15 has the following properties: $w = 20\%$; LL $= 28\%$; PL $= 12\%$. (*a*) Compute the porewater pressure at point A. (*b*) Compute the porewater pressure at point A if the water table drops to 10 ft below the ground surface. Assume that the clay above the water table remains saturated. (*c*) Does the drop in water table effect the stresses between the solid particles of the soil? Explain.
Answer: 312 psf; -312 psf; increases it by 624 psf

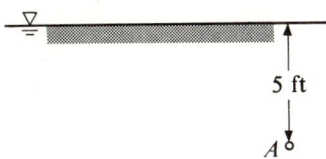

FIGURE 2.15

2.5 An excavation (Fig. 2.16) is to be made in a deposit of clay 30 ft thick underlain by a layer of sand. The water in the sand has a head of 25 ft above the top of the sand layer (artesian water). The depth of the cut is 20 ft. Calculate the height of water h in the excavation that would be necessary to prevent boiling.
Answer: 4.5 ft

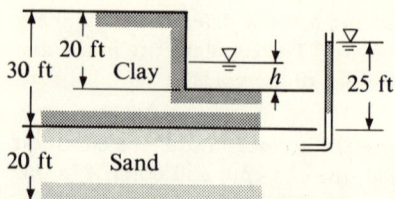

	Clay	Sand
Saturated weight	128 pcf	135 pcf
Water content	28%	18%

FIGURE 2.16

2.6 It is reported that the permeability coefficients of two soils are 1 cm/sec and 0.01×10^{-4} cm/sec, respectively. One soil is a clay, the other a sand. (a) Which coefficient applies to the sand? (b) If the porosity of the clay is twice that of the sand, would this change your answer to part (a)? Why?

2.7 A deposit of clay lies between two layers of sand as shown in Fig. 2.17. The lower sand layer is under artesian pressure. Calculate: (a) The seepage through the clay in ft³/day/ft² of area. (b) The porewater pressure and effective stress at the bottom of the clay. (c) The head H that would cause boiling.

	Sand	Clay
Sat. unit wt.	120 pcf	110 pcf
Dry unit wt.	105 pcf	100 pcf
k	10^{-3} cm/sec	10^{-6} cm/sec

Answer: 0.04 ft³/day; 1875 psf; 350 psf; 15.6 ft

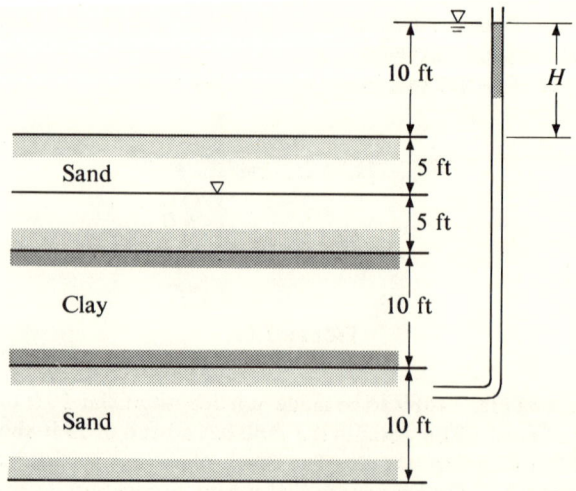

FIGURE 2.17

2.8 Compute the porewater pressure and effective stress in the silty clay (Fig. 2.18) at a point 10 ft below the ground surface when (a) the water table is at the ground surface and (b) the water table drops to the surface of the sand layer. The silty clay has the following properties:

> Water content = 28%   Sp. gr. of solids = 2.65
> LL = 21%; PL = 10%   Capillary rise, 18 ft

Answer: 625, 605 psf; −313, 1542 psf

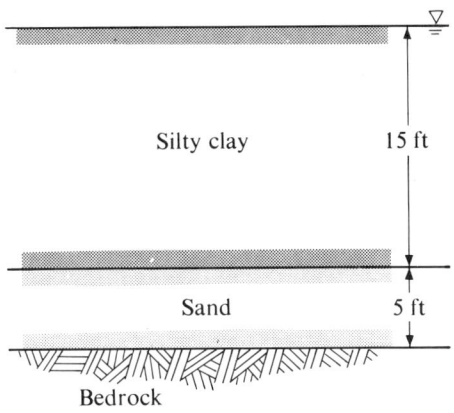

FIGURE 2.18

3

Seepage Problems

Seepage caused by flow under a pressure gradient is a common problem in engineering. Seepage forces in the soil may cause deformation or shear failure. The topic is therefore of fundamental importance in the realm of soil mechanics.

The measurement of the permeability coefficient is discussed in the first part of the chapter and the analysis of seepage and seepage forces is presented in the second part.

PERMEABILITY

3.1 Permeability tests

It is shown in Chapter 2 that the rate of flow through a porous medium under a pressure gradient can be described by Darcy's law:

$$q = vA = kiA \qquad (2.7)$$

The coefficient k cannot be calculated and is a soil property depending on the size distribution of the voids and the shape of the flow channels. Therefore it must be determined experimentally for a given soil if we are to calculate the velocity of flow v under a given gradient i.

The measurement of the permeability coefficient k is based on Darcy's law. To determine k we apply a known gradient i to a soil specimen of known cross-sectional area A and measure the rate of flow q. These quantities then make it possible to calculate k from Eq. (2.7).

The simplest permeability test is the constant-head permeability test. It is illustrated in Fig. 3.1(a). The sample is placed in a cylindrical container of length L and cross-sectional area A. Water flows through the sample under a head difference h, which is kept constant. The quantity of flow Q, during a time interval t, is determined by weighing the water collected in the cup C. The rate of flow q is then

$$q = \frac{Q}{t} = kA\frac{h}{L}$$

From this we get

$$k = \frac{QL}{Aht} \tag{3.1}$$

All the terms on the right-hand side are known or can be measured in the experiment. Thus k can be evaluated.

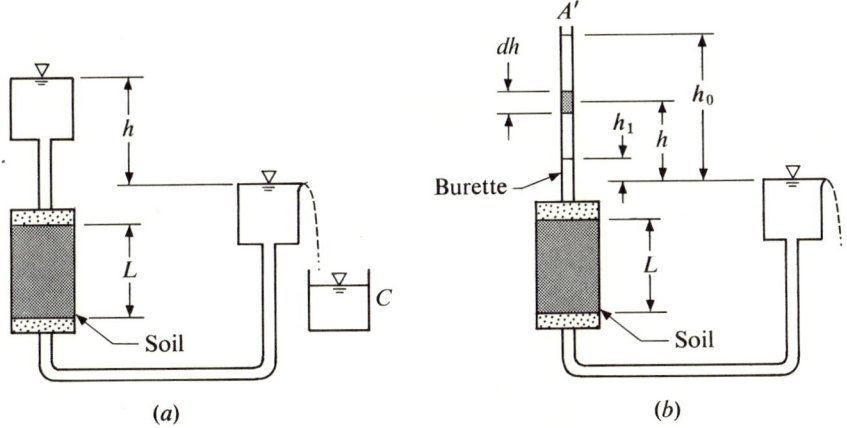

FIGURE. 3.1 *Permeability tests: (a) constant head; (b) falling head.*

Figure 3.1(b) illustrates the falling-head permeability test. Water flows through the sample under an initial head difference h_0. The water level in the burette drops as flow progresses and at the end of the time interval t the head difference is h_1. The quantity of flow Q is measured by the burette, whose inside cross-sectional area is A'. Since the head varies throughout the duration of the test, Darcy's law can be written only as a differential equation for a particular head difference h. If during a time interval dt the water level in the burette drops through dh, then the rate of flow during this interval is

$$\frac{dQ}{dt} = -\frac{A'dh}{dt} = k\frac{h}{L}A$$

or

$$-\frac{dh}{h} = \frac{kA}{A'L}dt$$

Integrating this equation from h_0 to h_1 we get

$$\log_e\frac{h_0}{h_i} = \frac{kA}{A'L}t$$

This allows us to solve for k:

$$k = \frac{A'L}{At} \log_e \frac{h_0}{h_i} \tag{3.2}$$

3.2 Pumping from wells

If a field pumping test is performed, the results enable one to calculate the average permeability coefficient of the soil deposit in the direction of flow. We consider the flow through a permeable layer (the aquifer) located between two impermeable deposits. The pumping is usually made from a well extending through the depth of the aquifer (Fig. 3.2), and flow can enter the well only from the aquifer. Pumping from the well is maintained at a constant discharge rate q and the pumping depresses the piezometric surface around the well as shown. The piezometric surface is determined by means of observation wells located at distances r_1 and r_2 from the pumping well. The piezometric levels are h_1 at r_1 and h_2 at r_2.

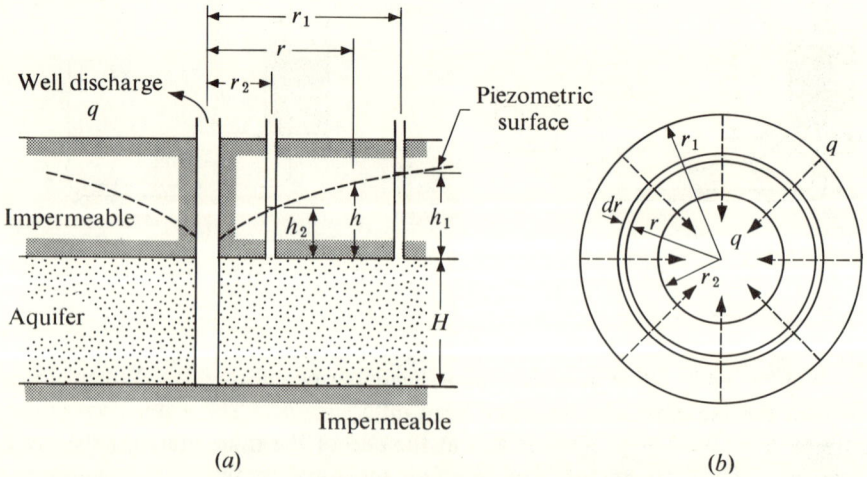

FIGURE 3.2 *Radial flow toward a well.*

After the piezometric surface has reached the equilibrium position,* we may consider the steady radial flow toward the well. The flow q passes through the thick-walled cylinder with radii r_1 and r_2. Darcy's law may be written in differential form for the flow through an element of this cylinder with a radius r and a thickness dr. If the change in head through the element is dh, then

$$q = k \frac{dh}{dr} 2\pi r H \tag{3.3}$$

---

* Equilibrium is taken as the stage when the rate of change in piezometric level is very small compared with the rate of flow and the values of h_1 and h_2. Theoretically the level is never stationary during the pumping process.

Equation (3.3) can be integrated between the limits h_1 and h_2 and r_1 and r_2 to give

$$q \log_e \frac{r_1}{r_2} = 2\pi k(h_1 - h_2)H$$

Solving for k, we get

$$k = \frac{q \log_e (r_1/r_2)}{2\pi(h_1 - h_2)H} \tag{3.4}$$

For additional information on laboratory and field permeability measurements, consult Wenzel (1942).

3.3 Permeability of nonhomogeneous soils

Nonhomogeneity, which is the mark of most natural soil deposits, has a very profound effect on permeability. Its effect can be illustrated by considering a simplified case in which the aquifer consists of a series of layers of different soils with permeabilities equal to $k_1, k_2, k_3, \ldots$ and thicknesses equal to $H_1, H_2, H_3, \ldots$ (Fig. 3.3). Let us consider flow in the x direction under a gradient i between ab and cd. The total quantity of flow is the sum of the quantities in each layer. If the average permeability of the series of layers is k_x in the x direction, we have

$$k_x i H = (k_1 H_1 + k_2 H_2 + k_3 H_3 + \cdots)i$$

or

$$k_x = \frac{1}{H}(k_1 H_1 + k_2 H_2 + k_3 H_3 + \cdots) \tag{3.5}$$

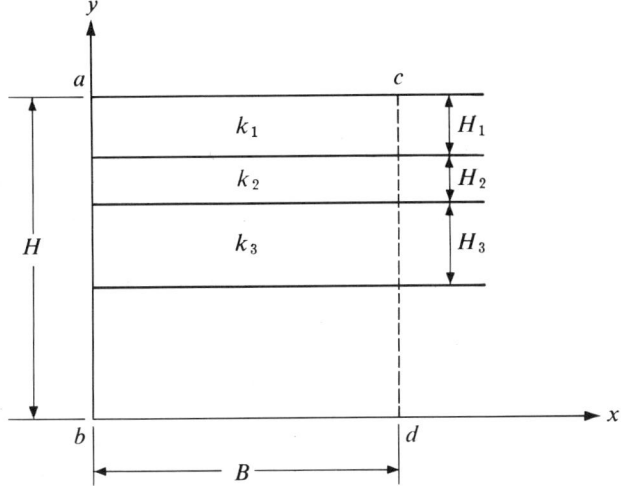

FIGURE 3.3 *A nonhomogeneous medium consisting of layers with different permeabilities.*

If the flow is in the y direction, a particle of water must pass through successively the layers 1, 2, 3, ... Hence the quantity and velocity of flow through each individual layer must be the same. If the overall gradient is i and the gradients across the individual layers are $i_1, i_2, i_3, \ldots$, we have

$$v_y = k_y i = k_1 i_1 = k_2 i_2 = k_3 i_3 \cdots$$

in which k_y is the average permeability of the entire aquifer in the y direction. Also, the total head loss must be the sum of the head losses through each individual layer. Hence

$$iH = i_1 H_1 + i_2 H_2 + i_3 H_3 + \cdots$$

or

$$i = \frac{1}{H}(i_1 H_1 + i_2 H_2 + i_3 H_3 + \cdots)$$

Then we have

$$k_y = \frac{v_y}{i} = \frac{H}{(H_1/k_1) + (H_2/k_2) + (H_3/k_3) + \cdots} \tag{3.6}$$

Equations (3.5) and (3.6) lead to some important conclusions. First we note that if the subsoil consists of strata of different soils and if the values of $k_1, k_2, \ldots$ are very different from each other, k_x may be approximated by $(1/H)(k'H')$, in which k' and H' represent the permeability and thickness of the most permeable layer. On the other hand, k_y is controlled by the least permeable layer and approaches $(H/H'')k''$, in which k'' and H'' are the permeability and thickness of the least permeable layer. Hence for a stratified soil the horizontal permeability k_x is always greater than the vertical permeability k_y. The ratio k_x/k_y for stratified natural soil deposits may range from 2 to 10 or more.

SEEPAGE

3.4 Equation of continuity

The conditions of flow studied in Secs. 3.1 and 3.2 are rather simple, and it is possible to write Darcy's law for flow across a given area or section. In more complicated problems this cannot be done, as the conditions of flow (velocity, gradient, etc.) vary throughout the medium and can be expressed only in the form of a differential equation for a particular point in the medium.

We consider an element of soil with sides dx, dy, and dz, as shown in Fig. 3.4. The rate of flow in the x direction is q_x across the plane $x = 0$ and $q_x + dq_x$ across the plane $x = dx$. The hydraulic gradients in the x direction at these two planes are, respectively, i_x and $i_x + di_x$. Similar notations are adopted for the flow in the y and z directions. From Darcy's law we can write

$$\begin{aligned} q_x &= k_x i_x \, dy \, dz & q_x + dq_x &= k_x(i_x + di_x) \, dy \, dz \\ q_y &= k_y i_y \, dx \, dz & q_y + dq_y &= k_y(i_y + di_y) \, dx \, dz \\ q_z &= k_z i_z \, dx \, dy & q_z + dq_z &= k_z(i_z + di_z) \, dx \, dy \end{aligned} \tag{3.7}$$

in which k_x, k_y, and k_z are the permeabilities in the x, y, and z directions, respectively. If the volume of the voids in the element remains constant, and if the fluid is incompressible, then the total rate of flow into the element must equal that coming out of the element, or

$$q_x + q_y + q_z = (q_x + dq_x) + (q_y + dq_y) + (q_z + dq_z)$$

From this we obtain

$$k_x di_x\, dy\, dz + k_y di_y\, dx\, dz + k_z di_z\, dx\, dy = 0 \tag{3.8}$$

Since $i_x = \partial h/\partial x$, $i_y = \partial h/\partial y$, $i_z = \partial h/\partial z$, then

$$di_x = \frac{\partial^2 h}{\partial x^2} dx \qquad di_y = \frac{\partial^2 h}{\partial y^2} dy \qquad di_z = \frac{\partial^2 h}{\partial z^2} dz \tag{3.9}$$

in which h represents the total head. Equation (3.8) may now be written as

$$\left[k_x \frac{\partial^2 h}{\partial x^2} + k_y \frac{\partial^2 h}{\partial y^2} + k_z \frac{\partial^2 h}{\partial z^2}\right] dx\, dy\, dz = 0 \tag{3.10}$$

If there is a change in volume at the rate dV/dt in the element during the flow, then we have

$$\left[k_x \frac{\partial^2 h}{\partial x^2} + k_y \frac{\partial^2 h}{\partial y^2} + k_z \frac{\partial^2 h}{\partial z^2}\right] dx\, dy\, dz = \frac{dV}{dt} \tag{3.11}$$

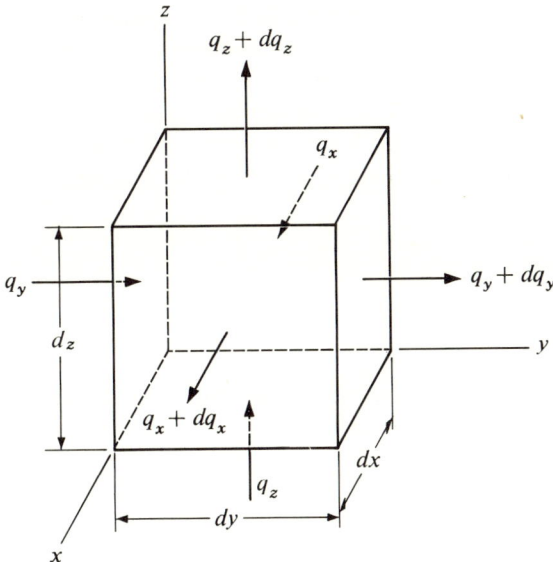

FIGURE 3.4 *Three-dimensional flow through an element.*

Equations (3.10), (3.11) are called *equations of continuity*, because continuous flow must satisfy one of the two equations. In the case of two-dimensional flow, the equations simplify into

$$\left(k_x \frac{\partial^2 h}{\partial x^2} + k_y \frac{\partial^2 h}{\partial y^2}\right) dx\, dy = 0 \qquad (3.12)$$

and

$$\left(k_x \frac{\partial^2 h}{\partial x^2} + k_y \frac{\partial^2 h}{\partial y^2}\right) dx\, dy = \frac{dV}{dt} \qquad (3.13)$$

If the *soil* is *isotropic* ($k_x = k_y = k$), we have, instead of Eq. (3.12),

$$\frac{\partial^2 h}{\partial x^2} + \frac{\partial^2 h}{\partial y^2} = 0 \qquad (3.14)$$

Since $v = ki$, this can also be written

$$\frac{\partial v_x}{\partial x} + \frac{\partial v_y}{\partial y} = 0 \qquad (3.15)$$

When the fluid in the voids is compressible, as in the case of an unsaturated material, then the volume change due to changes in the pressure must be taken into account. For a treatise on flow of compressible fluids see Muskat (1937).

3.5 Solutions of the continuity equation

Equation (3.14) describes the rate of change of h with x and y within the porous medium. To solve seepage problems we need to know the velocity at every point in the medium; or, if the head and piezometric level at any point is known, the velocity may be calculated from the gradient. To obtain the head, we must solve the differential equation (3.14) for prescribed boundary conditions.

We illustrate the solution by taking the simple problem of one-dimensional flow, as shown in Fig. 2.5. The cross-sectional area through which flow occurs is uniform within the length L. If we choose the direction of the x axis along the length of the rectangle then there is no flow in the y direction and Eq. (3.14) reduces to

$$\frac{\partial^2 h}{\partial x^2} = 0$$

Integrating this twice, we get

$$h = C_1 x + C_2 \qquad (3.16)$$

in which C_1 and C_2 are constants of integration to be determined from the boundary conditions. These are

$$\text{at} \quad x = 0, h = H \quad \text{at} \quad x = L, h = 0$$

Putting these values into Eq. (3.16) we get

$$C_1 = -\frac{H}{L} \quad C_2 = H$$

and the solution is

$$h = -\frac{H}{L}x + H$$

The solution shows that the head h decreases uniformly from H at $x = 0$ to 0 at $x = L$. Since the velocity is proportional to dh/dx, it follows that the velocity is constant throughout the rectangle. Other examples of direct solution are given by Scott (1963).

3.6 Potential and stream functions

For more complicated problems, it is convenient to introduce two functions $\Phi(x,y)$ and $\Psi(x,y)$, called the *potential function* and the *stream function*, respectively. We consider here the two-dimensional problem for an isotropic material ($k_x = k_y$). The functions Φ and Ψ are chosen to be

$$\frac{\partial \Phi}{\partial x} = v_x = -k\frac{\partial h}{\partial x} \qquad \frac{\partial \Phi}{\partial y} = v_y = -k\frac{\partial h}{\partial y} \qquad (3.17a)$$

and

$$\frac{\partial \Psi}{\partial y} = v_x = -k\frac{\partial h}{\partial x} \qquad -\frac{\partial \Psi}{\partial x} = v_y = -k\frac{\partial h}{\partial y} \qquad (3.17b)$$

We can combine Eqs. (3.17a) and (3.17b) and get

$$\frac{\partial \Phi}{\partial x} = \frac{\partial \Psi}{\partial y} \qquad -\frac{\partial \Phi}{\partial y} = \frac{\partial \Psi}{\partial x}$$

Equation (3.17a) can be integrated to yield

$$\Phi(x,y) = -kh(x,y) + C_1$$

in which C_1 is a constant depending on the boundary conditions. This equation can be studied as follows. If Φ is assigned a constant value equal to say Φ_1, then we have

$$h(x,y) = \frac{1}{k}(C_1 - \Phi_1) = \text{constant}$$

An equation of the form $h(x,y) = \text{constant}$ represents a curve in the xy plane. on this curve the value of Φ is constant and equal to the assigned value Φ_1. Since $\Phi_1 = -kh + C_1$, this is also a curve of equal h. If we assign a series of values $\Phi_1, \Phi_2, \ldots$ to Φ we get a family of curves which are called *equipotential lines*. Along each of these lines h is a constant equal to $h_1, h_2 \ldots$.

If we now assign constant values $\Psi_1, \Psi_2, \ldots$ to the function $\Psi(x,y)$, we also get a family of curves in the xy plane. The slope along such a curve is $(dy/dx)_\Psi$. This value can be obtained from the relation

$$d\Psi = \frac{\partial \Psi}{\partial x}dx + \frac{\partial \Psi}{\partial y}dy \qquad (3.18)$$

Along a curve on which Ψ is a constant, $d\Psi$ is 0. Hence Eq. (3.18) gives

$$\left(\frac{dy}{dx}\right)_\Psi = -\frac{\partial \Psi/\partial x}{\partial \Psi/\partial y}$$

From Eq. (3.17b) this is equal to

$$\left(\frac{dx}{dy}\right)_\Psi = \frac{v_y}{v_x}$$

Since v_x and v_y are the x and y components of the velocity, we see that the slope is also equal to the direction of the resultant velocity. Hence the curves for $\Psi_1, \Psi_2, \ldots$ represent directions of flow and are called *flow lines* or *stream lines*. Similarly, the slope of the equipotential lines $(dy/dx)_\Phi$ is equal to

$$\left(\frac{dy}{dx}\right)_\Phi = -\frac{\partial\Phi/\partial x}{\partial\Phi/\partial y} = -\frac{v_x}{v_y}$$

We see that it is perpendicular to that of the flow lines. Thus for an isotropic material flow and equipotential lines intersect each other at right angles.

Another useful characteristic of the flow lines is shown in Fig. 3.5. The flow q that occurs between two flow lines Ψ_1 and Ψ_2 can be written as

$$q = \int_{\Psi_2}^{\Psi_1} v_x\, dy$$

But from Eq. (3.17b) we see that v_x is equal to $\partial\Psi/\partial y$; so

$$q = \int_{\Psi_2}^{\Psi_1} \frac{\partial\Psi}{\partial y}\, dy = \Psi_1 - \Psi_2 \tag{3.19}$$

Therefore, the flow between any two flow lines is the difference between the stream functions.

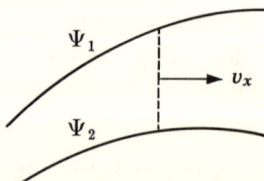

FIGURE 3.5 *Flow between two stream lines.*

The objective of the solution is to find the functions $\Phi(x,y)$ and $\Psi(x,y)$ for the given boundary conditions. The following example illustrates the solution of a seepage problem by the use of complex variables. Figure 3.6(a) shows an impervious dam over a pervious foundation extending to infinite depth. The foundation material is assumed to be homogeneous and isotropic.

Let the complex potential $\omega = \Phi + i\Psi$ be an analytical function of w so that

$$\omega = \Phi + i\Psi = g(w) = g(x + iy)$$

From the theory of complex variables [see, for example, Hildebrande (1962), pp. 496–98]

$$\frac{d\omega}{dw} = \frac{\partial\Phi}{\partial x} + i\frac{\partial\Psi}{\partial x}$$

SEC. 3.6   Potential and stream functions

From Eq. (3.17) this is also

$$\frac{d\omega}{dw} = v_x - iv_y$$

and this is called the *complex velocity*. For the boundary conditions of the problem, we have the following. The horizontal velocity is 0 along AB and CD, whereas along the base of the dam v_y is 0, as flow must be parallel to the impermeable boundary. So

$$y = 0, \quad x < -b \text{ and } x > b: \quad v_x = 0$$
$$y = 0, \quad -b < x < b: \quad v_y = 0$$

The first of these conditions means that $d\omega/dw$ is imaginary when $x < -b$ or $x > b$; the second condition means that $d\omega/dw$ is real if $-b < x < b$. Consider now the function $(b - w)^{1/2}(b + w)^{1/2}$. It meets both the above requirements.

A second boundary condition is that the velocity should approach 0 at infinite depth.

$$y \to \infty: \quad v_x \to 0, \quad v_y \to 0$$

So we let the function $d\omega/dw$ take the form

$$\frac{d\omega}{dw} = \frac{A}{(b^2 - w^2)^{1/2}}$$

in which A is a constant. Integrating this yields

$$\omega = A \sin^{-1}\left(\frac{w}{b}\right) + B \qquad (3.20)$$

and B is the constant of integration.

The xy plane and $\Phi\Psi$ plane are shown in Fig. 3.6(a) and (b). On AB the head is H. It is an equipotential line and we assign the value $\Phi = H$. On CD the head is 0 and so $\Phi = 0$. On CB we let $\Psi = 0$. Thus the area $ABCD$ in the $\Phi\Psi$ plane corresponds to the pervious foundation in the xy plane. To evaluate A and B we have

$$\text{at } C: \quad \omega = \Phi + i\Psi = 0 \quad w = b$$
$$\text{at } B: \quad \omega = H \quad\quad\quad\quad\quad w = -b$$

Substitution of the boundary conditions into Eq. (3.20) gives

$$A = -\frac{H}{\pi}, \quad B = \frac{H}{2}$$

and

$$\omega = -\frac{H}{\pi}\sin^{-1}\left(\frac{w}{b}\right) + \frac{H}{2} = -\frac{H}{\pi}\cos^{-1}\left(\frac{w}{b}\right)$$

or

$$w = b\cos\frac{\pi\omega}{H}$$

Since $w = x + iy$ and $\omega = \Phi + i\Psi$, we get

$$x + iy = -b\cos\left[\frac{\pi}{H}(\Phi + i\Psi)\right]$$

$$= -b\left(\cos\frac{\pi\Phi}{H}\cos\frac{\pi i\Psi}{H} - \sin\frac{\pi\Phi}{H}\sin\frac{\pi i\Psi}{H}\right)$$

$$= b\left(\cos\frac{\pi\Phi}{H}\cosh\frac{\pi\Psi}{H} - i\sin\frac{\pi\Phi}{H}\sinh\frac{\pi\Psi}{H}\right)$$

Equating the real and imaginary parts, we have

$$x = b\cos\frac{\pi\Phi}{H}\cosh\frac{\pi\Psi}{H} \quad (3.21a)$$

$$y = -b\sin\frac{\pi\Phi}{H}\sinh\frac{\pi\Psi}{H} \quad (3.21b)$$

From which we obtain

$$\cos\frac{\pi\Phi}{H} = \frac{x}{b\cosh(\pi\Psi/H)}$$

$$\sin\frac{\pi\Phi}{H} = -\frac{y}{b\sinh(\pi\Psi/H)}$$

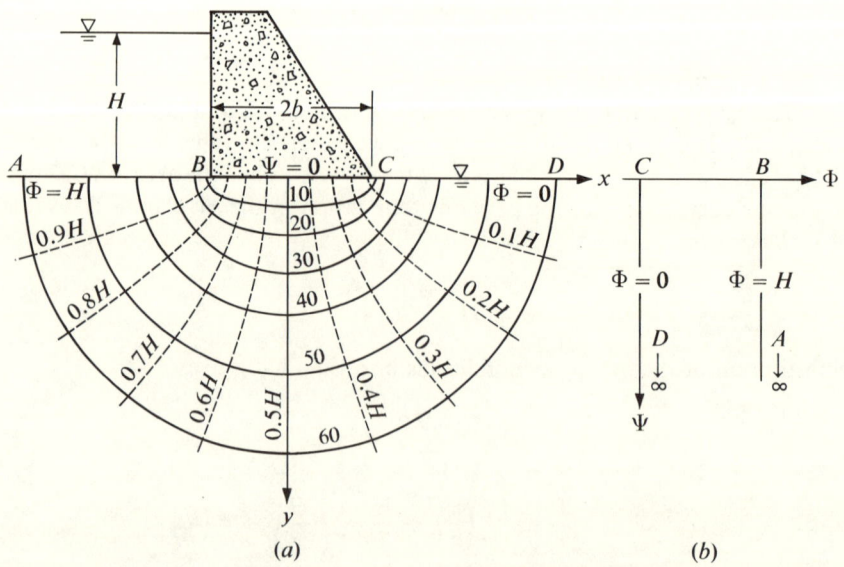

FIGURE 3.6 *Flow beneath a dam on a permeable foundation.* [After Zanger (1953).]

Since $\cos^2 x + \sin^2 x = 1$
and $\cosh^2 x - \sinh^2 x = 1$
we have

$$\frac{x^2}{[b \cosh (\pi \Psi / H)]^2} + \frac{y^2}{[b \sinh (\pi \Psi / H)]^2} = 1 \qquad (3.22a)$$

$$\frac{x^2}{[b \cos (\pi \Phi / H)]^2} - \frac{y^2}{[b \sin (\pi \Phi / H)]^2} = 1 \qquad (3.22b)$$

Thus we have obtained Φ and Ψ as functions of x, y and the problem is solved.

We may plot the results graphically as equipotential lines and flow lines. By taking Ψ equal to a sequence of constants $\Psi_1, \Psi_2, \ldots$ we can plot a series of ellipses by using Eq. (3.22a), which are the flow lines. The equipotential lines may be plotted by taking Φ equal to $\Phi_1, \Phi_2, \ldots$. The curves obtained with Eq. (3.22b) are hyperbolas. The flow lines and equipotential lines are shown in Fig. 3.6(a). (This solution may also be found in the following references: Harr, 1962, pp. 88–91; Muskat, 1937, pp. 183–86; Scott, 1963, pp. 100–102.)

The solutions may be utilized to determine a number of quantities of practical importance as shown in the following examples. Along the base of the dam $\Psi = 0$. Substituting this condition into Eq. (3.21a), we obtain

$$x = b \cos \frac{\pi \Phi}{H}$$

and

$$\Phi = \frac{H}{\pi} \cos^{-1} \left(\frac{x}{b}\right) \qquad (3.23)$$

Since $\Phi = -kh$, Eq. (3.23) gives the distribution of water pressure along the base of the dam. To obtain the velocity along the base, Eq. (3.23) may be differentiated to give

$$\frac{d\Phi}{dx} = k \frac{H}{\pi b} \left(\frac{1}{1 - (x/b)^2}\right)$$

3.7 Flow net

A set of flow lines and equipotential lines as shown in Fig. 3.6(a) is called a *flow net*. Another example, given in Fig. 3.7, illustrates the flow through the pervious subsoil around an impermeable barrier. The equipotential lines are shown as dashed lines and the flow lines are shown as solid lines. As defined in Sec. 3.6, the equipotential lines are lines connecting points of equal total head h. In other words, any two points such as b and c on the same equipotential line possess the same piezometric level.

If we use the elevation of point b as datum, the total head at b is equal to the hydrostatic head $h'_b + h_b$. At point c the total head is the same and consists of the hydrostatic head plus the elevation difference between b and c. It is obvious that the hydraulic gradient between two points b and c on the same equipotential

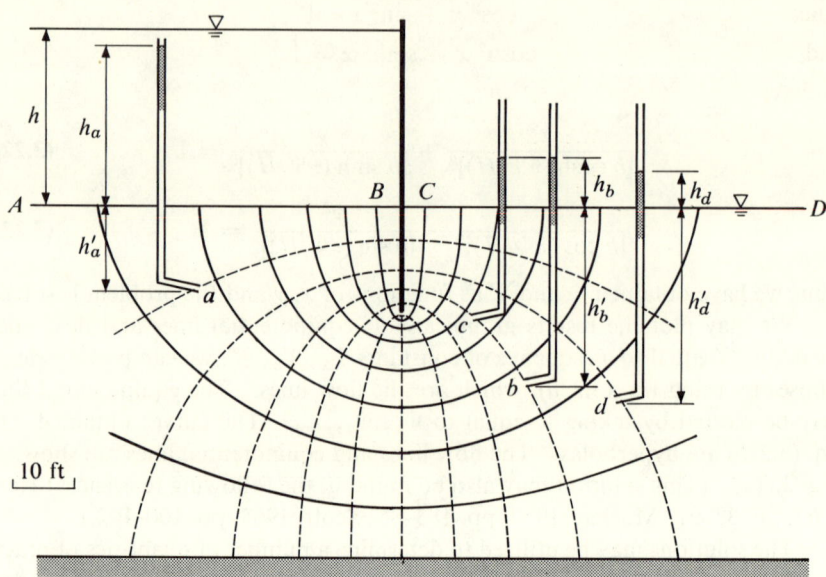

FIGURE 3.7 *Flow net around a barrier wall.*

line is zero. A point d on another equipotential line has a different piezometric level. Therefore, flow takes place from one equipotential line toward another.

The flow net can also be used for two-dimensional flow in the horizontal plane. Such is the case of the confined flow in the aquifer shown in Fig. 3.2(a). Since the aquifer is of uniform thickness, there is no vertical component v_z in the seepage velocity. At any distance r_1 from the well, the piezometric level throughout the entire thickness of the aquifer is at h_1 above the top of the aquifer. Hence

$$\frac{\partial h}{\partial z} = 0 \quad \text{and} \quad v_z = 0$$

Therefore the flow is two-dimensional in the horizontal plane. In the plan [Fig. 3.2(b)] the equipotential lines appear as concentric circles and the flow lines are radial lines. As long as the flow is entirely in the horizontal direction, the flow net can be drawn for the plan. Figure 3.8(a) is an example of flow from a river into a drained area. The flow takes place through a layer of permeable sand that is connected to the river and is bounded at the upper and lower surfaces by impermeable clayey soils [Fig. 3.8(b)]. The wells used for drainage completely penetrate the sand [as in Fig. 3.2(a)] and flow occurs only in the horizontal direction.

GRAPHICAL CONSTRUCTION. In many soil mechanics problems, the boundary conditions are often so complex that the solution cannot be obtained analytically. Numerical methods are then used to obtain the flow net. For the problem illustrated in Fig. 3.7 the boundary conditions are described as follows. First,

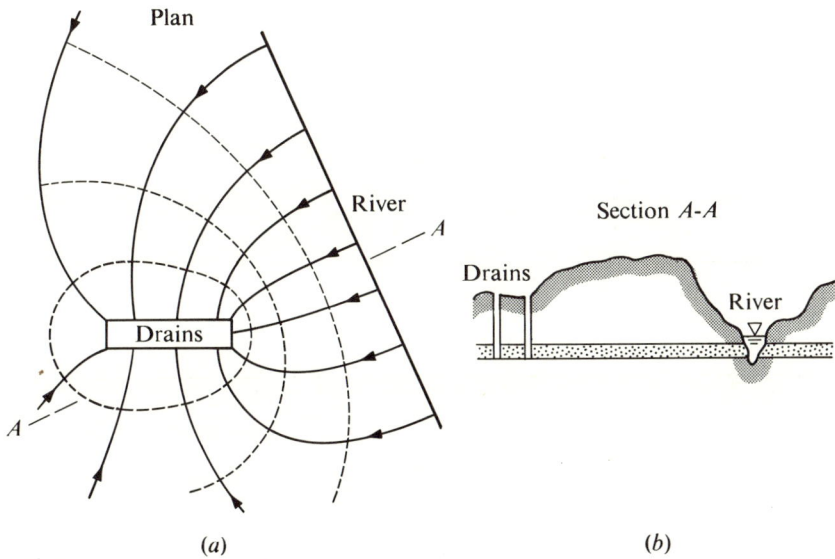

FIGURE 3.8 *Seepage from a river into a drained area.*

the ground surfaces AB on the upstream side and CD on the downstream side are equipotential lines. Also, flow must move along the barrier, because flow cannot go through it. Hence the barrier is a flow line, and so is the bottom of the previous stratum. It should be realized that the boundary conditions can be satisfied by only one pattern of flow and equipotential lines, which constitutes the correct solution. Starting from these boundaries we construct the flow net by trial and error. In addition to meeting the boundary conditions the flow and equipotential lines should intersect each other at right angles (Sec. 3.6). Second, the area enclosed between two adjacent pairs of flow lines and equipotential lines should be square. Trial flow nets are sketched in and revised until these requirements are met.

The flow lines obtained by construction of the flow net can be checked by laboratory experiments on models. Figure 3.9(*a*) shows the flow net for seepage of groundwater underneath the slopes of a valley into a river. Figure 3.9(*b*) is a photograph of the flow lines marked by dye color introduced into the water at a number of points along the boundaries.

SEEPAGE AND VELOCITY. After the flow net has been constructed we may proceed to calculate the hydraulic gradient, velocity of flow, and seepage force at any point in the medium, as well as the total rate of flow. We consider a series of squares between two flow lines (Fig. 3.10). The equipotential lines represent piezometric levels h_1, h_2, h_3, and h_4. Since there is no flow across any flow line we have, for any "lane" between two flow lines,

$$\Delta q_1 = \Delta q_2 = \Delta q_3 = \Delta q$$

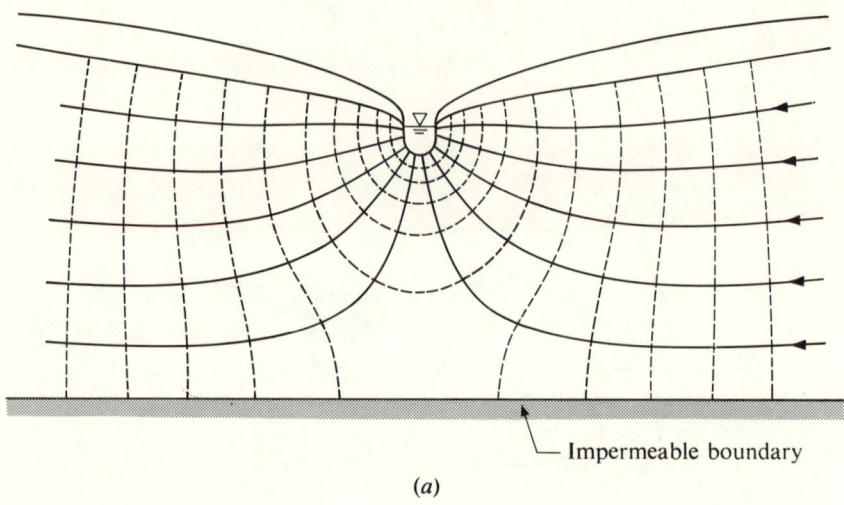

(a)

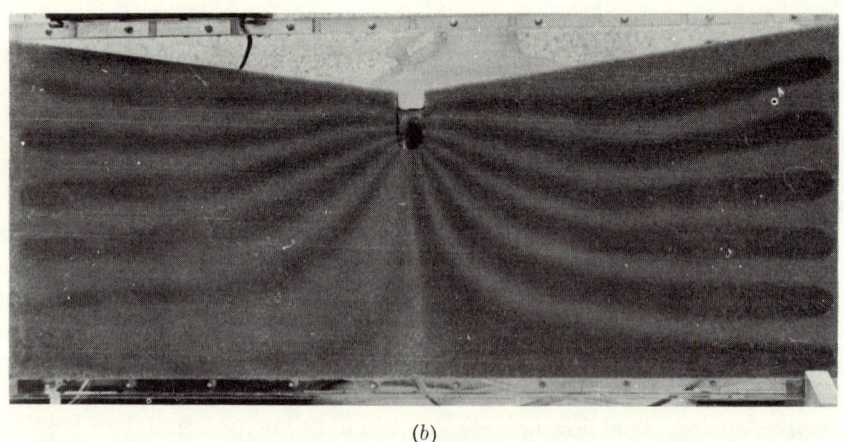

(b)

FIGURE 3.9 *Seepage of groundwater into a river.* [(b) courtesy of J. H. Lehr.]

For an isotropic material k is the same in all directions, and the flow is equal to kiA. It follows that

$$k\frac{h_1 - h_2}{l_1}b_1 = k\frac{h_2 - h_3}{l_2}b_2 = k\frac{h_3 - h_4}{l_3}b_3$$

For squares,

$$\frac{b_1}{l_1} = \frac{b_2}{l_2} = \frac{b_3}{l_3} = \frac{b}{l} = 1$$

Hence, we have

$$h_1 - h_2 = h_2 - h_3 = h_3 - h_4 = \Delta h = \frac{h}{N} \qquad (3.24)$$

in which N is the number of equipotential increments.

SEC. 3.7   Flow net   63

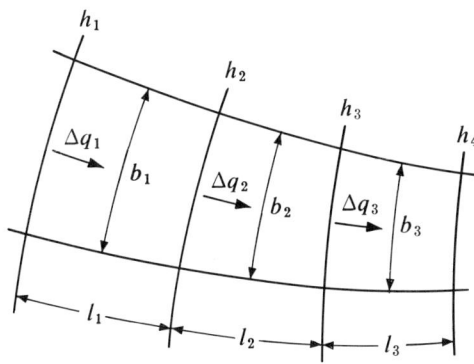

FIGURE 3.10 *The flow net.*

From Eq. (3.24) we see that the drop in head is the same between any two successive equipotential lines, and the rate of flow Δq between any two adjacent flow lines is also a constant. Therefore, the total rate of flow q in the flow net is

$$q = F\,\Delta q = Fk\,\Delta h\frac{b}{l} = k\frac{F}{N}h \tag{3.25}$$

in which F represents the number of flow lanes and h is the total difference in piezometric level. In Fig. 3.7, F and N are equal to 8 and 12, respectively. The head difference Δh between two successive equipotential lines is therefore $h/12$. At point a we have $h - h_a = h/12$ or $h_a = \frac{11}{12}h$. At point b we have $h - h_b = \frac{9}{12}h$, since there are nine decrements beginning at the equipotential line AB. The head at b is $\frac{3}{12}h$. This is the head difference between b and the 0 equipotential line CD.

The rate of flow q is the same in all flow lanes, but the velocity varies inversely with the width of the flow lane as given by

$$v = \frac{\Delta q}{b} = \frac{q}{F}\frac{1}{b} \tag{3.26a}$$

and the hydraulic gradient is

$$i = \frac{\Delta h}{l} \tag{3.26b}$$

For any square (Fig. 3.10), the value of b and l may be measured from the flow net.

Example 3.1. Calculate the piezometric heads and velocities at points a and b, Fig. 3.7. The value of h is 28 ft and the permeability coefficient is 10^{-5} ft/hr.

Solution: The equipotential line that passes through a represents a piezometric level that is one decrement lower than that of AB. Since the decrement of head between two adjacent equipotential lines is h/N, we have

$$\Delta h = \frac{h}{N} = \frac{28}{12} = 2.33 \text{ ft}$$

Hence the piezometric level at a is 2.33 ft below the free water surface. The piezometric head (or hydrostatic head) is equal to $h'_a + h_a$:

$$h_a = h - \Delta h = 28.00 - 2.33 = 25.67 \text{ ft}$$

By scaling we get

$$h'_a = 13.00 \text{ ft}$$

Therefore,
$$h_a + h'_a = 38.67 \text{ ft}$$

The equipotential line through b represents a piezometric level that is nine decrements lower than that of AB. Thus

$$h_b = h - 9\,\Delta h = 28.00 - (9 \times 2.33) = 7.00 \text{ ft}$$

The piezometric level at b is also three increments higher than that of CD, or

$$h_b = 0 + 3\,\Delta h = 3 \times 2.33 = 7.00 \text{ ft}$$

The distance h'_b is scaled and we get

$$h'_b = 28.00 \text{ ft}$$

The piezometric head is

$$h_b + h'_b = 7.00 + 28.00 = 35.00 \text{ ft}$$

The velocity at any point is $k(\Delta h/l)$. The quantity Δh is 2.33 ft, while b, which is width of the squares in the flow net, varies. At point a, b is 12.5 ft according to the scale, and at point b it is 8.5 ft. Hence the velocities are

$$v = k\frac{\Delta h}{l} = 10^{-5} \times \frac{2.33}{12.50} = 1.87 \times 10^{-6} \text{ ft/hr at } a$$

$$= 10^{-5} \times \frac{2.33}{8.50} = 2.74 \times 10^{-6} \text{ ft/hr at } b$$

3.8 Anisotropic materials

When the material is anisotropic, it is necessary to modify the flow net and the relationships developed in Sec. 3.5. We consider only the case in which k_x and k_y are different. It is necessary to go back to the basic relationship defined by Eq. (3.12). It may be rewritten as

$$\frac{\partial^2 h}{(k_y/k_x)\,\partial x^2} + \frac{\partial^2 h}{\partial y^2} = 0$$

In this case, we can adopt a new coordinate scale x', such that $x' = \sqrt{k_y/k_x}\,x$. In terms of this coordinate, the above equation becomes

$$\frac{\partial^2 h}{\partial x'^2} + \frac{\partial^2 h}{\partial y^2} = 0$$

This equation is the same as (3.12) except that it has x' as the coordinate instead of x. Hence we can construct the flow net as if the material is isotropic after transforming the x dimensions to x'. The y scale remains unchanged.

3.9 Nonhomogeneous materials and transfer conditions

After the flow net has been drawn in the $x'y$ coordinate, the quantity of seepage can be calculated by

$$q' = k'\frac{F}{N}h$$

This is the same as Eq. (3.25) except that k' in the $x'y$ coordinate system is given by

$$k' = \sqrt{k_x k_y} \tag{3.27}$$

This relationship follows from the condition that flow in the x and x' directions for a given square (Fig. 3.11) must be the same. In the natural x scale the flow is $q_x = k_x(\Delta h/L)A$, while in the transformed x' scale it is $q_x = k'(\Delta h/L\sqrt{k_y/k_x})A$. Equating the two quantities, we obtain Eq. (3.27).

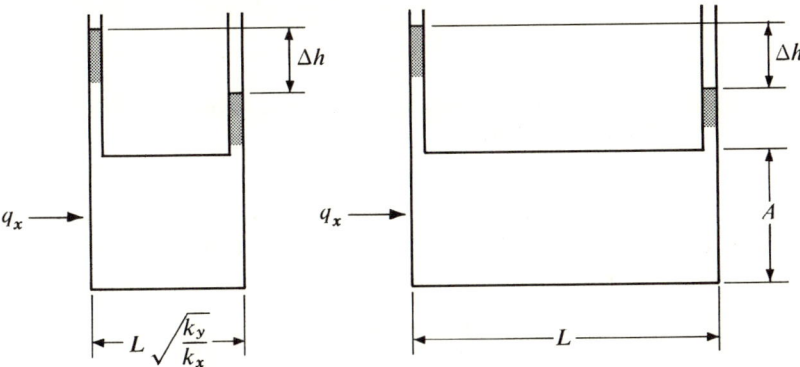

FIGURE 3.11 *Flow through anisotropic materials.*

3.9 Nonhomogeneous materials and transfer conditions

When flow crosses the boundary between two materials with different permeabilities, the flow line is deflected. This is called *transfer condition*. To analyze the transfer condition, we consider the portion of the flow net near the boundary, as drawn in Fig. 3.12(a). The condition that must be satisfied is that the quantity of flow Δq between two adjacent flow lines must be the same on both sides of the boundary. The quantity Δq on the left side is equal to $k_1 a(\Delta h/a)$, since the flow net consists of squares with sides a. It can be seen that for k_2 different from k_1 the ratio $c:b$ on the right-hand side cannot be equal to 1 if the flow Δq is to be the same on both sides of the boundary. This can be expressed as follows:

$$\Delta q = k_1 a \frac{\Delta h}{a} = k_2 c \frac{\Delta h}{b}$$

It follows that the ratio $c:b$ depends on the ratio between the two permeabilities, or

$$\frac{c}{b} = \frac{k_1}{k_2} \tag{3.28}$$

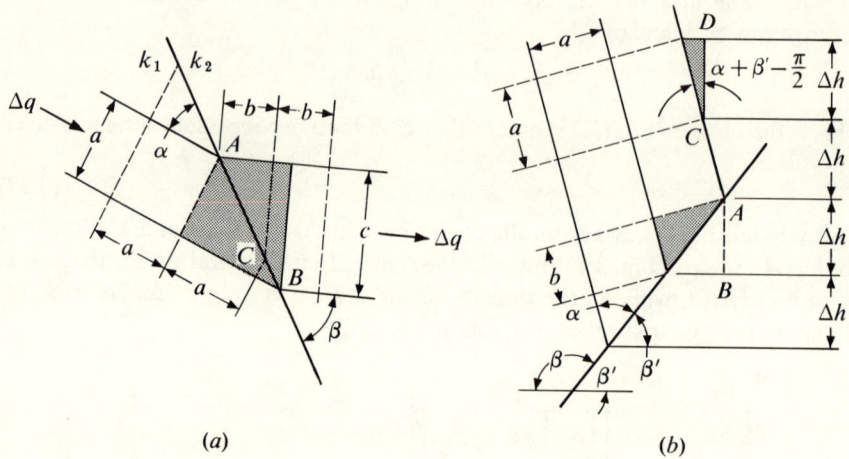

FIGURE 3.12 *Transfer conditions in nonhomogeneous materials.*
[After Casagrande (1937).]

The deflection of the flow line can now be evaluated from the dimensions of the two shaded triangles in Fig. 3.12(a). The distances AB and AC can be expressed as

$$\overline{AB} = \frac{\alpha}{\sin \alpha} = \frac{c}{\sin \beta} \qquad \overline{AC} = \frac{a}{\cos \alpha} = \frac{b}{\cos \beta}$$

Combining these, we obtain

$$\frac{\tan \beta}{\tan \alpha} = \frac{c}{b} = \frac{k_1}{k_2} \tag{3.29}$$

Therefore, the tangent of the intersecting angles is inversely proportional to the permeabilities. An example of the deflection of the flow lines is given in the flow net in Fig. 3.13.

Example 3.2. Construct a flow net for seepage through the nonhomogeneous subsoil shown in Fig. 3.13.

Solution. From Eq. (3.29),

$$\frac{\tan \beta}{\tan \alpha} = \frac{c}{b} = \frac{k_1}{k_2} = 2$$

Thus a flow line crossing the boundary is deflected so that $\tan \beta/\tan \alpha$ (Fig. 3.13) is equal to 2. If the part of the flow net in the top layer is made up of squares, the part in the lower layer should have a ratio c/b [Figs. 3.12(a) and 3.13] equal to 2. The flow net is constructed to meet these two additional requirements.

Next we consider flow from a material with permeability k_1 into a free-draining material. Its permeability is so large compared to k_1 that it can be

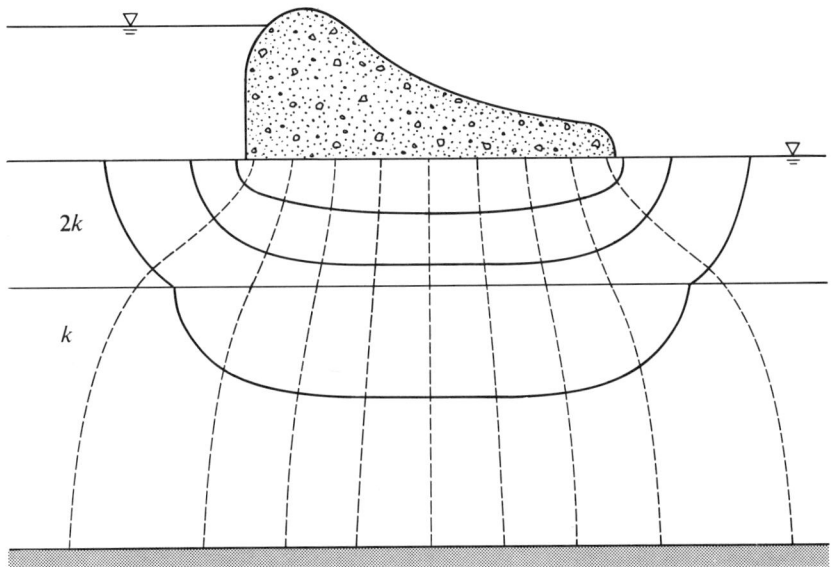

FIGURE 3.13 *Flow net in a subsoil that consists of two layers with different permeabilities.*

taken as infinity. Such a case is often called *discharge*. The opposite case, where flow passes from a free-draining material into a material with permeability k_2, is called *entrance*. Figure 3.12(b) shows the case of discharge into an overhanging slope. The angle α must be such that the flow net shall consist of squares throughout, or b should be equal to a. Considering the shaded triangles, this requirement can be written as

$$\overline{AB} = \frac{b}{\cos \alpha} \sin \beta'$$

and $\overline{CD} = a \cos(\alpha + \beta' - 90°)$

Since both AB and CD must be equal to Δh, we have

$$\frac{b}{\cos \alpha} \sin \beta' = a \cos(\alpha + \beta' - 90°)$$

To meet the requirement $a = b$, we must have

$$\alpha = 90 - \beta'$$

which means that the line of seepage must have a vertical slope at the boundary.

By similar analyses, other entrance and discharge conditions have been worked out and are given in Fig. 3.14. For a detailed treatment of this subject, see Casagrande (1937). Examples of flow through nonhomogeneous materials are also given by Shea and Whitsett (1958).

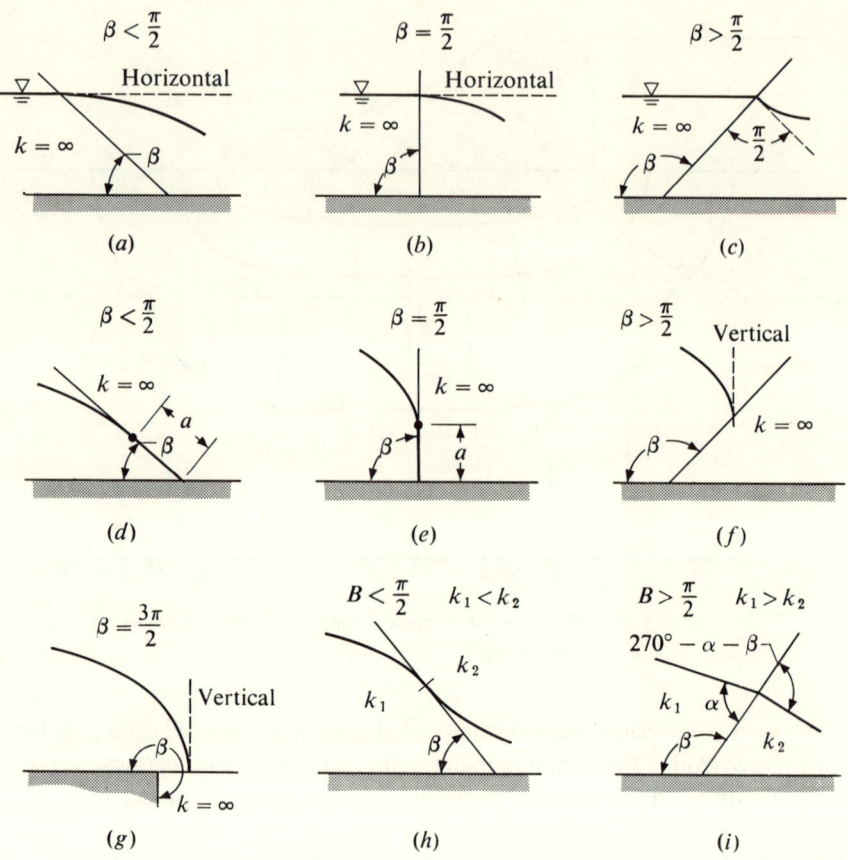

FIGURE 3.14 *Transfer conditions.* [After Casagrande (1937).]

3.10 Free surface flow and basic parabola

The examples of flow net we considered in Secs. 3.6 and 3.7 all have boundary conditions completely defined by the physical boundaries of the problem. Such is not always the case, and one important type of problem, involving free surface flow, is illustrated in Figs. 3.15(*a*) and 3.16(*a*). Figure 3.15(*a*) shows seepage through a pervious earth dam resting on an impervious foundation. Figure 3.16(*a*) shows flow under a head h into a drain. The uppermost flow line is called *phreatic surface* or *line of seepage* and is not determined by the physical boundaries. Therefore it should be located before the flow net can be constructed. Otherwise, the construction of flow net involves one additional item to be determined by trial and error and becomes very time-consuming.

The phreatic surface is subjected only to atmospheric pressure. Therefore it must obey one physical condition—the only head along this flow line is the

SEC. 3.10　　　　Free surface flow and basic parabola　　　　69

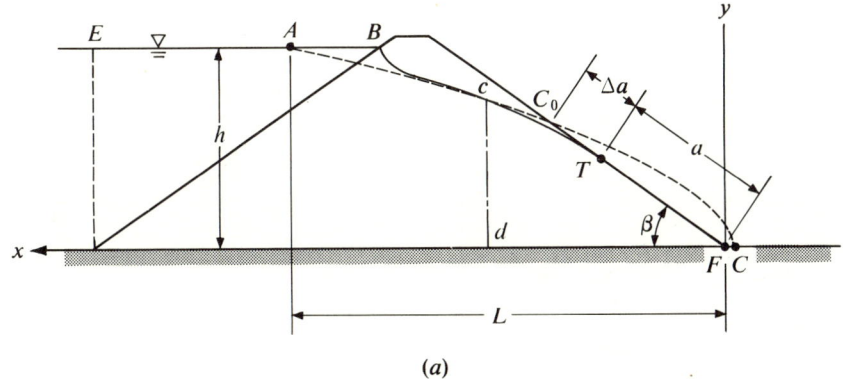

(a)

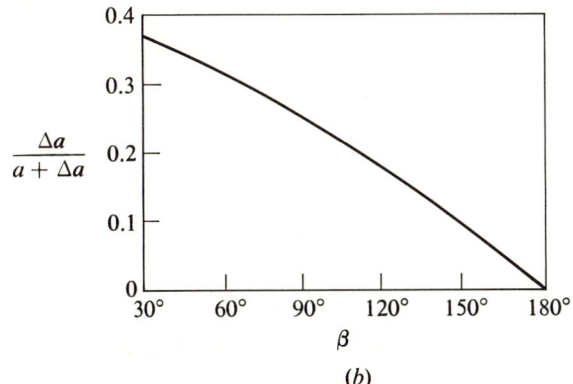

(b)

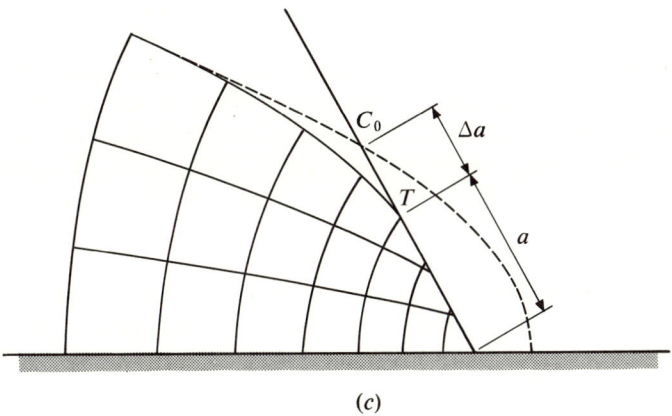

(c)

FIGURE 3.15 *Flow through a pervious earth dam.* [After Casagrande (1937).]

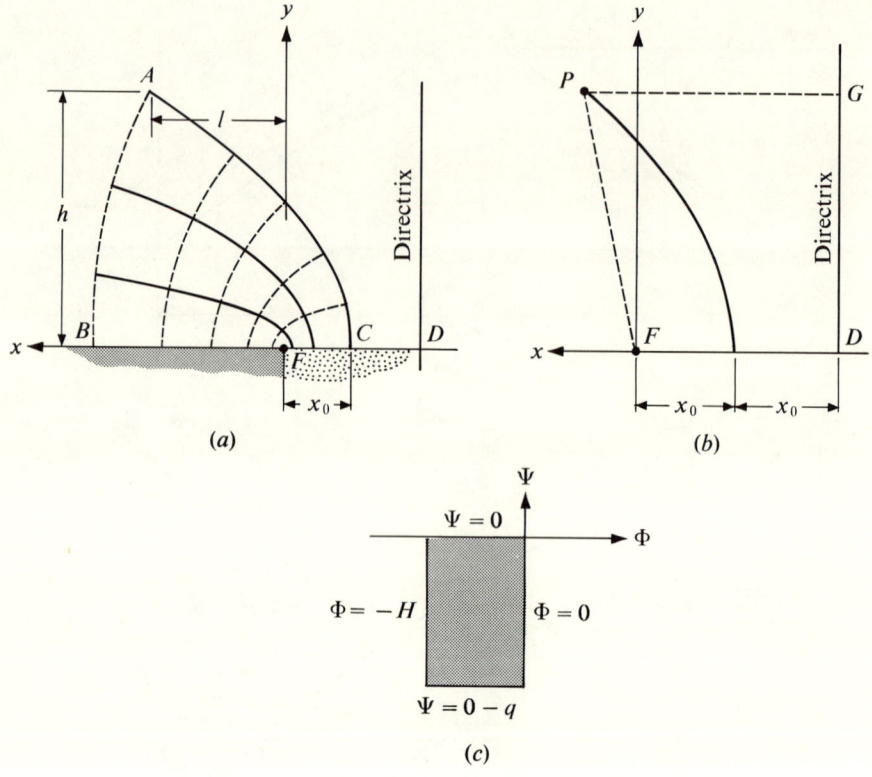

FIGURE 3.16 *Flow through a pervious soil into a drain.*

elevation head. Consequently, the head drop between successive equipotential lines must be constant and equal to the vertical distance between them.

We study this topic by beginning with a simple problem in which the flow lines are confocal parabolas with the same focus at F, and the equipotential lines are conjugate confocal parabolas. Figure 3.16(a) illustrates the physical conditions for such a flow net. It is a flow through an isotropic material over a horizontal, impervious surface BF, which at point F continues as a free-draining material with infinite permeability. The material below DCF is free-draining, so that DF is an equipotential line. The other boundary condition, AB, an equipotential surface, is a parabola with focus at F. Under these conditions the two sets of parabolas meet the requirements for the flow net (see Example 3.3). To locate the phreatic surface AC we make use of the property [Fig. 3.16(b)] that the distance FP from the focus to any point P is equal to the distance PG from the point P to the directrix. Hence

$$\sqrt{x^2 + y^2} = x + 2x_0 \tag{3.30}$$

or
$$x = \frac{y^2 - 4x_0^2}{4x_0} \tag{3.31}$$

SEC. 3.10   **Free surface flow and basic parabola**

The known point of this parabola is at A, because of the boundary conditions. At A we have $y = h$ and $x = l$. Equation (3.30) then leads to

$$x_0 = \tfrac{1}{2}(\sqrt{l^2 + h^2} - l) \tag{3.32}$$

With x_0 known, the various points on the parabola can be calculated by Eq. (3.31) and the flow net can be drawn.

Example 3.3. The basic parabola can be obtained by the following calculations. The upstream face is an equipotential line and we assign to it the value $\Phi = -H$. Along CF we have $\Phi = 0$. If we assign the value $\Psi = 0$ to the impervious base BF, then it follows from Eq. (3.19) that $\Psi = -q$ for the phreatic surface. The quantity q denotes the flow between the phreatic surface and the impermeable base. Thus the area of flow in the xy plane corresponds to the shaded area in the $\Phi\Psi$ plane [Fig. 3.16(c)].

Solution. We take the relationship between the $w = x + iy$ plane and the $\omega = \Phi + i\Psi$ plane as

$$w = C\omega^2$$

in which C is a constant. Along the phreatic surface

$$\Phi = -y^* \qquad \Psi = -q$$

Hence,

$$w = C\omega^2 = C(\Phi + i\Psi)^2$$

or $x + iy = C(-y - iq)^2 = C(y^2 + 2iqy - q^2)$

Equating real and imaginary parts,

$$x = C(y^2 - q^2) \qquad y = 2Cyq$$

From which we obtain

$$x = \frac{1}{2}\left(\frac{y^2}{q} - q\right)$$

This is the equation of the basic parabola. The other flow lines can be obtained by assigning different values to Ψ such as $0.8q$, $0.6q$......

Along the upstream face AB

$$\Phi = -H$$

Then $w = x + iy = C\omega^2 = C(-H + i\Psi)^2 = C(H^2 - 2iH\Psi - \Psi^2)$

Equating real and imaginary parts,

$$x = C(H^2 - \Psi^2) \quad \text{or} \quad \Psi^2 = \frac{x}{C} - H^2$$

$$y = -2CH\Psi \quad \text{or} \quad \Psi^2 = \frac{y^2}{4C^2H^2}$$

Thus we have

$$\frac{x}{C} - H^2 = \frac{y^2}{4C^2H^2}$$

---

* This is because the only head is the elevation head y.

Substitution of $C = 1/2q$ into the above yields

$$x = \frac{H^2}{2q} + \frac{qy^2}{H^2}$$

This is the equation of the upstream face and equipotential surface AB.

3.11 Examples of flow through earth embankments

The example of a flow through an earth dam, shown in Fig. 3.15, departs somewhat from the ideal example described in the preceding section. Most, if not all, dams and other earth structures have forms that do not lead to exact solutions. In many cases the flow conditions are close enough to the basic parabola so that approximate solutions can be obtained by introducing modifications to the basic parabola.

UPSTREAM FACE. In Fig. 3.15 the conditions depart from the basic parabola in that the upstream boundary is not a conjugate confocal parabola. Casagrande (1937) has found the line of seepage is close to the basic parabola through point A (dashed curve). The distance AB is approximately 0.3 of the distance EB. This was obtained from a study of "correct" flow nets constructed by trial. With AB known, the approximate seepage line can be constructed as a parabola. Other approximate methods are available, and for these the reader may consult Polubarinova-Kochina (1962) or Karpoff (1954), who translated some of Pavlov's works.

DOWNSTREAM FACE. A second departure is that the toe drain is omitted in Fig. 3.15(a). Thus we have to consider the changes at the downstream face. Since the water cannot enter the impermeable base, the line of seepage must intersect the downstream slope. According to the results given in Fig. 3.14, it should be tangent to the downstream slope. The point of tangency, represented by distance a, must then be determined. To obtain an approximate solution, we introduce the simplifying assumption that along any vertical line cd, the hydraulic gradient is constant with depth and equals the slope dy/dx of the line of seepage at c. This approximation is attributed to Dupuit. Under this assumption, we have

$$q = kiA = k\frac{dy}{dx}y$$

Since the continuity condition must hold, we have

$$\frac{\partial q}{\partial x} = k\frac{\partial^2(y^2)}{\partial x^2} = 0$$

Integrating twice we get the equation of the parabola:

$$y^2 = Ax + B$$

Given the dimensions, h, l, and β the boundary conditions are as follows:

at point A: $\quad x = l, \quad y = h$

at point T: $\quad x = a \cos \beta, \quad \dfrac{dy}{dx} = \tan \beta, \quad y = a \sin \beta$

The solution is

$$y^2 = 2a \frac{\sin^2 \beta}{\cos \beta} x + h^2 - 2a \frac{\sin^2 \beta}{\cos \beta} l \qquad (3.33)$$

For the point of tangency T, we have

at $x = a \cos \beta, \quad y = a \sin \beta$

Substituting in Eq. (3.33) we get

$$a = \frac{l}{\cos \beta} - \sqrt{\frac{l^2}{\cos^2 \beta} - \frac{h^2}{\sin^2 \beta}} \qquad (3.34)$$

The quantity of seepage is

$$q = ka \sin \beta \tan \beta \qquad (3.35)$$

For angles of β greater than 30 deg, the departure from Dupuit's assumption becomes more noticeable. Hence, use of Eq. (3.35) involves a considerable error. Dupuit's assumption may then be modified to

$$q = k \frac{dy}{ds} y$$

in which s is the distance along the line of seepage (Casagrande, 1937; Gilboy, 1934).

On the basis of many flow nets constructed by trial and error, Casagrande (1937) has found that for large angles of β, the actual line of seepage is close to the basic parabola and the difference is limited to the area at the toe [Fig. 3.15(c)]. Since the process of obtaining the correct flow net and line of seepage by trial and error is very time-consuming, he suggested that the line of seepage be determined first by application of a correction to the basic parabola. The basic parabola is constructed first and its intersection with the downstream face (point C_0) can be determined. A correction to the basic parabola is then applied to locate point T. By comparing the flow nets constructed by trial and error with the basic parabola as shown in Fig. 3.15(c), for example, Casagrande (1937) obtained the correction to the basic parabola. This correction expressed as a ratio, $\Delta a/(a + \Delta a)$, is given in Fig. 3.15(b) as a function of the slope angle β.

The flow net for a simple homogeneous dam (Fig. 3.15) may therefore be constructed by beginning with a section of the dam drawn according to the transformed scale

$$x' = \sqrt{\frac{k_y}{k_x}} x$$

After this the point A on the basic parabola can be located by setting AB equal to $\frac{1}{3}EB$. The focus of the parabola is at F and point C is located by use of Eq. (3.32). The parabola intersects the downstream surface at C_0, a point which is known after the parabola has been drawn. The point of tangency of the line of seepage with the downstream face is at T and the distance Δa is taken from Fig. 3.15(b). The line of seepage can then be sketched in. After this, the flow net can be drawn. If β is less than 30 deg, the point T may be located by calculating a from Eq. (3.34).

The approximate solutions given by Casagrande are in good agreement with the experimental work of Casagrande (1932), in which the line of seepage and flow lines were located by traces of dye injected into the water. Additional examples of flow through earth dams are given by Karpoff (1954), Reinus (1955), and Harr (1962).

Example 3.4. Locate the line of seepage for the earth-dam cross section shown in Fig. 3.25(a). Assume the material to be isotropic.

Solution. The first step is to locate the basic parabola. On the upstream face the point A [Fig. 3.25(b)] is determined by making

$$\overline{AB} = \tfrac{1}{3}\overline{BE} = \tfrac{1}{3} \times 250 = 83.3 \text{ ft}$$

To construct the basic parabola x_0 is computed with Eq. (3.32):

$$l = 250 + 12.5 + 30 + 12.5 + 83.3 = 388.3 \text{ ft}$$

$$x_0 = \tfrac{1}{2}[\sqrt{(388.3)^2 + (100)^2} - 388.3] = 12 \text{ ft}$$

The locations of the points on the parabola are then computed from Eq. (3.31), and point T is located as follows:

$$\beta = \tan^{-1}\frac{1}{2.5} = 21.8°$$

$$a = \frac{388.3}{\cos 21.8°} - \left[\frac{(388.3)^2}{\cos^2 21.8} - \frac{(100)^2}{\sin^2 21.8}\right]^{1/2}$$

$$= 418 - 320 = 98 \text{ ft}$$

The connecting sections BB' and TT' are then sketched in and the line of seepage is complete.

In the case where β exceeds 30 deg, point T may be located by taking the quantity Δa from Fig. 3.15(b).

3.12 Seepage force

The flow of water through soil exerts a force on the soil that is called the *seepage force*. To show that such a force exists, we consider an element of soil $abcd$ bounded by two equipotential lines and two flow lines [Fig. 3.17(a)]. The equipotential line ab has a potential surface Δh above that of the equipotential line cd, and the direction of flow is from ab to cd. The water pressure on ab is $h_1\gamma_w$

SEC. 3.12 Seepage force

and that on cd is $(h_1 - \Delta h + b \sin \alpha)\gamma_w$. The normal force on ab is P and it increases to $P + \Delta P$ on cd. We take the distances ab and cd as equal to b and sum the forces in the x direction, which consists of the forces exerted by the water on ab and cd plus the component of the weight of $abcd$ in the x direction. Thus

$$\Sigma F_x = h_1\gamma_w b - (h_1 - \Delta h + b \sin \alpha)\gamma_w b + \gamma b^2 \sin \alpha - \Delta P = 0$$

where ΔP is the difference between the reaction on face cd and that on ab. The value of ΔP is

$$\Delta P = \Delta h \gamma_w b + (\gamma - \gamma_w) b^2 \sin \alpha$$

We see that the second term is simply the component of the submerged weight of $abcd$ in the x direction. The first term shows that the reaction on cd contains

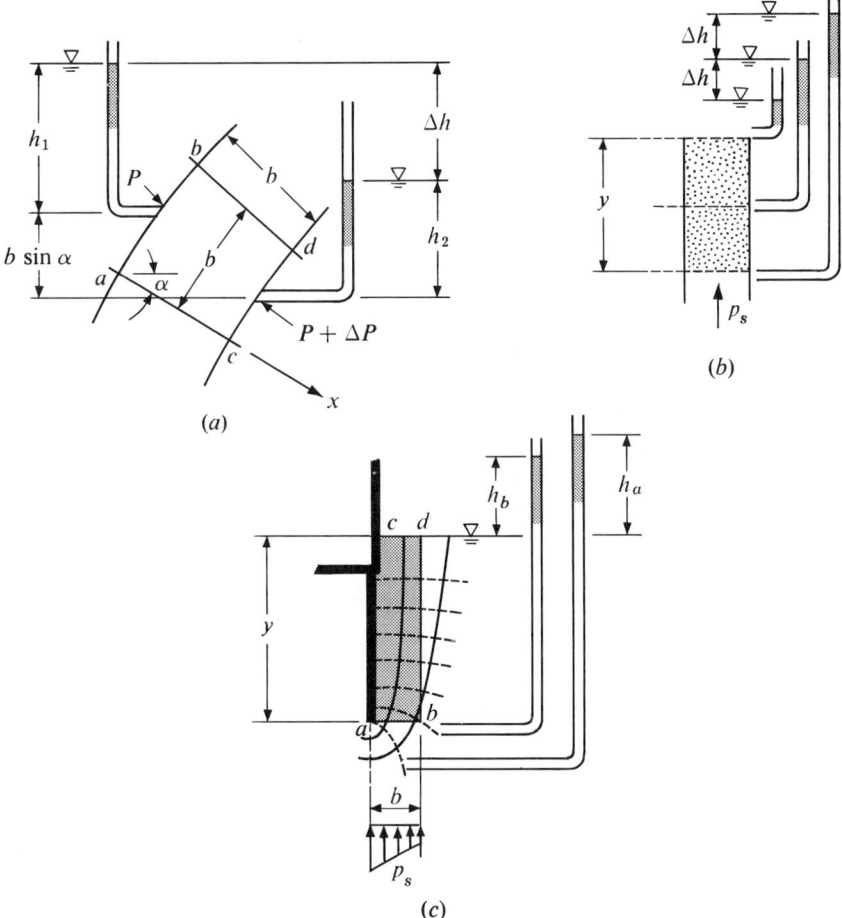

FIGURE 3.17 *Seepage force and failure by boiling.*

a quantity $\Delta h \gamma_w b$ in addition to the submerged weight of the element. This additional quantity is caused by seepage and is called the *seepage force*. It is sometimes convenient to express it in terms of force per unit volume of the soil, or

$$j = \frac{\Delta h \gamma_w}{b} = i\gamma_w \tag{3.36}$$

We see that it is equal to the hydraulic gradient times the unit weight of water. Since it has the unit of force per volume it may be treated as a body force. The derivation shows that the seepage force acts in a direction normal to the equipotential line.

CRITICAL GRADIENT. The seepage force is of great practical importance in soil mechanics because it is responsible for the phenomenon known as *boiling*. We consider the case in which the flow is directed upward [Fig. 3.17(*b*)] through a column of soil with height y and unit cross-sectional area. The hydraulic gradient i is equal to $2\Delta h/y$, since the section contains two equipotential increments. The seepage force on *ab* acts upward and is

$$\Delta P = i\gamma_w y \times 1$$

The only downward force is the submerged weight of the column of soil, which is

$$W' = y(\gamma - \gamma_w) \times 1$$

We can visualize what happens if the hydraulic gradient is gradually increased until

$$i\gamma_w y = y(\gamma - \gamma_w) \tag{3.37}$$

The seepage force is then equal to the submerged weight. Upon further increase in i, the seepage force exceeds the submerged weight and thereby lifts the whole column of soil upward. This is the phenomenon known as boiling or heave. If it occurs near the toe of a dam it undermines the structure and brings about the collapse of the dam. From Eq. (3.37) the hydraulic gradient required to produce boiling is equal to

$$i_{cr} = \frac{\gamma - \gamma_w}{\gamma_w} \tag{3.38}$$

and is called the *critical gradient*.

If we consider the flow underneath a dam, as shown in Fig. 3.18, we find that upward flow occurs in the soil at the downstream side of the dam. Consequently the danger of boiling exists on this side. Furthermore Eq. (3.36) tells us that the seepage force is directly proportional to the hydraulic gradient i. Hence the most dangerous areas are those where the hydraulic gradient is largest. This means that the areas where the equipotential lines are closest together are most susceptible to boiling. These are usually located immediately near the toe of the dam, as shown in Fig. 3.18.

To investigate the stability of a soil mass against boiling, we consider the

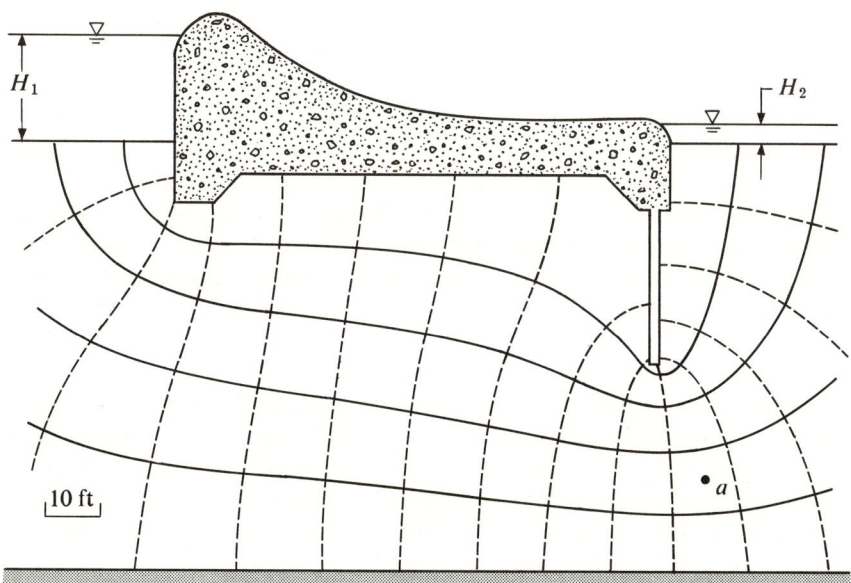

FIGURE 3.18 *Flow net beneath a dam.*

column of soil *abcd* shown in Fig. 3.17(*c*). Unlike the conditions shown in Fig. 3.17(*b*), the flow in this case is nonuniform, as shown by the curved flow and equipotential lines. We must make some approximations and use the average condition. Consider the soil column *ab*, in which the hydraulic gradient varies from a maximum at *a* to a minimum at *b*. For simplicity the hydraulic gradient is assumed to be uniform throughout the column. Terzaghi (1943) recommended that the distance *ab* be taken as $\frac{1}{2}$ the depth *y*. If we use the average of the heads at *a* and *b*, the hydraulic gradient is

$$i = \frac{1}{2y}(h_a + h_b)$$

The critical hydraulic gradient is computed from Eq. (3.38). If *i* is smaller than i_{cr}, a margin of safety exists. The factor of safety is defined as the ratio

$$F_S = \frac{i_{cr}}{i} \tag{3.39}$$

Because of the catastrophic nature of boiling, it is common practice to require a safety factor of at least 3.

3.13 Limitations of seepage theory

The theory of seepage and its applications are based on the equation of continuity. This, in turn, is based on Darcy's law. As noted in Chapter 2, Darcy's law is restricted to the case of laminar flow. The early experiments by Terzaghi (1925a) indicate that the law is valid over the range of velocities that are likely to be encountered in most practical problems. However, the transition from laminar to turbulent flow is not well established. Experiments with soils and other porous media have shown that the Reynold's number at which the flow changes from laminar to turbulent may range between very wide limits [see Scheidegger (1957)]. In connection with this Leonards (1962) has pointed out that the application of the Reynold's number, which is a characteristic of straight tubes, to porous media may itself be of questionable validity. For soils of very low permeability, Hansbo (1960) has suggested that a minimum hydraulic gradient must be exceeded before flow occurs. This minimum gradient is small and therefore does not impose serious limitations on most practical problems.

A most serious limitation is the nonhomogeneity of natural soil deposits. This was recognized by Terzaghi as early as 1929, and the major difficulties he emphasized at that time still remain. The quantity and velocity of seepage and the seepage force are governed to a very large extent by highly permeable layers in the soil deposit. They constitute zones of high seepage velocity and high seepage force, both of which are major concerns in design. An illustration of the effect of a highly pervious pocket on the piezometric levels is shown in Fig. 12.3(a). However, the extent of these permeable layers, or even their existence, is sometimes difficult to determine by the usual means of subsoil exploration. Even when careful permeability tests are performed on representative soil specimens, the permeability of a soil deposit may be quite different from that measured in the permeability tests.

An instructive example is the study by Mansur (1957). The laboratory and in situ permeabilities of a deposit of stratified coarse to fine sand were investigated. Specimens of sand were obtained and subjected to permeability tests in the remolded condition. This was necessary because of the great difficulty encountered in measuring horizontal permeability in vertical cylindrical specimens of cohesionless soils. The in situ permeability of the deposit in the horizontal direction was measured by pumping tests. The bar graph in Fig. 3.19 shows the in situ permeabilities (k_x) of the successive strata. The laboratory permeabilities of remolded specimens are plotted as circles and differ considerably from the in situ values. This example serves to emphasize the inadequacy of the available sampling and laboratory techniques. Second, the wide range in permeability within the soil deposit, as revealed by field tests, makes it extremely difficult to evaluate the average permeability of the entire deposit from laboratory tests, even if such tests could give reliable values. The number of permeability tests required to obtain a statistically significant average would be prohibitive for many projects. If the subsoil consists of mixtures of gravel, sand, and silt, the scatter of the in situ permeability would be many times greater than that in Fig. 3.19.

As the success or failure of a project often depends upon the magnitude of the seepage and seepage force, recognition of the above limitation is of the utmost importance. There is always an element of uncertainty concerning the

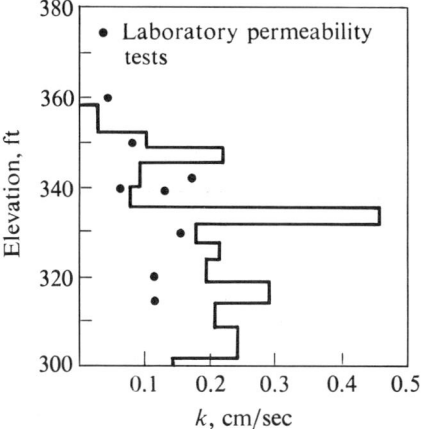

FIGURE 3.19 *Laboratory and in situ permeabilities of an alluvial sand deposit in the Mississippi Valley.* [After Mansur (1957).]

validity of the soil conditions on which the seepage calculations are based. This frequently requires that systematic field observations be made as the project progresses, so that any significant departures from the theoretical results can be detected immediately and corrective measures taken. The field observations may range from measurement of seepage quantity or water pressure to close surveillance for local springs and boils indicating locations of high seepage velocity. The corrective measures depend upon the structure and the problem, and some excellent examples are given by Esmiol (1957), Mansur and Kaufman (1956), and Terzaghi and Leps (1958).

PROBLEMS

3.1 It is necessary to estimate the seepage loss through the sandy silt layer beneath the earth dam shown in Fig. 3.20. A permeability test on the material showed that during an interval of 5 min, 0.010 ft³ of water under a head of 12 in. flowed

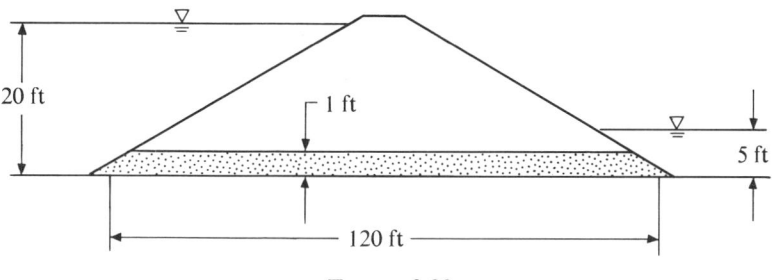

FIGURE 3.20

through the specimen. The dimensions of the specimen were $A = 15$ in.^{2}, $d = 8$ in. How much water would leak through a 1-ft section of the dam during a period of 24 hr?
Answer: 2.3 ft^3

3.2 The following data have been determined for two soils:

	Soil A	Soil B
Description	Silt	Medium sand
D_{10}	0.010 mm	0.10 mm
D_{60}	0.050 mm	0.20 mm
k	0.01 cm/sec	0.0001 cm/sec
e	0.50	0.56

(a) Do you suspect anything wrong with the data? (b) If so, why?

3.3 (a) Flow of water under gravitation occurs through a saturated silty sand layer at a rate of 1 ft^3/sec. What will be the rate of flow if the head is doubled and the distance of flow cut by one-half? (b) Flow of water under artesian pressure occurs through a saturated fine sand layer at a rate of 5 ft^3/sec. What will be the rate of flow if the head is doubled and the distance of flow cut by one-half?
Answer: will be four times as large

3.4 The quantity of flow through the sand layer in Fig. 3.21 was measured to be 30 ft^3/day/ft width. The piezometers were installed and the measured water pressures were as shown. What is the coefficient of permeability of the sand?
Answer: 0.833 ft/hr

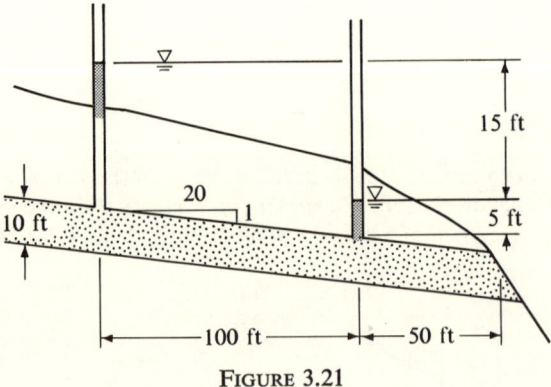

FIGURE 3.21

3.5 Shown in Fig. 3.22 are portions of flow nets. State "what is wrong" with each of the flow nets.

3.6 Sketch the boundary conditions which are responsible for two flow lines tending to intersect, and explain the nature of the flow that results as a consequence of this condition.

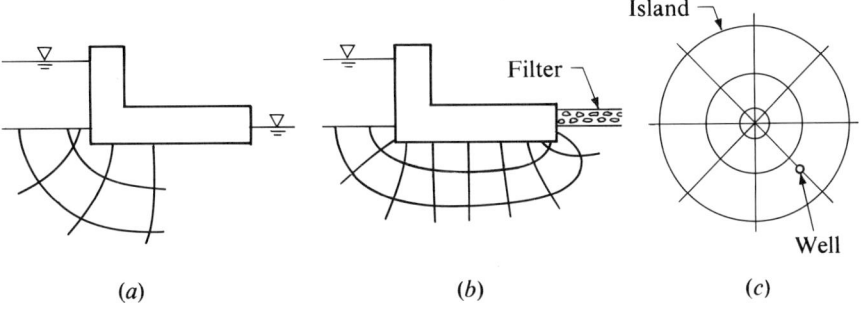

(a)   (b)   (c)

FIGURE 3.22

3.7 Figure 3.23 shows a dam constructed over pervious foundation material. The following properties were determined for the foundation material:

$$e = 0.56 \quad \text{sp. gr.} = 2.65 \quad k = 10^{-4} \text{ ft/sec}$$

At a point 18 ft left of the upstream face of the dam and 13 ft below the ground surface, compute the following: (a) Hydrostatic pressure. (b) Hydraulic gradient. (c) Velocity.
Answer: 1550 psf; 0.082; 0.82×10^{-5} ft/sec

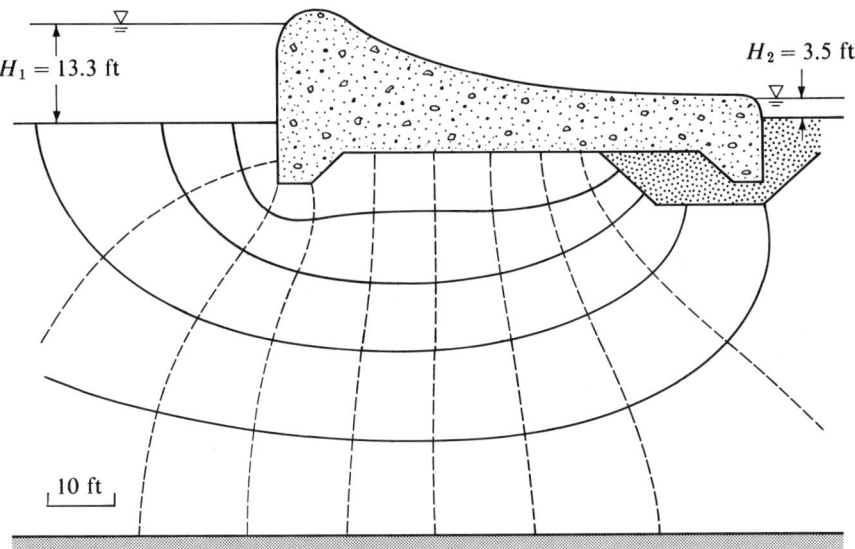

FIGURE 3.23 *Flow net beneath a dam with toe drain.*

3.8 The pressure exerted by the water on the base of a dam is called uplift pressure. Compute the resultant of the uplift pressure for the dam shown in Fig. 3.23.
Answer: 28,000 lb/ft length of dam.

82   Seepage problems   CHAP. 3

3.9 Using the flow net in Fig. 3.18 calculate: (*a*) The total quantity of seepage per hour underneath the dam. (*b*) The velocity at point *a*. (*c*) The factor of safety with respect to boiling.
Use the following values for calculations:
                 Scale: 1 in. = 40 ft
                 Unit wt. of satd. soil = 125 pcf
                 Sp. gr. of solids = 2.67
                 D_{10} = 0.01 mm
Answer: 7.1×10^{-2} ft^3/hr; 1.18×10^{-3} ft/hr; 10

3.10 The earth dam shown in Fig. 3.24 has a toe drain with length *L*. What is the minimum length of *L* (in terms of the slope angle β) in order that the line of seepage does not touch the downstream slope?
Answer: $L > x_0(1 + \cot^2 \beta)$

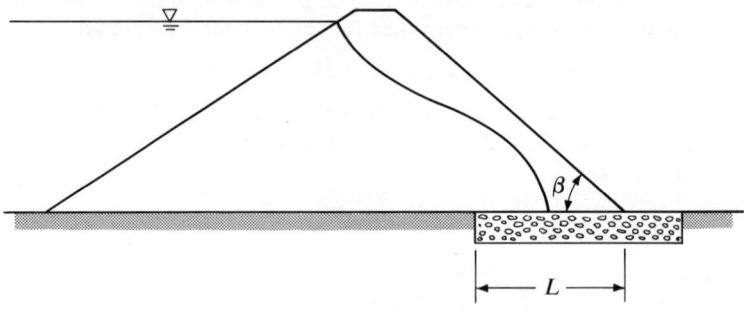

FIGURE 3.24

3.11 For the earth-dam whose cross section is shown in Fig. 3.25(*a*), determine the values of *a* and the quantities of seepage per foot of dam for the following two conditions: (*a*) $k_x = k_y = 0.8 \times 10^{-4}$ cm/sec. (*b*) $k_x = 5 \times 10^{-4}$ cm/sec, $k_y = 0.8 \times 10^{-4}$ cm/sec.
Answer: 0.44×10^{-4} ft^3/sec; 2.44×10^{-4} ft^3/sec

3.12 Consider the seepage underneath the dam shown in Fig. 3.26. The seepage loss and seepage pressure at the toe may be modified by the following construction features: (1) A sheet pile wall is driven at the toe to a depth of 25 ft below ground surface. (2) A drain filter is constructed at the toe. (3) An impervious clay blanket 100 ft long is placed on the upstream ground surface. Evaluate the three possible measures as to their effectiveness in (*a*) reducing seepage loss, (*b*) increasing safety against boiling at the toe.
Answer:

Condition	Seepage loss, %	Factor of safety
As is	100	1.3
Sheet pile only	86	2.7
Filter only	140	13.0
Clay blanket only	57	2.3

Problems

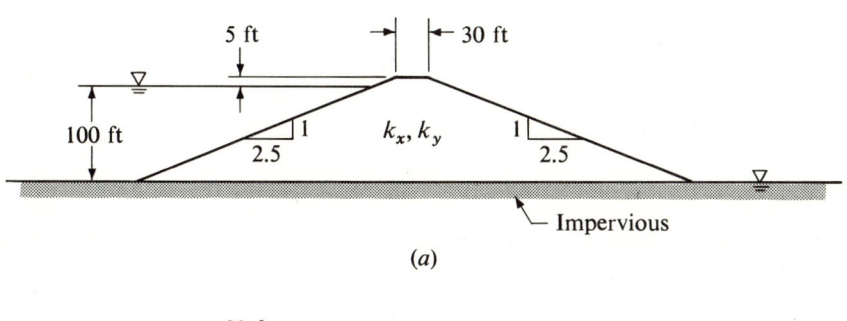

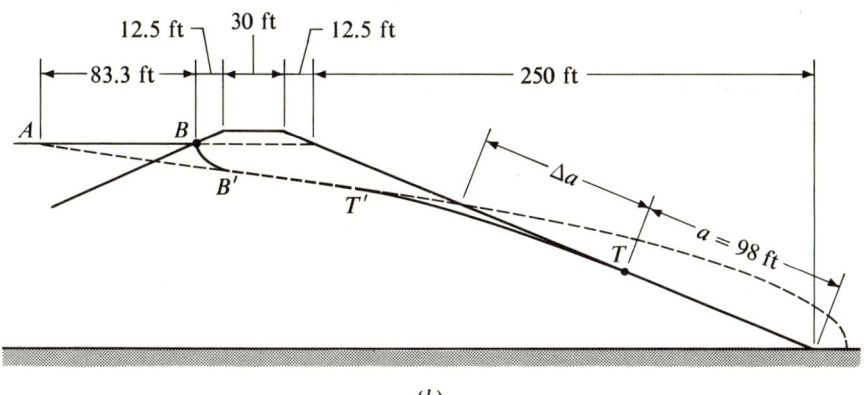

FIGURE 3.25

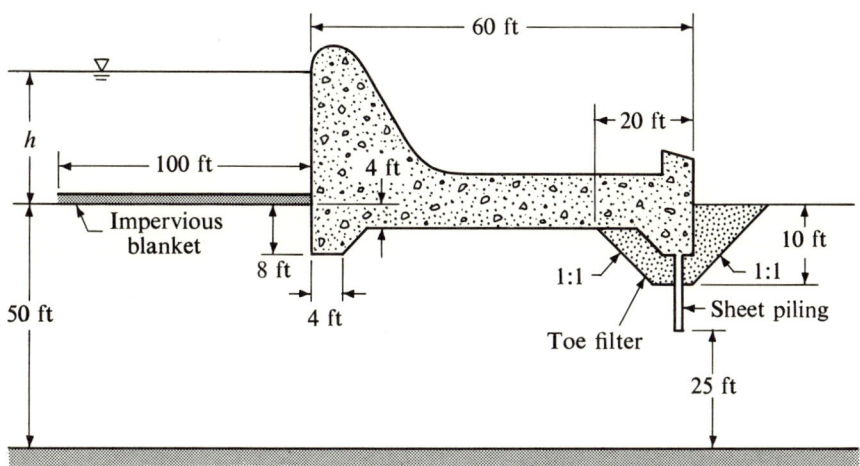

FIGURE 3.26

3.13 For the well problem described in Sec. 3.2 and Fig. 3.2: (*a*) Is the solution as given still valid if the head drops below the top of the aquifer? Why? (*b*) If the head is everywhere below the top of the aquifer, what simplifying assumption may be introduced in order to solve the problem as two-dimensional radial flow? (*c*) Find the solution for *k* under the conditions given in (*b*).
Answer: Flow is not two-dimensional; Dupuit's assumption;

$$k = \frac{q \log s(r_1/r_2)^*}{2\pi(h_1^2 - h_2^2)H}$$

3.14 Write the equations for the equipotential lines and flow lines for the problem described in Example 3.3.

Answer: $x = C\left(\dfrac{y^2}{4\Psi^2} - \Psi^2\right); \quad x = C\left(\Phi^2 - \dfrac{y^2}{4\Phi^2}\right)$

---

* h_1 and h_2 are measured to the bottom of aquifer.

4

Stresses and Strains

4.1 Stresses at a point

Figure 4.1 shows the generalized case of stresses at a point. If we consider a cubical element, there exist shear and normal stresses on each of the six faces [Fig. 4.1(a)]. In the derivation of the fundamental equations, tensile stress is taken as positive to conform with the sign convention in most mechanics books. For any given set of stresses there also exist three mutually perpendicular directions, along which there exist only normal stresses. These are called principal axes. The stresses along these axes are called principal stresses (denoted by $\sigma_1, \sigma_2, \sigma_3$) and the planes perpendicular to the principal axes are called principal planes [Fig. 4.1(b)].

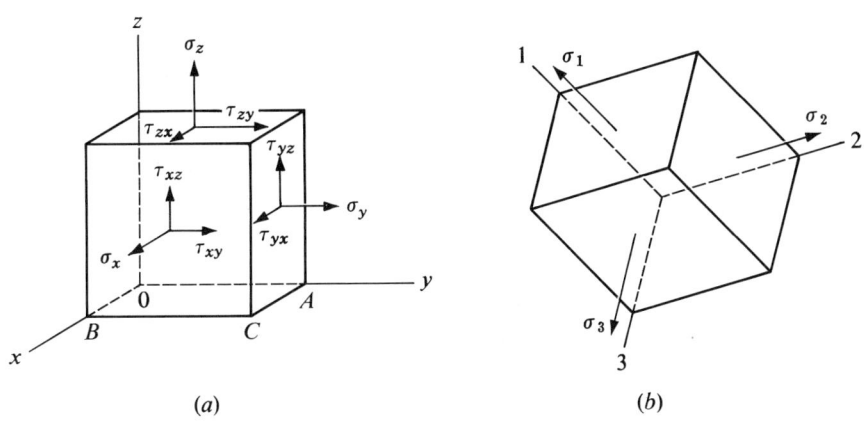

FIGURE 4.1 *Stresses on an element.*

We consider first the simple case in which the stresses σ_z, τ_{zx}, τ_{zy} are zero. This is called plane stress, and the stresses are in the x–y plane as shown in Fig. 4.2. An example of plane stress is a plate [Fig. 4.2(e)] loaded by forces in the plane of the plate. Another simple case is the plane strain in which the deformation (or strain) is zero in the z direction. This occurs if the body with which we are dealing is very long in the z direction and the section and loads do

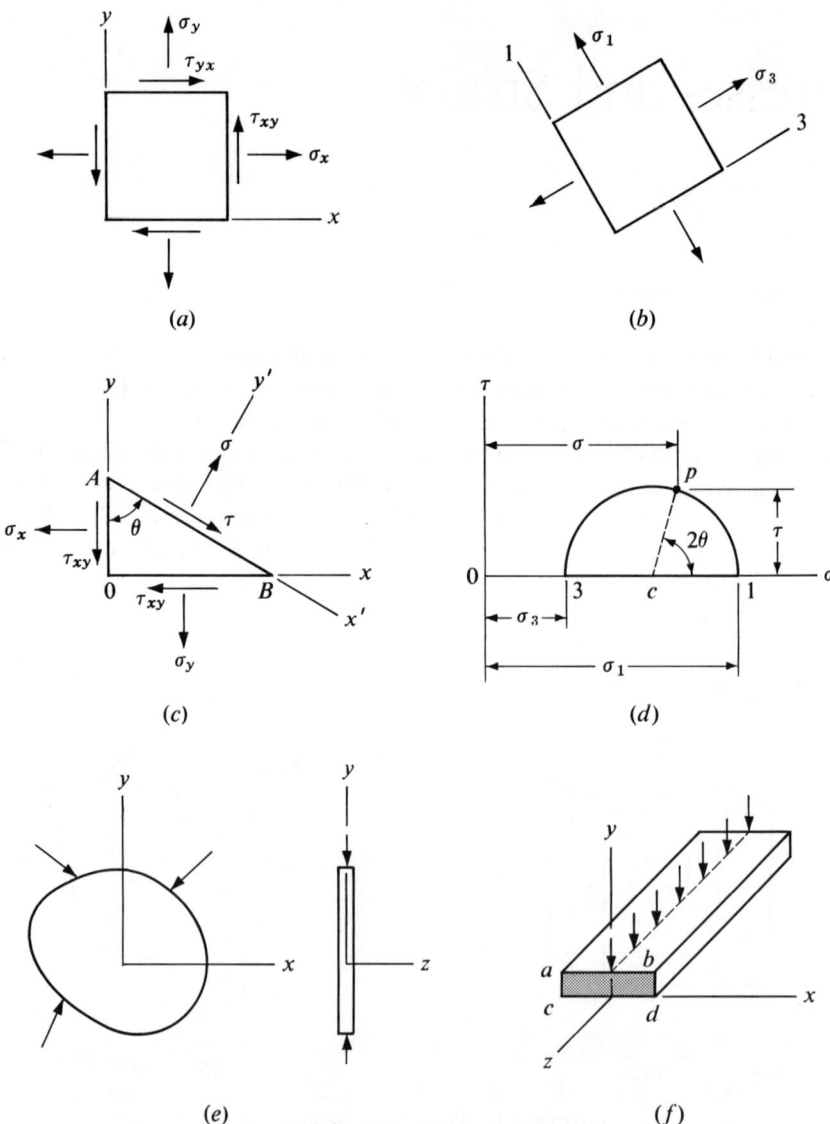

FIGURE 4.2 *Stresses on an element; a two-dimensional problem.*

4.1 Stresses at a point

not vary with z. Thus every section in the x–y plane is the same as every other and there are no shear stresses on the plane perpendicular to the z axis. There is a normal stress σ_z on the z plane [$abcd$ in Fig. 4.2(f)], but the shear stresses τ_{zx} and τ_{zy} are zero. Such situations are common in soil mechanics. A long foundation [Fig. 4.2(f)] may be treated as a plane-strain problem.

For both plane stress and plane strain the relationship between the stresses in the x and y directions (σ_x, σ_y, τ_{xy}) and those on any plane making an angle θ with the y axis [Fig. 4.2(c)] can be found by resolution of forces in the x–y plane. We consider the stresses on an element $AB0$ [Fig. 4.2(c)]. The stresses σ and τ that act on AB can be found by resolving forces in the x' and y' directions. If the length of AB is taken as unity, we have

$$A0 = \cos\theta \qquad B0 = \sin\theta$$

and

$$\Sigma F_{y'} = -(\sigma_x \cos\theta)\cos\theta - (\tau_{xy}\sin\theta)\cos\theta \\ -(\sigma_y \sin\theta)\sin\theta - (\tau_{xy}\cos\theta)\sin\theta + \sigma = 0$$

$$\Sigma F_{x'} = -(\sigma_x \cos\theta)\sin\theta + (\tau_{xy}\sin\theta)\sin\theta \\ +(\sigma_y \sin\theta)\cos\theta - \tau_{xy}\cos\theta\cos\theta + \tau = 0$$

From the first of these we get

$$\sigma = \sigma_x \cos^2\theta + \sigma_y \sin^2\theta + 2\tau_{xy}\sin\theta\cos\theta$$
$$\sigma = \tfrac{1}{2}(\sigma_x + \sigma_y) + \tfrac{1}{2}(\sigma_x - \sigma_y)\cos 2\theta + \tau_{xy}\sin 2\theta \qquad (4.1a)$$

From the second of these we get

$$\tau = \sigma_x \sin\theta\cos\theta - \sigma_y \sin\theta\cos\theta + \tau_{xy}(\cos^2\theta - \sin^2\theta)$$
$$\tau = \tfrac{1}{2}(\sigma_x - \sigma_y)\sin 2\theta + \tau_{xy}\cos 2\theta \qquad (4.1b)$$

It can be seen that the angle θ may be chosen such that τ is 0. This means, from Eq. (4.1b),

$$\tan 2\theta = \frac{\sin 2\theta}{\cos 2\theta} = -\frac{2\tau_{xy}}{\sigma_x - \sigma_y} \qquad (4.2)$$

Thus for every set of σ_x, σ_y, τ_{xy}, there are two values of θ that satisfy the above conditions. These are the directions of the principal planes. Substitution of Eq. (4.2) into (4.1a) gives the principal stresses:

$$\genfrac{}{}{0pt}{}{\sigma_1}{\sigma_3} = \frac{\sigma_x + \sigma_y}{2} \pm \left[\left(\frac{\sigma_x - \sigma_y}{2}\right)^2 + \tau_{xy}^2\right]^{1/2} \qquad (4.3)$$

If the x and y directions are the directions of the principal planes, then $\sigma_x = \sigma_1$, $\sigma_y = \sigma_3$, $\tau_{xy} = 0$, and Eq. (4.1) reduces to

$$\sigma = \frac{\sigma_1 + \sigma_3}{2} + \frac{\sigma_1 - \sigma_3}{2}\cos 2\theta$$
$$\tau = \frac{\sigma_1 - \sigma_3}{2}\sin 2\theta \qquad (4.4)$$

It can be seen from Eq. (4.4) that on planes making an angle $\pi/4$ with the principal planes ($\theta = \pi/4$), the shear stress attains its maximum value.

These stresses may also be calculated graphically by means of Mohr's circle of stress, as shown in Fig. 4.2(d). In this method of construction, the state of stress on any plane is represented by a point on the graph. The abscissa and ordinate of this point are equal to the normal and shear stresses, respectively, on this plane. The stresses on the principal planes are indicated by points 1 and 3, and a circle is drawn through these points with its center on the horizontal axis. Points on this circle designate the shear and normal stresses on planes between the principal planes. The stresses on a plane at an angle θ with the major principal plane are determined by locating point p on the circle so that angle $pc1$ is equal to 2θ.

4.2 Strains at a point

We consider the strains and displacements of the element shown in Fig. 4.1(a) and let u, v, w denote the displacements in the x, y, z directions, respectively. The side $AOBC$ is shown in Fig. 4.3. After the application of stress, the two perpendicular lines $A0$ and $B0$ are displaced to the new position $A''0'$ and $B''0'$. The displacement may be considered to consist of a translation from AOB to $A'0'B'$ plus a distortion from $A'0'B'$ to $A''0'B''$. The displacement of point 0 is u and v in the x and y directions, while that at B, located at $x = dx$ away, is $u + (\partial u/\partial x)\,dx$, $v + (\partial v/\partial x)\,dx$. Similarly, the displacement at A is $u + (\partial u/\partial y)\,dy$, $v + (\partial v/\partial y)\,dy$.

We see that the elongation of $B0$ in the x direction is $(\partial u/\partial x)\,dx$. The length of $B0$ is dx. Hence the strain of the element in the x direction can be written as

$$\epsilon_x = \frac{\partial u}{\partial x}$$

Similarly, we obtain

$$\epsilon_y = \frac{\partial v}{\partial y}$$

$$\epsilon_z = \frac{\partial w}{\partial z}$$

(4.5)

During the process of strain, $B0$ also undergoes a rotation. The angle change $B'0'B''$ is $(\partial v/\partial x)dx(1/dx)$. The angle change of $A0$ is $(\partial u/\partial y)dy(1/dy)$. Thus the total angular distortion of the element is $(\partial u/\partial y) + (\partial v/\partial x)$. This is the shear strain and is denoted by γ_{xy}. We have, therefore,

$$\gamma_{xy} = \frac{\partial u}{\partial y} + \frac{\partial v}{\partial x}$$

$$\gamma_{yz} = \frac{\partial v}{\partial z} + \frac{\partial w}{\partial y}$$

$$\gamma_{xz} = \frac{\partial u}{\partial z} + \frac{\partial w}{\partial x}$$

(4.6)

SEC. 4.2   Strains at a point   89

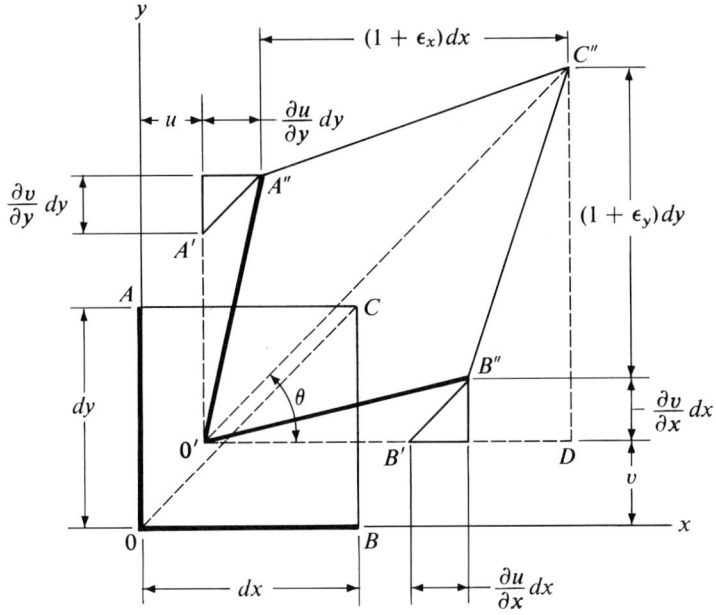

FIGURE 4.3   *Strain and displacement of an element.*

The shear strain is positive if the first and third quadrants become smaller. Given the strains ϵ_x, ϵ_y, γ_{xy} along the x and y directions, the strains and γ in any direction θ from the x axis can be found. Consider the case of plane strain. The geometric relations are shown in Fig. 4.3. If we denote the strain in the direction $0'C''$ as ϵ_θ, we get

$$0'C'' = 0C(1 + \epsilon_\theta)$$

Also

$$(0'C'')^2 = (0'D)^2 + (C''D)^2$$

Then

$$(0C)^2(1 + \epsilon_\theta)^2 = \left[(1 + \epsilon_x)dx + \frac{\partial u}{\partial y}dy\right]^2 + \left[(1 + \epsilon_y)dy + \frac{\partial v}{\partial x}dx\right]^2$$

Since

$$dx = 0C \cos \theta, \quad dy = 0C \sin \theta, \quad \gamma_{xy} = \frac{\partial v}{\partial x} + \frac{\partial u}{\partial y}$$

we get

$$\epsilon_\theta = \epsilon_x \cos^2 \theta + \epsilon_y \sin^2 \theta + \gamma_{xy} \sin \theta \cos \theta$$
$$= \tfrac{1}{2}(\epsilon_x + \epsilon_y) + \tfrac{1}{2}(\epsilon_x - \epsilon_y) \cos 2\theta + \tfrac{1}{2}\gamma_{xy} \sin 2\theta$$

By similar calculations

$$\gamma_\theta = (\epsilon_x - \epsilon_y) \sin \theta \cos \theta + \gamma_{xy}(\cos^2 \theta - \sin^2 \theta)$$
$$= \tfrac{1}{2}(\epsilon_x - \epsilon_y) \sin 2\theta + \tfrac{1}{2}\gamma_{xy} \cos 2\theta \tag{4.7}$$

It is seen that Eq. (4.7) for strain at a point has the same form as Eqs. (4.1) for stress. Similarly, at angles $2\theta = \tan^{-1} [2\tau_{xy}/(\epsilon_x - \epsilon_y)]$, the shear strain vanishes; these are the principal planes of strain. Mohr's circle can also be used to calculate the strain. It can also be shown that for an elastic, isotropic material, the principal axes of stress and strain coincide.

4.3 Invariants of stress and strain

For any given set of stresses at a point the principal stresses are known and Eq. (4.3) shows that

$$\sigma_x + \sigma_y = \sigma_1 + \sigma_3$$

Thus, the sum $(\sigma_x + \sigma_y)$ of the normal stresses on two perpendicular planes is a constant and this constant does not vary with the direction of the x, y axes which may be chosen arbitrarily. For three-dimensional stress we have

$$\sigma_x + \sigma_y + \sigma_z = \sigma_1 + \sigma_2 + \sigma_3 = 3\sigma_0 \qquad (4.8a)$$

This sum is called the stress invariant, and σ_0 as defined above is called the octahedral normal stress.

Similarly we find that

$$\sigma_x\sigma_y + \sigma_y\sigma_z + \sigma_z\sigma_x - \tau_{xy}^2 - \tau_{yz}^2 - \tau_{zx}^2 = \sigma_1\sigma_2 + \sigma_2\sigma_3 + \sigma_3\sigma_1$$

is another invariant. It follows that

$$\tfrac{2}{9}[(\sigma_1 - \sigma_2)^2 + (\sigma_2 - \sigma_3)^2 + (\sigma_3 - \sigma_1)^2] = \tau_0^2 \qquad (4.8b)$$

is an invariant and τ_0 as defined is called the octahedral shear stress.

In the same way we find the strain invariants

$$\epsilon_0 = \tfrac{1}{3}(\epsilon_1 + \epsilon_2 + \epsilon_3) \qquad (4.9a)$$

and
$$\gamma_0 = \tfrac{2}{3}[(\epsilon_1 - \epsilon_2)^2 + (\epsilon_2 - \epsilon_3)^2 + (\epsilon_3 - \epsilon_1)^2]^{1/2} \qquad (4.9b)$$

and ϵ_0 and γ_0 are called the octahedral normal strain and octahedral shear strains, respectively.

The stress and strain invariants have important physical significance. It should be noted that σ_0 is the mean of the normal stresses. The expression for τ_0 contains only the principal stress differences $(\sigma_1 - \sigma_2)$, etc. These are also the maximum shear stresses and therefore τ_0 is indicative of the magnitude of the shear stresses in the element. The term ϵ_0 is the mean normal strain and is also three times the volumetric strain of the element,

$$\epsilon_v = \epsilon_1 + \epsilon_2 + \epsilon_3 = 3\epsilon_0 \qquad (4.9c)$$

whereas γ_0 represents the shear distortion of the element.

4.4 Stress-strain relations

Materials deform when subjected to stress. Figure 4.1(b) illustrates and element subjected to stresses $\sigma_1, \sigma_2,$ and σ_3. As a result the material undergoes strains.

The stress-strain relationship of most real materials is usually complex. Hence, in order to describe their behavior mathematically, it is necessary to introduce simplified ideal materials as approximations to the real material. Several types of ideal material behavior that have been used in engineering problems are the following.

ELASTICITY. A simple ideal material is the elastic material that obeys Hooke's law, which says that the stress is proportional to the strain. The material properties that are needed to calculate the elastic stresses and strains are the modulus of elasticity E and Poisson's ratio μ. Figure 4.4(a) illustrates the stress-strain relationship of an elastic material.

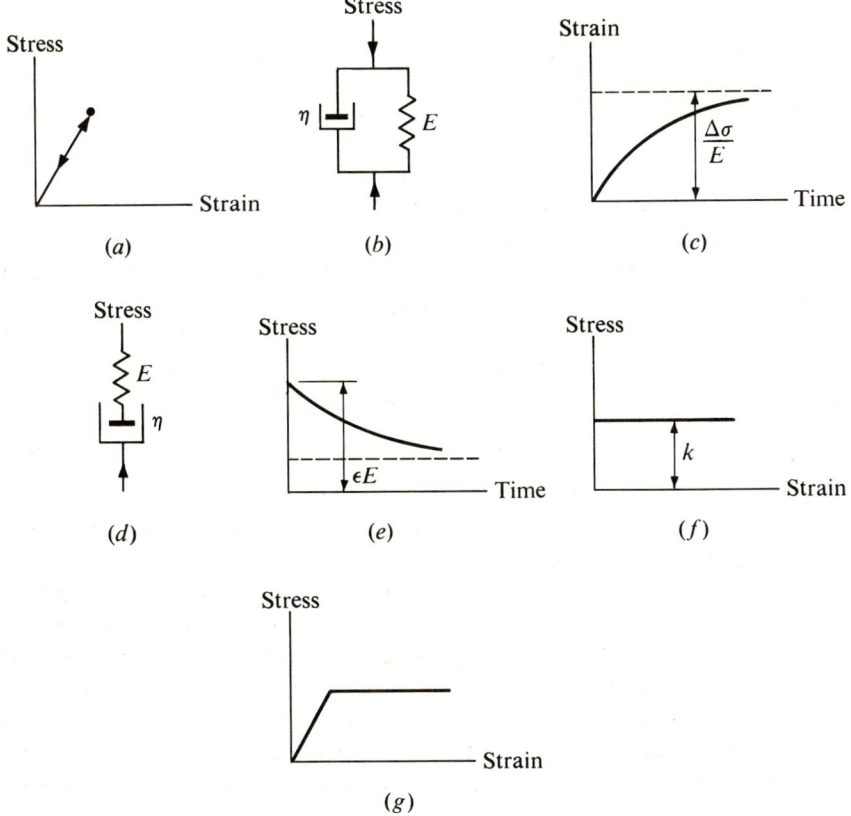

FIGURE 4.4 *Stress-strain relationships.*

When the element in Fig. 4.1 is subjected to the stresses σ_1, σ_2, and σ_3 the strains are

$$\epsilon_1 = \frac{1}{E}(\sigma_1 - \mu\sigma_2 - \mu\sigma_3) \tag{4.10a}$$

$$\epsilon_2 = \frac{1}{E}(\sigma_2 - \mu\sigma_3 - \mu\sigma_1) \qquad (4.10b)$$

$$\epsilon_3 = \frac{1}{E}(\sigma_3 - \mu\sigma_1 - \mu\sigma_2) \qquad (4.10c)$$

Equations (4.8a) and (4.9a) may be combined with Eq. (4.10) to give

$$\epsilon_0 = \frac{\sigma_0}{E}(1 - 2\mu) = \frac{\sigma_0}{K} \qquad (4.11a)$$

and K is called the bulk modulus. Similarly Eqs. (4.8b) and (4.9b) may be combined with Eq. (4.10) to give

$$\gamma_0 = \frac{2\tau_0}{E}(1 + \mu) = \frac{\tau_0}{G} \qquad (4.11b)$$

and G is called the shear modulus. Equations (4.11a) and (4.11b) show that for an ideally elastic material the volumetric strain and shear strain are dependent only on the octahedral normal stress and octahedral shear stress, respectively.

VISCOELASTICITY. Many materials exhibit creep under a constant applied stress, and stress relaxation under a constant applied strain. This can be described by viscoelastic models which are combinations of elastic behavior with viscous flow phenomenon. In Fig. 4.4(b) the elastic behavior as represented by the elastic spring E is coupled in parallel to a dashpot with a viscosity η. As a simple illustration, consider an element loaded axially to a given strain ϵ_y at a strain rate $d\epsilon_y/dt = \dot{\epsilon}_y$. The stresses in the spring element and the dashpot element are respectively $\epsilon_y E$ and $\eta\dot{\epsilon}_y$. Hence for this model, we have

$$\sigma_y = \epsilon_y E + \eta\dot{\epsilon}_y \qquad (4.12)$$

As in elasticity, viscoelastic behavior can be considered in terms of the volumetric and shear components. However, at present, such general formulations are not available for soils because of inadequate experimental data.

If a given stress is applied, it is resisted first by the fluid pressure in the dashpot. Since the fluid is incompressible, there is no strain immediately after the application of stress. Strain develops as the fluid flows past the piston in the dashpot. This gradually compresses the spring, which then takes up an increasing portion of the stress. The strain ceases to increase when the stress in the sping becomes equal to the applied stress. Hence, under a given stress increment, the strain increases with time and approaches an asymptote, as shown in Fig. 4.4(c). The increase of strain with time is called creep.

In Fig. 4.4(d), the spring and dash-pot are connected in series. If the system is loaded rapidly to a given strain ϵ and this strain is held constant, the strain is supplied by compression of the spring immediately after loading. The compression produces a high stress in the spring equal to $\epsilon_y E$. However, with the passage of time, the fluid in the dashpot flows past the piston. This reduces the compression in the spring and thus relieves part of the stress in the spring. In this case the stress decreases with time as shown in Fig. 4.4(e); this is called stress relaxation. Other models of creep and relaxation are available, but these two suffice as examples.

PLASTICITY. If the stress on a material is increased progressively, it sooner or later reaches a value that is equal to the yield strength of the material and plastic yielding occurs. At this stage, the strain increases indefinitely at the same stress [Fig. 4.4(f)], and plasticity* results. The yield stress k is the maximum shear stress that the material can sustain and is called the shear strength. The property of shear strength is treated in Chapter 8.

It follows that the above models may be used in combination to approximately describe a real material. Thus a material may behave elastically up to a stress equal to the yield strength. At that point it behaves as a plastic material. The stress-strain relationship is then a combination, as shown in Fig. 4.4(g). All these models have been used to describe soil behavior under various kinds of loading.

4.5 Stress-strain relationships of soils

The preceding section shows that stresses and strains may be resolved into the normal and shear components. Also in ideal elasticity, the volumetric and shear strains are related directly to the normal and shear stresses. Although soil behavior is far from elastic, it is nevertheless convenient to study volumetric compression and shear distortion separately.

VOLUMETRIC STRAIN. To measure the volumetric compression, we may perform the test shown in Fig. 4.5(a). The stresses on the specimen are increased by $\Delta \bar{\sigma}_0$ in all directions. The term $\Delta \bar{\sigma}_0$ denotes effective stress. To compress the soil it is necessary to compress the soil skeleton. No shear stresses are introduced. The consequent volumetric strain may be plotted against the all-around stress as shown in Fig. 4.5(b). It shows that the material becomes stiffer as the stress is increased. For a given range ab, the stress-strain curve may be assumed to be a straight line and its slope is equal to the bulk modulus K. If the soil is loaded to c and then unloaded, the curve cd is obtained. Only a small portion of the volume compression is recovered.

SHEAR STRAIN. An ideal way to measure the shear distortion is to subject a specimen to pure shear as illustrated in Fig. 4.6. The specimen is initially confined by an all-around stress $\bar{\sigma}_0$ [Fig. 4.6(a)]. Then the shear stress τ is applied [Fig. 4.6(b)]. The principal stresses are equal to $\sigma_0 \pm \tau$. The stress-strain relationship under shear is illustrated in Fig. 4.6(c). In the region $0a$, the material

---

* Plasticity here refers to a type of stress-strain behavior and should not be confused with the definition in Sec. 1.8.

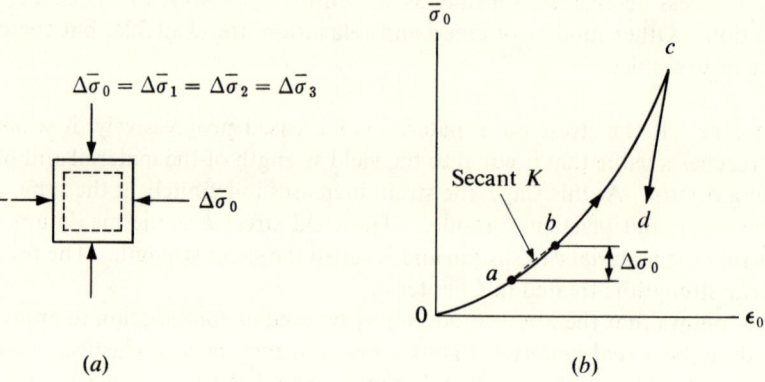

FIGURE 4.5 *Volumetric compression.*

response may be considered as approximately elastic. The slope of 0a may be taken to be the shear modulus G for this stress range. However, if unloaded to b, only a small fraction of the strain is recovered. With increasing τ, plastic strain becomes more important. At c, the material behavior may be considered to be essentially plastic flow. The stress cannot be increased further and this state is also called shear failure. The shear stress at this point is called the shear strength. It is important to note that the stress-strain curve is strongly dependent on the all-around pressure $\bar{\sigma}_0$. Figure 4.6(c) shows two curves. The larger the all-around confining pressure, the greater the shear strength, and also the greater the shear modulus G.

4.6 Excess hydrostatic porewater pressure

THE UNDRAINED AND DRAINED CONDITIONS. In many practical problems, it is necessary to study the undrained condition. The compression of the soil under the applied stresses (see Fig. 4.5) would force some water out of the voids. However, consider a soil with low permeability, containing water in the voids. If the stresses are applied very rapidly, there is no drainage of water from this soil immediately after loading because there is insufficient time for the water to flow out. This is the undrained condition and leads to the development of excess hydrostatic porepressures. On the other hand, if the stresses are increased very slowly, the water is forced out of the voids and this is the drained condition.

PORE PRESSURE PARAMETER B. Let us consider the soil element illustrated in Fig. 4.7(a) in equilibrium under the initial stresses σ_1, σ_2, and σ_3 in the three directions 1, 2, 3 and with an initial pore pressure u_0. The soil is loaded so that the element is subjected to an all-around stress increase of the amount $\Delta\sigma_3$ [Fig. 4.7(b)]. Following the derivation by Skempton (1954), the porewater pres-

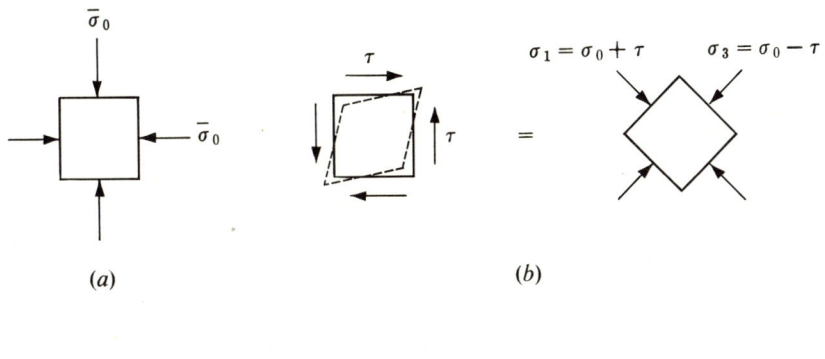

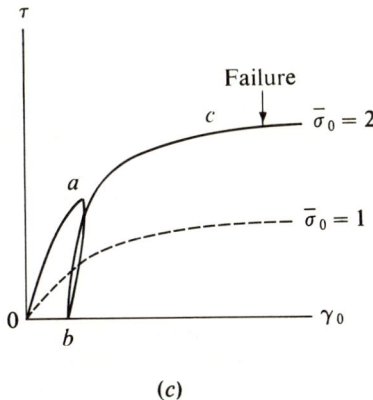

FIGURE 4.6 *Shear distortion.*

sure Δu_b may be determined from the following considerations. The rise in pore pressure Δu_b results in a compression of the pore space and the volume change is

$$\Delta V_n = m_n \, \Delta u_b$$

in which m_n is the compressibility of the pore space (air and water) due to an all-around increase in pressure. The increase in effective stress $\Delta \sigma_3 - \Delta u_b$ causes a compression of the soil structure equal to

$$\Delta V_s = 3\epsilon_0 = 3m_c(\Delta \sigma_3 - \Delta u_b)$$

The symbol m_c is the volume compressibility and is equal to $1/K$. If we assume the solid particles to be incompressible, then for the undrained condition ΔV_n must equal ΔV_s; hence

$$\Delta u_b = \Delta \sigma_3 \, \frac{1}{1 + (m_n/3m_c)} = B \, \Delta \sigma_3 \qquad (4.13)$$

where B is a coefficient dependent on the soil properties m_n and m_c.

For a saturated soil, m_n is the compressibility of water. The value of m_n is very small compared with m_c, as water is almost incompressible; so the coefficient B may be taken as equal to 1.0. For unsaturated soils the compressibility

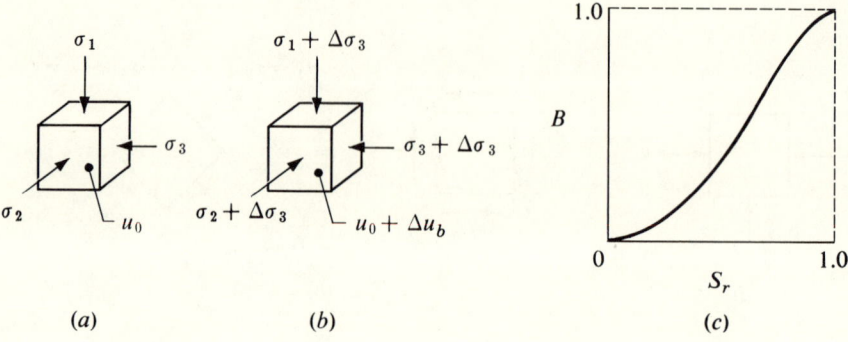

FIGURE 4.7 *Element of soil subjected to hydrostatic stress increase* $\Delta\sigma_3$.

of the pore air is large, and B is less than 1.0. The coefficient B varies with the degree of saturation. An example is Fig. 4.7(c). The pore pressure u_b for an unsaturated soil is a combination of the pore air and porewater pressures as stated in Eq. (2.29). The magnitudes of the pore air pressure and porewater pressure can be derived by considering the compressibility of the air and the capillary forces [see Hilf (1956)].

PORE PRESSURE PARAMETER A. We consider next the case shown in Fig. 4.8. The stress σ_1 is increased to $\sigma_1 + \Delta\sigma_1$ and the pore pressure in the soil undergoes a corresponding change Δu_a. To evaluate the quantity Δu_a, we consider the deformations of the element under the stress increase. We again assume the undrained state.

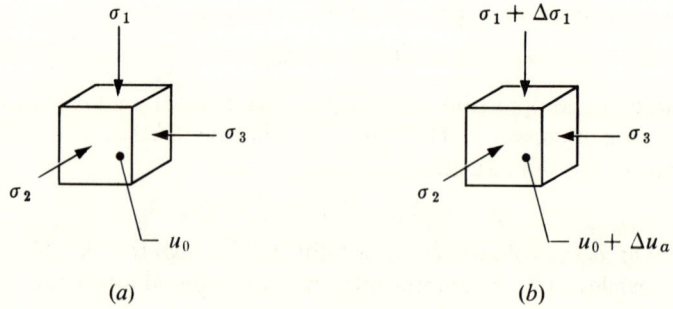

FIGURE 4.8 *Element of soil subjected to stress increase* $\Delta\sigma_1$.

The effective stress in the direction 1 is increased by $\Delta\sigma_1 - \Delta u_a$, while stresses in directions 2 and 3 are decreased by Δu_a. Increase in stress $\Delta\sigma_1 - \Delta u_a$ produces a compression equal to $m_c(\Delta\sigma_1 - \Delta u_a)$. Similarly, the reduction of the effective stress in directions 2 and 3 tends to produce an expansion equal to $2m_e(\Delta u_a)$ and m_e is the volume expansibility. We analyze first a saturated soil. In the undrained state the total volume change is 0, and we have

$$m_c(\Delta\sigma_1 - \Delta u_a) = 2m_e \Delta u_a \qquad (4.14)$$

It follows from Eq. (4.14) that

$$\Delta u_a + \frac{1}{1 + (2m_e/m_c)} \Delta\sigma_1 = A \Delta\sigma_1 \qquad (4.15)$$

It is thus seen that a change in porewater pressure always accompanies a stress change and that its magnitude is governed by the compressibility and expansibility of the soil. For highly compressible soils (such as soft clays), the compressibility is large compared to the expansibility, and A may approach 1.0. For soils of low compressibility, such as stiff clays and dense sands, A is usually very small. It should also be noted that should the soil be restrained so that expansions in the 2 and 3 directions are prevented, A becomes 1.0.

If the soil is unsaturated, then there is a volume change due to the compression of the void space. This amount is

$$\Delta V_n = m_n \Delta u_a$$

and Eq. (4.14) must be modified. The compression of the soil $m_c(\Delta\sigma_1 - \Delta u_a)$ must exceed the expansion by an amount equal to the compression of the void space, or

$$m_c(\Delta\sigma_1 - \Delta u_a) = 2m_e \Delta u_a + m_n \Delta u_a$$

This yields

$$\Delta u_a = \frac{1}{1 + (m_n/m_c) + (2m_e/m_c)} \Delta\sigma_1 \qquad (4.16)$$

By comparison, the product AB [Eqs. (4.13) and (4.15)] is

$$AB = \frac{1}{1 + (m_n/m_c)[(m_c + 2m_e)/3m_c] + (2m_e/m_c)}$$

The amount $(m_c + 2m_e)/3m_c$ is not very different from 1, and Eq. (4.16) may be simplified to

$$\Delta u_a = AB \Delta\sigma_1 \qquad (4.17)$$

The general case of stress change can be obtained by suitable superpositions of the two preceding cases. For stress increases equal to $\Delta\sigma_1$, $\Delta\sigma_3$, and $\Delta\sigma_3$ (Fig. 4.9), we combine Eqs. (4.13), (4.15), and (4.17), to get

$$\Delta u = \Delta u_a + \Delta u_b = B[\Delta\sigma_3 + A(\Delta\sigma_1 - \Delta\sigma_3)] \qquad (4.18)$$

In the preceding derivations, the coefficients A and B are presented as constants. In reality this is far from the case, because in real soils the compressibility and expansibility are not constant, as assumed in the derivation, but depend on the stress level. As a result, coefficients A and B are also dependent upon the stress level.

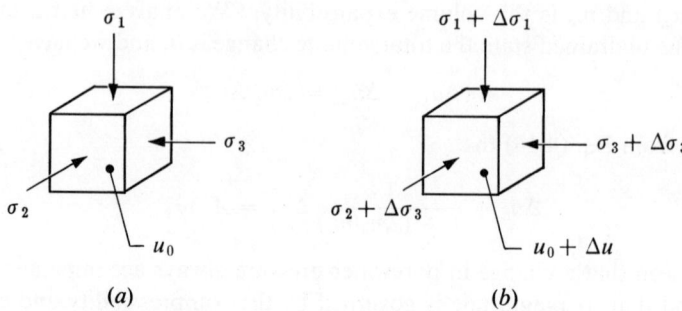

FIGURE 4.9 *Element of soil subjected to hydrostatic stress increase and a deviator stress ($\Delta\sigma_1 - \Delta\sigma_3$).*

The nature of the excess hydrostatic porewater pressure can be shown by the conditions in Fig. 4.10. A load Q on the surface increases the stresses at point A by $\Delta\sigma_1$, $\Delta\sigma_2$, and $\Delta\sigma_3$. This results in an excess hydrostatic porewater pressure Δu. If a piezometer is installed at point A, the piezometric surface rises to a level $\Delta u/\gamma_w$ above the water table. At point B a large distance away, the load Q does not effect the stress. There is no excess hydrostatic porewater pressure at B.

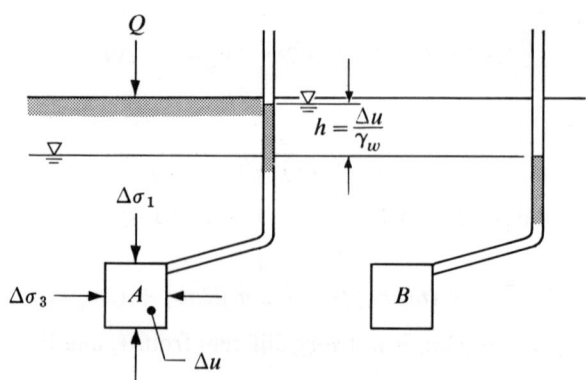

FIGURE 4.10 *Excess hydrostatic porewater pressure in a soil deposit due to a point load Q on the ground surface.*

4.7 Measurement of stress-strain properties

To obtain the stress-strain curves shown in Figs. 4.5(b) and 4.6(c), it would be desirable to use experimental devices that would apply the stress systems shown in Fig. 4.5(a) and 4.6(a) and (b) to a soil sample. Although devices have recently been developed to accomplish this, they are rather complicated to operate. For practical purposes it is common to use simpler devices and sacrifice uniformity in stress conditions. The principal features of several common soil tests are described here.

THE CONSOLIDATION TEST. This test is widely used to measure compressibility. Instead of using the hydrostatic compression shown in Fig. 4.5(a), it compresses the soil in one dimension only. A cylindrical sample is contained in a metal ring as shown in Fig. 4.11 and the stress $\bar{\sigma}_1$ is increased slowly so the drained condition is maintained. The metal ring allows no radial strain and $\epsilon_2 = \epsilon_3 = 0$. Setting $\epsilon_2 = 0$ and $\bar{\sigma}_2 = \bar{\sigma}_3$ in Eq. (4.10b), we get the ratio

$$K_0 = \frac{\bar{\sigma}_3}{\bar{\sigma}_1} = \frac{\mu}{1 - \mu} \qquad (4.19)$$

Thus, we see that the radial stress is K_0 times the axial stress and K_0 is called the coefficient of earth pressure at rest. Substituting Eq. (4.19) into (4.10a) we obtain the constrained modulus M:

$$M = \frac{\bar{\sigma}_1}{\epsilon_1} = E\frac{1 - \mu}{(1 - 2\mu)(1 + \mu)} = K\frac{1 - \mu}{1 + \mu} \qquad (4.20)$$

The constrained modulus increases with increasing $\bar{\sigma}_1$ in much the same manner as described in Fig. 4.5 for K.

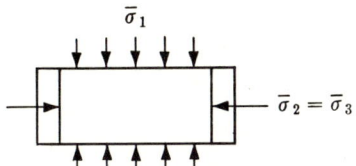

FIGURE 4.11 *The consolidation test.*

THE DIRECT SHEAR TEST. To measure the stress-strain behavior under shear, the test should follow the scheme shown in Fig. 4.6. This condition is often called pure shear or simple shear and several instruments have been built to accomplish this [Bjerrum and Landva (1966); Roscoe (1953); Roscoe et al. (1967); Ko and Scott (1967)]. However, these instruments are rather complicated and in practice the direct shear test is commonly used. The direct shear test is illustrated in Fig. 4.12(a). The soil specimen is enclosed in a box consisting of an upper and a lower half. The porous stones permit drainage of water from the specimen. The potential plane of failure is a-a. A normal stress σ is applied on plane a-a through a loading head, and the shear stress is increased until the specimen fails along the plane a-a. Loading is carried out slowly so that the drained condition is maintained. A stress deformation curve is obtained by plotting the shear stress against the shear displacement. Because the thickness of the shear zone a-a is not precisely known, the shear strain cannot be determined.

The test gives the value of τ_{xz} at failure. The stresses applied to the element are shown in Fig. 4.12(b). The vertical stress $\bar{\sigma}_z$ and the shear stress τ_{xz} are known, but $\bar{\sigma}_x$ is not. The direction of the principal stresses are approximately as shown in Fig. 4.12(c), and the Mohr's circle for the stresses at failure is shown in Fig. 4.12(d). However, only the point $(\bar{\sigma}_z, \tau_{xz})$ on the Mohr's circle is known; hence, the circle cannot be constructed for the direct shear test. The foregoing represents the common interpretation of the direct shear test. For more elaborate analyses, see Hill (1950) and Morgenstern and Tchalenko (1967a).

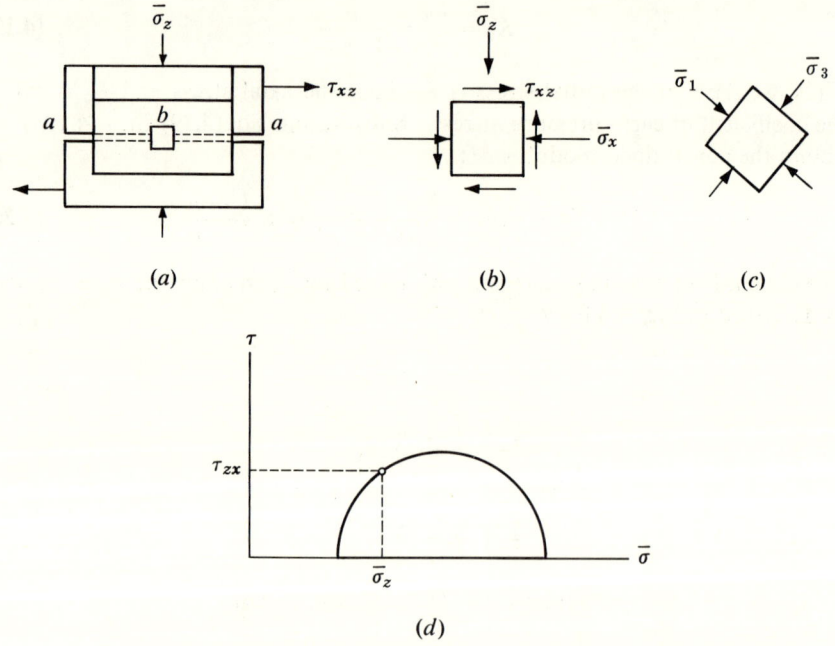

FIGURE 4.12 *The direct shear test.*

THE TRIAXIAL TEST. This is a highly versatile test, and a variety of stress and drainage conditions may be employed. Figure 4.13(a) is a schematic diagram describing the triaxial apparatus. The cylindrical soil specimen is enclosed within a thin rubber membrane and is placed inside a triaxial cell. The cell is then filled with a fluid. By applying pressure to the fluid in the cell, the specimen may be subjected to a hydrostatic compressive stress $\bar{\sigma}_3$. If the specimen undergoes a volume change, the drainage is provided by the porous stone at the bottom, which is connected to a burette. The volume change is measured by the change of water level in the burette. This allows the determination of the volume compressibility.

The axial stress may be increased by application of a load through the piston. This produces the stress state in Fig. 4.13(b). From the known stresses

SEC. 4.7 Measurement of stress–strain properties

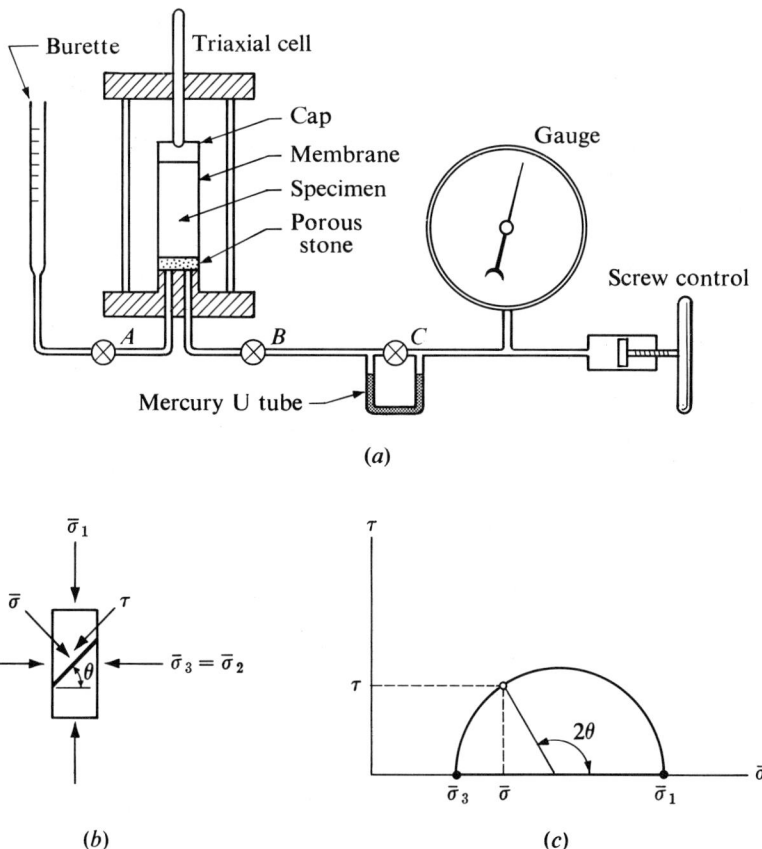

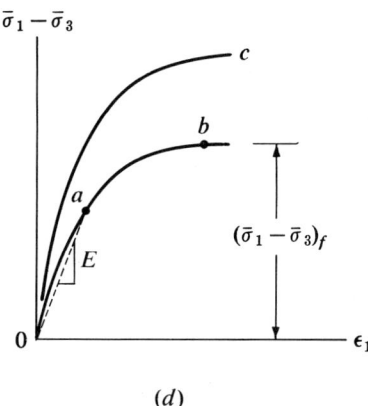

FIGURE 4.13 *The triaxial test.*

$\bar{\sigma}_1$ and $\bar{\sigma}_3 = \bar{\sigma}_2$, the Mohr's circle in Fig. 4.13(c) may be constructed, and the stresses $\bar{\sigma}$, τ on any plane may be found. If the principal stress difference $(\bar{\sigma}_1 - \bar{\sigma}_3)$ is plotted against the axial strain, the curve shown in Fig. 4.13(d) is obtained. If the sample is loaded to point a, the secant modulus E may be determined from the slope of $0a$. If the load is increased to point b, failure occurs under the shear stress $\frac{1}{2}(\bar{\sigma}_1 - \bar{\sigma}_3)_f$. If a larger value of $\bar{\sigma}_3$ is used the stress-strain curve $0c$ is obtained.

Valves A and B may be closed to prevent drainage from the specimen. Another line from the base leads to a mercury-filled U tube used for the measurement of porewater pressure. To measure the porewater pressure the porous stone and the lines leading to it are completely saturated with water. With this arrangement the undrained condition can be maintained by opening valve B, closing valve C, and keeping the mercury level fixed in the U tube. To do this the screw control is manipulated so that the pressure on the right side of the mercury U tube is equal to the porewater pressure in the soil, which acts on the left side. This amount is registered by the pressure gauge.

Thus, if a hydrostatic stress σ_3 is applied, the measured pore pressure Δu_b allows the determination of the pore pressure parameter B in Eq. (4.13). Similarly, if a stress σ_1 is applied, the measured pore pressure Δu_a allows the determination of the pore pressure parameter A in Eq. (4.15).

The secant modulus E may be computed for either the drained or undrained conditions. Consider first the drained condition:

$$\epsilon_1 = \frac{\Delta \bar{\sigma}_1}{E} \tag{4.21a}$$

$$\epsilon_2 = \epsilon_3 = -\mu \frac{\Delta \bar{\sigma}_1}{E} \tag{4.21b}$$

Since the volume change is measured, we also have

$$\epsilon_v = \epsilon_1 + 2\epsilon_3 \tag{4.21c}$$

Equation (4.21a) allows the determination of E and Eqs. (4.21b) and (c) allow the determination of μ. For the undrained condition

$$\epsilon_1 = \frac{\Delta \sigma_1 - \Delta u}{E} - 2\mu \frac{(-\Delta u)}{E} \tag{4.22a}$$

$$\epsilon_2 = \epsilon_3 = -\mu \frac{\Delta \sigma_1 - \Delta u}{E} + (1 - \mu) \frac{(-\Delta u)}{E} \tag{4.22b}$$

and if the soil is saturated

$$\epsilon_v = \epsilon_1 + \epsilon_2 + \epsilon_3 = 0 \tag{4.22c}$$

The quantities ϵ_v, $\Delta \sigma_1$ and Δu are measured and E, μ, and ϵ_3 are computed from the three equations. For a given soil the values of E and μ are dependent on the magnitudes of $\bar{\sigma}_3$ and $(\bar{\sigma}_1 - \bar{\sigma}_3)$. The stress and strain distributions in a real test are more complicated. For this topic, see Haythornthwaite (1960), Roscoe et al. (1963b) and Perloff and Pombo (1969).

PROBLEMS

4.1 Given the stresses shown in Fig. 4.14, use Mohr's circle to find (*a*) the major and minor principal stresses, (*b*) the directions of the principal planes, (*c*) the maximum shear stress, and (*d*) the direction of the plane of maximum shear.
Note: $\sigma_y = \tau_{xy} = \tau_y = 0$.
Answer: (*a*) $\sigma_1 = +110$ psi, $\sigma_3 = -1700$ psi; (*b*) σ_1 plane makes an angle of 58.5 deg counterclockwise from Z axis; (*c*) ± 900 psi; 45 deg from principal stresses.

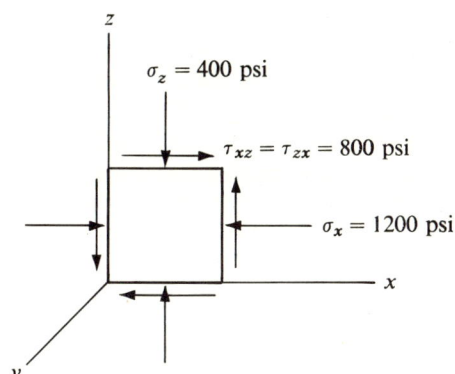

FIGURE 4.14

4.2 Given the stresses shown in Fig. 4.15, use the Mohr's circle to find (*a*) the maximum shear stress and its direction, and (*b*) the normal and shear stresses on the plane $\theta = 30°$.
Answer: (*a*) $\tau_{max} = \pm 250$ psi, 45 deg from X and Z axes; (*b*) $\sigma = -525$ psi, $\tau = -130$ psi

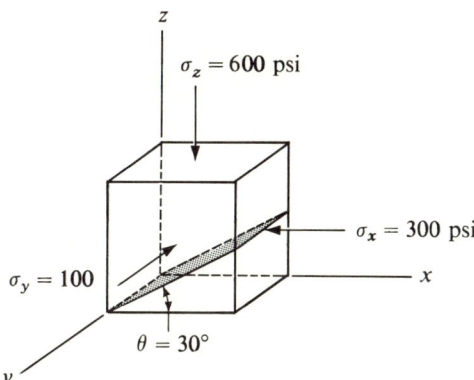

FIGURE 4.15

4.3 Calculate the maximum normal strain and maximum shear strain at a point whose stresses are given in (4.1). For this material $E = 10^4$ psi and $\mu = 0.35$.
Answer: $+0.0705$; 0.243

4.4 A material cannot withstand a shear stress greater than 200 psi, (a) Will it fail under the stresses shown in Fig. 4.15? (b) If so, along what planes will failure occur? Show this on a Mohr's circle.
Answer: (a) It will fail; (b) on planes where $\tau > 200$ psi

4.5 For a point whose stresses are given in Fig. 4.15, compute (a) the volumetric strain and (b) the maximum shear strain. For this material $E = 10^4$ psi and $\mu = 0.35$.
Answer: (a) 0.03; (b) 0.068

4.6 Consider the stresses shown in Fig. 4.15. How large a compressive stress can be put in the X direction before failure occurs, if the material has a shear strength equal to 300 psi? (σ_y and σ_z remain constant during this time.)
Answer: -700 psi

4.7 A cylindrical sample of clay was subjected to the unconfined compression test, as shown in Fig. 4.16(a). The stress-deformation curve obtained is shown in Fig. 4.16(b). (a) If the material is assumed to be elastic up to point A [Fig. 4.16(b)], what is E? (b) If point A is taken to represent failure of the sample, what is the maximum shear stress in the sample at failure? What is the direction of the plane on which the maximum shear stress acts?
Answer: (a) $E = 4 \times 10^4$ psf; (b) $T_{\max} = 10^3$ psf; it acts on planes inclined at 45 deg from the vertical

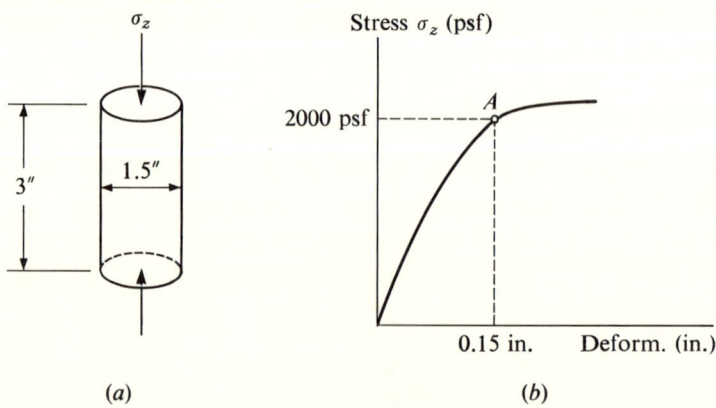

FIGURE 4.16

4.8 A clay sample was placed in a triaxial cell with the drainage valve closed, and a confining pressure of 1.0 kg/cm² was applied. The measured response of the pore pressure was 1.0 kg/cm². The sample was then loaded axially until failure occurred at an axial stress of 6 kg/cm². The resulting pore pressure in the sample at failure was 4 kg/cm². Compute the values of A and B.
Answer: $A = 0.8$; $B = 1.0$

4.9 A sample of dry sand is placed in a consolidation ring and loaded to an axial stress of 5 kg/cm². If the coefficient of earth pressure at rest, K_0, is 0.5, then we know that the vertical and horizontal effective stresses are 5 and 2.5 kg/cm², respectively. If an additional vertical stress of 4 kg/cm² is applied, what are the vertical and horizontal stresses?
Answer: $\bar{\sigma}_z = 9.0$ kg/cm²; $\bar{\sigma}_x = \bar{\sigma}_y = 4.5$ kg/cm²

4.10 The major, intermediate, and minor principal stresses at point A of the clay layer (Fig. 4.17) are increased by 3000, 1500, and 1500 psf, respectively, due to loads applied near the surface. (*a*) What is the porewater pressure at A before loading? (*b*) What is the porewater pressure immediately after loading? (*c*) What is the piezometric surface immediately after loading with respect to the groundwater table?

	Sand	Clay
Unit wt.	125 pcf	110 pcf
Permeability	2 in./hr	2×10^{-4} in./hr
Void ratio	0.72	1.32
Compress. (m_c)	0.01	0.10 tsf^{-1}
Expans. (m_e)	0.01	0.02 tsf^{-1}

Answer: 938 psf; 3488 psf; 30.8 ft above ground surface

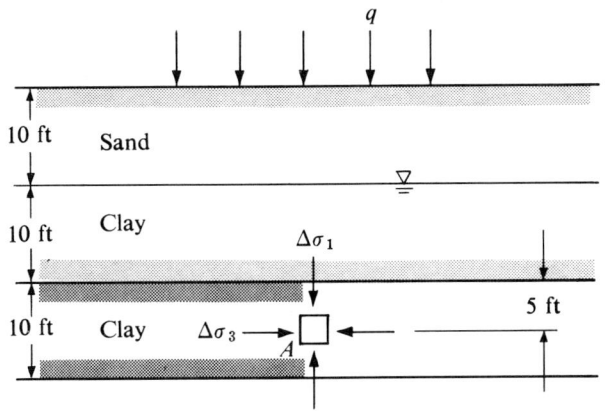

FIGURE 4.17

4.11 Derive an expression for the porewater pressure coefficient A for a specimen loaded as shown in Fig. 4.18. The pressure in two directions is increased by $\Delta\sigma_1$ simultaneously while it is kept constant in the third direction. Assume the material to be elastic in the stress range considered.
Answer: $\Delta u_b = A \cdot \Delta\sigma_3$; $A = 1/[1 + (m_e/2m_c)]$

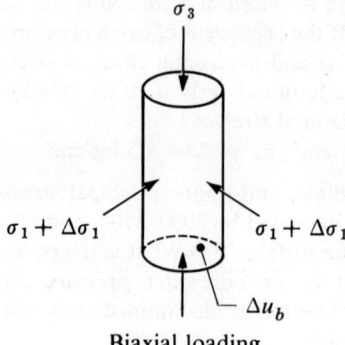

Biaxial loading
—Undrained

FIGURE 4.18

4.12 Prove that if the soil specimen in Fig. 4.9 is loaded by increasing $\Delta\sigma_1$ and holding the strains in 2 and 3 directions equal to 0, the porewater pressure is given by $\Delta u_b = \Delta\sigma_1$ or $A = 1$.

5

Volume Change and Compressibility

5.1 Introduction

A common example of a circular load applied to the ground surface is shown in Fig. 5.1(a). It increases the stresses on a soil element A by $\Delta\sigma_z$ and $\Delta\sigma_x = \Delta\sigma_y = \Delta\sigma_r$. If the soil is saturated there is no volume change immediately after the stresses are imposed because it takes time for the water to flow from the voids. Thus the porewater pressure in A increases by an amount Δu, as given by Eq. (4.18).

While there is no volume change at this stage, the shear stress increment $\frac{1}{2}(\Delta\sigma_z - \Delta\sigma_x)$ produces a shear distortion of the element as shown in Fig. 5.1(b). With the passage of time, the pore pressure Δu dissipates. The flow of water from the voids results in a volumetric compression. This process of compression by drainage of water from the soil is called consolidation. During consolidation the stress is transferred from the pore pressure to the soil skeleton. At the end of this process, the conditions at A are shown in Fig. 5.1(c).

For many clay soils the volumetric compression due to the applied loads leads to large settlements. The determination of the compressibility is the subject of this chapter.

5.2 Compressibility

The consolidation test is shown in Fig. 5.2(a). The soil specimen is placed inside a metal ring. Pressure is applied to the specimen through the loading head, and the porous stones allow free escape of the porewater as the voids in the soil are compressed. A dial indicator is used to measure the downward movement

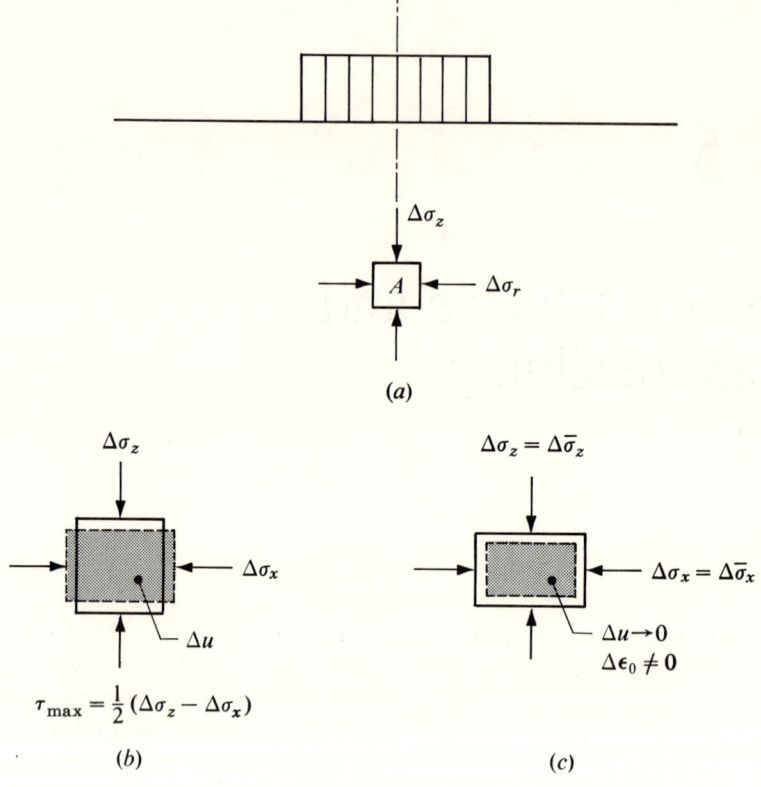

FIGURE 5.1 *Stresses and strains under the center of a circular loaded area.*

of the loading head; this allows the calculation of the volume change of the soil. Table 5.1 gives the final dial readings (compression) measured after the application of various pressure increments in a conventional consolidation test.

In conventional consolidation tests the pressure is doubled every 24 hr. The long time interval between load increments is necessary because the drainage of water from the voids does not occur instantaneously but requires considerable time (Sec. 5.5). Such an experiment gives us the empirical relationship between the pressure and the volume change that is called compressibility. Compressibility (m_v) is defined as the volume strain for a unit increment of stress:

$$m_v = \frac{\Delta \epsilon_z}{\Delta \bar{\sigma}_z} \tag{5.1}$$

Alternatively, the constrained modulus,

$$M = \frac{\Delta \bar{\sigma}_z}{\Delta \epsilon_z} = \frac{\Delta \bar{\sigma}_1}{\Delta \epsilon_1} \tag{4.20}$$

may also be used.

SEC. 5.2 Compressibility 109

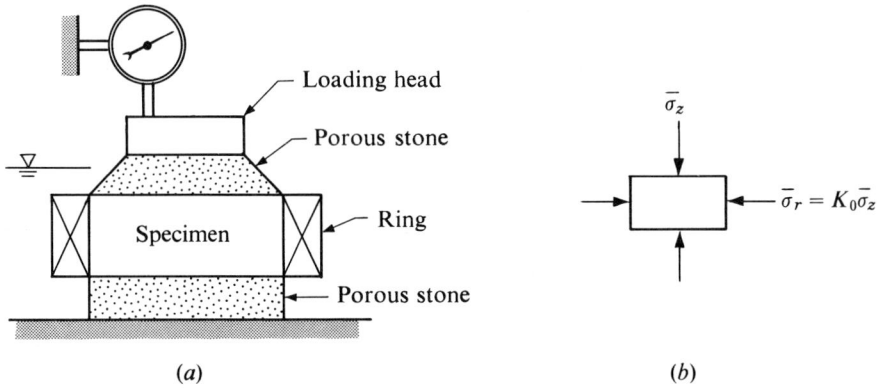

FIGURE 5.2 *Consolidation test.*

TABLE 5.1 *Consolidation Test Data*

Soil: Grey clay with silt laminations   Boring No. 7
Location: Detroit, Mich.   Specimen No. 1–3
Depth: 10 ft 10 in.
Existing overburden pressure = 1160 psf
Consolidation specimen: Height = 1.0 in.
                            Diameter = 2.5 in.
                            Drainage: both sides
                            Initial void ratio: 0.740

No.	Pressure, kg/cm^2	Dial reading, in.
	0	0
1	0.158	0.0026
2	0.316	0.0048
3	0.634	0.0094
4	1.26	0.0154
5	2.53	0.0254
6	5.06	0.0418
7	10.1	0.0726
8	20.2	0.1105
9	40.5	0.1633
10	10.1	0.1477
11	2.53	0.1277
12	0.634	0.1075
13	0.158	0.0904

PRESSURE VS. VOID RATIO RELATIONSHIP. The results of the consolidation test may be plotted as a stress-strain curve as shown in Fig. 4.6(b). For convenience in computations the void ratio is commonly used instead of the strain and a pressure-void ratio curve is plotted as shown in Fig. 5.3(a). For a particular pressure increment, the segment *ab* on the curve may be assumed to be a straight line. The slope of *ab* is

$$a_v = \frac{\Delta e}{\Delta \bar{\sigma}} = (1 + e_0)\frac{\Delta \epsilon_z}{\Delta \bar{\sigma}} = (1 + e_0)m_v \qquad (5.2)$$

where e_0 denotes the initial void ratio.

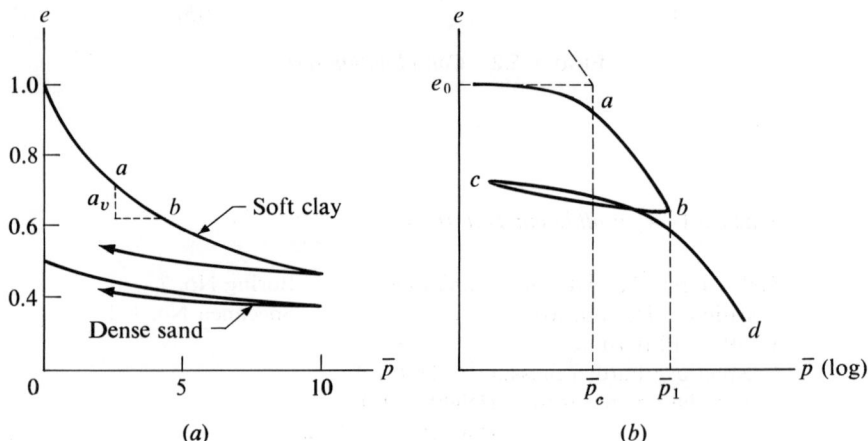

FIGURE 5.3 *Pressure vs. void ratio curves: (a) arithmetic plot; (b) semilogarithmic plot.*

The pressure-void ratio curve may also be plotted with the pressure on a logarithmic scale. The advantage is that for most clayey soils part of the curve is approximately a straight line, as shown by the curve *ab* in Fig. 5.3(b).

5.3 Field consolidation

STRESS HISTORY. Experiments have shown that the sharp curvature in the $e - \log \bar{p}$ curve [Fig. 5.3(b)] reflects the stress history of the sample. As an example we load the soil to pressure $\bar{p}_1$. Then it is unloaded and reloaded. The unloading and reloading curves form a hysteresis loop. When the pressure exceeds $\bar{p}_1$ the reloading curve approaches *bd*, which is an extension of the loading curve *ab*. Thus we see that the steep portion of the curve *cd* represents the compressibility of the soil when the pressure exceeds the maximum past pressure $\bar{p}_1$.

Following this reasoning, we may interpret the flat portion $e_0 a$ of the loading curve as a reloading, since the extraction of the soil sample from the ground relieves it of the overburden pressure. The pressure $\bar{p}_c$ is called the preconsolidation pressure and is equal to the maximum pressure under which the soil has previously been consolidated in the field.

NORMALLY CONSOLIDATED AND PRECONSOLIDATED CLAYS. In order to interpret rationally the $e - \log \bar{p}$ curve, it is necessary to consider the stress history of the soil. The soil is said to be normally loaded or normally consolidated if the value $\bar{p}_c$ is equal to its present effective overburden pressure $\bar{p}_0$. Point a [Fig. 5.4(a)] represents the condition of the soil as it exists in the field. If the vertical effective stress on a normally loaded soil is increased from $\bar{p}_0$ to $\bar{p}_0 + \Delta \bar{p}$ in the field, the void ratio changes by Δe, as shown in the figure. This amount can be measured directly from the $e - \log \bar{p}$ curve. Since the field consolidation or virgin curve is nearly a straight line, we can denote its slope by C_c, called the compression index.

The change in void ratio Δe can also be computed according to the relationship

$$\Delta e = C_c [\log (\bar{p}_0 + \Delta \bar{p}) - \log \bar{p}_0] = C_c \log \frac{\bar{p}_0 + \Delta \bar{p}}{\bar{p}_0} \tag{5.3a}$$

The volume strain is

$$\epsilon_v = \frac{\Delta e}{1 + e_0} \tag{5.3b}$$

A soil may have been consolidated under a large overburden in times past. Then the value $\bar{p}_c$ is much greater than $\bar{p}_0$. Figure 5.4(b) shows that the soil was first consolidated in the field to point a. Subsequent removal of part of the overburden (i.e., erosion) brought about an unloading and the conditions changed to that of point b. Such a soil is called a preconsolidated or overconsolidated soil. To compute the value of Δe for a preconsolidated soil, Eq. (5.3) cannot be used. Figure 5.4(b) shows that if $\bar{p}_0 + \Delta \bar{p}$ is less than $\bar{p}_c$, the slope of the curve is much smaller than C_c. Careful studies of $e - \log \bar{p}$ curves by Schmertmann (1953) led to the conclusion that the reloading branch (ba) has about the same slope as the unloading branch (cd) for good undisturbed specimens, and field consolidation would follow the curve bac. Hence to compute the value of Δe for a preconsolidated clay, the slope C_e of the reloading branch or unloading branch should be used instead of C_c in Eq. (5.3). As with normally consolidated clays, Δe can also be obtained directly from the $e - \log \bar{p}$ curve.

If $\bar{p}_0 + \Delta \bar{p}$ exceeds $\bar{p}_c$, the compression of the soil follows line bac [Fig. 5.4(b)]. Then we have

$$\Delta e = C_e \log \frac{\bar{p}_c}{\bar{p}_0} + C_c \log \frac{\bar{p}_0 + \Delta \bar{p}}{p_c} \tag{5.3c}$$

The first term on the right hand side denotes the compression from b to a and the second term denotes the compression beyond a. The accuracy and reliability of using laboratory consolidation data for settlement calculations is summarized in Sec. 7.6.

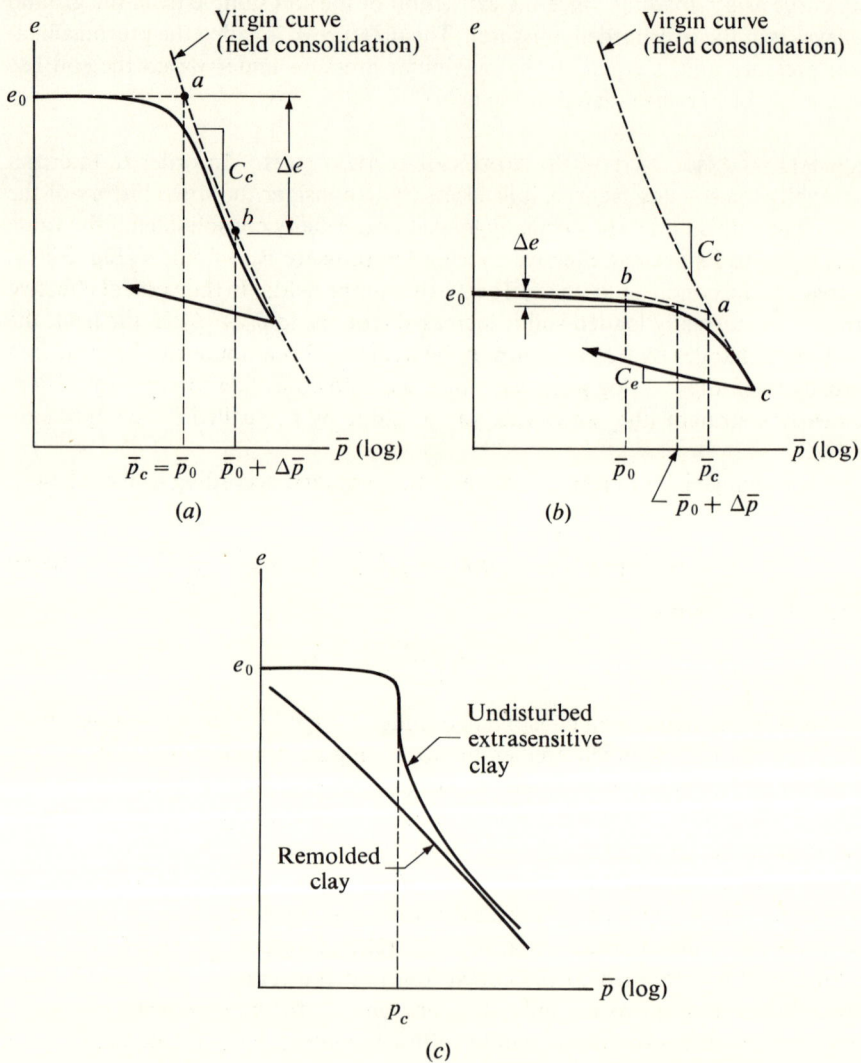

FIGURE 5.4 *Pressure vs. void ratio curves:* (a) *normally consolidated clay;* (b) *preconsolidated clay;* (c) *extra sensitive clay and remolded clay.*

Inasmuch as the compressibility is the result of the compression of the soil-skeleton structure, the shape of the $e - \log \bar{p}$ curve is a reflection of the changes in structure. If we consider the curve in Fig. 5.4(a), we note that little compression takes place when the pressure is less than the preconsolidation pressure. In this pressure range, the soil structure can withstand the stresses because it has previously been loaded to pressure $\bar{p}_c$. Only when the applied pressure exceeds $\bar{p}_c$ do significant changes in structure occur, yielding large compressions.

Some clay structures are very sensitive to disturbances; these are called sensitive clays. In a consolidation test on a highly sensitive clay, a curve such as the one shown in Fig. 5.4(c) is often obtained. The curve has an almost vertical slope at a pressure close to $\bar{p}_c$, indicating that the structure suffers a breakdown when the pressure is increased beyond $\bar{p}_c$. In contrast, a soil sample that is completely remolded has lost all the structural characteristics of the natural soil. When it is loaded in the consolidation tests, the particles are progressively forced closer together and the curve shows no prominent changes in curvature. Thus a remolded soil cannot be used to evaluate the preconsolidation pressure. Indeed the accuracy of the measured preconsolidation pressure is dependent on the quality of the specimen.

DETERMINATION OF PRECONSOLIDATION PRESSURE. The laboratory $e - \log \bar{p}$ curve undergoes a change in slope at pressures close to the preconsolidation pressure. If we examine a typical curve as shown in Fig. 5.4(a), we realize that the value of $\bar{p}_c$ is not precisely given by the curve. After a consolidation test is performed on a natural soil, the only information we have on its stress history, other than geologic deductions, is the $e - \log \bar{p}$ curve. To enable one to calculate $\bar{p}_c$ from the $e - \log \bar{p}$ curve, Casagrande (1936) studied the laboratory loading, unloading, and reloading curves. Since in his laboratory reloading the preconsolidation pressure [$\bar{p}_1$ in Fig. 5.3(b)] was known, he was able to establish an empirical method for the calculation of $\bar{p}_c$.

Casagrande's procedure is as follows. On the $e - \log \bar{p}$ curve [Fig. 5.5] pick out the point of greatest curvature and designate this as point c. From this point two lines are drawn, one tangent to the curve and another parallel to the p axis. The angle α between these two lines is bisected by line Ac, and A is the intersection of this line with the tangent to the steep part of the consolidation curve. The abscissa of point A is the preconsolidation pressure.

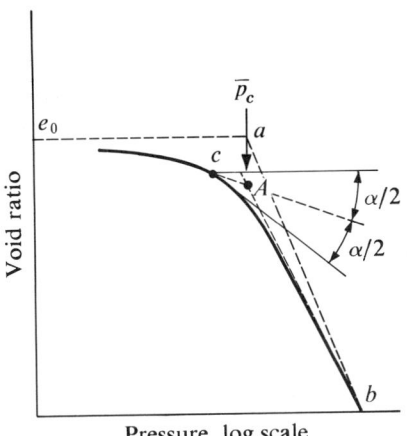

FIGURE 5.5 *Pressure vs. void ratio relationship.*

An interesting check on the method of calculation of the preconsolidation pressure is provided by some clay deposits. Clays deposited at the bottom of lakes or bays during recent geologic times are consolidated under the weight of the overburden. Many of these clays are normally consolidated, because geologic evidence indicates that no erosion has taken place since deposition. For such clays, the calculated preconsolidation pressure should equal the existing overburden pressure. Many examples showing good agreement between the two values have been presented [Casagrande and Fadum (1944); Skempton (1948a); Wu (1958), for example] which support, in general, the method for calculation of the preconsolidation pressure. On the other hand, Leonards and Ramiah (1959) found that certain loading sequences, such as very small pressure increments or very long time intervals, affect the value of the measured preconsolidation pressure.

It should also be noted that the measured preconsolidation pressure is not necessarily equal to the overburden pressure under which the soil has been consolidated in situ. Cementation between soil particles or desiccation of the soil deposits may produce $e - \log \bar{p}$ curves that indicate large preconsolidation pressures. In these cases the values of $\bar{p}_c$ are not related to past overburden pressures. For example, Hamilton (1964) has observed that ocean bottom sediments with little or no overburden pressure possess appreciable preconsolidation pressures.

FIELD $e - \text{LOG } \bar{p}$ CURVE. After $\bar{p}_c$ is evaluated, the approximate field curve is determined as follows. If the clay is normally consolidated, then the preconsolidation pressure $\bar{p}_c$ and the initial void ratio of the specimen e_0 determine the location of point a [Fig. 5.5]. This is the status of the clay in situ before any load has been applied. The virgin consolidation curve ab is slightly different from the straight portion of the laboratory consolidation curve and may be sketched in. Several empirical methods for locating the virgin curve are also available. Terzaghi and Peck (1967), for example, suggested that ab should intersect the extension of the laboratory consolidation curve at a void ratio of $0.4e_0$ (point b).

For an overconsolidated clay, the field consolidation curve is constructed by first establishing the in situ pressure and void ratio [point b, Fig. 5.4(b)]; the coordinates of this point are the existing overburden pressure and initial void ratio. Line ba is then drawn parallel to the rebound curve, with the abscissa of point a equal to the preconsolidation pressure. The virgin compression curve is located as described in the preceding paragraph. [See also Schmertmann (1953).]

5.4 Rate of consolidation

It has already been pointed out that the volume change during consolidation can be entirely attributed to the reduction of the volume of the voids. For saturated soils the rate of consolidation is governed by the rate at which the porewater can escape from the soil.

SEC. 5.4   Rate of consolidation

This process can be understood by an examination of the effective stress and porewater pressure and is illustrated graphically by the model in Fig. 5.6. In the figure, the effective stress and porewater pressure are represented, respectively, by the stress in the spring and the pressure of the water. Initially the system is in equilibrium, and the effective overburden pressure $\bar{p}_0$ is resisted by an effective stress of the same magnitude. The hydrostatic porewater pressure is u_0 [Fig. 5.6(a)]. Figure 5.6(b) illustrates the condition immediately after the application of the pressure Δp. Since water has not had time to drain, the volume change is 0. For the condition of no lateral strain, the porewater-pressure increase is equal to the applied vertical pressure, and the effective stress remains at the initial value. The applied pressure then is resisted entirely by an increase in porewater pressure equal to Δp. With the passage of time, volume changes occur as water is forced out of the container, and the effective stress increases. This is accompanied by a reduction in the porewater pressure of equal magnitude, since $u + \bar{\sigma}$ must equal $\Delta p + \bar{p}_0$. Finally, $\bar{\sigma}$ reaches a value equal to $\Delta p + \bar{p}_0$, and at this point the porewater pressure is again equal to the value u_0 [Fig. 5.6(c)]. Since the applied pressure and the effective stress are in equilibrium, there is no further compression, and consolidation is complete.

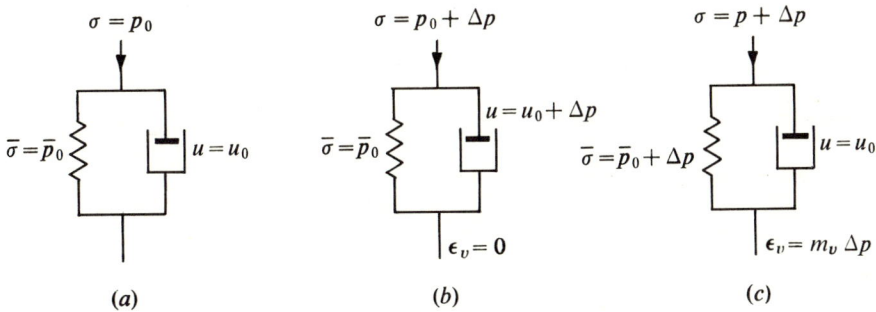

FIGURE 5.6   *Pore pressure and effective stress.*

The progress of consolidation after a particular time t is represented by a ratio

$$U_z = \frac{\Delta V}{\Delta V_{t=\infty}}$$

called the *degree of consolidation*. In this equation ΔV is the volume change after a time t and $\Delta V_{t=\infty}$ is the total volume change after the completion of the consolidation process. Frequently, this is also expressed as a percent.

As the volume change is the result of the increase in effective stress, we can write

$$\epsilon_v = \frac{\Delta V}{V} = m_v \Delta \bar{\sigma} = -m_v \Delta u$$

If m_v is assumed to be a constant within this stress increment, the degree of consolidation may be written

$$U_z = \frac{\Delta V}{\Delta V_{t=\infty}} = \frac{\Delta u}{\Delta u_{t=\infty}} = \frac{u_i - u}{u_i - u_0} \qquad (5.4)$$

in which Δu and $\Delta u_{t=\infty}$ are the changes in porewater pressure after time t and ∞, u_i is the porewater pressure immediately after the application of Δp (equal to the applied pressure plus the hydrostatic porewater pressure u_0), and u is the porewater pressure at time t.

5.5 Terzaghi consolidation theory

The relationship between the degree of consolidation U_z and time can be derived from the principles of hydraulics. This problem was solved by Terzaghi in 1925 for saturated soils. Figure 5.7(a) shows a layer of clay subjected to one-dimensional consolidation, and an element from this layer is shown in Fig. 5.7(b).

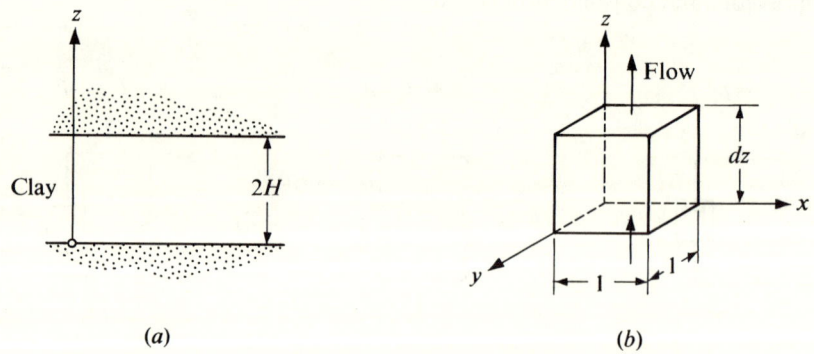

FIGURE 5.7 One-dimensional consolidation problem.

The applied pressure is Δp. If the hydraulic gradient is $-\partial h/\partial z$, the velocity of flow is

$$v = ki = -k\frac{\partial h}{\partial z} = -\frac{k}{\gamma_w}\frac{\partial u}{\partial z}$$

in which u is the porewater pressure. Following the derivation in Sec. 3.1, the difference between the rate of flow out of and into the element is

$$-\frac{k}{\gamma_w}\left(\frac{\partial u}{\partial z} + \frac{\partial^2 u}{\partial z^2}dz - \frac{\partial u}{\partial z}\right) = -\frac{k}{\gamma_w}\frac{\partial^2 u}{\partial z^2}dz$$

and this should be equal to the rate of change of volume dV/dt. Hence the continuity equation for one-dimensional flow is

$$-\frac{k}{\gamma_w}\frac{\partial^2 u}{\partial z^2}dz = \frac{dV}{dt} \qquad (5.5)$$

Terzaghi consolidation theory

For a consolidating clay layer dV/dt is given by

$$\frac{dV}{dt} = m_v \frac{\partial(\bar{\sigma})}{\partial t} dz = -m_v \frac{\partial u}{\partial t} dz \tag{5.6}$$

since dV is $m_v\, d\bar{\sigma}\, dz$. Combining Eqs. (5.5) and (5.6), we obtain

$$-m_v \frac{\partial u}{\partial t} = -\frac{k}{\gamma_w} \frac{\partial^2 u}{\partial z^2}$$

or

$$\frac{\partial u}{\partial t} = \frac{k}{m_v \gamma_w} \frac{\partial^2 u}{\partial z^2} = c_v \frac{\partial^2 u}{\partial z^2} \quad \left(c_v = \frac{k}{m_v \gamma_w}\right) \tag{5.7}$$

The factor c_v is called the *coefficient of consolidation*. It is a soil property, inasmuch as k and m_v are properties of the soil. This is the fundamental equation for the consolidation theory.

The differential equation (5.7) may be solved for the following simplified conditions. It is assumed that the compression of the soil is small compared to the thickness H. It is further assumed that the soil property c_v is constant for the pressure increment and is not affected by the compression under this pressure increment. Consequently, H and c_v are constants in Eq. (5.7). For simplicity we can take u_0 as the datum for porewater pressure, or $u_0 = 0$. The quantity u then consists entirely of the excess hydrostatic porewater pressure induced by the applied pressure Δp.

If the pressure increment Δp is applied instantaneously the initial and boundary conditions are as follows.

At zero time, the applied stress Δp is carried by the porewater pressure at every point in the clay layer. If the initial porewater pressure is designated by u_i, we have,

$$\text{when } t = 0, \quad u = u_i = \Delta p \tag{5.8a}$$

Since there is complete drainage at the top and bottom of the clay layer, there exists no excess hydrostatic porewater pressure on these planes. The porewater pressure along these two surfaces is equal to the hydrostatic pressure outside the clay. Therefore,

$$\text{when } z = 0, \quad u = 0 \tag{5.8b}$$
$$\text{when } z = 2H, \quad u = 0 \tag{5.8c}$$

MATHEMATICAL SOLUTION. To solve Eq. (5.7) for the above boundary conditions we let

$$u = F(z)G(t) \tag{5.9}$$

Equation (5.7) then becomes

$$c_v G(t) F''(z) = F(z) G'(t)$$

or

$$\frac{F''(z)}{F(z)} = \frac{G'(t)}{c_v G(t)}$$

The quantity on the left side of the equation is a function of z only and independent of t, while that on the right side is a function of t only and is independent of z. Hence both quantities must be equal to a constant, say, $-A^2$. Thus it follows that

$$F''(z) = -A^2 F(z)$$

and

$$G'(t) = -A^2 c_v G(t)$$

A solution to the first of the above equations is of the form

$$F(z) = C_1 \cos Az + C_2 \sin Az$$

and a solution to the second is of the form

$$G(t) = C_3 e^{-A^2 c_v t}$$

where C_1, C_2, and C_3 are constants. Thus Eq. (5.9) becomes

$$u = (C_1 \cos Az + C_2 \sin Az) C_3 e^{-A^2 c_v t}$$
$$= (C_4 \cos Az + C_5 \sin Az) e^{-A^2 c_v t}$$

The boundary condition (5.8b) requires that $C_4 = 0$. To satisfy Eq. (5.8c) requires that

$$C_5 \sin(2AH) = 0$$

or

$$2AH = n\pi$$

where n is any integer. Thus

$$u = \sum_{n=1}^{n=\infty} C_n \sin \frac{n\pi z}{2H} e^{-n^2 \pi^2 c_v t / 4H^2}$$

The boundary condition (5.8a) is satisfied if the coefficients C_n are such that

$$u_i = \sum_{n=1}^{n=\infty} C_n \sin \frac{n\pi z}{2H}$$

which is a Fourier sine series. The coefficients C_n are given by*

$$C_n = \frac{1}{H} \int_0^{2H} u_i \sin \frac{n\pi z}{2H} dz$$

The solution then becomes

$$u = \sum_{n=1}^{n=\infty} \left(\frac{1}{H} \int_0^{2H} u_i \sin \frac{n\pi z}{2H} dz \right) \left(\sin \frac{n\pi z}{2H} \right) e^{-n^2 \pi^2 c_v t / 4H^2} \qquad (5.10)$$

RESULTS. We consider the case in which u_i is a constant. Then the solution, Eq. (5.10), becomes

$$u = \sum_{n=1}^{n=\infty} \frac{2u_i}{n\pi} (1 - \cos n\pi) \left(\sin \frac{n\pi z}{2H} \right) e^{-n^2 \pi^2 c_v t / 4H^2}$$

or

$$u = \sum_{m=0}^{m=\infty} \left(\frac{2u_i}{M} \sin \frac{Mz}{H} \right) e^{-M^2 T_v} \qquad (5.11)$$

---

* See, for example, Churchill (1941), pp. 24–30.

SEC. 5.5 Terzaghi consolidation theory

in which $M = \frac{1}{2}\pi(2m + 1)$, m is an integer, and

$$T_v = \frac{c_v t}{H^2} \tag{5.12}$$

The term T_v is a dimensionless number called the *time factor*. The quantities c_v and H are known for a given problem. The expression therefore gives the porewater pressure u as a function of time t and depth z. The degree of consolidation at any point in the stratum is obtained by substituting the expression for u [Eq. (5.11)] in (5.3), or

$$U_z = 1 - \sum_{m=0}^{m=\infty} \frac{2}{M}\left(\sin\frac{Mz}{H}\right) e^{-M^2 T_v} \tag{5.13}$$

The progress of consolidation is illustrated graphically in Fig. 5.8(a), in which the excess hydrostatic porewater pressure in a clay layer $2H$ thick is plotted using Eq. (5.11). Immediately after loading ($T_v = 0$), the porewater pressure everywhere in the layer is equal to Δp. The variation of porewater pressure with z after different time intervals T_v is shown in Fig. 5.8(a). Such curves are called *isochrones*. The porewater pressure as calculated by the consolidation theory has been verified experimentally by Taylor (1942). An example of the porewater-pressure distribution is given in Fig. 5.8(b). The clay embankment of Selset Dam contained drainage layers of sand. The porewater-pressure distribution in the clay located between two sand layers was measured at time intervals after loading and had the characteristic shape of the isochrone.

The degree of consolidation U_z at any time also varies with the depth z. Figure 5.8(c) contains plots of U_z versus z for various time intervals T_v. It is frequently necessary to find the average degree of consolidation for a layer of clay. This is equal to

$$U_H = \frac{u_i - (1/2H)\int_0^{2H} u\,dz}{u_i} = 1 - \sum_{m=0}^{m=\infty} \frac{2}{M^2} e^{-M^2 T_v} \tag{5.14}$$

Figure 5.9 is the graphical plot of Eq. (5.14).

We note in Fig. 5.8(a) that the porewater-pressure distribution is symmetrical about the center of the consolidating layer. If the clay layer is only half as thick and drains on one side only (as would be the case if the clay lies over an impermeable base) then the progress of consolidation is exactly the same as that shown in the upper half of Fig. 5.8(c). Hence if a layer has only one free-draining face, the value of H [in Eqs. (5.11), (5.13), and (5.14)] should be taken as the thickness of the layer.

The theory of consolidation can also be extended to problems of radial flow and of two- and three-dimensional flow. A common example of radial flow is the consolidation of a compressible layer by drainage toward vertical wells. These wells frequently consist of columns of highly permeable sand (hence the name *sand drain*) and are often used as a means of accelerating the process of consolidation. The consolidation theory for sand drains was solved by Barron

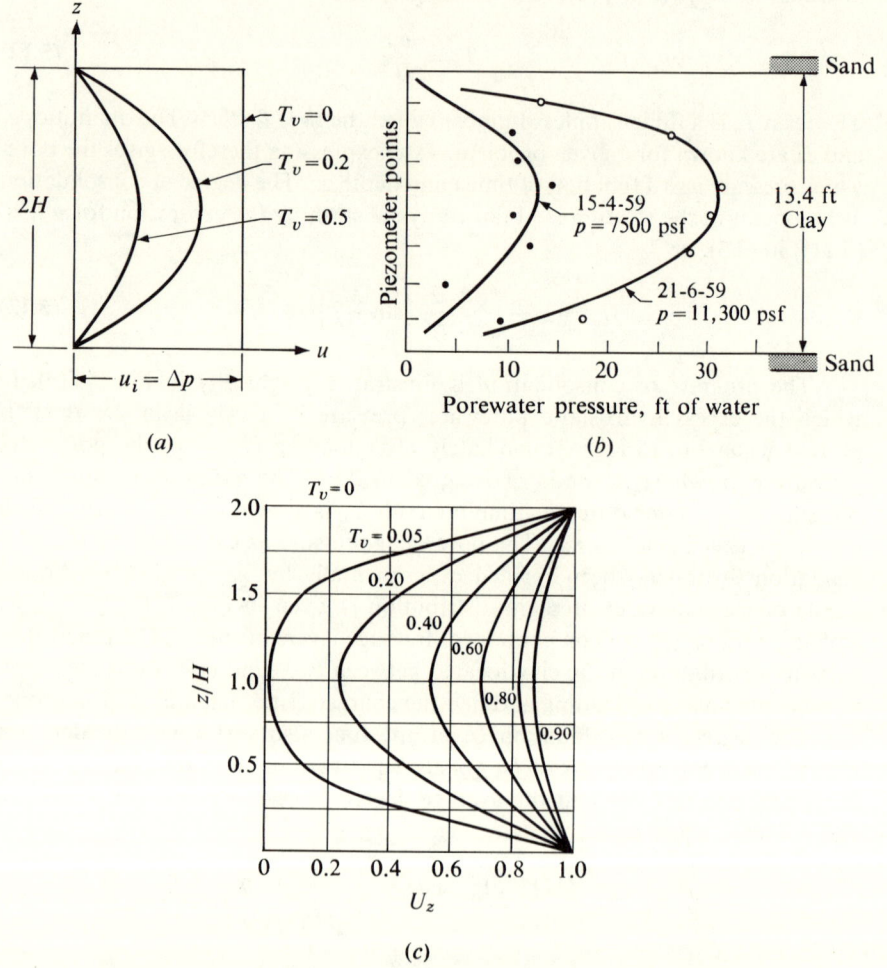

FIGURE 5.8 Distribution of pore pressure in a clay layer: (a) isochrones; (b) pore pressures in Selset Dam. [After Bishop et al. (1960 b)]; (c) degree of consolidation. [After Taylor (1948).]

(1948) and reviewed by Richart (1957). For solutions of two-dimensional problems, the reader might consult the papers by Gibson and Lumb (1953) and McNamee and Gibson (1960).

5.6 Experimental time-consolidation curves

Figure 5.10 compares an experimental time-consolidation curve (a) with a theoretical one (b). The shapes of the curves are similar up to about 90 percent

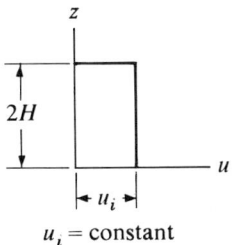

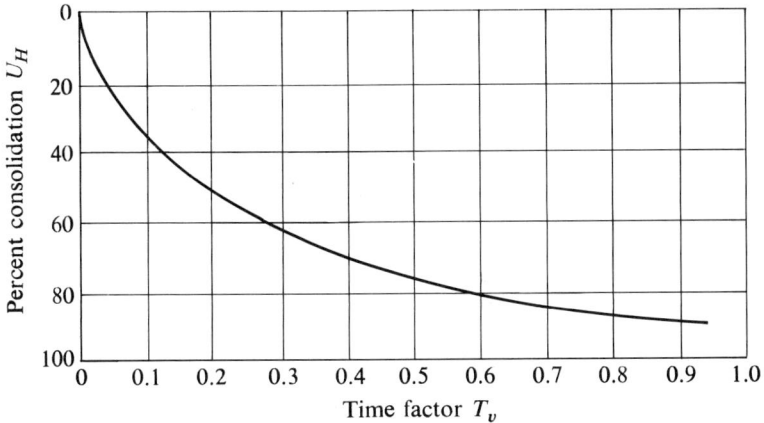

FIGURE 5.9 *Time-consolidation relationship.* [After Taylor (1948).]

consolidation. Beyond this point, the theoretical curve flattens out to a horizontal asymptote. In the experiment, the compression continues for a long period of time. This part of the compression continues under very small excess hydrostatic porewater pressures and is not contolled by hydraulic flow. It is called *secondary consolidation*. The part of the volume change that is accounted for by the theory of consolidation is referred to as *primary consolidation*. For many clays the secondary consolidation curve is approximately a straight line on the semilogarithmic plot until a final or ultimate compression is attained [Buisman (1936), for example]. However, this is not always the case and other secondary consolidation versus time relationships have also been observed. Experimental factors that affect the rate of secondary consolidation include the size of pressure increment Δp, preconsolidation, and temperature [Leonards and Ramiah (1959); Leonards and Girault (1961); Lo (1962a)].

In clays, the magnitude and time duration of secondary consolidation may vary over a wide range. If the magnitude of the secondary consolidation is large, a consolidation test in which the specimen is loaded every 24 hr would underestimate the amount of compression, since only a fraction of the secondary consolidation would have taken place during this interval [see Zeevaert (1957), for example]. However, because of the strong influence of test conditions on the

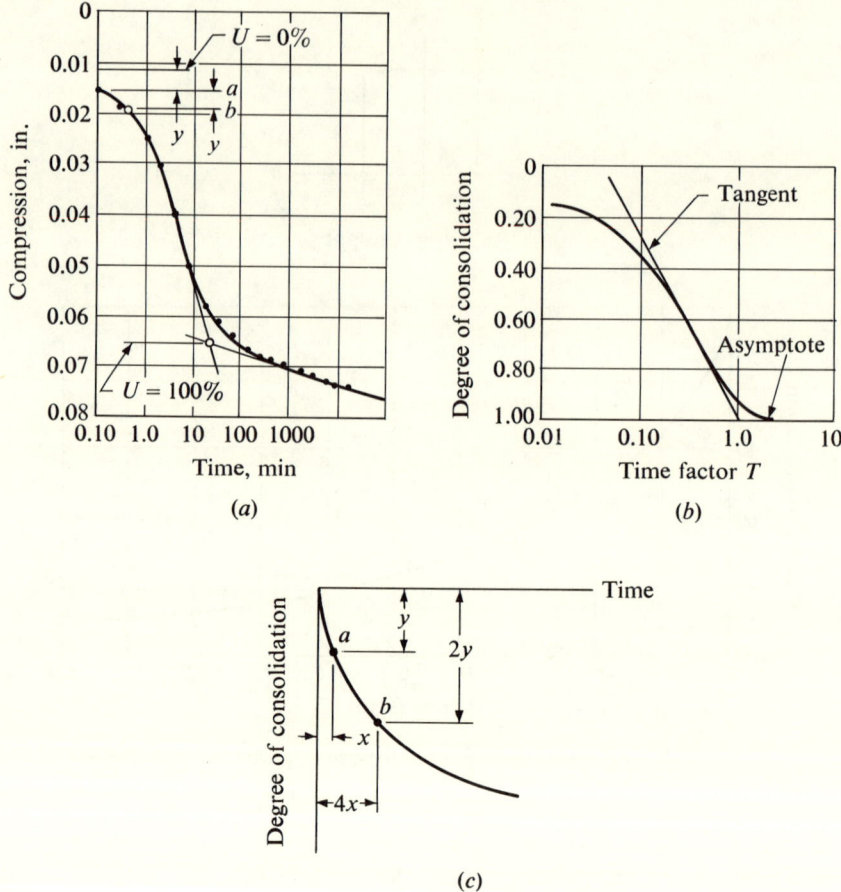

FIGURE 5.10 *Time-consolidation curves.*

rate of secondary consolidation, extrapolations from laboratory data to estimate field behavior involve many uncertainties.

Example 5.1. The vertical stress on a normally consolidated clay layer 5 ft thick is increased from 1500 to 2500 psf. The clay is confined laterally so that only one-dimensional compression is possible, and is drained on the top only. Its compression index is 0.15, its consolidation coefficient is 1.5×10^{-2} in.^{2}/min, and its initial void ratio is 0.80. Calculate the compression due to consolidation.

Solution

$$\Delta e = C_c \log \frac{p_0 + \Delta p}{p_0} = 0.15 \log \frac{2500}{1500} = 0.033$$

$$\epsilon_v = \frac{\Delta V}{V} = \frac{0.033}{1 + 0.80} = 0.018$$

The compression of a layer 5 ft thick would be $5 \times 0.018 = 0.090$ ft.

SEC. 5.6 Experimental time-consolidation curves

Example 5.2. What is the time required to attain 70 percent of the void ratio change calculated in Example 5.1 if the clay layer is twenty feet thick and drained on both sides?

Solution. For 70 percent consolidation $T_v = 0.40$ (Fig. 5.9):

$$T_v = \frac{c_v t}{H^2} \qquad t = \frac{T_v H^2}{c_v} = \frac{(0.40)(10 \times 12)^2}{1.5 \times 10^{-2}} = 3.84 \times 10^5 \text{ min}$$

$$= 2.7 \times 10^2 \text{ days}$$

Example 5.3. What is the porewater pressure and vertical effective stress at the middle of the clay layer at the time calculated in Example 5.2?

Solution. At a given time, the porewater-pressure distribution in the clay is represented by an isochrone. For $T_v = 0.40$ and $z/H = 1.0$, the value of U_z is found to be 0.53 [Fig. 5.8(c)]. The initial excess hydrostatic porewater pressure u_i is 1000 psf. Hence

$$U_z = \frac{u_i - u}{u_i - u_0} = \frac{1000 - u}{1000} = 0.53 \qquad u = 470 \text{ psf}$$

To find the effective stress we note that before loading $\bar{\sigma}_y = p_0 = 1500$ psf. After loading the total stress is increased to 2500 psf. At any time t,

$$\bar{\sigma} = \sigma - u$$

At $t = 2.7 \times 10^2$ days,

$$\bar{\sigma}_y = 2500 - 470 = 2030 \text{ psf}$$

Example 5.4. Why is the value of U_z calculated in Example 5.3 not equal to 0.70 as given in Example 5.2 at a time equal to 2.7×10^2 days?

Solution. It is because the degree of consolidation is not uniform throughout the clay layer [see Fig. 5.8(c)]. The value of 0.70 is the average degree of consolidation for the entire layer, whereas the value of 0.53 is the degree of consolidation at the middle of the layer.

Example 5.5. What is the void ratio after 2.7×10^2 days at the middle of the clay layer?

Solution. Example 5.3 shows that $U_z = 0.53$ at 2.7×10^2 days. Hence

$$U_z = \frac{\Delta V}{\Delta V_{t=\infty}} = \frac{\Delta e}{\Delta e_{t=\infty}}$$

$$\Delta e = U_z \Delta e_{t=\infty} = 0.53 \times 0.033 = 0.017$$

$$e = e_{t=0} - \Delta e = 0.80 - 0.017 = 0.0783$$

This may also be found by means of the relationship between pressure and void ratio. At $t = 2.7 \times 10^2$ days, Example 5.3 shows that the effective vertical stress has increased to 2030 psf. Hence

$$\Delta e = C_c \log \frac{p_0 + \Delta p}{p_0} = 0.15 \log \frac{2030}{1500} = 0.020$$

We note that this value differs slightly from that of 0.017 obtained from the degree of consolidation. This is because in the definition of the degree of consolidation [Eq. (5.3)] it is assumed that the volume compressibility m_v is a constant. The value of 0.020 is obtained from Eq. (5.1), which is based on the assumption that C_c is constant. This implies that the relationship between void ratio and the logarithm of pressure is linear. The different assumptions result in slightly different answers.

Example 5.6. If a laboratory consolidation test is performed on a specimen of the clay described in Example 5.1 and the specimen is 1 in. thick, drained on both sides, what will be the time required to attain 70 percent consolidation?

Solution. Since the clay is the same, c_v is 1.5×10^{-2} in.²/min. At 70 percent consolidation, T_v is equal to 0.40. Hence

$$t = \frac{T_v H^2}{c_v} = \frac{(0.40)(0.5)^2}{1.5 \times 10^{-2}} = 6.7 \text{ min}$$

To compare the time required for 70 percent consolidation in these two cases, we note that T_v equals 0.40 for both cases. Hence t is proportional to the square of H. Thus we can obtain the time by

$$t_1 = \left(\frac{H_1}{H_2}\right)^2 t_2 = \left(\frac{0.5}{10 \times 12}\right)^2 (3.84 \times 10^5) = 6.7 \text{ min}$$

CURVE FITTING. The theory of consolidation given in Sec. 5.5 may be used to determine the value of the consolidation coefficient c_v from experimental data. Under a given load increment, if the experimental time for a certain percentage of consolidation, say 50 percent, is known, then the value of c_v may be evaluated by means of Eq. (5.12). The coefficient of permeability k may be calculated from c_v with Eq. (5.7).

The direct comparison between the experimental and theoretical time-consolidation curves is complicated by the presence of secondary consolidation, described in the preceding section. Hence empirical methods have been developed to calculate the value of c_v based on the characteristics of the time-consolidation curves. These are called *curve-fitting methods*. One such method, the logarithm-of-time method, is described below. Another widely used method is the square-root-of-time method [see Taylor (1948)].

Figure 5.10(*b*) shows that the asymptote to the theoretical consolidation curve intersects the tangent to the steep part of the curve at 100 percent consolidation. Therefore, Casagrande suggested that the point of 100 percent consolidation in the experimental curve be represented by the intersection of the two tangents [Fig. 5.10(*a*)]. The point of zero consolidation is obtained by assuming the early portion of the curve to be a parabola on the natural scale. This geometric relationship is shown in Fig. 5.10(*c*). One locates two points *a* and *b* on the curve with times in the ratio 1:4. The property of the parabola is such that the vertical distance between zero consolidation and point *a* is equal to that between points *a* and *b*. Therefore, the point of zero consolidation on the semilogarithmic plot [Fig. 5.10(*a*)] may be determined by locating two points *a* and *b* in the early part of the consolidation curve at times equal to t and $4t$. If point a is located at a time of 0.10 min, point b is located at 0.40 min. Then according to the geometrical relationship shown in Fig. 5.10(c), the zero consolidation point must be at a distance y above point a and y is also equal to the vertical distance between a and b.

After the zero and 100 percent consolidation points have been determined, the 50 percent point is simply located at one-half the vertical distance between these two points. This gives us the time t for U_H equal to 0.50. The value of T_v for 50 percent consolidation may be obtained from the theoretical solution in Fig. 5.9. The thickness of the specimen drained on both sides is $2H$. Then c_v can be calculated from Eq. (5.12).

5.7 Visco-elastic models for consolidation

The solution described in Sec. 5.5 is based on the assumption that the effective stress vs. volumetric strain relation may be described by linear elasticity. More sophisticated analyses have used viscoelastic models for the stress-strain behavior [Taylor (1942); Tan (1958); Lo (1962); Schiffman et al. (1964)]. The soil-skeleton–porewater system is represented by the model in Fig. 5.11. The

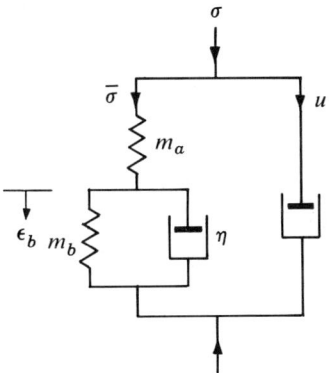

FIGURE 5.11   Rheologic model.

soil skeleton is described by the compressibilities m_a and m_b and viscosity η; m_a represents the immediate response, while m_b and η the delayed response of the soil skeleton to a change in effective stress. If ϵ_b denotes the compression of the spring m_b then [see Eq. (4.12)]

$$\bar{\sigma}(t) = \frac{\epsilon_b}{m_b} + \eta \frac{d\epsilon_b}{dt}$$

For the initial condition $t = 0$, $\epsilon_b = 0$, the solution is*

$$\epsilon_b = \int_0^t \bar{\sigma}(\tau) e^{-(\eta/m_b)(t-\tau)} \, d\tau \qquad (5.15)$$

---

* See for example Hildebrand (1962), Chapter 1.

The rate of compression of the soil element in Fig. 5.7(b) is

$$\frac{\partial V}{\partial t} = \left(\frac{\partial \epsilon_a}{\partial t} + \frac{\partial \epsilon_b}{\partial t}\right) dz = \left(m_a \frac{\partial \bar{\sigma}}{\partial t} + \frac{\partial \epsilon_b}{\partial t}\right) dz \tag{5.16}$$

Equations (5.15) and (5.16) are substituted into Eq. (5.5) to yield

$$-\frac{k}{\gamma_w} \frac{\partial^2 u}{\partial z^2} = m_a \frac{\partial \bar{\sigma}}{\partial t} + \eta \bar{\sigma} - \frac{\eta^2}{m_b} \int_0^t \bar{\sigma} e^{-(\eta/m_b)(t-\tau)} d\tau \tag{5.17}$$

Solutions of the differential equation may be obtained by numerical methods. The nature of the time compression curve depends strongly on the relative values of m_a, m_b, and η. Some examples are shown in Fig. 5.12 as solid curves. For a given ratio of $(H^2/m_b c_v)$ $(=0.10)$, entirely different time compression curves may be obtained by using different ratios of $m_a/(m_a + m_b)$. The terms U and T_v are defined as

$$U = \frac{\Delta V}{\Delta V_{t=\infty}}$$

$$T_v = \frac{c_v t}{4H^2} \tag{5.18}$$

$$c_v = \frac{k}{m_a \gamma_w}$$

Thus, c_v and T_v are equivalent to the same terms in the Terzaghi theory (Eq. 5.12). The dashed curves in Fig. 5.12 represent the solution of the Terzaghi model in which the soil compressibility is represented by m_a only.

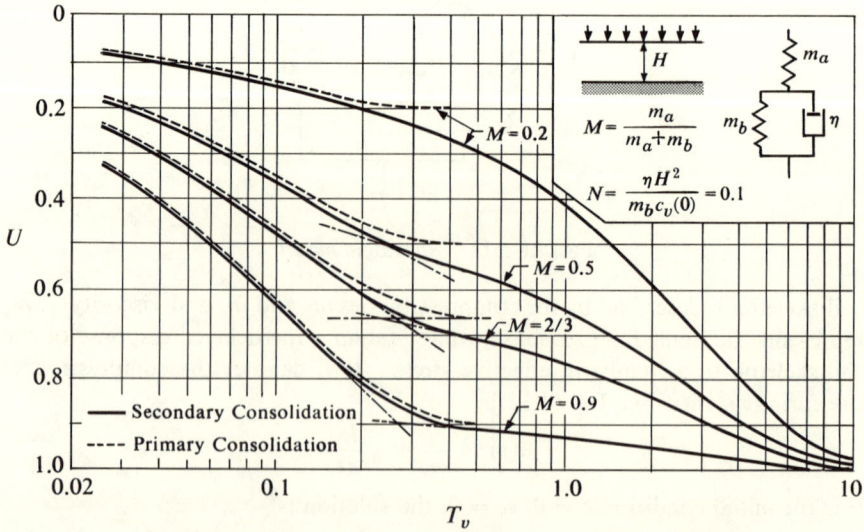

FIGURE 5.12 *Effect of compressibility factor on time-consolidation relationships; a three-parameter system.* [After Schiffman et al. (1964).]

PROBLEMS

5.1 A compressible soil layer is 30 ft thick and its initial void ratio is 1.038. Laboratory tests indicate that the final void ratio of this soil under the weight of a projected building will be 0.981. What will be the probable total settlement of the building over a long period of time?
Answer: 0.84 ft

5.2 The clay layer in Fig. 5.13 has been consolidated under the weight of the overburden as shown on the left side. The valley on the right side is the result of erosion. (*a*) What is the preconsolidation pressure of the clay at point A? (*b*) If a load is placed so that the vertical stress on the clay layer is increased by 1000 psf, what will be the settlement?

The following soil properties are known:

	Clay	Sand
Unit wt.	115	130 pcf
Moisture content	37%	20%
c_v (coeff. of consol.)	2×10^{-4} cm²/sec	—
C_c (normally consol. branch)	0.20	—
C_e (unloading and reloading)	0.05	—
σ'_u (unconfined comp.)	1.75 tsf	—

Answer: 4313 psf; 0.06 ft

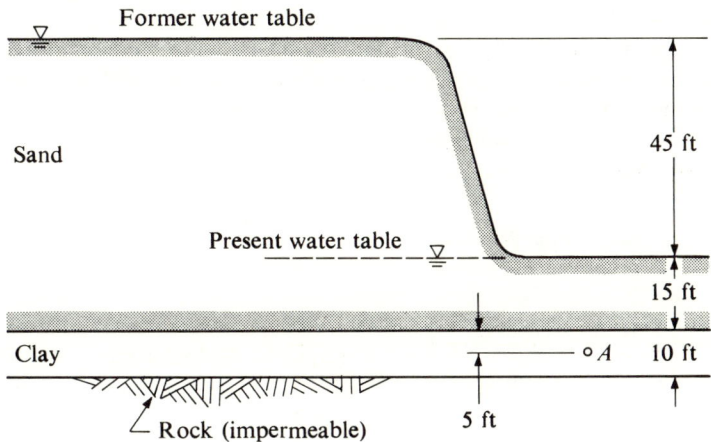

FIGURE 5.13

5.3 A laboratory consolidation test was performed on a specimen of clay. The time-volume change curve was obtained and calculations gave the times for 50% and 90% consolidation as 20 min and 40 min, respectively. (*a*) Is anything wrong with the data? (*b*) Why?
Answer: yes; $T_{50}:T_{90}$ should be 2:9

5.4 A laboratory consolidation test was performed on a specimen of clay. The time required for 50% consolidation is equal to 20 min. The specimen is 1 in. thick and its initial moisture content is 40%. (*a*) Calculate the coefficient of consolidation. (*b*) What time would be required to reach 90% consolidation for a clay layer in situ that is 10 ft thick and drained on both sides?
Answer: 0.0025 in./min; 950 days

5.5 A fully saturated clay specimen is placed in a consolidometer and subjected to a loading of 2 tsf. After a period of time it is determined that the average pore pressure in the specimen is 10 psi. What percentage of consolidation has been reached at this time?
Answer: 64%

5.6 For the consolidation test described in Problem 5.4, calculate the velocity of flow across the face of the specimen at a time of 20 min after a pressure increment of 1000 psf has been applied. Under this pressure increment the volumetric compression of the specimen is 0.0434.
Answer: 2.72×10^{-4} in./min

5.7 For the test described in Problem 5.6, calculate the velocity of flow across the midplane of the specimen. If this differs from the answer to Problem 5.6, give the reason for the difference.
Answer: 0

5.8 A large storage tank with a radius of 140 ft will be constructed at a site where subsoil conditions are as shown in Fig. 5.14. The sand and the stiff clay may be considered as incompressible. It is required that the settlement due to compression of the soft silty clay be estimated. Laboratory tests have shown that the silty clay has the following properties: $C_c = 0.20$, $C_r = 0.02$; $\bar{p}_c$ lies between 2300 and 2800 psf. *Note:* A range is given for $\bar{p}_c$ because the graphical construction described in Sec. 5.3 is not very precise. Assuming that the vertical stress at the middle of the silty clay layer is increased by 1400 psf, compute the settlement (*a*) if the soil is normally consolidated; (*b*) if the soil has a preconsolidation pressure of 2400 psf.
Answer: 0.55 ft; 0.34 ft

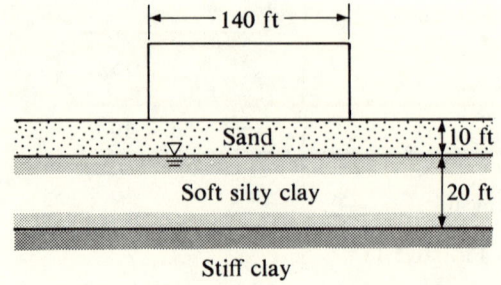

FIGURE 5.14

5.9 Based on your answers to Problem 5.8 (*a*) what do you think is the significance of the preconsolidation pressure? (*b*) What would happen if you used assumption (*b*) in Problem 5.8 and the soil is in fact normally consolidated?

5.10 Assume that the soil in Problem 5.8 is normally consolidated. The following scheme is proposed. A fill 12 ft high will be placed on the ground and the silty clay will be allowed to consolidate. After this, the fill will be removed and the tank will be constructed. Estimate the settlement if this scheme is followed.
Answer: 0.11 ft

5.11 What would happen to the operation described in Problem 5.10 if the fill is removed prematurely before the silty clay is consolidated? Design a scheme of measurements to help ascertain that consolidation is complete (say $U = 90\%$ at least) before removing the fill.

6

Elastic Equilibrium

6.1. Introduction

This chapter deals with problems involving stresses in soil masses and the accompanying deformations, such as settlement of foundations. The stresses considered are substantially less than the yield strength of the material, hence problems involving failure are not included here. Figure 6.1 illustrates the simplified assumption of the soil behavior adopted for our analysis. When the ideal material is stressed, it first behaves as an elastic material through the range $a0$. When the stress reaches the value represented by point a, the material becomes perfectly plastic and there is no increase in stress with further strain. This second stage is equivalent to yielding or plastic flow in the material.

Most problems in soil mechanics may be divided into two categories. In the first category, the stresses in the soil are in the elastic range and substantially lower than those required to cause plastic flow. These problems may be solved

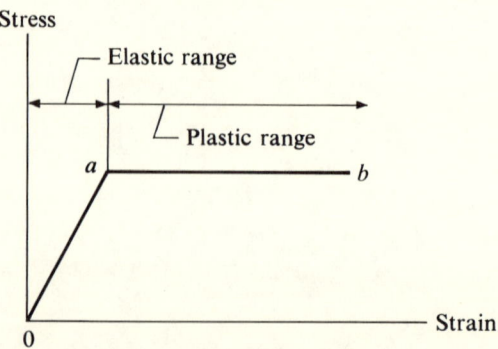

FIGURE 6.1 *Idealized elastic-plastic behavior.*

SEC. 6.2   Fundamental relationships of elasticity

by the theory of elasticity and they are called problems of elastic equilibrium. The second group of problems deals with stresses in the plastic range. These problems may be solved on the basis of the assumption that the material is perfectly plastic and are called problems of plastic equilibrium (Chapter 9). The assumptions of elasticity and plasticity are of course gross simplifications of real soil behavior. One may obtain a fair idea of the degree of approximation by comparing the ideal stress-strain curve (Fig. 6.1) with those measured from real soils (Fig. 8.6). The inelastic behavior of soils at even low stresses is pointed out in Chapter 4.

The employment of the methods of elasticity are simplified if we introduce two additional assumptions regarding the material properties—the material is considered to be homogeneous and isotropic. Like the first requirement, these assumptions are almost never satisfied in real soils. Despite these limitations, calculations based on elasticity are invaluable in providing a guide to the behavior of soil under stress. For example, the Boussinesq equations (Sec. 6.4) are widely used to estimate stresses in soils and settlement under load (Chapter 7). In the subsequent sections the basic principles of elasticity are outlined, followed by a discussion of their applications to soil mechanics problems.

6.2 Fundamental relationships of elasticity

The theory of elasticity is based on the elastic behavior of the material and the law of equilibrium.

HOOKE'S LAW. For an isotropic elastic material the stress-strain relation is defined by Eq. (4.10) or

$$\epsilon_x = \frac{1}{E}[\sigma_x - \mu(\sigma_y + \sigma_z)] \tag{6.1a}$$

$$\epsilon_y = \frac{1}{E}[\sigma_y - \mu(\sigma_x + \sigma_z)] \tag{6.1b}$$

$$\epsilon_z = \frac{1}{E}[\sigma_z - \mu(\sigma_x + \sigma_y)] \tag{6.1c}$$

Equation (6.1) can be rearranged so that the stresses are expressed in terms of the strains as follows:

$$\sigma_x = \frac{\mu E}{(1+\mu)(1-2\mu)}(\epsilon_x + \epsilon_y + \epsilon_z) + \frac{E}{1+\mu}\epsilon_x$$

$$\sigma_y = \frac{\mu E}{(1+\mu)(1-2\mu)}(\epsilon_x + \epsilon_y + \epsilon_z) + \frac{E}{1+\mu}\epsilon_y$$

$$\sigma_z = \frac{\mu E}{(1+\mu)(1-2\mu)}(\epsilon_x + \epsilon_y + \epsilon_z) + \frac{E}{1+\mu}\epsilon_z$$

For simplicity the equations may be written

$$\sigma_x = \lambda\epsilon + 2G\epsilon_x$$
$$\sigma_y = \lambda\epsilon + 2G\epsilon_y \qquad (6.2a)$$
$$\sigma_z = \lambda\epsilon + 2G\epsilon_z$$

in which

$$\epsilon = \epsilon_x + \epsilon_y + \epsilon_z \qquad \lambda = \frac{\mu E}{(1+\mu)(1-2\mu)} \qquad G = \frac{E}{2(1+\mu)} \qquad (6.2b)$$

The relationship between shear stress and shear strain is obtained by combining Eqs. (4.4), (4.7), and (6.1), giving

$$\gamma_{xy} = \frac{\tau_{xy}}{G} \qquad (6.3a)$$

$$\gamma_{yz} = \frac{\tau_{yz}}{G} \qquad (6.3b)$$

$$\gamma_{zx} = \frac{\tau_{zx}}{G} \qquad (6.3c)$$

Equations (6.1) or (6.2) plus (6.3) comprise six equations that define the elastic stress-strain relationship.

EQUILIBRIUM. In addition, the stresses must obey the law of equilibrium. Since the stresses vary from point to point, we can only write down the equilibrium condition for an element inside the soil mass. Consider the cubic element shown in Fig. 6.2. The stresses on the three faces $x = 0$, $y = 0$, $z = 0$ are, respectively, σ_x, τ_{xy}, τ_{xz}; σ_y, τ_{yz}, τ_{yx}; σ_z, τ_{zx}, τ_{zy}. On the face $x = dx$ the stresses are $\sigma_x + (\partial\sigma_x/\partial x)\,dx$, $\tau_{xy} + (\partial\tau_{xy}/\partial x)\,dx$, $\tau_{xz} + (\partial\tau_{xz}/\partial x)\,dx$. The stresses on $y = dy$ and $z = dz$ are also shown in the figure. In addition, there exists the body force, which has components X, Y, Z and is expressed as force per unit volume. Common body forces include the weight of the material and seepage force.

Taking the summation of forces in the x direction, we get

$$F_x = -\sigma_x\,dy\,dz - \tau_{xz}\,dx\,dy - \tau_{xy}\,dx\,dz + X\,dx\,dy\,dz$$
$$+ \left(\sigma_x + \frac{\partial\sigma_x}{\partial x}dx\right)dy\,dz + \left(\tau_{xz} + \frac{\partial\tau_{xz}}{\partial z}dz\right)dx\,dy$$
$$+ \left(\tau_{xy} + \frac{\partial\tau_{xy}}{\partial y}dy\right)dx\,dz = 0^*$$

or

$$\frac{\partial\sigma_x}{\partial x} + \frac{\partial\tau_{xy}}{\partial y} + \frac{\partial\tau_{xz}}{\partial z} + X = 0 \qquad (6.4a)$$

---

* $\tau_{xy} = \tau_{yx}$, $\tau_{yz} = \tau_{zy}$, $\tau_{zx} = \tau_{xz}$.

SEC. 6.2   Fundamental relationships of elasticity   133

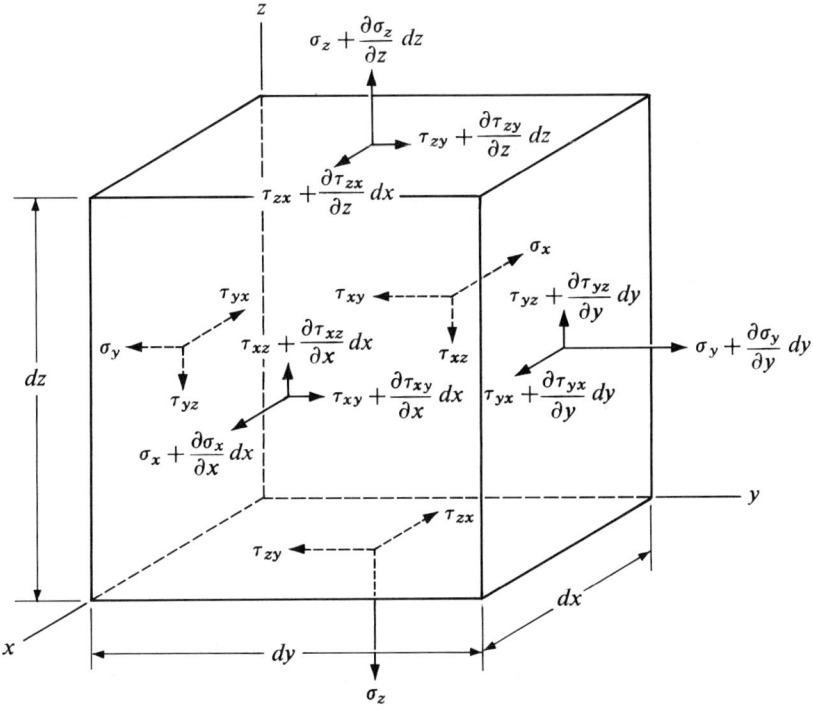

FIGURE 6.2   *Stresses on an element.*

Similarly, summations of forces in the y and z directions yield

$$\frac{\partial \sigma_y}{\partial y} + \frac{\partial \tau_{yx}}{\partial x} + \frac{\partial \tau_{yz}}{\partial z} + Y = 0 \qquad (6.4b)$$

$$\frac{\partial \sigma_z}{\partial z} + \frac{\partial \tau_{zx}}{\partial x} + \frac{\partial \tau_{zy}}{\partial y} + Z = 0 \qquad (6.4c)$$

If the stress-strain relationships [Eqs. (6.2) and (6.3)] and the strain-displacement relationships [Eqs. (4.5) and (4.6)] are substituted in Eqs. (6.4), we obtain equilibrium equations in terms of the displacements u, v, w. They are

$$(\lambda + G)\frac{\partial \epsilon}{\partial x} + G \nabla^2 u + X = 0$$

$$(\lambda + G)\frac{\partial \epsilon}{\partial y} + G \nabla^2 v + Y = 0 \qquad (6.5)$$

$$(\lambda + G)\frac{\partial \epsilon}{\partial z} + G \nabla^2 w + Z = 0$$

in which ∇^2 denotes the operator

$$\nabla^2 = \frac{\partial^2}{\partial x^2} + \frac{\partial^2}{\partial y^2} + \frac{\partial^2}{\partial z^2}$$

The problem of theory of elasticity is to determine the stress distribution inside a body. Alternatively we may want to determine the strain and displacement inside a body. The body may be subjected to certain loads on the boundary, or certain displacements such as the settlement underneath a rigid footing may be enforced along part of the boundary. These constitute the boundary conditions that must be satisfied by the solution.

COMPATIBILITY. If we seek the solution in terms of stresses, the equilibrium and boundary conditions are insufficient. In addition, the stress solution must result in displacements that conform to certain physical conditions. The displacements u, v, w must bear definite relationships, as stated in Eqs. (4.5) and (4.6). For example, the relationship

$$\frac{\partial^2 \epsilon_x}{\partial y^2} + \frac{\partial^2 \epsilon_y}{\partial x^2} = \frac{\partial^2 \gamma_{xy}}{\partial x \, \partial y} \tag{6.6}$$

is obtained by differentiating $\epsilon_x = \partial u/\partial x$ twice with respect to y, differentiating $\epsilon_y = \partial v/\partial y$ twice with respect to x, and differentiating $\gamma_{xy} = (\partial u/\partial y) + (\partial v/\partial x)$ once with respect to x and once with respect to y. This equation is called the compatibility equation.

In three-dimensional problems there are six compatibility equations, whereas in plane strain there is only one. The compatibility equation can also be expressed in terms of stress if Hooke's law [Eq. (6.1)] is substituted in Eq. (6.6).

6.3 Stress and displacement functions

PLANE STRESS. To obtain solutions to problems in elasticity, it is necessary to integrate the differential equations of equilibrium together with the compatibility condition.

We shall first consider the case of plane stress ($\sigma_z = \tau_{yz} = \tau_{zx} = 0$). For this case the equilibrium equations consist of Eqs. (6.4a) and (6.4b), and the compatibility condition is given by Eq. (6.6). To relate the stresses and strains we have Hooke's law as defined by Eqs. (6.1a), (6.1b), and (6.3a). There are six equations and the unknowns are σ_x, σ_y, τ_{xy}, ϵ_x, ϵ_y, and γ_{xy}.

The six equations may be combined to yield*

$$\left(\frac{\partial^2}{\partial x^2} + \frac{\partial^2}{\partial y^2}\right)(\sigma_x + \sigma_y) = -(1 + \mu)\left(\frac{\partial X}{\partial x} + \frac{\partial Y}{\partial y}\right) \tag{6.7a}$$

If the body forces are zero, we have

$$\left(\frac{\partial^2}{\partial x^2} + \frac{\partial^2}{\partial y^2}\right)(\sigma_x + \sigma_y) = 0 \tag{6.7b}$$

---

* For details of the algebraic manipulation see Timoshenko and Goodier (1951).

SEC. 6.3   Stress and displacement functions

The solution for stress must satisfy Eqs. (6.4a), (6.4b), and (6.7). A common way of solving these equations is to introduce a stress function Φ such that

$$\sigma_x = \frac{\partial^2 \Phi}{\partial y^2}, \quad \sigma_y = \frac{\partial^2 \Phi}{\partial x^2}, \quad \tau_{xy} = -\frac{\partial^2 \Phi}{\partial x\, \partial y} \tag{6.8}$$

These conditions satisfy Eqs. (6.2a) and (6.2b). When substituted in Eq. (6.7) we obtain

$$\frac{\partial^4 \Phi}{\Phi x^4} + 2 \frac{\partial^4 \Phi}{\partial x^2\, \partial y^2} + \frac{\partial^4 \Phi}{\partial y^4} = 0 \tag{6.9}$$

The problem is then reduced to finding a function that satisfies (6.9) and the boundary conditions. Equation (6.9) was first used by Airy and is called the Airy stress function.

PLANE STRAIN. The condition of plane strain means that $\epsilon_z = \gamma_{yz} = \gamma_{zx} = 0$. Substitution of this condition into Eq. (6.1c) yields respectively

$$\sigma_z = \mu(\sigma_x + \sigma_y) \tag{6.10}$$

If this is substituted into Eqs. (6.1a) and (6.1b) we get

$$\epsilon_x = \frac{1 - \mu^2}{E} \left(\sigma_x - \frac{\mu}{1 - \mu} \sigma_y \right)$$

$$\epsilon_y = \frac{1 - \mu^2}{E} \left(\sigma_y - \frac{\mu}{1 - \mu} \sigma_x \right) \tag{6.11}$$

We now compare these two equations with Eqs. (6.1a) and (6.1b). They are similar in form except that the terms $1/E$ and μ in Eqs. (6.1a) and (6.1b) are now replaced by $(1 - \mu^2)/E$ and $\mu/(1 - \mu)$ respectively. We may then write Eq. (6.11) as

$$\epsilon_x = \frac{1}{E^*}(\sigma_x - \mu^* \sigma_y)$$

$$\epsilon_y = \frac{1}{E^*}(\sigma_y - \mu^* \sigma_x) \tag{6.12a}$$

where $\quad E^* = \dfrac{E}{1 - \mu^2} \quad$ and $\quad \mu^* = \dfrac{\mu}{1 - \mu} \quad$ (6.12b)

With this substitution the equilibrium and compatibility equations for plane stress, Eqs. (6.7) and (6.9), may also be used for plane strain. The solution for plane strain differs from that for plane stress only in that E^* and μ^* should be used instead of E and μ.

In most soil mechanics publications, the normal stress σ is taken as positive for compression. The fundamental equations derived in this section remain unchanged if all the stresses shown in Fig. 6.2 are reversed in direction. In all subsequent calculations, compressive stress and strain are taken as positive to conform with the sign convention in soil mechanics.

LOADS ON THE BOUNDARY

In this section we analyze stresses and strains arising from loads applied at or near the surface of soil masses. A frequently encountered engineering problem in this category is the settlement or surface displacement.

6.4 Stresses in a soil mass due to surface loads

STRESSES DUE TO CONCENTRATED LOAD. In Fig. 6.3(*a*) we illustrate the application of elasticity to the problem of stress distribution inside a soil mass, caused by a concentrated line load on the surface.

If we take the line load to be infinitely long, the problem is one of plane strain. We solve the problem by using the stress function

$$\Phi = Cx \tan^{-1}\left(\frac{y}{x}\right)^z \tag{6.13}$$

Applying Eq. (6.8) we have

$$\sigma_z = \frac{\partial^2 \Phi}{\partial x^2} = \frac{\partial}{\partial x}\left[C \tan^{-1}\left(\frac{z}{x}\right) - Cx\frac{1}{1 + (z^2/x^2)}\frac{z}{x^2}\right]$$

$$\sigma_z = -\frac{Cz}{x^2 + z^2} - \frac{Cz}{x^2 + z^2} + \frac{Cxz}{(x^2 + z^2)^2}2x = -\frac{2Cz^3}{(x^2 + z^2)^2} \tag{6.14}$$

We see that the stress function chosen gives us the general pattern of σ_z that we would expect. Immediately beneath the load $x^2 + z^2 = 0$ the stress is infinite, and at large distances the stress becomes very small. To evaluate C, we apply the boundary condition that the summation of σ_z over any horizontal plane must be equal to Q. Hence

$$\int_{-\infty}^{\infty} \sigma_z \, dx = Q$$

or

$$\int_{-\infty}^{\infty} \sigma_z \, dx = -\int_{-\infty}^{\infty} \frac{2Cz^3}{(x^2 + z^2)^2} \, dx$$

$$= -2Cz^3 \times \left\{\left[\frac{x}{2z^2(x^2 + z^2)}\right]_{-\infty}^{\infty} + \frac{1}{2z^2}\int_{-\infty}^{\infty}\frac{dx}{x^2 + z^2}\right\}$$

$$= -2Cz^3\left\{0 + \left[\frac{1}{2z^3}\tan^{-1}\left(\frac{x}{z}\right)\right]_{-\infty}^{\infty}\right\} = -C\pi = Q$$

$$C = -\frac{Q}{\pi}$$

Thus

$$\sigma_z = \frac{2Q}{\pi}\frac{z^3}{(x^2 + z^2)^2}$$

SEC. 6.4 Stresses in a soil mass due to surface loads

The solution for σ_x and τ_{xz} may be obtained in a similar way. If we let

$$\cos \theta = \frac{z}{(x^2 + z^2)^{1/2}}$$

[Fig. 6.3(a)], then the solutions for σ_x, σ_z, τ_{xz} become

$$\sigma_z = \frac{2Q}{\pi z} \cos^4 \theta$$

$$\sigma_x = \frac{2Q}{\pi z} (\cos^2 \theta \sin^2 \theta) \qquad (6.15)$$

$$\tau_{xz} = \frac{2Q}{\pi z} (\cos^3 \theta \sin \theta)$$

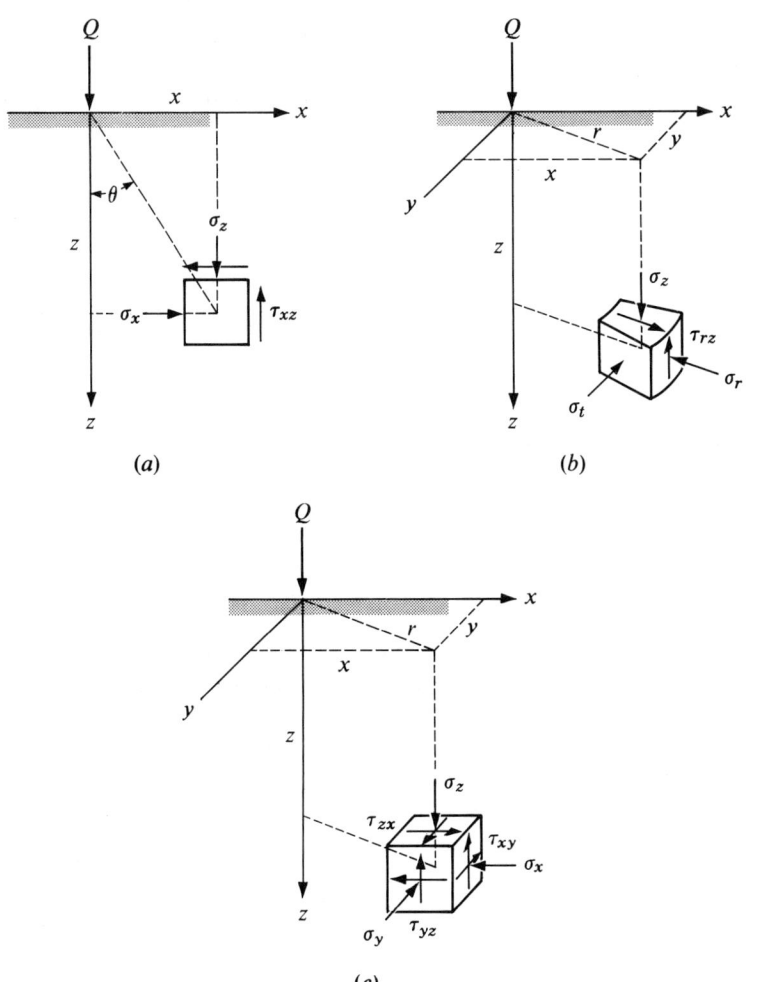

FIGURE 6.3 *Stresses due to a concentrated load.*

For the three-dimensional problem of a point load on the surface [Fig. 6.3(b)], we have axial symmetry, and the solution in terms of polar coordinates is

$$\sigma_z = \frac{Q}{2\pi} \frac{3z^3}{(r^2 + z^2)^{5/2}} = \frac{Q}{2\pi z^2} (3 \cos^5 \theta) \tag{6.16a}$$

$$\sigma_r = \frac{Q}{2\pi} \left[\frac{3r^2 z}{(r^2 + z^2)^{5/2}} - \frac{1 - 2\mu}{r^2 + z^2 + z(r^2 + z^2)^{1/2}} \right]$$

$$= \frac{Q}{2\pi z^2} \left[3 \sin^2 \theta \cos^3 \theta - \frac{(1 - 2\mu) \cos^2 \theta}{1 + \cos \theta} \right] \tag{6.16b}$$

$$\sigma_t = \frac{Q}{2\pi} (1 - 2\mu) \left[\frac{z}{(r^2 + z^2)^{3/2}} - \frac{1}{r^2 + z^2 + z(r^2 + z^2)^{1/2}} \right]$$

$$= -\frac{Q}{2\pi z^2} (1 - 2\mu) \left[\cos^3 \theta - \frac{\cos^2 \theta}{1 + \cos \theta} \right] \tag{6.16c}$$

$$\tau_{rz} = \frac{Q}{2\pi} \frac{3rz^2}{(r^2 + z^2)^{5/2}} = \frac{Q}{2\pi z^2} (3 \sin \theta \cos^4 \theta) \tag{6.16d}$$

When expressed in terms of rectangular coordinates [Fig. 6.3(c)] the stresses are

$$\sigma_z = \frac{3P}{2\pi} \frac{z^3}{r^5} \tag{6.17a}$$

$$\sigma_x = \frac{3P}{2\pi} \left\{ \frac{x^2 z}{r^5} + \frac{1 - 2\mu}{3} \left[\frac{1}{r(r + z)} - \frac{(2r + z)x^2}{r^3(r + z)^2} - \frac{z}{r^3} \right] \right\} \tag{6.17b}$$

$$\sigma_y = \frac{3P}{2\pi} \left\{ \frac{y^2 z}{r^5} + \frac{1 - 2\mu}{3} \left[\frac{1}{r(r + z)} - \frac{(2r + z)y^2}{r^3(r + z)^2} - \frac{z}{r^3} \right] \right\} \tag{6.17c}$$

$$\tau_{zx} = -\frac{3P}{2\pi} \frac{xz^2}{r^5} \tag{6.17d}$$

$$\tau_{yz} = -\frac{3P}{2\pi} \frac{yz^2}{r^5} \tag{6.17e}$$

$$\tau_{xy} = \frac{3P}{2\pi} \left\{ \frac{xyz}{r^5} - \frac{1 - 2\mu}{3} \frac{(2r + z)xy}{r^3(r + z)^2} \right\} \tag{6.17f}$$

This problem was first solved in 1883 by Boussinesq, and the equations are called Boussinesq equations. These are the stresses produced by the load Q and do not include the stresses due to the weight of the overburden.

STRESSES DUE TO LOADED AREA. The stresses at a point due to an area loaded uniformly to intensity q at the surface can be obtained by dividing the loaded area into many small elements, each with an area of dA and a small concentrated load at its center equal to

$$dQ = q \, dA$$

From Eq. (6.16a) the vertical stress at a due to dq is

$$d\sigma_z = \frac{dQ}{2\pi} \frac{3z^3}{(r^2 + z^2)^{5/2}} = \frac{q}{2\pi} \frac{3z^3}{(r^2 + z^2)^{5/2}} dA \tag{6.18a}$$

SEC. 6.5 Influence charts 139

The stress σ_z can be found by integrating Eq. (6.18a) over the loaded area. If we consider the stress underneath the center of a circular loaded area of radius R, we have [Fig. 6.4]

$$dA = r \cdot d\psi \cdot dr$$

in which ψ ranges from 0 to 2π and r ranges from 0 to R. Thus

$$\sigma_z = \frac{3q}{2\pi} \int_0^{2\pi} \int_0^R \frac{z^3}{(r^2 + z^2)^{5/2}} r \cdot d\psi \cdot dr = q\left[1 - \frac{z^3}{(R^2 + z^2)^{3/2}}\right]$$

$$= q\left[1 - \frac{1}{\left(\frac{R^2}{z^2} + 1\right)^{3/2}}\right] \quad (6.18b)$$

Similarly we get

$$\sigma_r = \frac{q}{2}\left[1 + 2\mu - \frac{2(1+\mu)z}{(R^2+z^2)^{1/2}} + \frac{z^3}{(R^2+z^2)^{3/2}}\right]$$

$$= \frac{q}{2}\left[1 + 2\mu + \frac{2(1+\mu)}{\left(\frac{R^2}{z^2}+1\right)^{1/2}} + \frac{1}{\left(\frac{R^2}{z^2}+1\right)^{3/2}}\right] \quad (6.18c)$$

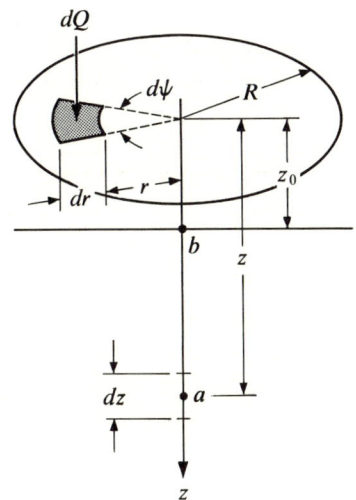

FIGURE 6.4 *Stresses due to a loaded area.*

6.5 Influence charts

Equations (6.18b) and (6.18c) show that the stresses σ_z and σ_r may be written as

$$\sigma_z = qN_z \quad \text{and} \quad \sigma_r = qN_r \quad (6.18d)$$

where N_z and N_r depend only on the ratio R/z for a given value of μ. Thus the values of N_z and N_r, called influence numbers, may be calculated and presented in the form of tables or charts to facilitate computations. Numerical procedures have been used to calculate the influence numbers for rectangular areas [Newmark (1935), for example]. A widely used influence chart is the one developed by Newmark (1942).

For convenience in dealing with irregular loads, the stresses in rectangular coordinates [Fig. 6.3(c)] are used. The influence chart for the vertical stress σ_z is given in Fig. 6.5. The chart consists of a number of elements called influence

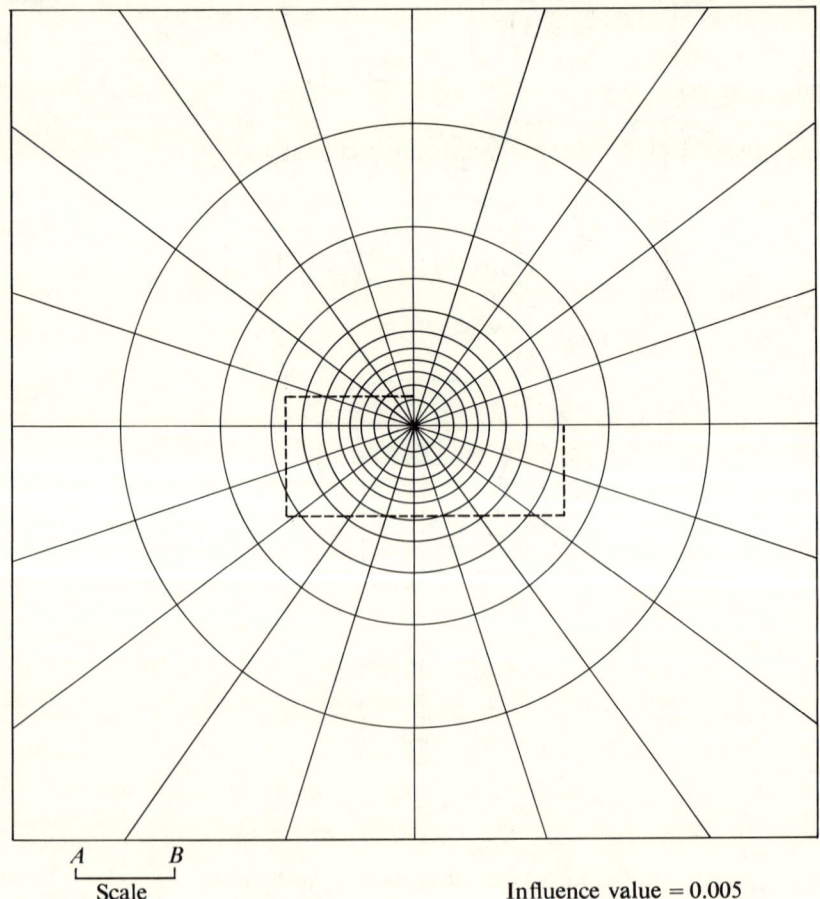

FIGURE 6.5 *Influence chart for σ_z.* [After Newmark (1942).]

areas. It is so constructed that a uniform load q that covers one influence area produces a stress of $0.005q$ at a point beneath the center of the chart. To compute σ_z at a depth z directly beneath point P of the loaded area [Fig. 6.6(a)], the loaded area is drawn to a scale so that the depth z is equal to the distance AB

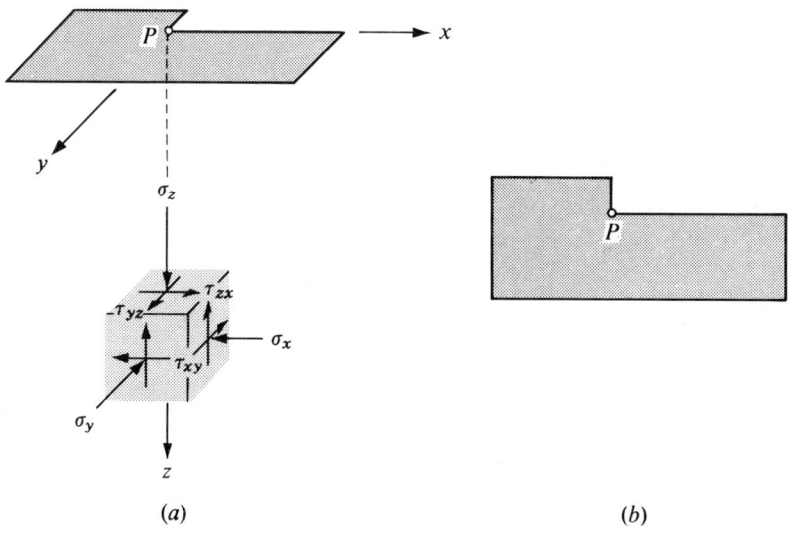

FIGURE 6.6 *Stress computation.*

shown on the chart [Fig. 6.6(b)]. The outline of the loaded area is superimposed on the chart with point P directly over the center of the chart. This is shown by the dashed outline in Fig. 6.5. The number of influence areas N enclosed by the outline is counted and the stress σ_z is given as

$$\sigma_z = 0.005\ Nq$$

The value of 0.005 is the influence value of the chart. [See Newmark (1940), for the derivation.]

Figure 6.7 is the influence chart for the horizontal stress σ_x. Its use is the same as that for the vertical stress σ_z. However, we note that the influence chart for σ_x is not symmetrical radially. Hence it is necessary to consider the directions X and Y in relation to the loaded area. For example, to obtain σ_x beneath the center of the loaded area in Fig. 6.6, the area is drawn to the appropriate scale and placed in the position shown in Fig. 6.7. The X axis of the load is in the same direction as that of the influence chart. To obtain σ_y, the outline of the loaded area is placed in a position 90 degrees to that shown in Fig. 6.7, so that the X axis of the load is in the Y direction of the influence chart.

The influence charts for the shear stresses τ_{xy}, and τ_{zx} are shown in Figs. 6.8(a) and 6.8(b) respectively. The consideration for the directions X and Y are the same as that for σ_x. The influence areas in the shaded zones are counted as negative. We note that along the centerline of a symmetric loaded area, the shear stresses τ_{zx} and τ_{zy} are zero. Hence the minimum and maximum values of σ_x are the minor and intermediate principal stresses, respectively. The major principal stress is the vertical stress σ_z.

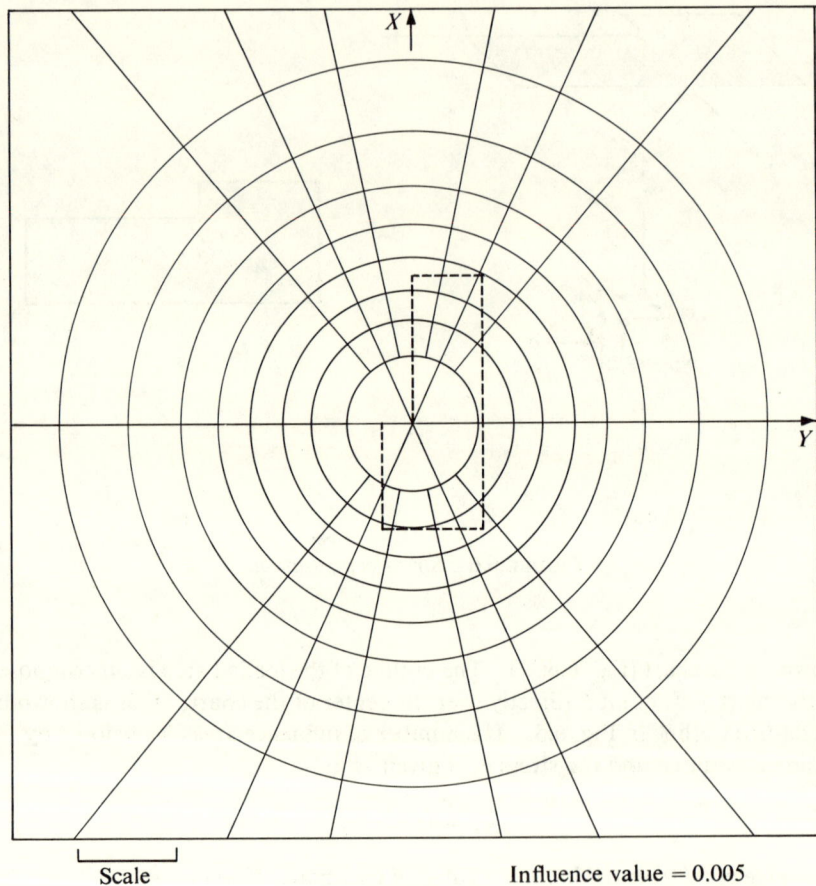

Scale | Influence value = 0.005

FIGURE 6.7 *Influence chart for σ_x or σ_y.* [After Newmark (1942).]

6.6 Stress distribution

It is important to appreciate the stress distribution in a mass of soil under load. The results of calculations by the use of elastic theory are presented in Figs. 6.9, 6.10, and 6.11. Figure 6.9 shows the principal stresses and maximum shear stresses in the soil beneath a uniform continuous strip load. The directions of the vectors indicate the directions of the stresses; the lengths of the vectors represent the magnitudes of the stresses. Curves connecting points of equal stress are shown in Fig. 6.10. Figure 6.11(a) shows the distributions of vertical stress on horizontal planes at various depths beneath the surface, and Fig. 6.11(b) shows the relationship between depth and the vertical stress beneath the center of the loaded area. The graphs point out certain characteristic features of stress distribution which are summarized qualitatively as follows. The application of pressure over a portion of the surface results in shear and normal stresses at every

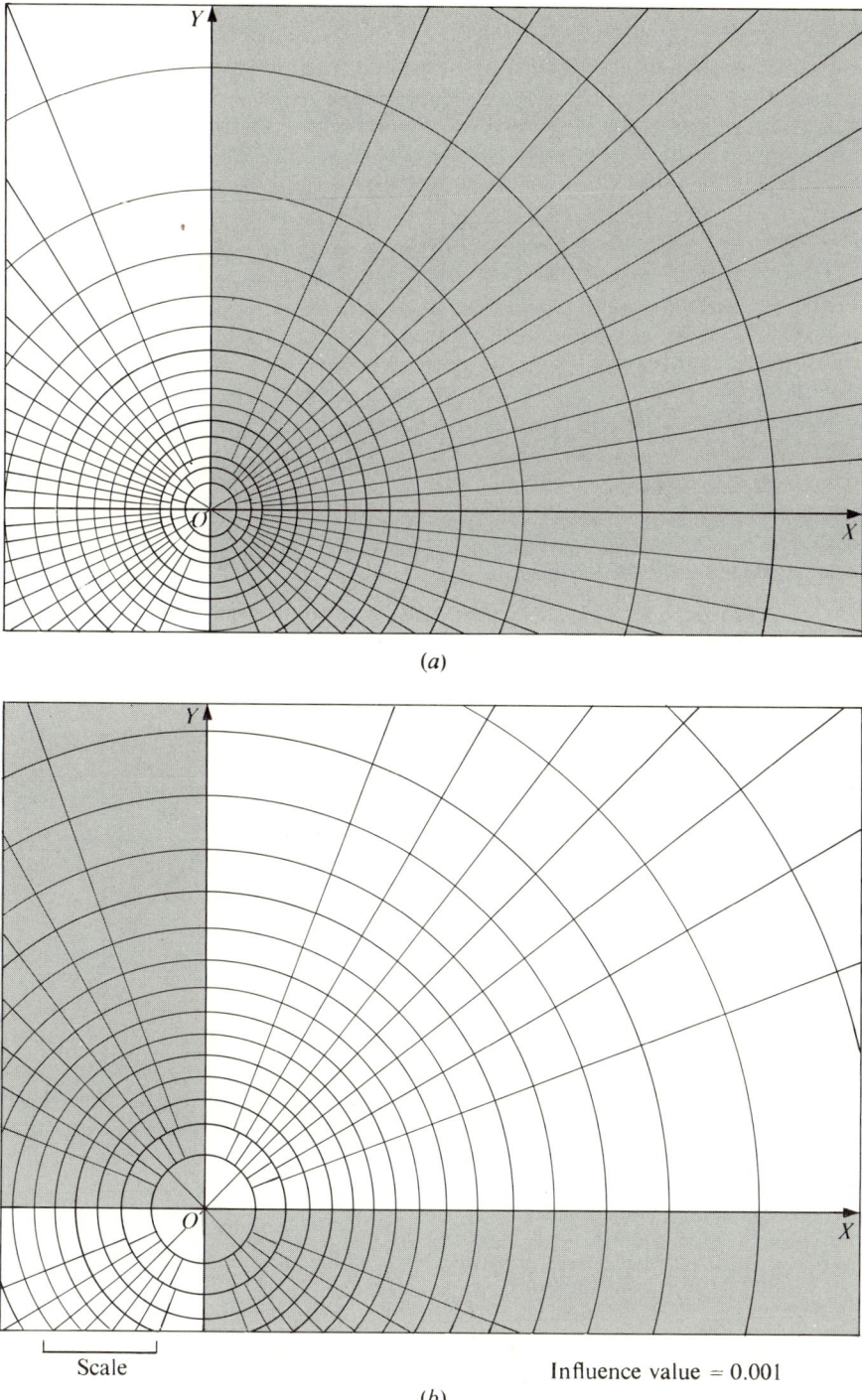

FIGURE 6.8 *Influence charts for* (a) τ_{xy} *and* (b) τ_{zx}. [After Newmark (1942).]

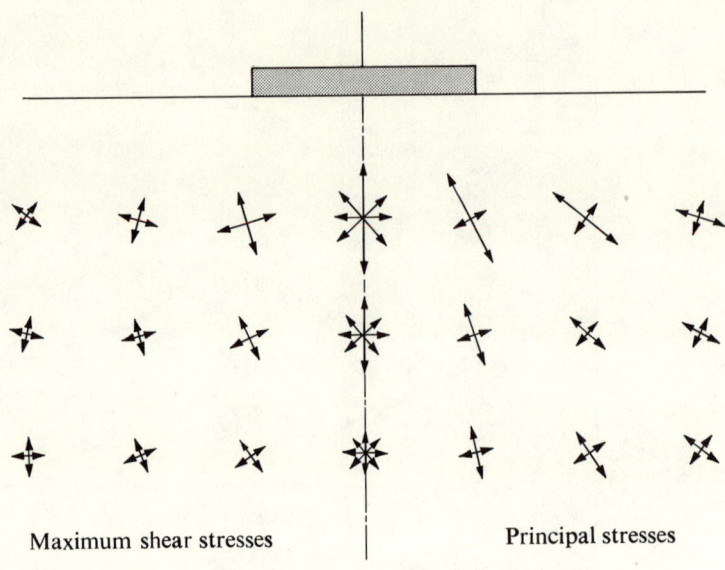

FIGURE 6.9 *Calculated shear and principal stresses beneath a loaded strip.* [After Jurgenson (1934).]

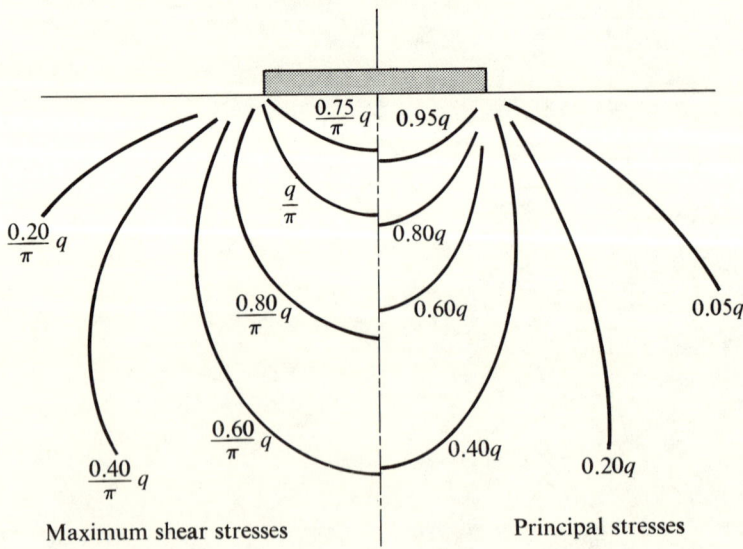

FIGURE 6.10 *Curves of equal shear and normal stresses beneath a loaded strip.* [After Jurgenson (1934).]

point in the soil mass. The stresses are highest immediately underneath the loaded area, but decrease rapidly with increasing distance from the load. The direction of the principal axes, hence also that of maximum shear stress, varies from point to point throughout the mass.

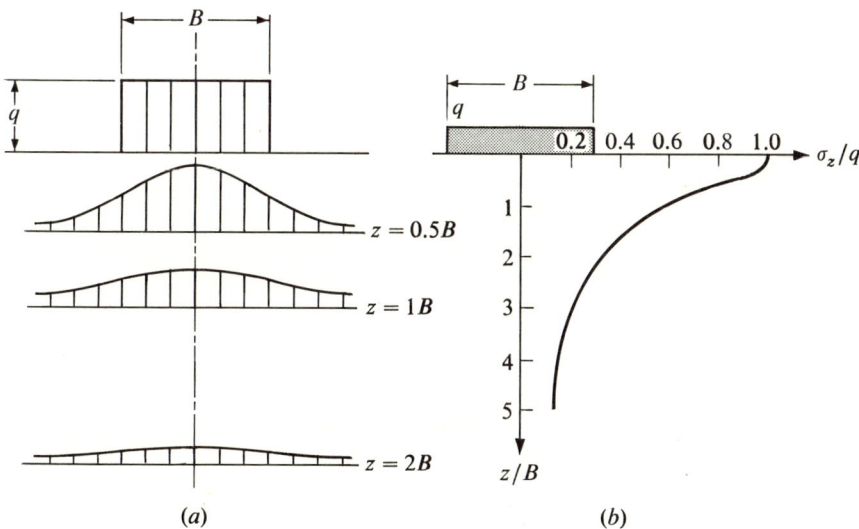

FIGURE 6.11 *Stress distribution beneath a loaded strip.*

LAYERED MEDIUM. The preceding treatment considers the soil mass as a homogeneous material—which is usually not the case. A comparatively simple model of nonhomogeneity is one in which the soil mass consists of layers with different moduli of elasticity. This case can be solved by the elastic theory, and results have been published by Burmister (1943), Acum and Fox (1951), and Jones (1961), among others. A condition commonly encountered in natural soil deposits is a layer of stiff clay or dense sand overlying a deposit of softer material. The effect of the stiff surface layer is to reduce the stress concentration in the underlying soil. Figure 6.12 shows the stresses in a three-layer medium under the center of a circular loaded area with a radius R. The stresses for three different ratios of E_1/E_2 and E_2/E_3 are plotted as curves A, B, and C. Curve C represents the case of a homogeneous material. The reduction in stress beneath a stiff layer also reduces the settlements considerably. This is the principle employed in the design of pavements, in which a comparatively rigid pavement is used to reduce the stress concentration in the subgrade soil.

6.7 Deformation

The elastic deformation of a soil element may be calculated from the stress changes and the modulus of elasticity and Poisson's ratio. The unit strain in the z direction at any point (point a in Fig. 6.13) is expressed as

$$\epsilon_z = \frac{\sigma_z}{E} - \frac{\mu\sigma_r}{E} - \frac{\mu\sigma_t}{E} \qquad (6.19a)$$

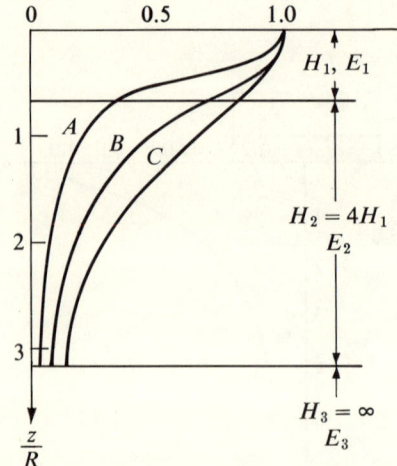

(a)

$R = 1.6H_1 =$ radius of loaded area
A: $E_1/E_2 = 20$, $E_2/E_3 = 2$
B: $E_1/E_2 = 2$, $E_2/E_3 = 2$
C: $E_1/E_2 = 1$, $E_2/E_3 = 1$

(b)

FIGURE 6.12 *Stress distribution in layered media.*

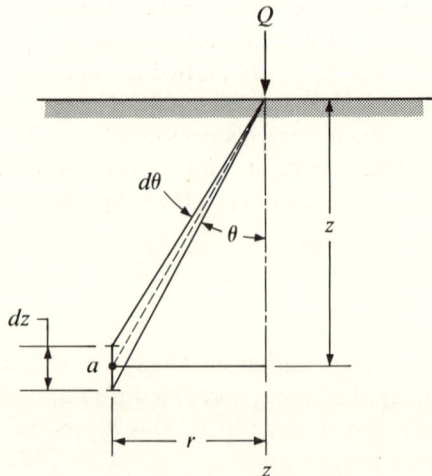

FIGURE 6.13 *Strains at a point in a semi-infinite mass.*

SETTLEMENT DUE TO CONCENTRATED LOAD. To calculate the displacement at the surface due to a concentrated load Q we substitute Eq. (6.16) in (6.19a) and obtain

$$\epsilon_z = \frac{Q}{2\pi z^2}\frac{1+\mu}{E}\cos^3\theta(3\cos^2\theta - 2\mu) \tag{6.19b}$$

The vertical displacement at the surface ρ may be found by integration:

$$\rho = \int_0^\infty \epsilon_z\, dz = \int_0^{\pi/2} \epsilon_z \frac{r\, d\theta}{\sin^2\theta} = \frac{Q}{2\pi r}\frac{1+\mu}{E}\left[\cos^2\theta + 2(1-\mu)\right]\sin\theta\bigg]_0^{\pi/2}$$

$$= \frac{Q}{\pi r}\frac{1-\mu^2}{E} \tag{6.20}$$

SETTLEMENT DUE TO LOADED AREA. As an example of a distributed load we consider a circular loaded area with radius R acting on the ground surface [Fig. 6.3(d)]. The strain ϵ_z at point a on the center line is obtained by substituting Eq. (6.18) in Eq. (6.19a). Since $\sigma_r = \sigma_t$ along the centerline,

$$\epsilon_z = \frac{1}{E}\left\{q\left[1 - \frac{z^3}{(R^2+z^2)^{3/2}}\right] - \mu q\left[1 + 2\mu - \frac{2(1+\mu)z}{(R^2+z^2)^{1/2}} + \frac{z^3}{(R^2+z^2)^{3/2}}\right]\right\}$$

For an incompressible material, $\mu = 0.5$. Then

$$\epsilon_z = \frac{q}{E}\left[\frac{1.5z}{(R^2+z^2)^{1/2}} - \frac{1.5z^3}{(R^2+z^2)^{3/2}}\right] \tag{6.19c}$$

The settlement at the surface, sometimes called elastic settlement, is

$$\rho = \int_0^\infty \epsilon_z\, dz = \frac{1.5qR}{E} \tag{6.21a}$$

The settlement at the edge of a circular loaded area is

$$\rho_r = \frac{3}{\pi}\frac{qR}{E} \tag{6.21b}$$

The displacement of a point b at a depth z_0 beneath the ground surface [Fig. 6.3(d)] may be found by integrating between z_0 and ∞, or

$$\rho_{z_0} = \int_{z_0}^\infty \epsilon_z\, dz = \frac{1.5qR}{E}\frac{1}{\left[1 + \left(\frac{z_0}{R}\right)^2\right]^{1/2}} \tag{6.21c}$$

This equation may be used to calculate the settlement of the subgrade soil underneath a pavement of thickness z_0 [Palmer and Barber (1940)]. Then it is assumed that the stress distribution is the same as that in a homogeneous material.*

The settlement due to a distributed load over an irregular area can be obtained by numerical integration of Eq. (6.21) in the same manner that the stress

---

* Corrections to Eq. (6.21c) to include the effect of the pavement stiffness have been suggested by Hogentogler (1940). See also Worley (1943).

under a distributed load is obtained. The results obtained by Schleicher (1926) for the settlement ρ_b at the corners of rectangular areas under a uniform load q are given as

$$\rho_b = qB \frac{1 - \mu^2}{E} N_\rho \qquad (6.22a)$$

in which B is the width and N_ρ is an influence number dependent on the length to width ratio L/B. Values of N_ρ are given in Table 6.1 for various ratios L/B.

TABLE 6.1   Influence Values N_ρ

L/B	N_ρ
1.0	0.56
2.0	0.76
3.0	0.88
4.0	0.96
5.0	1.00

For a square area and μ of 0.5, the settlement at the center and that at the corner are, respectively,*

$$\rho_0 = 0.84 \frac{qB}{E} \qquad (6.22b)$$

and

$$\rho_b = 0.42 \frac{qB}{E} \qquad (6.22c)$$

For a uniformly loaded area the settlement is shown in Fig. 6.14(a). The settlement at the center is much larger than the settlement at the edge of the loaded area.

In many soil mechanics problems, the loads are applied through a rigid foundation. In such cases the foundation enforces a uniform settlement ρ_d over the entire loaded area. The distribution of the contact pressure is not uniform, and according to the theory of elasticity has the shape shown in Fig. 6.14(b). The settlement of a rigid circular footing is equal to*

$$\rho_d = \frac{Q}{2ER}(1 - \mu^2) \qquad (6.22d)$$

If the loads are applied at some depth below the ground surface (for example, a deep foundation) then the preceding equations are not strictly applicable. Fox (1948) calculated the settlement due to a load with width B applied at a depth D below the surface. If the depth D is equal to B, the settlement is about 0.75 times that of a load on the surface. This ratio approaches 0.50 as a limit for an infinitely deep foundation.

---

* For proof, see Timoshenko and Goodier (1951), pp. 366–72.

SEC. 6.7   Deformation

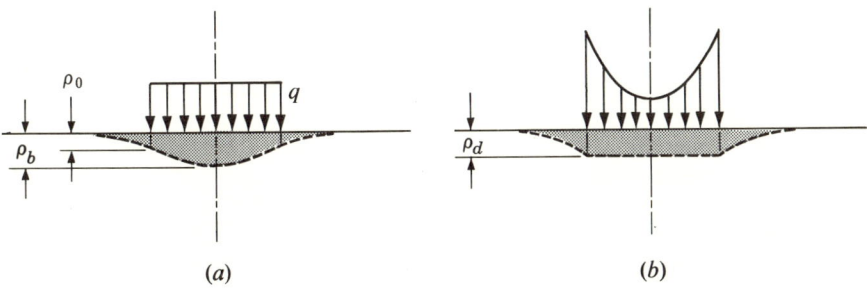

FIGURE 6.14 *Elastic settlement and contact pressure.*

Example 6.1. It is required to calculate the settlement at the center of a circular loaded area shown in Fig. 6.15. The calculated stress changes $\Delta\sigma_1$ and $\Delta\sigma_3$ on the element A at the middle of the layer are 440 and 185 psf, respectively. The modulus of elasticity is measured in a laboratory test which duplicates the field loading condition as follows. The specimen is consolidated in the K_0 condition under a vertical pressure equal to 500 psf which is the average overburden pressure on the clay layer. This represents the initial condition of the soil element A. The vertical pressure is then increased rapidly allowing no drainage. This approximates the application of the load. The measured vertical strain of the specimen at a value of $\Delta\sigma_1$ equal to 255 psf is 0.5 percent. Note that in this test we keep $\Delta\sigma_3$ equal to zero. This is because in a saturated soil the porewater-pressure coefficient B is 1. Hence, applying a $\Delta\sigma_3$ in the undrained state does not change the effective stress and has no effect on the value of E.

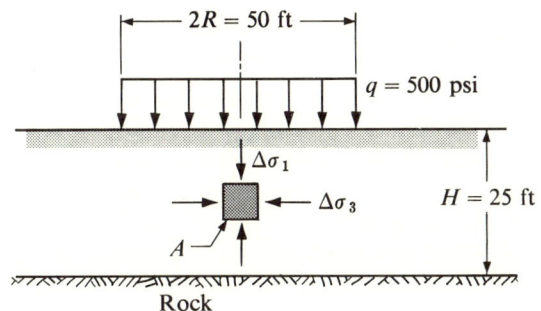

FIGURE 6.15 *Elastic deformation of a clay layer.*

Solution. The value of E may be obtained from the test data as

$$E = \frac{255}{0.005} = 51{,}000 \text{ psf}$$

If we calculate the settlement with Eq. (6.22d) we have

$$\rho = \frac{500 \times 25\pi \times 0.75}{51{,}000 \times 2} = 0.286 \text{ ft}$$

We may also use the elementary relationship

$$\rho = \epsilon_z H \tag{6.23}$$

in which H is the thickness of the compressible layer and ϵ_z the average strain. Then, $\rho = 0.005 \times 25 = 0.125$ ft

It is interesting to note the difference between the quantities calculated from Eqs. (6.22d) and (6.23). Equation (6.22d) is based on the assumption that strain occurs throughout the soil mass—from the surface to infinite depth. In the present case, the strains exist in a layer only 25 ft thick, since the rock is assumed to be rigid. Hence Eq. (6.23) gives a more realistic estimate. However, Eq. (6.23) involves an inaccuracy in that we have used the average stress changes $\Delta\sigma_1$ and $\Delta\sigma_3$ for the entire 25 ft. Its accuracy can be improved if we divide the 25 ft into several layers and calculate the deformation for each layer using the stress changes in each layer.

DISPLACEMENTS ON THE BOUNDARY

A different kind of boundary-value problem is presented by a soil mass subjected to certain displacements along part of the boundary. Such would be the case if the construction of a tunnel or underground structure introduces certain deformations in the soil. Another example would be an earth dam resting on a compressible subsoil that settles a substantial amount. There exists a wide variety of such problems. Two problems are illustrated in the subsequent sections because the availability of observational data makes it possible to evaluate the applicability of the solutions.

6.8 Beam subjected to deflection

A beam with a rectangular cross section is subjected to certain arbitrary deflections along its length [Fig. 6.16(a)]. It is required to find out the stress and strain distribution inside the beam under the assumption that the material is elastic and isotropic. In soil mechanics this is used to represent the case of an earth dam [Fig. 6.16(b)] subjected to given settlements as a result of compression of the foundation material [Leonards and Narain (1963)]. Of course we realize that an earth dam differs considerably from a beam of constant depth and rectangular cross section.

This is a two-dimensional problem that is solved by means of a stress function Φ which satisfies Eq. (6.9). In the case of an earth dam, the known boundary condition is the measured settlement along the crest of the dam. If the dam is being designed, the settlement may be estimated according to the principles outlined in the preceding sections. Denoting the horizontal and vertical displacements by u and v, we represent the irregular settlement curve by a Fourier series. Along the crest,

$$y = 0 \qquad v = \frac{a_0}{2} + \sum_{n=1}^{m} \left(a_n \cos \frac{n\pi}{L} x + b_n \sin \frac{n\pi}{L} x \right)$$

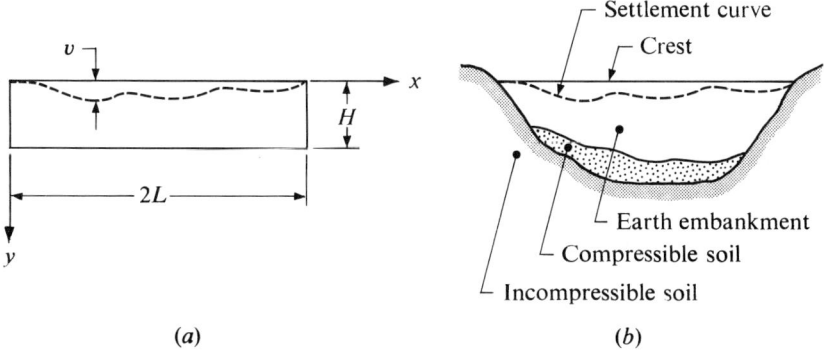

FIGURE 6.16 *Settlement of an earth dam on a compressible foundation.*

in which a_0, a_n, and b_n are constant coefficients that can be evaluated from the given settlement curve. This may be accomplished numerically if the length of the dam (2L) is divided into a finite number of $2m$ increments. The value of v is known at these points from the settlement data. For $2m$ coefficients the values of a_n and b_n can be obtained by solving $2m$ simultaneous equations. This can be readily accomplished on the computer.

It is assumed that the abutments are rigid and unyielding, so

$$\text{at } x = 0 \text{ and } x = 2L: \quad u = 0, \quad v = 0$$

Since there are no loads on the crest of the dam, we have

$$\text{at } y = 0: \quad \sigma_y = 0, \quad \tau_{xy} = 0$$

To simplify the mathematics it is further assumed that the shear stress along the base of the dam is negligible and at $y = H$, $\tau_{xy} = 0$. A solution to Eq. (6.9) is of the form

$$\Phi = \sum_{n=1}^{m} \left(A_n \cos \frac{n\pi}{L} x + B_n \sin \frac{n\pi}{L} x \right) F(y)$$

in which A_n and B_n are constants. Substituting into Eq. (6.9) yields

$$\sum_{n=1}^{\infty} \left(A_n \cos \frac{n\pi}{L} x + B_n \sin \frac{n\pi}{L} x \right)$$

$$\left[\frac{\partial^4}{\partial y^4} F(y) - 2 \left(\frac{n\pi}{L}\right)^2 \frac{\partial^2}{\partial y^2} F(y) + \left(\frac{n\pi}{L}\right)^4 F(y) \right] = 0$$

which means

$$\left[\frac{\partial^4}{\partial y^4} F(y) - 2 \left(\frac{n\pi}{L}\right)^2 \frac{\partial^2}{\partial y^2} F(y) + \left(\frac{n\pi}{L}\right)^4 F(y) \right] = 0$$

Thus the function $F(y)$ has the form

$$F(y) = C_1 \sinh \frac{n\pi}{L} y + C_2 \cosh \frac{n\pi}{L} y + C_3 y \sinh \frac{n\pi}{L} y + C_4 y \cosh \frac{n\pi}{L} y$$

in which C_1, C_2, C_3, and C_4 are constants.

The boundary conditions stated above lead to the following results.

$$A_n = -\frac{a_n E}{2C_1}\left(\frac{L}{n\pi}\right) \qquad B_n = -\frac{b_n E}{2C_1}\left(\frac{L}{n\pi}\right)$$

$$C_2 = 0 \qquad C_3 = \frac{C_1}{\nu}\left(\frac{n\pi}{L}\right) \qquad C_4 = -C_1 \left(\frac{n\pi}{L}\right)$$

Therefore,

$$\Phi = \frac{E}{2} \sum_{n=1}^{m} \left(a_n \cos \frac{n\pi}{L} x + b_n \sin \frac{n\pi}{L} x\right)$$

$$\left(-\frac{L}{n\pi} \sinh \frac{n\pi}{L} y + y \cosh \frac{n\pi}{L} y - \frac{1}{\nu} y \sinh \frac{n\pi}{L} y\right)$$

in which

$$\nu = \frac{\sinh \frac{n\pi}{L} H + \frac{n\pi}{L} H \cosh \frac{n\pi}{L} H}{\frac{n\pi}{L} H \sinh \frac{n\pi}{L} H}$$

With Φ determined, the longitudinal strain ϵ_x is, according to Eqs. (6.8) and (6.1),

$$\epsilon_x = \sum_{n=1}^{m} \left(a_n \cos \frac{n\pi}{L} x + b_n \sin \frac{n\pi}{L} x\right) \left[-\frac{n\pi}{\nu L} \cosh \frac{n\pi}{L} y + \frac{1-\mu}{2} \frac{n\pi}{L} \sinh \frac{n\pi}{L} y \right.$$

$$\left. + \frac{1+\mu}{2}\left(\frac{n\pi}{L}\right)^2 y \cosh \frac{n\pi}{L} y - \frac{1+\mu}{2\nu}\left(\frac{n\pi}{L}\right)^2 y \sinh \frac{n\pi}{L} y\right]$$

We note that the tensile strain is a maximum at the top of the dam. Hence, for $y = 0$,

$$\epsilon_x = -H \sum_{n=1}^{m} \left[\frac{\left(\frac{n\pi}{L}\right)^2 \sinh \frac{n\pi}{L} H}{\sinh \frac{n\pi}{L} H + \frac{n\pi}{L} H \cosh \frac{n\pi}{L} H}\right]$$

$$\left(a_n \cos \frac{n\pi}{L} x + b_n \sin \frac{n\pi}{L} x\right) \quad (6.24)$$

Leonards and Narain (1963) computed the maximum tensile strains in a number of dams with known settlements and compared these with the measured strains along the crest. In addition, tensile strains at which the soils developed tension cracks were measured by laboratory tests. Table 6.2 gives the calculated and measured tensile strains, the strains at which the soils crack, and the observed behaviors of the dams. We see that for the two dams whose strains were measured, the calculated and measured strains are in reasonable agreement. Thus, despite the simplifications introduced in the solution, the answers are still of the right order of magnitude. The field observations show that, except for the Woodcrest Dam, cracks develop in dams where the tensile strain exceeds that required to produce cracking in laboratory specimens.

TABLE 6.2 *Comparison of Calculated and Measured Strain in Earth Dams*[a]

	Tensile strain, %			
Dam	Calculated	Measured in dam	Cracking of lab. spec.	Field observ. of dam behavior
Portland, Colo.	0.30	—	0.12	Cracked
Rector Cr., Calif.	0.29	0.24	0.14	Cracked
	0.23			
Woodcrest, Calif.	0.12	0.09	0.20	Cracked
	0.09			
Shell Oil, Calif.	0.20	—	0.08	Cracked
	0.17			
Willard, Utah	0.17	—	0.24	Uncracked

[a] After Leonards and Narain (1963).

6.9 Pressure against yielding restraints

In this group belong a host of problems such as pressure on flexible underground structures and conduits and pressure on retaining walls. The general condition is that certain displacements are imposed on parts of a soil mass due to the yielding of the structure. The simplest problem is that of a yielding base, such as is shown in Fig. 6.17(a). The bottom of the soil mass yields by an amount ρ along the strip ab. The remainder of the base is rigid. This condition may be used as an approximation of the deformations introduced by the excavation of a tunnel with a yielding roof.

The boundary conditions are as follows. The base is assumed to be perfectly smooth, so that $y = 0$, $\tau_{xy} = 0$. Along the surface of the soil mass there are no vertical stresses and the vertical displacement is assumed to be 0:

$$\text{at } y = H: \quad \sigma_y = 0, \quad v = 0$$

At the base the yield is ρ along the strip ab and 0 everywhere else:

$$\text{at } y = 0: \quad x > B, \quad v = 0$$
$$\text{at } y = 0: \quad x < B, \quad v = -\rho$$

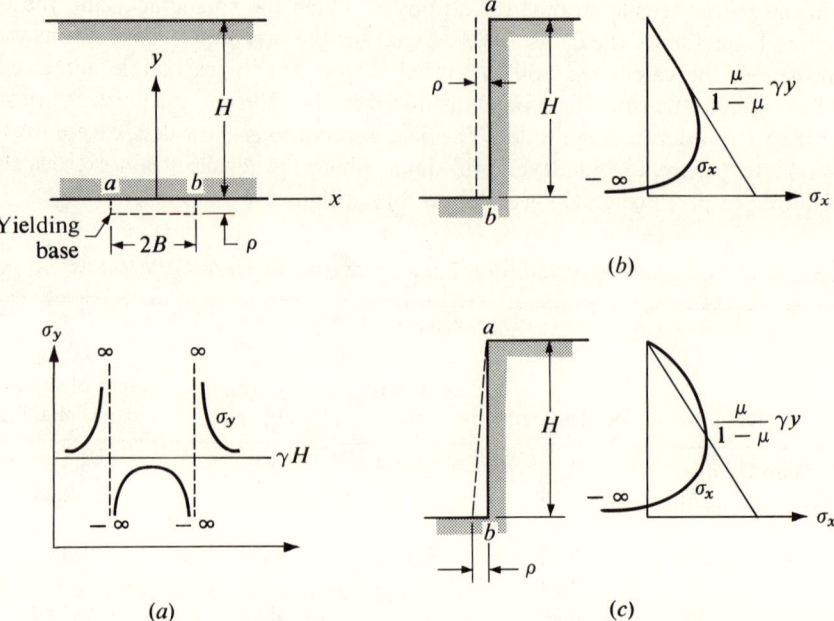

FIGURE 6.17 *Earth pressures on yielding boundaries.* [After Finn (1963).]

This condition is expanded into a Fourier integral of the form

$$v(x) = \frac{2}{\pi} \int_0^\infty \cos \xi x \, d\xi \int_0^\infty v(\nu) \cos \xi \nu \, d\nu$$

in which ξ and ν are parameters. For the boundary conditions just stated, the Fourier integral can be evaluated and we have [see Hildebrand (1962), pp. 236–39]

$$y = 0: \quad v(x) = \frac{2\rho}{\pi} \int_0^\infty \frac{\sin \xi (B/2)}{\xi} \cos \xi x \, d\xi$$

To obtain the stresses we assume a stress function Φ of the form

$$\Phi = \int_0^\infty \frac{1}{\xi^2} (C + D\xi y) e^{-\xi y} \cos \xi x \, d\xi$$

in which C and D are constants and Φ is the stress function as given by Eqs. (6.8) and (6.9).

To satisfy the boundary conditions, we have

$$y = 0, \quad \tau_{xy} = -\frac{\partial^2 \Phi}{\partial x \partial y} = \int_0^\infty (C - D) \sin \xi x \, d\xi = 0$$

or,

$$C = D$$

SEC. 6.9 **Pressure against yielding restraints** 155

Also,
$$\sigma_x = \frac{\partial^2 \Phi}{\partial y^2} = -\int_0^\infty D(1 - \xi y)e^{-\xi y} \cos \xi x \, d\xi$$

$$\sigma_y = \frac{\partial^2 \Phi}{\partial x^2} = -\int_0^\infty D(1 + \xi y)e^{-\xi y} \cos \xi x \, d\xi$$

In plane strain,
$$\epsilon_z = \frac{\sigma_z}{E} - \mu \frac{\sigma_x}{E} - \mu \frac{\sigma_y}{E} = 0$$

or
$$\sigma_z = \mu(\sigma_x + \sigma_y)$$

Hence we may write
$$\epsilon_y = \frac{\partial v}{\partial y} = \frac{\sigma_y}{E} - \mu \frac{\sigma_x}{E} - \mu \frac{\sigma_z}{E} = \frac{1 - \mu^2}{E} \sigma_y - \frac{\mu(1 + \mu)}{E} \sigma_x$$

$$= -\frac{1}{E}\int_0^\infty D[(1 - \mu - 2\mu^2) + (1 + \mu)\xi y]e^{-\xi y} \cos \xi x \, d\xi + g(x)$$

$$v = \frac{1}{E}\int_0^\infty \frac{D}{\xi}[2(1 - \mu^2) + (1 + \mu)\xi y]e^{-\xi y} \cos \xi x \, d\xi + g(x)$$

For $y = H = \infty$, $v = 0$. Hence $g(x) = 0$.
At $y = 0$,
$$v = \frac{2}{E}\int_0^\infty \frac{D}{\xi}(1 - \mu^2) \cos \xi x \, d\xi = \frac{2\rho}{\pi}\int_0^\infty \frac{1}{\xi} \sin \xi \left(\frac{B}{2}\right) \cos \xi x \, d\xi$$

or
$$D = \frac{\rho E}{(1 - \mu^2)\pi} \sin \xi \left(\frac{B}{2}\right)$$

Substituting this into the expressions for σ_x, σ_y, and τ_{xy}, we get

$$\sigma_x = -\frac{\rho E}{2\pi(1 - \mu^2)}\left\{\frac{x + (B/2)}{[x + (B/2)]^2 + y^2} - \frac{x - (B/2)}{[x - (B/2)]^2 + y^2}\right.$$
$$\left. - \frac{2[x + (B/2)]y^2}{\{[x + (B/2)]^2 + y^2\}^2} + \frac{2[x - (B/2)]y^2}{\{[x - (B/2)]^2 + y^2\}^2}\right\}$$

$$\sigma_y = -\frac{\rho E}{2\pi(1 - \mu^2)}\left\{\frac{x + (B/2)}{[x + (B/2)]^2 + y^2} - \frac{x - (B/2)}{[x - (B/2)]^2 + y^2}\right. \quad (6.25)$$
$$\left. + \frac{2[x + (B/2)]y^2}{\{[x + (B/2)]^2 + y^2\}^2} - \frac{2[x - (B/2)]y^2}{\{[x - (B/2)]^2 + y^2\}^2}\right\}$$

$$\tau_{xy} = \frac{E}{2\pi(1 - \mu^2)}\left\{y\left(\frac{y^2 - [x - (B/2)]^2}{\{[x + (B/2)]^2 + y^2\}^2} - \frac{y^2 - [x + (B/2)]^2}{\{[x - (B/2)]^2 + y^2\}^2}\right)\right\}$$

Of special interest is the vertical pressure on the base. Putting $y = 0$ in Eq. (6.25) we get

$$\sigma_y = -\frac{\rho B E}{2\pi(1 - \mu^2)}\left[\frac{1}{x^2 - (B/2)^2}\right] \quad (6.26)$$

This is the stress due to the displacement ρ only. To this we must add the overburden pressure γH. The calculated pressure distribution is shown in Fig. 6.17(a). The above solution was given by Finn (1963), who also solved the case of a rough base and a base rotating about one edge. The stress distribution shown in Fig. 6.17(a) is qualitatively in agreement with experimental results reported by Terzaghi and Peck (1967).

The above solution can be extended to the problem of pressures on a yielding wall (Finn, 1963). The wall may move away from the soil through a distance ρ [Fig. 6.17(b)] or rotate about the top [Fig. 6.17(c)]. The calculated stress distributions for these two cases are also shown in Fig. 6.17(b) and (c), respectively. Considerable information regarding pressures on yielding walls is available from model tests and field observations on actual structures. Under certain conditions the observed pressure distribution has the general shape shown in Fig. 6.17 (see Chapter 10).

The application of the theoretical solutions to design problems implies that the material behaves elastically. Some of the departures of soil behavior from ideal elasticity are pointed out in this chapter and in Chapter 4. The elasticity solutions are good approximations at small displacements. At large displacements, the stresses in certain regions reach the yield stress, and the stress distribution may be quite different from the elastic stress distribution. For example, the elastic solution gives a stress equal to infinity at the bottom of the yielding wall and also at the edge of the yielding base (see Fig. 6.17). This is obviously untenable in a real material. Plastic yielding would occur at these points as soon as the stress reaches the yield stress. Hence, the theoretical stress distribution is not valid in this zone, even at small deflections. The difference between the actual and the computed stress distribution can be expected to increase as the deflection increases, owing to the spread of the plastic zone. The practical solution of earth-pressure-distribution problems often falls in between the theories of elasticity and plasticity and is treated in Chapter 10.

PROBLEMS

6.1 An area 40 ft × 40 ft on the ground surface is loaded to a uniform pressure of 2000 psf. (a) Construct the vertical stress increase vs. depth relationship beneath the center and corner of the loaded area. (b) If stress increases less than 200 psf do not contribute significantly to the settlement, what is the maximum depth to which stress calculations should be made in a settlement analysis?
Answer: 80 ft

6.2 In Problem 6.1 the saturated unit weight of the soil is 120 psf and its water content is 24%. The water table is at a depth of 5 ft below ground surface. (a) Construct the overburden pressure vs. depth relationship. Assume the soil above the water table to be saturated by capillary action. (b) Construct the vertical stress vs. depth relationship after the load in Problem 6.1 is applied.

Problems

(c) Are the stresses calculated in (a) and (b) total stresses or effective stresses?
Answer: (a) is effective stress; (b) is total stress immediately after application of load becomes effective stress long time after application of load

6.3 A footing 8 ft square supports a load of 100 tons. Calculate the vertical stress due to the load at a depth of 12 ft below the center of the footing.
Answer: 533 psf

6.4 Another footing of the same size and carrying the same load is placed at a distance of 12 ft from the first one: (a) Calculate the vertical stress due to both loads at a depth of 12 ft below the center of the footing. (b) What conclusions can be drawn regarding the effect of the second footing on the settlement of the first footing?
Answer: 656 psf; increases the settlement

6.5 A pavement is to support airplane wheel loads of 40 tons at a tire pressure equal to 150 psi. The modulus of elasticity of the pavement material is equal to 20,000 psi and that of the base course which supports the pavement is equal to 4000 psi. (a) Calculate the thickness of the pavement if the stress on the base material is not to exceed 100 psi, assuming the stress distribution equal to that in a homogeneous material. (b) What is the effect of the stiff pavement on the stress distribution?
Answer: 15 in.; reduces the stress in the base and requires thinner pavement

6.6 On the basis of the settlement equations for homogeneous material, derive an expression for the thickness of a pavement if the settlement of the subgrade cannot exceed 1 in. and the wheel load is Q.

Answer: $\sqrt{\left(\dfrac{3Q}{2\pi E}\right)^2 - a^2}$

6.7 The load from a building foundation is represented as a line load in Fig. 6.18. (a) Compute the principal stresses at points (1) and (2) and their directions. (b) Compute the vertical strains at points (1) and (2) if the soil has $E = 2 \times 10^4$ psi and $\mu = 0.50$.
Answer: (a) $-3190, 0$ psf; $-1596, 0$ psf; vertical and horizontal; 45 deg. with vertical; (b) 11×10^{-4}, 5.5×10^{-4}

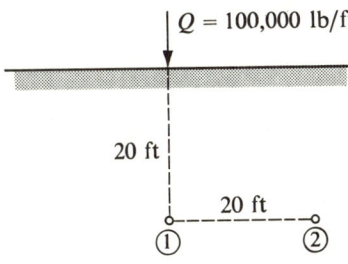

FIGURE 6.18

6.8 A sample was taken from the clay layer shown in Fig. 6.19(a). In the triaxial test it was found that the pore pressure parameters are $B = 1.0$ and $A = 0.3$. The sample was loaded as shown in Fig. 4.8(b). (a) If a point load is applied to the ground surface as shown in Fig. 6.19(b), compute the excess hydrostatic pore pressure at points (1) and (2) at the middle of the clay layer. (b) Show the piezometric level if piezometers are installed at points (1) and (2). Assume the soil to be homogeneous.
Answer: (a) 130; 48 psf

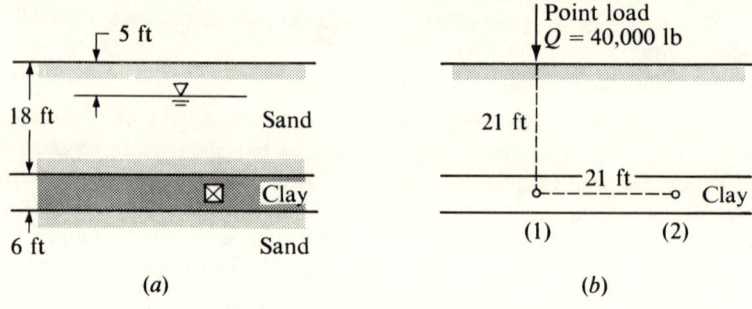

FIGURE 6.19

6.9 Consider the clay layer in Problem 6.8. (a) Compute the excess hydrostatic pore pressure at point (1) for the loading shown in Fig. 6.20(a). (b) Compute the excess hydrostatic pore pressure at point (1) for the loading shown in Fig. 6.20(b). (c) Give the reasons why you expect or do not expect Eq. (4.18) to hold for the conditions in Figs. 6.19(b), 6.20(a), and 6.20(b).
Answer: (a) 1000 psf; (b) 45 psf

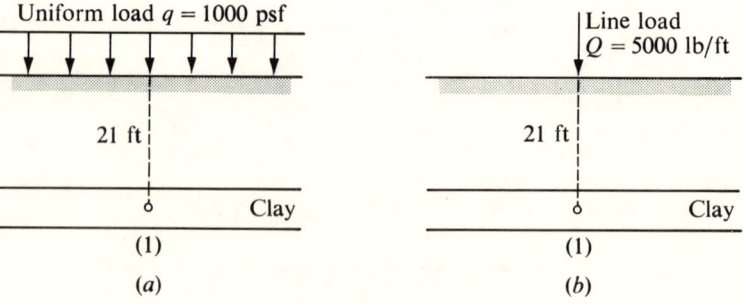

FIGURE 6.20

6.10 Compute the principal stresses and their directions at points a, b, and c in Fig. 6.21.
Answer: 970, 610, 440 psf, 0° at a; 810, 440, 200 psf, 0° at b; 630, 300, 70 psf, 31° at c

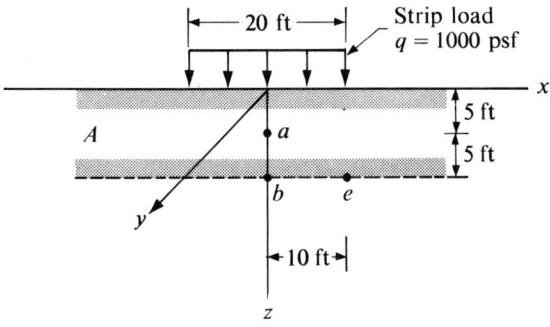

FIGURE 6.21

6.11 If the soil is saturated and has a pore pressure parameter A equal to 0.5, what are the pore pressures at points a, b, and c (Fig. 6.21) for the undrained condition?
Answer: 790, 630, 470 psf

7

Settlement Analysis

7.1 Introduction

Since no material is perfectly rigid, the imposition of a load (or stress) always produces a deformation (or strain). The construction of buildings and other structures imposes stresses on the supporting soil mass and causes deformations in the soil mass and settlement of the structures. In foundation engineering it is often necessary to estimate the settlement in advance. If it exceeds some tolerable value, it would be necessary to revise the foundation design.

If the soil behavior is elastic, the equations developed in Sec. 6.7 may be used in settlement analysis. The departure of soil behavior from ideal elasticity requires some modifications. However, it is generally accepted that at least the stresses from elasticity theory may be used for settlement calculations without unduly large errors. Since elasticity solutions of stress distribution are readily available, they are very useful in providing approximate answers.

Most common settlement problems may be simulated by one of the two simple loading conditions—the drained and the undrained-drained. The first is widely used for foundations on cohesionless soils. These soils usually have such high permeability that no excess hydrostatic porewater pressure develops during normal construction operations, so the drained condition holds. With clay soils the permeability is so small that very little dissipation of excess hydrostatic porewater pressure can occur during construction. Hence, the loading corresponds to the undrained condition. After the end of construction, the load may remain nearly constant, and the excess hydrostatic porewater pressure dissipates by consolidation (Sec. 5.4). This sequence is called the undrained-drained condition.

In soil mechanics literature the terms immediate settlement* and consolidation (or long-term) settlement are commonly used. This distinction based on

---

* Terzaghi and Peck (1967) used the term contact settlement.

rate has great practical advantage. It is, however, important to identify these terms with the loading conditions. For cohesionless soils, immediate settlement corresponds to the drained condition, whereas for clay soils it corresponds to the undrained condition and consolidation settlement corresponds to the drained condition. This identification is maintained throughout this book.

7.2 Settlement of loads on sand

The problem of settlement under the drained condition is illustrated in Fig. 7.1. The distribution of the stresses σ_z and σ_r below the center of the circular loaded area may be obtained with Eq. (6.18). If the material is assumed to be elastic, then the vertical strain ϵ_z may be calculated with Eq. (6.19c). This is shown in Fig. 7.1 as curve A. For simplicity we plot only the quantity inside the brackets of Eq. (6.19c). This is denoted by the influence number $I = \epsilon_z E/q$. Since the strain also depends on the value of Poisson's ratio, two curves are shown, one for $\mu = 0.40$ and one for $\mu = 0.50$.

As shown in Figs. 4.5 and 4.6 both the shear modulus G and the bulk modulus K (and hence, E and μ) of soils depend on the confining pressure. To account for this property we conduct laboratory experiments in which soil samples are subjected to stress paths that simulate the field stress path. These are called stress path tests. For the soil element a in Fig. 7.1, we calculate the initial stresses $\bar{p}_0$ and $K_0\bar{p}_0$ and the stress increases σ_z and σ_r due to the load q. We then subject a soil specimen to the stresses $\bar{p}_0$ and $K_0\bar{p}_0$ which simulates the initial condition. Then we increase the stresses by σ_z and σ_r and measure the strain ϵ_z brought about by this change. We may assume that the soil element a would suffer the same strain when subjected to the same stress change in-situ. If a series of tests are made for soil elements at various depths below the center of the loaded area, these give the strain distribution and may be plotted in Fig. 7.1 as curve B [D'Appolonia et al. (1968)]. Since the value of E varies with depth, an average value is used to obtain I. It is important to note the significant difference between the curves A and B, a difference caused by the stress dependence of G and K. Tests of model footings on sand have given results that fall within the zone D [Eggstad (1963)]. To account for these differences Schmertmann (1970) has recommended that curve C, Fig. 7.1, be used for settlement calculations.

The use of stress path tests provides one method for the prediction of settlement. To calculate the settlement we divide the subsoil into a number of layers with thickness Δz. For the layer shown in Fig. 7.1, the measured strain ϵ_z for element a is taken as the average strain. The compression of this layer is $\epsilon_z \Delta z$. The settlement is then obtained by summation of $\epsilon_z \Delta z$ for all layers, or

$$\rho = \sum \epsilon_z \Delta z \qquad (7.1a)$$

This is frequently called the stress path method [Lambe (1964)]. Alternatively, if the stress-strain curves and their relationship to the confining pressure (Fig.

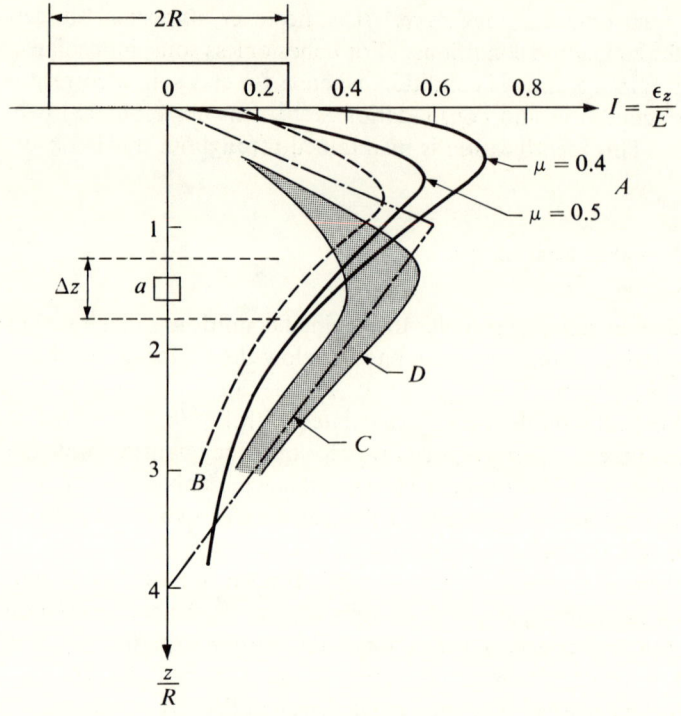

FIGURE 7.1 *Distribution of strain beneath a loaded area.*

4.6) are established, numerical methods such as the finite element method may be employed [i.e., Duncan and Chang (1970)].

One serious problem with cohesionless soils is the difficulty in securing undisturbed samples for the laboratory tests. For many design problems an expedient procedure is provided by field sounding tests such as the cone penetration and standard penetration tests (see Sec. 13.13). The results of sounding tests have been widely used to estimate the settlements on an empirical basis. Data collected by Schmertmann (1970) indicate that the value of E is approximately given by

$$E = 2q_c \qquad (7.2a)$$
and
$$E = 8N \qquad (7.2b)$$

where q_c is the cone penetration resistance (kg/cm²) and N is the number of blows in the standard penetration test. If E is known, curve C in Fig. 7.1 may be used to calculate the strains and Eq. (7.1a) may be used to calculate the settlement.

7.3 Settlement of loads on clay

Consider a circular loaded area on a deposit of saturated clay, as shown in Fig. 7.2. During the application of the load, the clay behavior is approximated

SEC. 7.3 Settlement of loads on clay

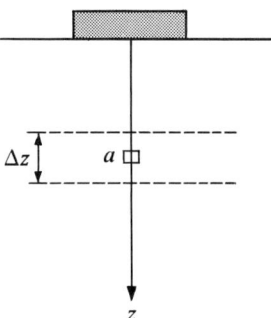

FIGURE 7.2 *Surface load on a clay deposit.*

by the undrained state, and the excess hydrostatic pore pressure at *a* may be given by Eq. (4.18). If there is no volume change, we have $\mu = 0.5$ and only shear distortion occurs. This stage is described by Fig. 7.3(*a*). If the stress path method is used, the specimen should be first consolidated under the stresses $\bar{p}_0$ and $K_0 \bar{p}_0$ in the triaxial cell. Then the stresses σ_z and σ_r are applied with the drainage valve *A* closed (see Fig. 4.13). The axial strain ϵ_{zu} is then the strain due to undrained loading. The stress path is shown as *ab* in Fig. 7.3(*c*).

With time, the excess hydrostatic pore pressure dissipates by consolidation. The all-around effective stress increases by an amount

$$\Delta \bar{\sigma} = -\Delta u$$

and the soil element undergoes a volumetric compression [Fig. 7.3(*b*)]. This stage can be simulated in the triaxial test by opening the drainage valve *A*. After complete consolidation, an additional vertical strain ϵ_{zc} is observed in the test. The stress path for this stage is *bc* in Fig. 7.3(*c*). To calculate the settlement we divide the subsoil into a number of layers as before, and obtain the average strains ϵ_{zu} and ϵ_{zc} for each layer. Then the settlement is

$$\rho = \rho_i + \rho_c = \sum (\epsilon_{zu} + \epsilon_{zc}) \Delta z \tag{7.1b}$$

We see that throughout the loading and consolidation process, $\bar{\sigma}_1$ progressively increases, while $\bar{\sigma}_3$ is first reduced and then increases. Therefore in the vertical direction the material is compressed. In the horizontal direction the soil first swells as $\bar{\sigma}_3$ decreases, and then is recompressed.

METHOD OF SKEMPTON AND BJERRUM. The settlement computation may be simplified considerably by using the method proposed by Skempton and Bjerrum (1957). The calculation of the undrained settlement is based on the assumption that the material is elastic. Then Eq. (6.21) may be used. It is, however, necessary to determine E and μ. Since there is no volume change, $\mu = 0.5$ and E may be measured by uniaxial loading in the undrained condition. As shown in Sec. 4.5, E is dependent on the confining pressure. Studies by

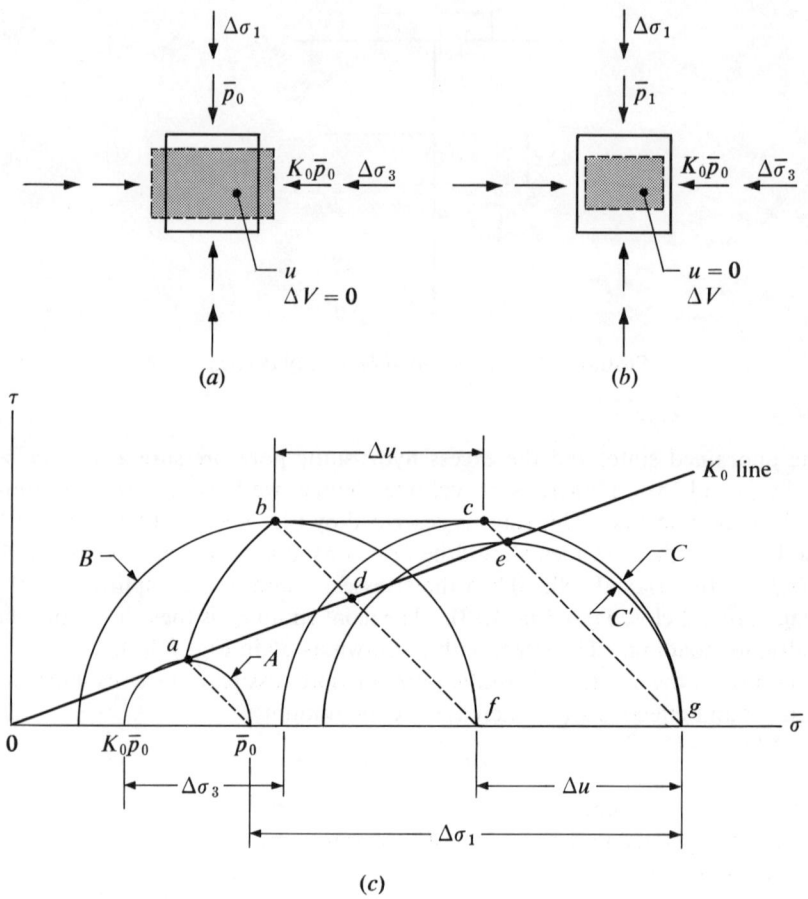

FIGURE 7.3 *Deformation and consolidation of an element.*

Ladd (1964) have shown that in undrained loading, E is approximately proportional to the initial confining pressure. Hence, the tests should be performed on samples consolidated under $\bar{p}_0$ and $K_0\bar{p}_0$, which represent the average initial stresses in the ground.

To compute the settlement due to consolidation, we first determine the volumetric strain, which is equal to

$$\epsilon_v = \frac{\Delta\bar{\sigma}_0}{K} = -\frac{\Delta u}{K}$$

where K is the bulk modulus measured by subjecting a sample to all-around stress increase. For simplification it is assumed that

$$K = \frac{1}{m_v}$$

where m_v is the compressibility measured in the consolidation test (Sec. 5.2).

Settlement of loads on clay

Since the compression in the horizontal direction is smaller than that in the vertical direction it is helpful to introduce the simplifying assumption that the settlement is due to compression in the vertical direction only. According to Skempton and Bjerrum (1957), the error introduced by this simplification is less than 20 percent.

At this point we should note that in the consolidation test in which there is no lateral yield, the major and minor principal stresses maintain a certain ratio K_0. For a constant ratio of $\bar{\sigma}_3/\bar{\sigma}_1$ the stress path is a straight line passing through the origin [from a to e, Fig. 7.3(c)]. This line is designated as the K_0 line. The assumption that the soil element compresses in the vertical direction only is thus equivalent to the assumption that the volume change is the same as that in a consolidation test. In other words, the compression produced by the stress path bc is assumed to be equal to that produced by the stress path de along the K_0 line. The quantity Δu is the same in both cases, and the effective major principal stress changes from point f to g.

The compression in the vertical direction due to the dissipation of the excess hydrostatic pcrewater pressure Δu (or an increase in effective stress of the same magnitude) for an element dz thick is

$$d\rho_c = m_v \, \Delta u \, dz$$

The compression of a layer with thickness H is the integral

$$\rho_c = \int_0^H m_v \, \Delta u \, dz$$

When the expression for Δu [Eq. (4.18)] is substituted in the above, we have

$$\rho_c = \int_0^H m_v \, \Delta\sigma_1 \left[A + \frac{\Delta\sigma_3}{\Delta\sigma_1}(1 - A) \right] dz \quad (7.3)$$

By comparison, the compression of a specimen in the one-dimensional consolidation test is

$$\rho_c' = \int_0^H m_v \, \Delta\sigma_1 \, dz \quad (7.4)$$

We can relate the compression of the soil layer to that of the consolidation test specimen by writing

$$\rho_c = C_\rho \rho_c' \quad (7.5a)$$

in which C_ρ is a factor given by

$$C_\rho = \frac{\int_0^H m_v \, \Delta\sigma_1 \left[A + \frac{\Delta\sigma_3}{\Delta\sigma_1}(1 - A) \right] dz}{\int_0^H m_v \, \Delta\sigma_1 \, dz} \quad (7.5b)$$

For a given soil layer, m_v and A are soil constants, and the equation reduces to

$$C_\rho = A + (1 - A) \frac{\int_0^H \Delta\sigma_3 \, dz}{\int_0^H \Delta\sigma_1 \, dz} \quad (7.5c)$$

The second term in the equation is dependent only upon the variation of $\Delta\sigma_1$ and $\Delta\sigma_3$ with the depth z, which is a geometrical property of the problem. For a given loading, such as a circular load or a strip load, the quantity

$$\frac{\int_0^H \Delta\sigma_3 \, dz}{\int_0^H \Delta\sigma_1 \, dz}$$

can be calculated from elasticity (Figs. 6.5 and 6.7) and the results are plotted in Fig. 7.4(a). This allows the determination of the value C_ρ for various values of A. Figure 7.4(b) shows the calculated values of C_ρ. The compressible layer has a thickness denoted by the ratio z/B, in which B is the diameter of the circular area or the width of the strip load. Of course, other loads and depth conditions can also be worked out. Strictly speaking, this method cannot be used to calculate the settlement of points located away from the axis of symmetry of the load. This is because, away from the axis, the principal stresses due to the load are no longer in the vertical and horizontal directions (see Fig. 6.9). Hence the stress conditions of Fig. 7.3 are no longer valid.

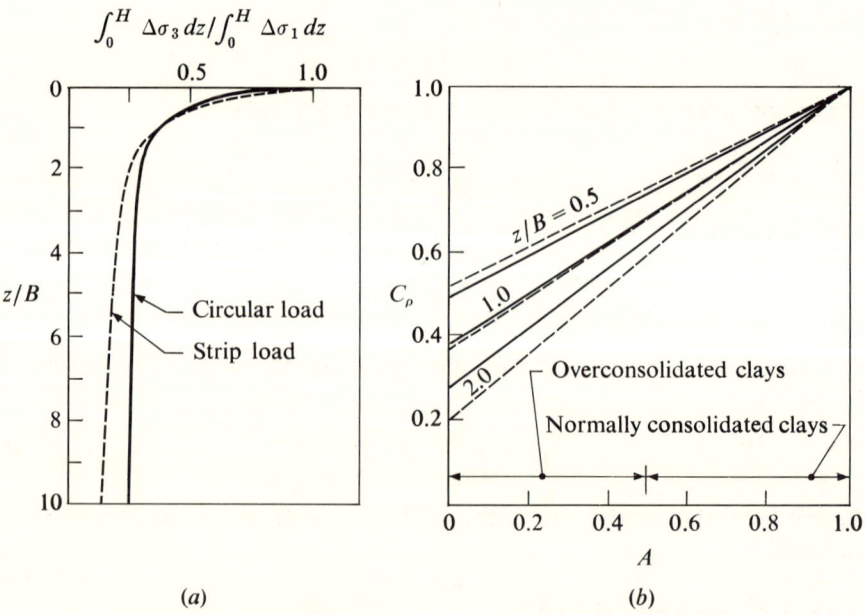

FIGURE 7.4 *Calculation of consolidation settlement.* [After Skempton and Bjerrum (1957).]

We can draw a number of useful conclusions from the results in Fig. 7.4. We note that the value of C_ρ approaches 1.0 for large values of A (normally consolidated clays) and small values of z/B (thin layers). At the other extreme, C_ρ may be as small as 0.2 for a thick layer of overconsolidated clay. For the former case we may calculate the settlement with Eq. (7.4) without serious error.

ONE-DIMENSIONAL METHOD. A practical consideration is that many compressible clays often occur between layers of stiff soils. Their presence restricts the lateral deformation of the compressible layer under the applied stresses. If the compressible clay cannot expand laterally at all it is under the same condition as the soil in a consolidation test. Hence Terzaghi and Peck (1967) and Taylor (1948) have recommended that Eq. (7.4) be used to calculate the settlement. In terms of the stress path, the settlement is calculated for the stress path ae [Fig. 7.3(c)]. The initial major principal stress is $\bar{p}_0$ and the final major principal stress is $\bar{p}_0 + \Delta\sigma_1$. The vertical compression of the soil element under the stress increment $\Delta\sigma_1$ is calculated from the pressure versus void ratio curve. If the reduction in void ratio is Δe, then

$$d\rho'_c = \frac{\Delta e}{1 + e_0} dz \qquad (7.6a)$$

For a normally consolidated clay, Δe is given by Eq. (5.3) and we may also use

$$d\rho'_c = \frac{C_c}{1 + e_0} [\log(\bar{p}_0 + \Delta\sigma_1) - \log \bar{p}_0] dz \qquad (7.6b)$$

The vertical settlement at a point on the ground surface is the result of compression of all the soil directly beneath this point. It is therefore equal to the summation of $d\rho_c$ over the entire depth H of the compressible layer beneath this point, or

$$\rho'_c = \frac{1}{1 + e_0} \int_0^H \Delta e \, dz \qquad (7.7)$$

In settlement calculations the integration is usually done numerically. The compressible stratum is divided into a finite number of layers. The increase in stress $\Delta\sigma_z$ at the middle of each layer is calculated by means of the influence chart (Fig. 6.5), and the value of Δe for each layer can be determined from the consolidation test data. The compression of each layer is computed with Eq. (7.6), and their summation gives the consolidation settlement at that point. This method further simplifies the settlement calculations. In addition, under the assumption that $\Delta\sigma_z = \Delta\sigma_1$ at all points in the compressible layer, we may also calculate the settlement at any point away from the centerline of the loaded area. The vertical stress increment at these points may be calculated by means of the influence chart.

Since the vertical stress on a horizontal plane beneath a loaded area varies as shown in Fig. 6.11(a), the consolidation settlement varies in a like manner and attains the largest value at the center of the loaded area.

7.4 Size effects

The relative behavior of areas of different sizes loaded to the same pressure intensity deserves attention for several reasons. As an illustration, the foundation of a structure may consist of footings of widely different sizes. If the design is to be based on a certain acceptable settlement, then the soil pressures should be such that the settlements of the individual footings are as nearly equal as possible. Hence the relative values of the settlement are important. As a second example, it is often necessary to use field load tests to predict the immediate settlement. Load tests are usually limited to test areas of small size. An understanding of the size effect is therefore essential if the load test results are to be extrapolated to predict the behavior of full-sized foundations.

Let us consider two footings with sizes B and nB loaded to equal intensity q (Fig. 7.5). We can assume without serious error that only stresses greater than

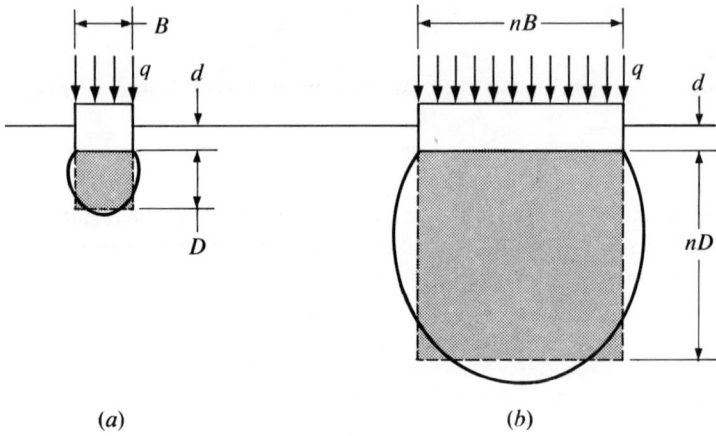

FIGURE 7.5 *Size effects.*

a certain value, say $0.2q$, produce any significant strains in the soil. Hence the soil that is strained lies within the stress bulbs representing that particular stress level. We may further simplify the problem by replacing the area inside the stress bulb with a rectangle whose dimensions are B and D [Fig. 7.5(a)]. For the footing whose width is nB, the size of the stress bulb and rectangle is proportionately larger, as shown in Fig. 7.5(b). If the soil modulus is constant with depth, we have for the two footings,

$$\rho_1 = \frac{Cq}{E} D \qquad \rho_2 = \frac{Cq}{E} nD$$

in which C is a constant. Hence,

$$\frac{\rho_2}{\rho_1} = n \tag{7.8}$$

We see that the settlement is proportional to the size of the footing. If the soil modulus increases in proportion to the depth, we may write $E = C'(d + z)$ as the soil modulus at any point z below the bottom of the footing. The average

SEC. 7.4 Size effects

modulus within the compressible rectangle is therefore $C'[d + (D/2)]$ for footing 1 and $C'[d + (nD/2)]$ for footing 2. The settlements for the two footings then become

$$\rho_1 = \frac{Cq}{C'[d + (D/2)]} D \qquad \rho_2 = \frac{Cq}{C'[d + (nD/2)]} nD$$

It follows that

$$\frac{\rho_2}{\rho_1} = n \frac{d + (D/2)}{d + (nD/2)} \tag{7.9}$$

This relationship is an approximation because of the simplifications introduced. However, it has an important advantage in that it allows the calculation of the ratio ρ_2/ρ_1 without having to determine the soil modulus.

Terzaghi and Peck (1967) on the basis of field load tests recommended the following size-effect relationship for sand:

$$\frac{\rho}{\rho_1} = \left(\frac{2B}{B + 1}\right)^2 \tag{7.10}$$

in which ρ_1 is the settlement of a footing 1 ft square and ρ is the settlement of a square footing of width B.

The size effects as defined by Eqs. (7.8) and (7.9) are compared in Fig. 7.6. Equation (7.8) is plotted as curve 1; curves 2 and 3 were obtained by using different values of d and D in Eq. (7.9). Thus, we see that Eq. (7.9) is useful only

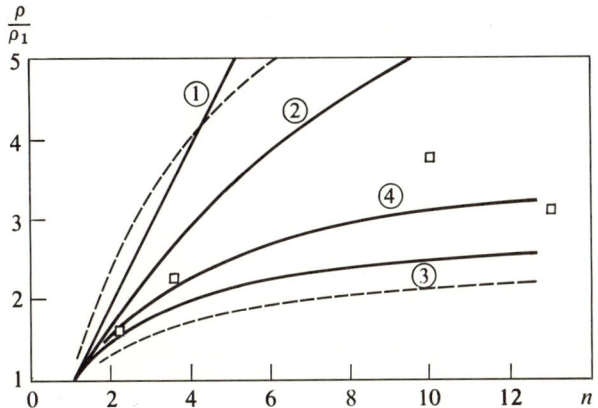

FIGURE 7.6 *Relationship between settlement and footing size.*

in the qualitative sense. The actual size-effect relationship may vary over a wide range depending on the relative values of d and D. Equation (7.10) is shown as curve 4 in Fig. 7.7, and the results of load tests used by Terzaghi and Peck to obtain Eq. (7.10) are shown in Fig. 7.6 as square points. Additional field data collected by Bjerrum and Eggstad (1963) show that the range of variation may be very large, as indicated by the dashed lines. The reason for this is not completely understood at present.

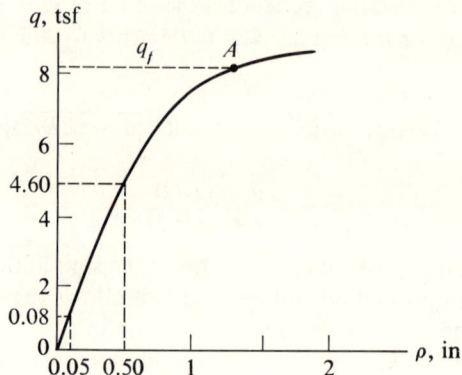

FIGURE 7.7 *Typical pressure settlement curve from a load test.*

As an illustration, Fig. 7.7 shows a typical pressure versus settlement curve obtained from a load test on a plate 1 ft square. Point A represents the ultimate bearing pressure. At greater pressures, the settlement increases rapidly and bearing capacity failure occurs. The settlement may be taken as being composed entirely of immediate settlement, since the process of consolidation requires much more time than the duration of the test. If the soil consists of clays, Eq. (7.8) shows that the immediate settlement of a footing 10 ft square will be 5.00 in. if loaded to a pressure of 4.60 tsf; the same pressure produced a settlement of 0.50 in. in the load test. If it is necessary to limit the immediate settlement of the footing to 0.50 in., then the soil pressure cannot be greater than the value that corresponds to a 0.05-in. settlement in the load test, which is equal to 0.80 tsf.

Example 7.1. The subsoil at a site consists of 10 ft of sand with a unit weight of 135 pcf. Beneath the sand is a layer of normally consolidated clay 10 ft thick. The clay rests on impermeable bedrock. The clay has a water content of 30 percent, a compression index of 0.15, and a consolidation coefficient of 1×10^{-3} in.^{2}/min. The water table is at the ground surface.

If the vertical stress at the middle of the clay layer is increased by 1500 psf by the construction of a structure, what will be the settlement due to consolidation of the clay?

Solution. To calculate the consolidation settlement assuming zero lateral strain we use Eq. (7.6b). For the 10-ft clay layer, we may take dz as being equal to 10 ft. The void ratio and unit weight may be calculated from the water content of the clay, assuming that the specific gravity is 2.70. We get $e = 0.81$ and $\gamma_{\text{sat}} = 121$ pcf. The initial overburden pressure (effective stress) at the middle of the clay layer is

$$\bar{p}_0 = (10)(135.0 - 62.4) + (5)(121.0 - 62.4) = 1019 \text{ psf}$$

Hence
$$p'_c = \frac{0.15}{1 + 0.81}\left(10 \log \frac{1019 + 1500}{1019}\right) = 0.33 \text{ ft}$$

If we wish to check the consolidation settlement by Eq. (7.5), we need to know the value of the porewater-pressure coefficient A. For a normally consolidated clay we estimate A to be about 0.7. If the loaded area is approximately circular in shape, Fig. 7.5(b) shows that the value of C_ρ is about 0.7, and

$$\rho_c = 0.7 \times 0.33 = 0.22$$

To calculate the change in the vertical effective stress during consolidation [points *of* and *og*, Fig. 7.2] we have

$$of = \bar{p}_1 + \Delta\sigma_1 - \Delta u = 1019 + 1500 - [600 + 0.7(1500 - 600)] = 1289$$
$$og = \bar{p}_1 + \Delta\sigma_1 = 1019 + 1500 = 2519.$$

Thus the stress changes from 1289 psf to 2519 psf. The settlement is

$$\rho_c = \frac{0.15}{1 + 0.81} \times 10 \log \frac{1519}{1289} = 0.24 \text{ ft}$$

Example 7.2. A large structure 100 ft square rests on a raft foundation at a depth of 6 ft below the ground surface as shown in Fig. 7.8. The design soil pressure is 2300 psf, assumed to be uniformly distributed over the entire foundation. The stiff clay is heavily overconsolidated by desiccation and may be considered as incompressible as far as consolidation is concerned. The soft clay is normally consolidated and has the same properties as the clay layer in Example 7.1. The average values of E measured by uniaxial compression with no lateral confinement are 450 tsf and 50 tsf for the stiff clay and soft clay, respectively.

Calculate the total settlement at the corner of the structure.

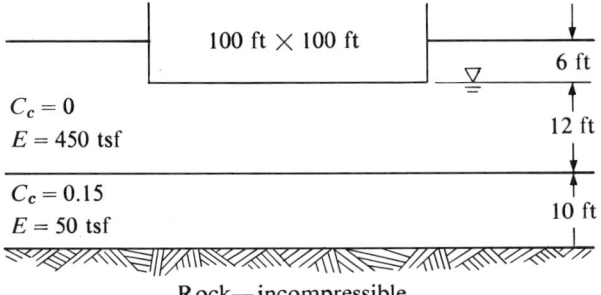

FIGURE 7.8 *Example 7.2.*

Solution. We calculate first the consolidation settlement of the soft clay layer. The initial overburden pressure is

$$\bar{p}_0 = 6 \times 138 + (12)(138 - 62.4) + (5)(121 - 62.4) = 2027 \text{ psf}$$

The net increase in pressure at the base of the foundation is the soil pressure of 2300 psf minus the weight of soil excavated for the basement:

$$q = 2300 - 6 \times 138 = 1472 \text{ psf}$$

The pressure increase at the middle of the clay layer is found by means of the influence chart (Fig. 6.5) to be

$$\Delta p = 1472 \times 50 \times 0.005 = 370 \text{ psf}$$

The consolidation settlement is

$$\rho'_c = \frac{0.15}{1 + 0.81}\left(10 \log \frac{2027 + 370}{2027}\right) = 0.063 \text{ ft}$$

The immediate settlement in the stiff clay may be computed by Eq. (6.22):

$$\rho_i = 0.42 \left(\frac{1472 \times 100}{900,000}\right) = 0.068 \text{ ft}$$

However, this is based on the assumption that the stiff clay extends to infinite depth. The presence of the soft clay layer at a distance of 12 ft below the base of the foundation is certain to increase the elastic settlement above 0.068 ft. This effect is difficult to calculate accurately. As an estimate we may consider the pressure exerted on the soft clay layer by the structure.

The vertical stress σ_z and the minimum horizontal stress σ_x at the middle of the soft clay layer are 370 psf and 89 psf, respectively. If we assume that these are the major and minor principal stresses, respectively, then the immediate settlement due to deformations in the 10-ft layer is

$$\rho_i = \frac{(370 - 89)}{100,000} \times 10 = 0.028 \text{ ft}$$

If the subsoil consists entirely of stiff clay, the contribution to the immediate settlement from the layer between depths of 18 and 28 ft may be estimated in the same way, or

$$\rho_i = \frac{(370 - 89)}{900,000} \times 10 = 0.003 \text{ ft}$$

Hence the difference 0.025 ft is the additional settlement due to the presence of the soft clay. The total settlement then becomes

$$\rho = 0.063 + 0.068 + 0.025 = 0.156 \text{ ft}$$

Example 7.3. Assuming the figures calculated in Example 7.2 to be correct, what would be the settlement after the structure is completed if the construction is considered to be rapid?

Solution. For rapid loading we may make the approximation that consolidation does not occur during loading. Hence at the end of construction, only the immediate settlement has occurred and the settlement is

$$\rho_i = 0.093 \text{ ft}$$

7.5 Settlement considerations in design

Settlement analyses are frequently made for projected construction. The objective is to predict whether excessive settlement is likely to occur and cause damage.

The deformation that is permissible is often controlled by the type of construction. If we consider the foundation of a structure, then the stresses introduced in the structure by the deformations constitute one of the criteria for determining the allowable settlement. Since the settlement of a loaded area is usually nonuniform, it is necessary to distinguish between the maximum settlement and the differential settlement. The latter is often more important as it produces distortions of the structure and other surface installations.

One can appreciate that the allowable settlement depends on the problem under consideration. In many cases the criteria for establishing this settlement are outside the realm of soil mechanics. Some common examples of allowable settlement are given in Table 7.1. For structures, the allowable differential settlement between two points may be expressed as some fraction of the distance L between the points (2, 3, and 4 in Table 7.1). Terzaghi and Peck (1967) have recommended that a maximum differential settlement of 0.75 in. be adopted for ordinary structures.

TABLE 7.1 *Values of Allowable Settlement*

Structure	Max. settl.	Diff. settl.	Ref.
1. Ordinary structures	1 in.	0.75 in.	Terzaghi and Peck (1967)
2. Framed structures	4 in.	$0.004L$[a]	Skempton and McDonald (1956)
3. Brick buildings		$0.002L$	Sowers (1962)
4. Turbogenerator foundations		$0.0002L$	Sowers (1962)
5. Flexible pavement	0.1 in.		State Hwy. Comm. Kansas (1947)

[a] L = distance between two points.

It is well to note that the available procedures for settlement computation are based on many assumptions that are often not satisfied in reality. Some of the complications that are frequently encountered in practice are outlined below.

The computation of the consolidation settlement with Eq. (5.3) assumes that the soil is compressed one dimensionally in the vertical direction. This condition is approached in the field if the loaded area is very large compared to the thickness of the compressible layer. If the Boussinesq equations are used to compute the stress distribution in the soil, the nonhomogeneity of the natural soil mass is ignored. Frequently, a stiff layer overlies the compressible clay and it may exert considerable influence on the stress distribution (Sec. 6.6). When necessary, numerical methods may be used to evaluate the effects of a stiff layer and some examples have been given by Mitchell and Gardner (1971).

If the settlement of a structure is calculated from the design column loads, the tacit assumption is that the settlement of the foundations, which is usually nonuniform, has no effect on the column loads. This is only correct if the structure is perfectly flexible. For a structure that possesses some degree of rigidity, unequal settlement of the foundation brings about a redistribution of the column loads. The effect of structural rigidity is to reduce the differential settlement. This problem has been investigated on the assumption that the structure is elastic [Taylor (1948), Chamecki (1956), DeJong and Morgenstern (1971)]. Table 7.2 shows the calculated settlements with the structural rigidity taken into consideration compared with that calculated for a perfectly flexible structure.

TABLE 7.2 *Effect of Structural Rigidity on Calculated Settlement*

A. Four-story reinforced concrete structure, 20 ft × 20 ft bays (Taylor, 1948):
    Max. ρ_c = 3.2 in.; min. ρ_c = 1.8 in.   No structural rigidity
    Max. ρ_c = 2.9 in.; min. ρ_c = 1.9 in.   Considering structural rigidity
B. Three-story framed structure, 6 m × 4 m bays (Chamecki, 1956):
    Max. ρ_c = 11.7 cm; min. ρ_c = 5.5 cm   No structural rigidity
    Max. ρ_c = 10.4 cm; min. ρ_c = 6.0 cm   Considering structural rigidity

In view of the complications outlined above and other uncertainties, such as those introduced by sample disturbance and natural variations in soil properties, it is important that we realize the approximate nature of settlement computations. We may assess our design methods by systematic comparison of predicted performance against actual performance observed in the field. Through such comparisons the designer obtains a perspective on the reliability of the tools at his disposal. Our experiences with settlement predictions for foundations on clay and on sand are described in the sections that follow.

7.6 Field observations of foundations on clay

A good overall picture of the settlement problem has developed from many careful observations on the behavior of structures in the field, pioneered by the early studies of Terzaghi (1937). Of great interest is the measured settlement of eight structures located over a thick deposit of soft clay in downtown Chicago (Peck and Uyanik, 1955). The extensive program of borings and laboratory tests and the high degree of uniformity of the subsoil conditions reduced to a minimum the uncertainties concerning the soil properties. The consolidation settlement was calculated by Eq. (7.7) and the immediate settlement was not included. The ratio of calculated to measured settlements has an average of 0.85. Thus under favorable conditions the errors introduced by the various factors are of the order of 15 percent.

The measured and computed settlement contours for the Masonic Temple Building are shown in Fig. 7.9(a) and (b). The magnitude of computed and measured settlements are in good agreement. Note also that the observed pattern of settlement distribution over the entire building is very well duplicated

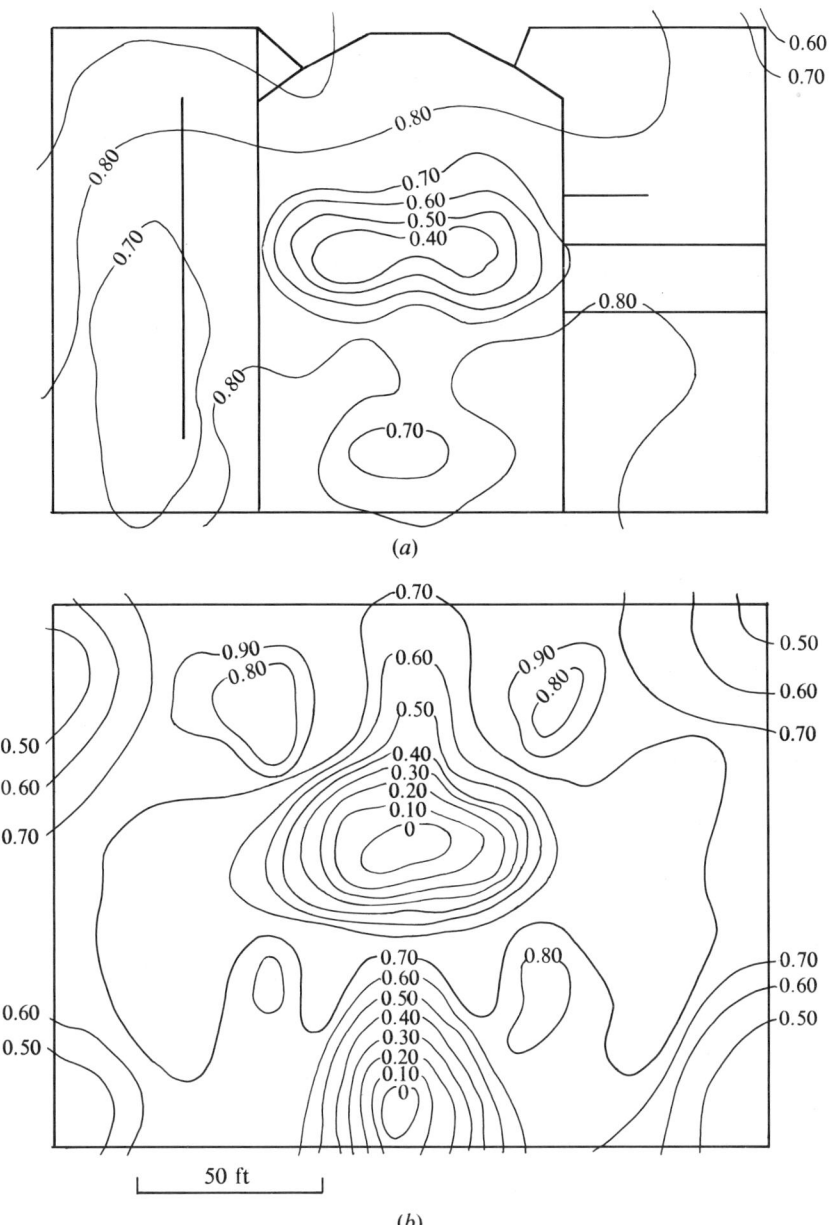

FIGURE 7.9 *Calculated and measured settlements (ft), Masonic Temple, Chicago.* [After Peck and Uyanik (1955).]

by the calculations (compare the area of small settlement near the center of the building, for example). The differences are relatively minor in nature. Figure 7.10 shows the calculated and measured settlement profile along a section of the Chicago Auditorium. It illustrates clearly the effect of the stiff upper clay and the structural rigidity. The effect is to smooth out the settlement profile, although the maximum and minimum settlements are close to the calculated values.

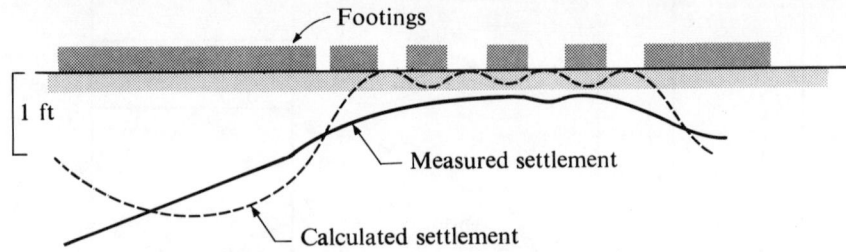

FIGURE 7.10 *Calculated and measured settlements, Chicago Auditorium.* [After Peck and Uyanik (1955).]

Skempton and Bjerrum (1957) calculated the settlements of four structures on overconsolidated clays—three of the Chicago buildings and an oil tank on soft clay—by Eqs. (7.5). In addition, the immediate settlement was calculated with Eq. (6.22a), using the elastic modulus measured in unconfined compression tests. The results are summarized in Table 7.3. Where the immediate settlement was measured, it is in reasonable agreement with the calculated value. In six of the structures it can not be separated from the consolidation settlement. However, in these cases the total calculated and measured settlements do not differ by more than 15 percent.

TABLE 7.3 *Settlement of Structures on Clays*

	Calculated			Measured		
Structure	ρ_i, in.	ρ_c, in.	Total, in.	ρ_i, in.	ρ_c, in.	Total, in.
Isle of Grain, oil tank	3.0	15.0	18.0	2.0	19.0	21.0
Masonic Temple, Chicago	3.0	8.0	11.0	—	—	10.0
Monadnock Block, Chicago	6.0	14.5	20.5	—	—	22.0
Auditorium, Chicago	6.5	19.5	26.0	—	—	24.0
Fire testing station, Elstree	0.25	0.35	0.60	—	—	0.7
Chelsea Bridge, London	0.80	1.65	2.45	—	—	2.1
Waterloo Bridge, London	0.90	2.60	3.50	—	—	3.7
Peterborough grain silo	0.35	0.55	0.90	0.25	0.65	0.90

While the cases described in the preceding sections and many others show good agreement between predicted and observed settlement, there are also a number of exceptions where the predicted and observed consolidation settlements are very different. In several cases [Elias and Storch (1970), Rico et al. (1969), for example] the predicted settlement is much larger than the observed value. This seems to happen more frequently with small settlements and has been attributed to sampling disturbance [Raymond (1972), Bjerrum (1972)].

For the prediction of the immediate settlement, the use of the modulus from unconfined compression tests has led to an overestimate of the settlement in several cases [Crawford and Burn (1962), Bozozuk (1963), Serota and Jennings (1959)]. A better estimate of the modulus is obtained if the soil is first consolidated in the triaxial cell under the initial overburden pressures and then loaded [Ladd (1964), Bozozuk and Leonards (1972)].

There remain various situations in which the loadings are not represented by the stress path described in Fig. 7.4 and for which the equations developed in Sec. 7.3 are not suitable. Such an example was given by Lambe (1964) for the consolidation of the clay beneath the Amuay Dam. The embankment of the Amuay Dam was constructed in stages and the field stress path for point B in the clay layer is shown in Fig. 7.11(b). This differs considerably from the K_0-line. The stress path used in a laboratory test is shown as a dashed curve in the figure (the details are given in the paper by Lambe). Under this loading, the vertical

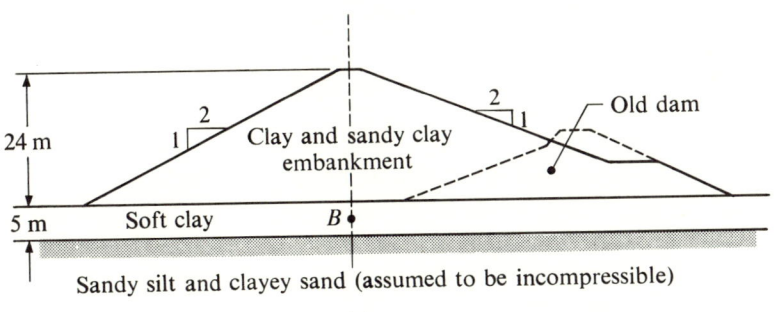

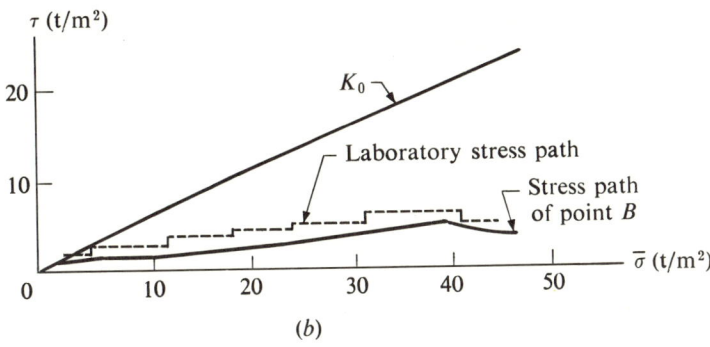

FIGURE 7.11 *Settlement of Amuay Dam.* [After Lambe (1963).]

strain in the clay layer should be close to the vertical strain measured in the laboratory specimen. Then the settlement can be calculated by Eq. (6.23) in which ϵ_z is the strain of the laboratory specimen. Table 7.4 shows the settlements calculated by the three methods and the measured settlement of the Amuay Dam. In this case the stress path method gives a better estimate of the settlement. Thus we should note that while in most foundation problems, the K_0-line is a reasonably good approximation to the field stress path, other loading conditions may depart from the K_0 condition to an extent that this assumption becomes unsatisfactory.

TABLE 7.4 *Settlement of Amuay Dam, Venezuela*[a]

Method	Elastic settlement	Consolidation settlement	Total settlement
Eq. (6.21)	—	94 cm	—
Eq. (6.23)	—	95 cm	—
Eq. (6.27)	14 cm	—	—
Stress path	5 cm	41 cm	46 cm
Measured	—	—	45 cm

[a] After Lambe (1964).

RATE OF SETTLEMENT. The prediction of the rate of settlement from consolidation test data is more complicated. If the load stresses a small part of a clay layer (as is usually the case), then the stress distribution in the compressible layer is nonuniform in the horizontal direction [Fig. 6.11(a)]. Immediately under the load the stress is largest, and so is the excess hydrostatic porewater pressure. With increasing distance from the load, the stress and excess hydrostatic porewater pressure decrease. Thus there is radial drainage within the compressible layer. Water movement takes place horizontally from the zone of high porewater pressure immediately beneath the load to surrounding parts, where the porewater pressure is much smaller. This occurs in addition to the vertical flow toward free-draining layers analyzed in Sec. 5.4 (see Sec. 12.6). The effect is to increase the rate of settlement above that given by Eq. (5.13). Thus the accuracy of the one-dimensional consolidation theory in predicting actual settlement rate in the field depends on the amount of radial drainage. The calculated and observed settlement rate may be in good agreement, as in the case of the Chicago Auditorium (Fig. 7.12). In the case of the Hayden Library [Aldrich (1952)] the value of c_v for field consolidation (assuming one-dimensional consolidation) is 15 times the value measured in the laboratory. Although the calculations are rather involved, it is possible to analyze combined radial and vertical drainage. Three-dimensional consolidation was first studied by Biot (1941). For results of numerical solutions see McNamee and Gibson (1960), Schiffman et al. (1967), and Christian et al. (1972).

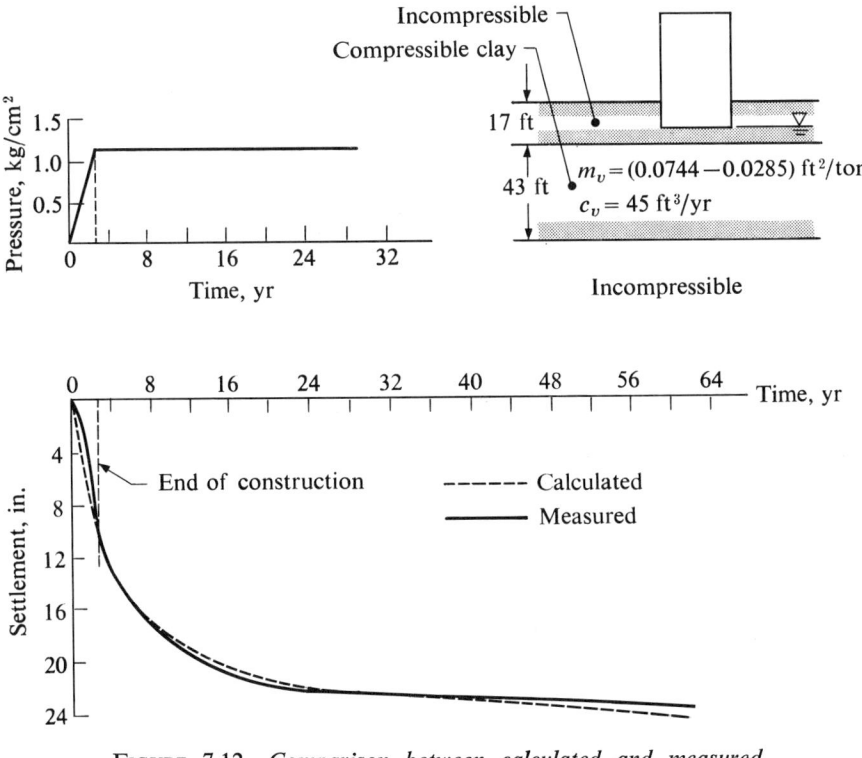

FIGURE 7.12 *Comparison between calculated and measured rates of settlement, Chicago Auditorium.* [After Skempton and Bjerrum (1957).]

Opportunities for comparison of predicted and observed settlement due to secondary consolidation are less frequent. However, from the few reported cases [Su and Prysock (1972), Crawford and Sutherland (1971)] the observed rates of settlement are comparable to the rates of secondary consolidation measured in the laboratory consolidation test.

7.7 Field observations of foundations on cohesionless soils

Because of difficulties encountered in the sampling and testing of undisturbed samples of cohesionless soils there have been only a few cases in which the immediate settlement could be predicted on the basis of theory. From a practical viewpoint a very important characteristic of immediate settlement is that the strains in the soil leading to the settlement occur in the zone immediately underneath the loaded area. This may be understood by a study of the calculated stresses shown in Figs. 6.10 and 6.11. The stresses are highest immediately beneath the load and decrease rapidly with depth. At a depth equal to twice the width B of the load, the vertical stress σ_z is only about one-fourth of the applied pressure. Hence, for general discussion, we may say that the settlement is primarily a result of the strains in a zone down to a depth of $2B$. It follows then that the soil properties in this zone exert the largest influence on the immediate settlement.

Figure 7.13 shows the measured settlements along the wall of a building supported by a continuous footing foundation on sand. In sharp contrast to the consolidation settlement, this profile shows no regular patterns. Since the load is uniform along the length of the wall, the variations in settlement may be attributed to variations in soil properties. Studies on natural soil deposits confirm the fact that most deposits of sand are highly variable in their properties (Chapter 13). Hence our second observation is that in the case of an erratic subsoil, such as a deposit of sand, the variations in soil properties are likely to be very great. Under such conditions settlement calculations become meaningless until the general pattern of the variations in soil property can be determined.

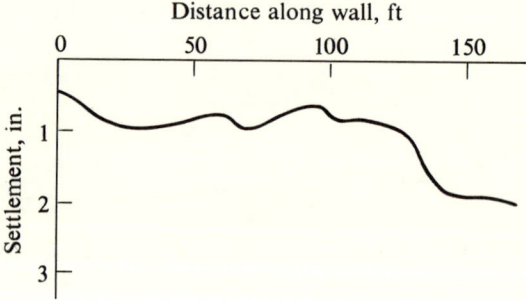

FIGURE 7.13 *Measured settlement along a continuous foundation resting on a deposit of sand.*

EMPIRICAL CORRELATIONS. Difficulties in calculating the immediate settlement in cohesionless soils has led to the development of a number of empirical methods. To be meaningful, the empirical method should circumvent the obstacles to the theoretical solution and yet correlate the settlement with the relevant soil properties. The greatest unknown is the variation in the soil properties. Hence, if the load is to cover a considerable area, some means must be used to establish the pattern of the variations. The most important soil property is the soil modulus. For cohesionless soils investigations have shown that it is closely related to the relative density [Chen (1948), Domasckuk and Wade (1969), for example]. In the United States, the spoon penetration test described in Chapter 13 is commonly used to measure the relative density in the field. Since the penetration resistance of granular soils is closely related to the relative density and soil modulus, it is possible to establish a correlation between penetration resistance and immediate settlement. Furthermore, the penetration test is inexpensive, and large numbers of them can be made to develop a general picture of density variations at the site. Thus it meets both requirements stated earlier in this paragraph.

On the basis of measured settlements of footings on sands of various relative densities, Terzaghi and Peck (1967) arrived at the empiricial correlation in Fig. 7.14. It gives the bearing pressure as a function of the penetration resistance and the size of the footing for a settlement of 1 in. The empirical data also provided information on the variability of soil properties. It was found that for footings of equal size and loaded to equal pressure, the differential settlement caused by variations in the density does not exceed one-half the largest settlement of an individual footing. The application of Fig. 7.14 to design is illustrated as follows.

The values of penetration resistance in each boring to a depth of $2B$ below the base of the foundation are averaged and the lowest value among the averages is used for design. Since the largest footing will settle the most, the width of the largest footing is entered in Fig. 7.14 and the pressure corresponding to the minimum average penetration resistance of the soil is determined. This means that the largest footing when located over the most unfavorable spot would not settle more than 1 in. A footing of same size located over the densest sand would settle at least 0.5 in., resulting in a differential settlement of at most 0.5 in. If in addition the smallest footing happens to be located over the densest sand, its settlement would be less than 0.5 in. This settlement may be estimated by the size-effect relationship given by Eq. (7.9) and Fig. 7.7. Considering an extreme range in foundation size between 5 ft and 20 ft, the settlement of the small footing is about two-thirds that of the large one. Hence, in this case, the smallest footing would settle at least 0.3 in. The differential settlement therefore should not exceed 0.70 in. Field data on the relation between maximum settlement and maximum differential settlement have been presented by Bjerrum (1963).

Figure 7.14 is designed for footings on unsaturated sand. If the groundwater is located closer than a distance B beneath the base of the footing, its influence on settlement should be taken into account. The effect of saturation is to reduce the vertical and horizontal effective stresses by approximately one-half. Consequently, the compressibility of the soil and also the settlement increase by a factor of approximately 2. The pressure for a settlement of 1 in. as given in the figure should therefore be reduced by one-half.

Recent observations [Meyerhoff (1956), D'Appolonia et al. (1968), and others] have shown that the Terzaghi-Peck method overestimates the settlement in the majority of cases. The ratio of predicted to measured settlement ranges between 1 and 9 with an average of about 3.

Such empirical correlation is less successful for foundations at large depths below ground surface. However, several interesting cases of field tests have been recorded [see Terzaghi and Peck (1967), pp. 565–66]. They show that for sands of equal density the settlement of a deep foundation is about one-half that of a shallow foundation. On the basis of these observations, the design pressure for deep foundations on sand may be taken as twice the values given in Figure 7.14 if a settlement of 1 in. is acceptable.

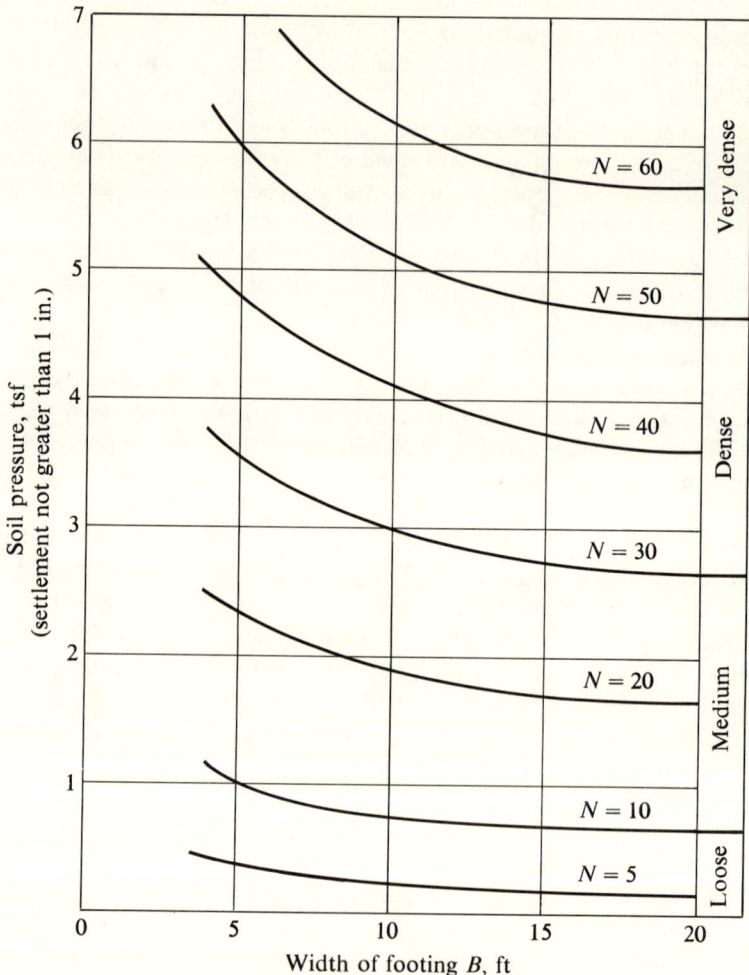

FIGURE 7.14 *Design pressures for footing foundations on sand, with a maximum tolerable settlement of 1 in.* [After Peck et al. (1974).]

It is obvious that other methods may also satisfy the requirements for empirical correlation stated earlier in this section. A common method in Europe is the cone penetration described in Sec. 13.13, and there is also data correlating cone penetration resistance with spoon penetration resistance [see Meyerhoff (1956) and Schmertmann (1970)]. For the purpose of settlement computation the correlation between the cone resistance and standard penetration resistance is given approximately as

$$q_c = 4N \tag{7.11}$$

PROBLEMS

7.1 A large foundation 30 ft square carrying a fluctuating load is to be constructed over a deposit of compressible clay. The steady load will produce a vertical stress of 1000 psf at the middle of the clay layer. Once in every 9 months the stress is expected to rise to 3000 psf. The duration of the higher stress will not exceed 3 days. If the clay layer is 10 ft thick and the clay has the properties given in Problem 5.2, what will be the probable settlement of the foundation by consolidation?
Answer: 0.06 ft

7.2 For the foundation in Problem 7.1, which of the stresses should be used to calculate the immediate settlement? Why?
Answer: 3000 psf

7.3 For the foundation in Problem 7.1, which of the stresses should be used to calculate the immediate settlement if the subsoil consists of cohesionless sand? Why?
Answer: 3000 psf

7.4 Heavy machinery and equipment with a total weight of 500 tons is to be supported on a rigid concrete mat 16 ft square located at a depth of 8 ft below ground surface. The subsoil consists of dense sand to a depth of 24 ft. The sand has a unit weight of 130 pcf. The water table is at a depth of 24 ft. Below this is a layer of clay 16 ft thick overlying bedrock. The clay has the following properties: LL = 45%; PL = 20%; w = 30%; γ_{sat} = 121 pcf; C_c = 0.16. (a) Calculate the maximum and differential settlement due to consolidation of the clay assuming the foundation to be perfectly flexible. (b) Estimate the settlement due to consolidation of the clay considering the rigidity of the foundation.
Answer: 1.57 in.; 0.50 in.; 1.30 in. approximately

7.5 At the site described in Problem 7.4 a footing 6 ft square is loaded to the same pressure as the mat (4000 psf) and is also located at a depth of 8 ft. (a) Calculate the maximum and differential settlement due to consolidation of the clay assuming the foundation to be flexible. (b) Comparing the answer in (a) with that in Problem 7.4(a), what conclusion can be drawn regarding the settlement of two footings with different sizes but loaded to equal pressures due to a soft stratum located at some depth below the ground surface?
Answer: 0.31 in.; 0.05 in.; settlement of a small foundation is much less even though the two footings have the same soil pressure.

7.6 A proposed structure has column loads ranging between 50 to 200 tons. The subsoil consists of sand of medium density with an average penetration resistance equal to 20. The water table is at a depth of 25 ft. (a) What should be the design pressure on the footings if the maximum and differential settlements are not to exceed 1 in. and 0.75 in., respectively? (b) What should be the design pressure on the footings if the maximum and differential settlements are not to exceed 0.75 in. and 0.50 in., respectively?
Answer: 3800 psf; 2500 psf

7.7 (a) Calculate the immediate and consolidation settlements for the foundation in Fig. 7.15 under a net pressure increase of 2000 psf at the foundation level.

(b) Explain why the more compressible clay 2 contributes less to the settlement than the stiffer clay 1.

7.8 In Problem 7.7, if the pressure is applied instantaneously: (a) What would be the settlement immediately after the pressure is applied? (b) What would be the settlement 3 years after the pressure is applied? (c) Explain the difference.
Answer: approx. 0.025 ft; approx. 0.447 ft; consolidation requires time

7.9 In Fig. 7.15, if clay 1 is 60 ft thick and clay 2 is 10 ft thick: (a) Will the contribution of clay 1 to the settlement be greater or less than that computed in Problem 7.7? Why? (b) Will the contribution of clay 2 to the settlement be greater or less than that computed in Problem 7.7? Why?
Answer: greater; less

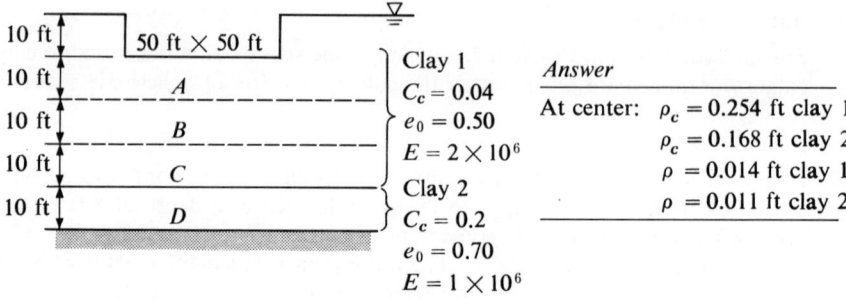

FIGURE 7.15

7.10 In Fig. 7.15, if it is necessary to estimate the settlement to the closest 0.05 ft, at what depth must clay 2 be located if its contribution to the settlement is to be negligible?
Answer: approx. 140 ft

7.11 A large machine foundation, carrying a load of 200 tons, is to be supported on a spread footing located at a depth of 9 ft below ground surface. The subsoil consists of moist medium to coarse sand. In order to predict the settlement of the foundation, a load test was made with the following results:

Load, lb	Settlement, in.
1,000	0.075
2,000	0.15
4,000	0.30
6,000	0.75
8,000	2.00
10,000	6.00

The load test was made on a plate 1 ft square inside a test pit 10 ft square and at a depth of 9 ft: (a) What should be the design soil pressure on the foundation if the settlement is not to exceed 1 in.? (b) What condition (if any) in the load test differs from that of the actual foundation? (c) Is the error resulting from

this difference on the safe or unsafe side? (d) If the density of the sand decreases below a depth of 17 ft, how does it affect the results? Explain.
Answer: 4000 psf: load test has no surcharge; safe side; actual foundation will settle more than calculated value

7.12 Determine the design soil pressure for footing foundations to be constructed at a site with subsoil conditions as shown in Fig. 7.16. The footings are to support column loads that range from 50 to 100 tons.
Answer: 1 tsf for a maximum settlement of 1 in. or less

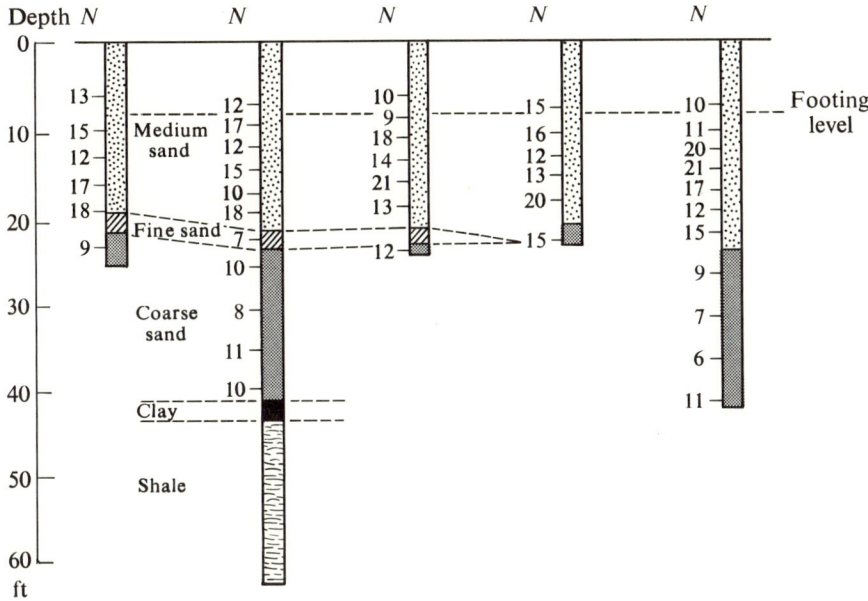

FIGURE 7.16

7.13 If the soil in Problem 6.10 is a saturated clay with the water table at the ground surface, construct the stress paths for points *a*, *b*, and *c*, assuming that the load is applied rapidly so that no consolidation occurs during loading. The unit weight of the soil is 125 pcf and the coefficient of earth pressure at rest is 0.6.
Answer: Fig. 7.17

7.14 If it is required to determine the settlement in Problem 7.13 by the stress path method, what stresses should be used in a triaxial test to simulate the stress conditions at *a*?
Answer: for a point *a*, consolidate sample under $\bar{\sigma}_1 = 310$ psf, $\bar{\sigma}_3 = 190$ psf, then increase stresses to 970 psf and 440 psf

7.15 If the specimen is loaded as dexcribed in Problem 7.14, do you expect the measured pore pressure to be greater or less than that computed in Problem 6.11? Why?

7.16 If the sample in Problem 7.14 suffered an axial strain of 0.03 under the applied stresses, what is the compression of layer A in Fig. 6.22?
Answer: 0.3 ft

7.17 What difficulties are encountered when one uses the triaxial test to duplicate the stress conditions at point a and point c (Fig. 7.17)?

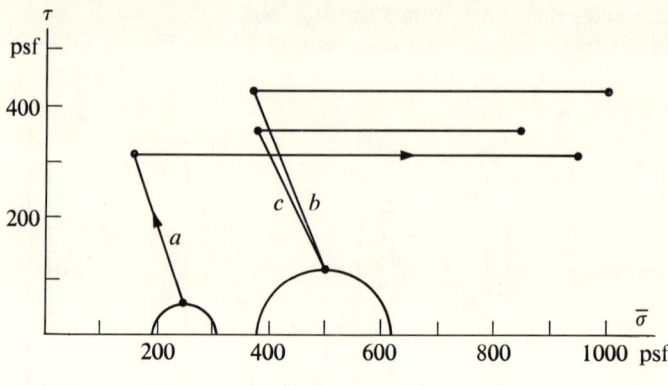

FIGURE 7.17

8

Shear Strength

8.1 Failure theories

A theory of failure is in essence a definition of the conditions that will produce failure in the material. We are accustomed to thinking of failure taking place when a material is subjected to a certain stress usually called strength (i.e., tensile strength). However, other criteria may also be valid. For instance, a material may fail after it has been deformed to some specific strain or after it has absorbed a certain amount of energy from the loading process.

Failure can occur in the form of fracture, in which the material disintegrates at a certain stress or strain. An example of this is the stress-strain curve shown in Fig. 13.12. Failure may also take place as yielding or plastic flow. In this case, the strain increases indefinitely at constant stress, as illustrated by the curves in Fig. 4.13(d).

In soil mechanics the most successful failure theories to date define failure in terms of stress. In a continuous medium the material is subjected to three principal stresses (see Sec. 4.1); hence, the failure criterion should be expressed as a function of all three principal stresses. It is reasonable to expect that each of the three principal stresses would contribute toward failure (contrast this with the case of simple tension, in which a single value defines the failure stress which is the tensile strength).

8.2 Mohr-Coulomb theory of failure

The Mohr-Coulomb failure theory has been found to be very successful in defining failure in soils. The theory states that failure in a material occurs if the shear stress on any plane equals the shear strength of the material. Furthermore, the shear strength s, along any plane is a function of the normal stress σ on that plane, or

$$s = f(\sigma) \tag{8.1}$$

This function is shown as a curve on the normal versus shear stress plot [Fig. 8.1(a)]. Coulomb (1776) defined the function f as a linear function of the normal stress. Equation (8.1) then becomes

$$s = c + \sigma \tan \phi^* \tag{8.2}$$

The Coulomb criterion is shown as a straight line in Fig. 8.1(b), with an intercept on the τ axis equal to c and a slope equal to $\tan \phi$. The quantities c and ϕ are material properties frequently called *cohesion* and *angle of internal friction*, respectively. The shear strength as defined by Eqs. (8.1) and (8.2) represents the maximum shear stress that may be sustained on any plane in a given material. Any combination of shear and normal stresses that plots as a point below the strength function [points a and b, for instance, in Fig. 8.1(b)] represents a safe state of stress, whereas a point on the line [point c, Figure 8.1(b)] represents stresses that will result in failure of the material. Stresses that plot above the line [point d, Figure 8.1(b)] cannot exist in this material because failure will have taken place before such stresses can be attained. Hence the strength function defines the limiting stress and it is often called the *failure envelope*.

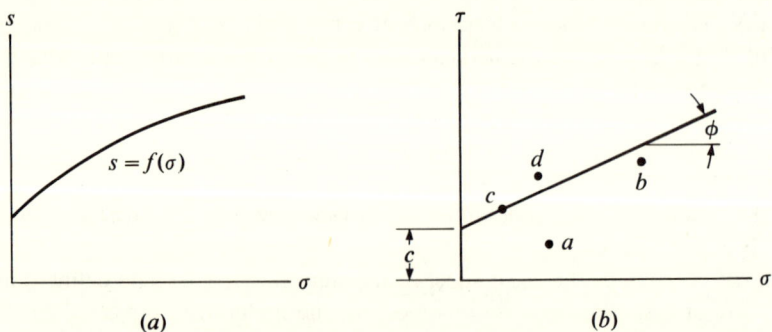

FIGURE 8.1 *Failure envelopes.*

If an element of material is acted on by major and minor principal stresses equal to σ_1 and σ_3 [Fig. 4.1(a)], the normal and shear stresses on any plane making an angle θ with the major principal plane are given by equations (4.4). These stresses may also be calculated graphically by means of Mohr's circle of stress. A Mohr's circle with principal stresses σ_1 and σ_3 is plotted in Fig. 8.2. It is seen that the values of the principal stresses are such that the circle is tangent

---

* This failure theory is given detailed treatment because it forms the basis of the analysis of plastic equilibrium (Chapter 9). It is not the only failure theory for soils. Others are presented in Sec. 8.11.

to the curve at point A. Therefore, on this particular plane, the shear stress τ is equal to the shear strength as defined by the strength function, and failure occurs. On the other hand, if the principal stresses are σ_1' and σ_3', every point on the Mohr's circle lies below the curve. On no plane does the shear stress equal the shear strength, and failure does not occur. Since the stress in a material can never exceed its strength, it follows that no part of a Mohr's circle can project above the envelope. All circles representing stresses at failure are tangent to the envelope.

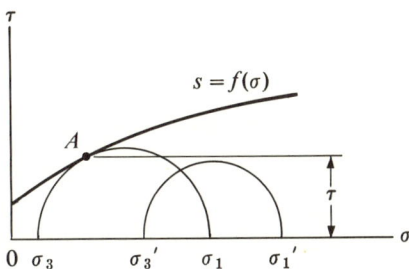

FIGURE 8.2 *Mohr's circle and failure envelope.*

It should be noted that according to the Mohr-Coulomb theory, the shear strength is independent of the intermediate principal stress σ_2.

8.3 Direct shear test

DRAINED SHEAR STRENGTH. The primary objective of strength measurement is to determine the failure envelope, which is the relationship between τ and σ at failure. (See also Sec. 4.7.) In the drained test, the loads are applied so slowly that the excess hydrostatic porewater pressure developed by the loads (Sec. 4.6) is dissipated by consolidation. Hence all the stresses are effective stresses.

In the direct shear test a normal stress $\bar{\sigma}_z$ is applied to the soil sample which is then subjected to an increasing shear stress τ_{xz} [Fig. 4.12(a)] until failure is reached. This gives the stresses $\bar{\sigma}_z$ and τ_{xz} at failure [Fig. 8.3(a)]. Consider a test in which the normal stress is $\bar{\sigma}_z$ [point a, Fig. 8.3(b)]. As the shear stress is steadily increased to failure, the normal and shear stresses on the surface ab change from that of point a to that of point b. The line ab which describes the change in the stress state during the test is called the stress path.

To obtain the failure envelope, several tests utilizing different normal stresses are performed on specimens of the same soil. The values of τ_{xz} at failure are plotted against the values of $\bar{\sigma}_z$. The samples shown in Fig. 8.4(a) simulate a soil consolidated in the field under an overburden pressure of 30 psi. The added loads and shear stresses at failure are shown in (b) and (c). The stresses at failure are shown as points in Fig. 8.4(f). The line that connects the points expresses the relation between $\bar{\sigma}$ and τ at failure and is the failure envelope.

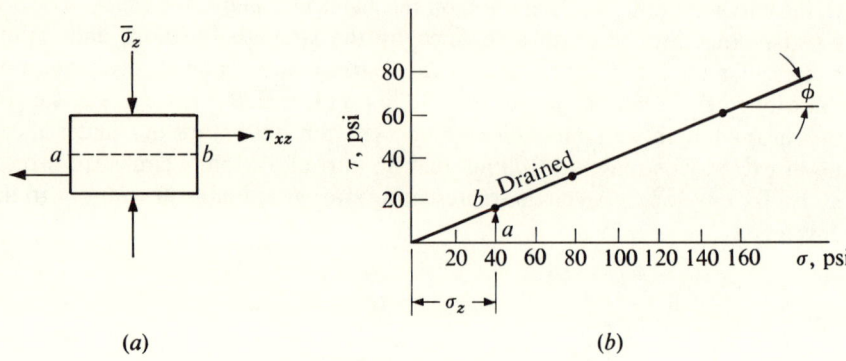

FIGURE 8.3   *Direct shear test.*

The drained condition has two important applications. For coarse-grained, cohesionless soils such as sands and gravels, the permeability is large enough so that under ordinary rates of loading (such as the construction of a building) the soil is always fully consolidated. Hence the drained condition applies. When a sample is tested under the drained condition, the excess hydrostatic porewater pressure is zero and the measured stresses represent effective stresses.*

For cohesive soils such as clay, the permeability is low and the dissipation of the excess hydrostatic porewater pressure requires some time. Hence, the drained condition is attained only if the load is applied very slowly or at a considerable time after loading. Under these two conditions the shear strength as defined by

$$\bar{s} = \bar{c} + \bar{\sigma} \tan \bar{\phi} \qquad (8.3a)$$

may be used. The bar above c and ϕ denotes the condition that the stress is $\bar{\sigma}$.

UNDRAINED SHEAR STRENGTH. To prevent volumetric strain in saturated clayey soils the direct shear test may be performed at a rapid rate so that the time duration is too short for any appreciable amount of water to flow into or out of the sample. This is the undrained state and excess hydrostatic porewater pressures are usually developed in the soil.

Consider the following example. The initial field condition is represented by the effective stresses shown in Fig. 8.4(a). The loading due to construction [Fig. 8.4(d) and (e)] is applied rapidly so that the samples fail without undergoing any volume change. We note that immediately after the application of a normal stress of 10 psi, an excess hydrostatic porewater pressure of 10 psi develops in the soil (Sec. 5.4). Hence the effective stress $\bar{\sigma}_z$ remains at 30 psi (point a'). When the shear stress is applied, the excess hydrostatic porewater pressure is further increased (Sec. 4.6), and at failure the pore pressure is 13 psi. The effective

---

* Note that in the laboratory test the hydrostatic pore pressure is almost zero because the pressure head is very small.

stress path is line A [Fig. 8.4(*f*)]. Note that point *a* which denotes the effective stresses at failure falls on the failure envelope.

Following the same argument, the effective stress $\bar{\sigma}_z$ remains at 30 psi after application of a normal stress of 48 psi. Hence the effective stress path for this test is also A. All the tests in Fig. 8.4(*d*) and (*e*) have the same effective stress path and the shear strength is the same.

In practical problems it is sometimes difficult to predict the pore pressure

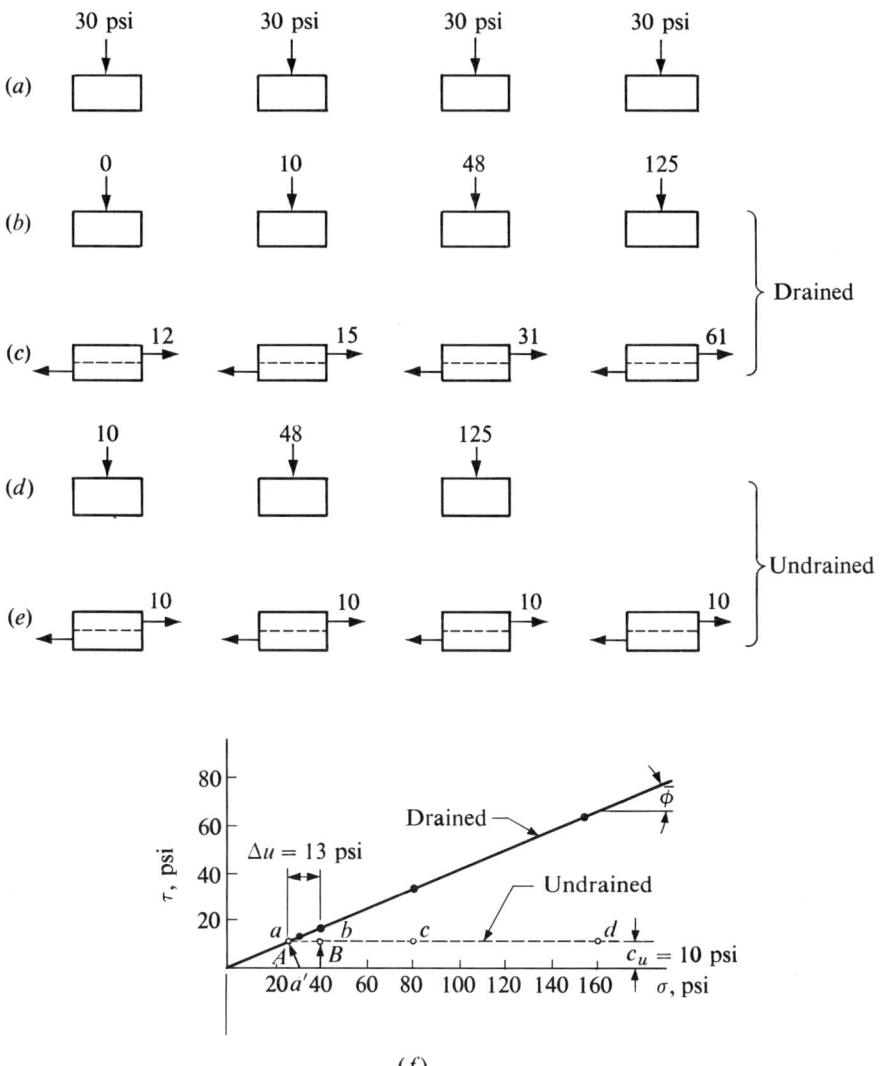

FIGURE 8.4 *Typical results of direct shear tests on a saturated clay.*

Δu. For simplicity the analysis may be made in terms of total stress. Consider again the examples in Fig. 8.4(a), (d), and (e). The measured shear strength in all three specimens is equal to 10 psi. This is the same as the value obtained in a specimen consolidated under the initial overburden pressure of 30 psi. Thus the shear strength is constant irrespective of the stress σ_z. If the total stresses of the three tests in Fig. 8.4(d) and (e) are plotted on the Mohr diagram, the points b, c, and d are obtained. The undrained shear strength may be expressed as

$$s_u = c_u, \quad \phi_u = 0 \tag{8.4}$$

The subscript u denotes the undrained state.

For saturated soils that contain an appreciable amount of clay, almost no consolidation takes place during construction with ordinary rates of loading. Hence the undrained shear strength is applicable to conditions during and immediately after construction.

8.4 Triaxial test

DRAINED SHEAR STRENGTH. The triaxial cell is shown in Fig. 4.13. In the first step [Fig. 8.5(a)] the fluid pressure acts as a hydrostatic compression. Then the stress ($\sigma' = \bar{\sigma}_1 - \bar{\sigma}_3$) is slowly increased until failure of the specimen occurs [Fig. 8.5(b)]. The excess hydrostatic porewater pressure is allowed to dissipate during loading and at failure the principal stresses are $\bar{\sigma}_1$ and $\bar{\sigma}_3$. The shear stress on the plane of failure is not measured directly. A photograph of a failure plane is shown in Fig. 8.5(d). Frequently shear displacement is widespread throughout a zone instead of being concentrated along a single plane. Then the specimen bulges.

To obtain Mohr's envelope, several triaxial tests are usually performed on specimens of the same material utilizing different values of cell pressure $\bar{\sigma}_3$. A Mohr's circle is drawn for the principal stresses at failure for each specimen. These circles are shown in Fig. 8.5(c); the envelope to these circles constitutes the failure envelope. The stresses on the failure surface are represented by the point of tangency a. From the geometry of Mohr's circle we know that this plane makes an angle of $\frac{1}{2}(\pi/2 + \phi)$ with the major principal plane.

We may trace the stress change that takes place on the failure plane as the specimen is gradually loaded to failure. Under the hydrostatic pressure $\bar{\sigma}_3$, the normal and shear stresses $\bar{\sigma}$ and τ on this plane are $\bar{\sigma}_3$ and 0, respectively. This state is represented by point b [Fig. 8.5(c)]. When the axial stress is increased to any value $\bar{\sigma}_1$, the normal and shear stresses on the failure plane may be calculated by means of Eq. (4.4) with θ equal to (45° + $\phi/2$). The τ and $\bar{\sigma}$ on this plane may be calculated for a series of values of $\bar{\sigma}_1$ as $\bar{\sigma}_1$ is gradually increased to failure. If these values of $\bar{\sigma}$ and τ are plotted on the Mohr diagram we obtain the stress path shown as line A in Fig. 8.5(c). The terminal point of the stress path is point a, which represents the stresses at failure.

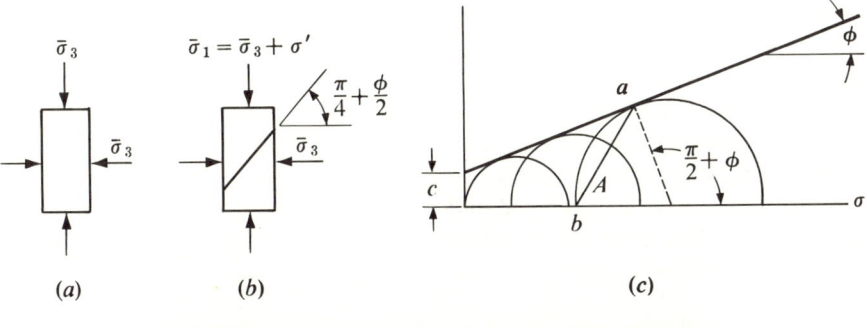

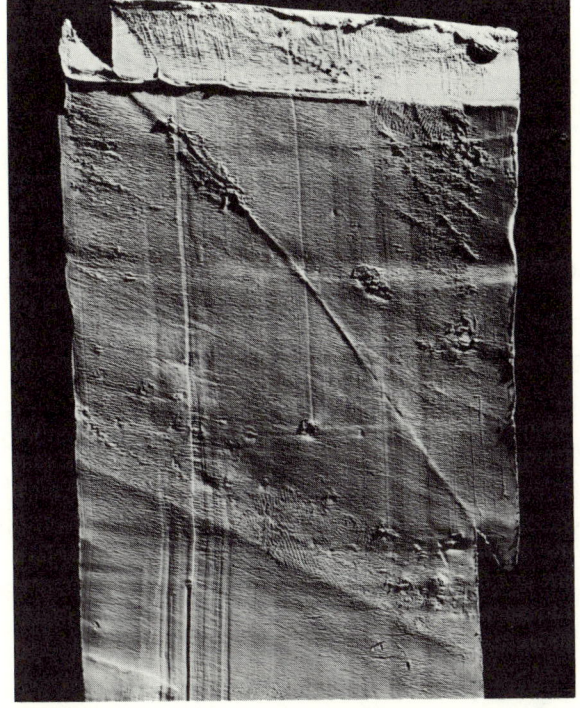

FIGURE 8.5 *Triaxial test: (a) and (b) stresses; (c) Mohr's circles and envelope; (d) failure plane in a specimen of undisturbed Detroit clay.*

DRAINAGE CONDITIONS IN THE TRIAXIAL TEST. Because of the capability of controlling drainage and measuring pore pressure during the test, the triaxial test is a very versatile tool for the measurement of shear strength. The following tests can be made with the triaxial apparatus. The porewater pressures and volume changes that occur in the various tests are illustrated in Fig. 8.6. The unconsolidated-undrained triaxial test, designated as the *UU test* (sometimes simply called the undrained test or quick test), allows no drainage throughout the test. Drainage can be prevented by closing valves A and C in Fig. 4.13. The pore pressure is measured by adjusting the screw control to maintain the mercury in the U tube at a fixed level. For no drainage the all-around pressure σ_3 induces a porewater pressure u_b in the soil. An additional porewater pressure u_a develops with the application of the stress σ'. In a consolidated-undrained triaxial test (CU test) the soil is consolidated under σ_3, but σ' is applied without drainage. Therefore, it involves a change of water content Δw_b under σ_3 and a porewater pressure u_a under σ'.

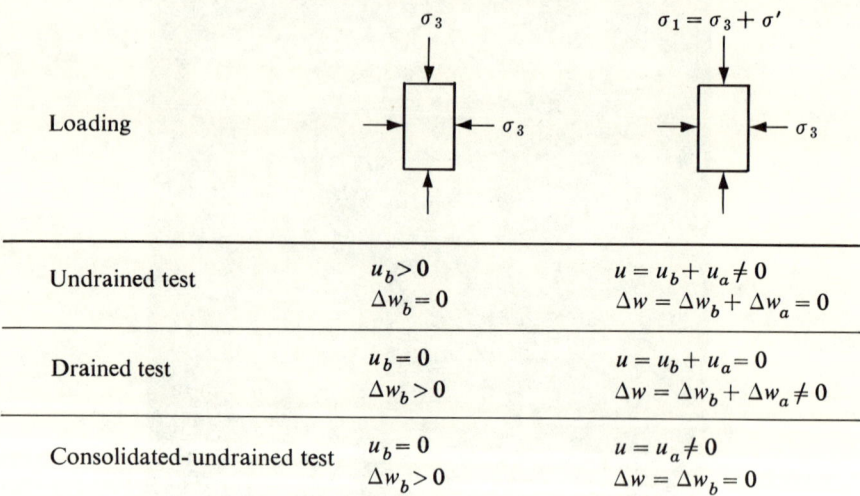

FIGURE 8.6 *Porewater pressures and volume changes in triaxial tests.*

In a consolidated-drained triaxial test, designated as the CD test (sometimes simply called a drained test or slow test), the soil is allowed to consolidate completely under the pressure σ_3, so that at the end of consolidation the excess hydrostatic porewater pressure in the soil is equal to 0. The consolidation results in a reduction of the water content equal to Δw_b. The stress σ' is applied at a very slow rate, with drainage permitted during loading, so that the soil consolidates completely as it is loaded. Under these circumstances, there is no excess hydrostatic porewater pressure developed during the application of the stress σ'. Instead, the soil undergoes a change in water content equal to Δw_a.

8.5 Porewater pressure and volume change

Consider two specimens of a clay both consolidated initially in a triaxial cell under an all-around stress σ_3 [Fig. 8.5(a)]. This represents the initial condition

SEC. 8.5   Porewater pressure and volume change

in the field where the soil is consolidated under the overburden pressure. The subsequent loading due to construction is represented by the addition of σ'. If the loading is carried out under the drained condition, the stress-strain curves in Fig. 8.7(a) are obtained. For undrained loading, the measured stress-strain

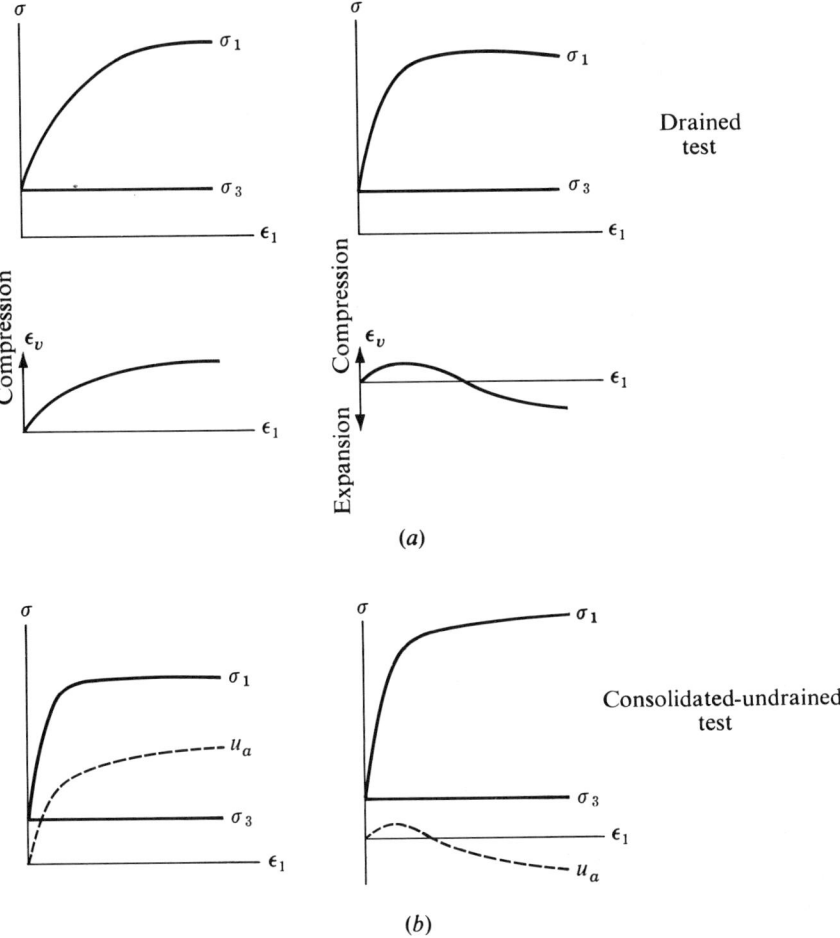

FIGURE 8.7 *Porewater pressures and volume changes in triaxial tests. Left side: normally consolidated clay or loose sand. Right side: over consolidated clay or dense sand.*

curve and pore pressure-strain curve are shown in Fig. 8.7(b). We may compare the soil behavior observed under the drained and undrained conditions. During undrained loading of a normally consolidated clay or a loose sand, the porewater pressure u_a increases steadily as the vertical stress σ_1 is increased. Since such soils are comparatively compressible, the porewater-pressure coefficient A [Eq. (4.15)] is high and may be close to 1 for soft clays. In the drained condition the porewater pressure is dissipated through drainage and the soil is compressed. The volume change increases (positive for compression) as σ_1 is increased to failure. In a heavily overconsolidated clay or dense sand only a small porewater pressure is observed during the initial part of loading. Owing to the low compressibility, the value of A is small. At larger strains as the material approaches failure, the soil tends to expand, and the porewater pressure decreases and becomes negative at failure. In the drained condition the dissipation of the negative porewater pressure results in a swelling of the soil. Henkel (1956) has observed that the value of A is closely related to the degree of overconsolidation of the clay represented by the overconsolidation ratio $\bar{p}_c/\bar{\sigma}_3$. The preconsolidation pressure of the soil is designated by $\bar{p}_c$, and $\bar{\sigma}_3$ is the cell pressure in the triaxial test. Typical values of A at failure for several soil types are given in Table 8.1.

TABLE 8.1 *Examples of the Porewater Pressure Coefficient A*

Soil		A at failure
Detroit clay	Normally consolidated	0.7
Detroit clay	Slightly overconsolidated	0.3
Weald clay, England	Overconsolidated	-0.3
Lilla Edet clay, Sweden	Normally consolidated; quick	1.2
Mississippi loess	Compacted; partially saturated	-0.2
Fine sand, Troudheim	Loose	4.0
Fine sand, Troudheim	Medium	0
Fine sand, Troudheim	Dense	-0.3

8.6 Effective-stress and total-stress Mohr envelopes

Porewater pressures of varying amounts may exist in shear and triaxial tests depending on the drainage condition during the test. The importance of porewater pressure in the evaluation of shear strength was emphasized as early as 1936 by Terzaghi (1936c). From Fig. 8.7 it can be seen that the effective stresses at the end of the test in the undrained, drained, and consolidated-undrained triaxial tests are, respectively,

$$\bar{\sigma}_1 = \sigma_1 - (u_a + u_b) \qquad \bar{\sigma}_3 = \sigma_3 - (u_a + u_b) \qquad (8.5a)$$

$$\bar{\sigma}_1 = \sigma_1 \qquad \bar{\sigma}_3 = \sigma_3 \qquad (8.5b)$$

$$\bar{\sigma}_1 = \sigma_1 - u_a \qquad \bar{\sigma}_3 = \sigma_3 - u_a \qquad (8.5c)$$

It is seen that since the porewater pressure is 0 in the drained tests, the applied stresses σ_1 and σ_3 are equal to the effective stresses in the soil. In all other tests, the applied stresses represent total stresses, since porewater pressures exist in the soil.

Figure 8.8(a) illustrates the stresses in the consolidated-undrained test. The sample is consolidated first under a hydrostatic stress indicated by point a. As the stress σ' is applied without drainage, a pore pressure u_a develops. The effective stress path is ab. The stresses on the failure plane are shown as point b, and the Mohr's circle at failure is B. This circle is tangent to the failure envelope defined by $\bar{c}$ and $\bar{\phi}$. The total stresses are shown as dashed lines in Fig. 8.8(a). If needed, a failure envelope in terms of total stress can also be drawn tangent to the total stress circles.

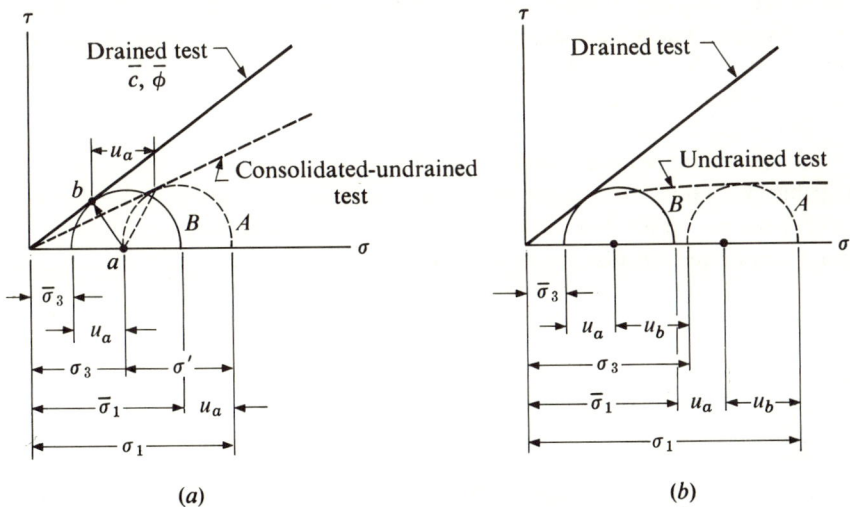

FIGURE 8.8 *Effective-stress and total-stress envelopes.*

The stresses in an undrained test are plotted in Fig. 8.8(b). The total stresses at failure are represented by circle A, the effective stresses by circle B. The latter is tangent to the drained test envelope. The effective-stress Mohr envelope may then be written as

$$s = \bar{c} + \bar{\sigma} \tan \bar{\phi} = \bar{c} + (\sigma - u) \tan \bar{\phi} \tag{8.3b}$$

Different Mohr envelopes are obtained from the different tests when total stresses are used, since different porewater pressures are developed in these tests.

SHEAR-STRENGTH PROPERTIES OF SOME COMMON SOILS

8.7 Shear strength of cohesionless soils

Granular soils such as sand and silt are usually called cohesionless soils. The Mohr's envelope of a cohesionless soil is approximately a straight line passing through the origin, and the shear strength of such soils may be expressed by

Eq. (8.3) with $\bar{c} = 0$. The value of $\bar{\phi}$ for cohesionless soils commonly ranges from about 28 to 42 deg. Generally the value of $\bar{\phi}$ increases with increasing density [Taylor (1948), pp. 347–51]. Extremely loose sands with an unstable structure may have a $\bar{\phi}$ as low as 10° [Bjerrum et al. (1961)].

The behavior of porewater pressure during a consolidated-undrained test on a saturated loose sand is described in Sec. 8.5. In a drained test a loose sand undergoes a compression. With saturated dense sands, negative porewater pressure is obtained at failure in a CU test, and an expansion occurs in the CD test. Typical results from drained direct shear tests on loose and dense sands are shown in Fig. 8.9. For a sand subjected to a given confining pressure, there exists a critical void ratio. At this void ratio the volume change is zero when the soil is loaded to the failure stress in the drained condition.

At the same relative density the value of $\bar{\phi}$ is affected by the particle size distribution and the particle shape. The value of $\bar{\phi}$ for a well-graded soil may be several degrees larger than that of a uniform soil. The same is true when a soil composed of angular grains is compared with a soil composed of rounded grains

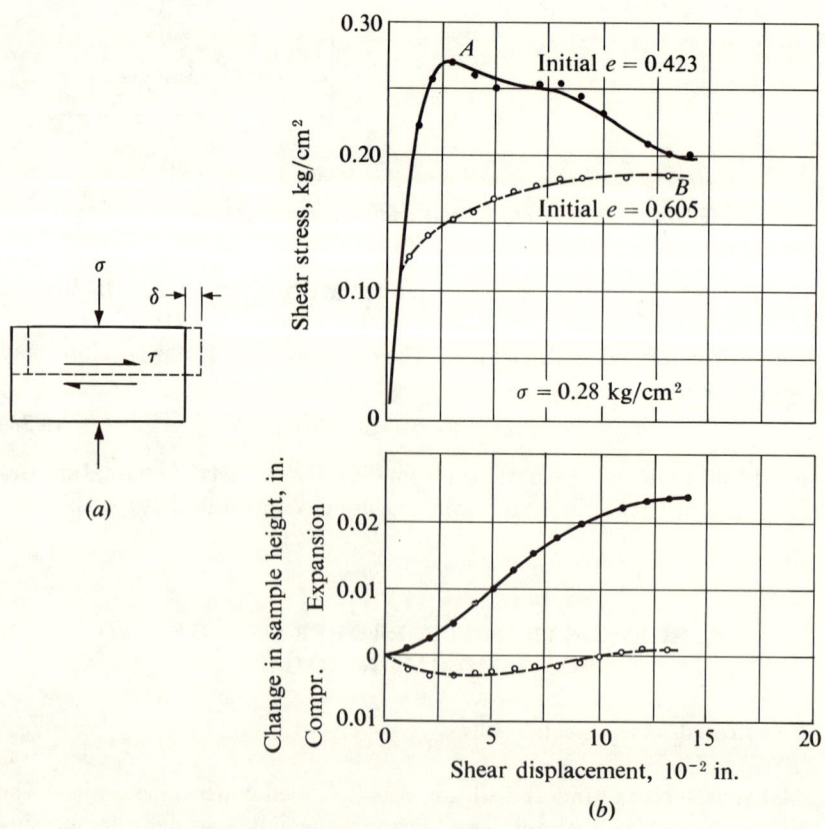

FIGURE 8.9 *Results of direct shear tests on a sand.*

[Holtz and Gibbs (1956)]. The value of $\bar{\phi}$ for a saturated sand is approximately the same as that for the same sand in a dry state.

The use of Eq. (8.3) and $\bar{\phi}$ measured in ordinary laboratory tests cannot be extended to cases where high confining pressures are encountered. Triaxial tests with $\bar{\sigma}_3$ equal to 100 psi or more have shown that the Mohr's envelope is curved at high confining pressures, as shown in Fig. 8.10 [see Bishop (1966) for examples]. In the high pressure range crushing of the grains occurs at the contact points.

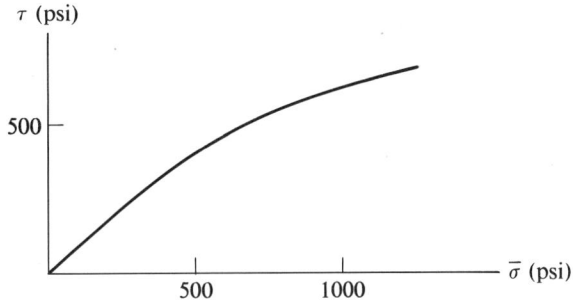

FIGURE 8.10 *Mohr's envelope for a cohesionless soil.*

8.8 Shear strength of saturated clays

Soils containing substantial amounts of clay are cohesive in nature. To study their shear-strength characteristics, we should distinguish between normally consolidated and overconsolidated clays. A normally consolidated clay is one in which the value of $\bar{\sigma}_3$ just prior to application of $\bar{\sigma}_1 - \bar{\sigma}_3$ is equal to or greater than the preconsolidation pressure. If the clay had previously been consolidated under a pressure $\bar{p}_c$ greater than $\bar{\sigma}_3$, then it is overconsolidated.

NORMALLY CONSOLIDATED CLAY. An unconsolidated clay slurry possesses no strength and gives a point at the origin of the Mohr diagram. Thus the Mohr's envelope passes through the origin. This means $\bar{c}$ is 0. The increase in strength with increase in consolidation pressure is linear, and the slope of the effective stress envelope is $\bar{\phi}$. This is illustrated by the test results shown in Fig. 8.3(b).

For a more detailed example, the results of a series of triaxial tests on normally consolidated Weald clay [Henkel (1956, 1959)] are given in Table 8.2. Many similar data are available, but the data on Weald clay has the advantage of having been presented and analyzed in a number of publications. The table lists the final water content w, the consolidation pressure σ_3, the stress $(\sigma_1 - \sigma_3)$ at failure, and the porewater pressure u at failure. From σ_3, $\sigma_1 - \sigma_3$, and u, the values in the subsequent columns are computed. The stresses on the failure plane are $\bar{\sigma}$ and τ. We first plot the results of the CU tests on the Mohr diagram

TABLE 8.2 *Triaxial Test Results on Normally Consolidated Weald Clay*

Spec. No.	Type	w, %	σ_3, psi	$\sigma_1 - \sigma_3$, psi	u, psi	$\bar{\sigma}_3$, psi	$\frac{1}{2}(\bar{\sigma}_1 + \bar{\sigma}_3)$, psi	A	$\bar{\sigma}$,[a] psi	τ,[a] psi
1	CU	23.1	30.0	17.0	16.5	13.5	22.5	0.97	18.2	7.4
2	CU	20.7	60.0	34.0	33.0	27.0	44.0	0.97	35.5	14.7
3	CU	18.3	120.0	68.0	66.0	54.0	88.0	0.97	71.0	28.5
4	CD	20.6	30.0	35.0	0	30.0	47.5	—	39.0	15.2
5	CD	18.3	60.0	70.0	0	60.0	95.0	—	77.5	30.3
6	CD	16.1	120.0	140.0	0	120.0	190.0	—	155.0	60.6

[a] Calculated for $\alpha = 60°$.

[Fig. 8.11(*a*)]. To avoid congestion we plot only the point representing the stresses on the failure plane rather than the Mohr circle. Also shown is the stress path for specimen No. 3. The stress path A is in terms of total stress. Because of the large increase in porewater pressure during the loading of a normally consolidated clay, the effective stress $\bar{\sigma}$ on the failure plane actually decreases during loading, as shown by the stress path B for effective stress.

Comparison between the results of CU tests and CD tests is given in Fig. 8.11(*b*). It is seen that the failure stresses of the two series of tests fall on the same envelope if the effective stress is used. We may compare the stress paths in CD and CU tests by considering Specimens 3 and 6. Both were consolidated under a pressure of 120 psi. Hence, just prior to the application of the deviator stress, the two specimens were identical. The stress path of the CU test is B. In a drained test there is no excess hydrostatic porewater pressure. Hence, the value of $\bar{\sigma}$ increases throughout the test as indicated by the stress path A. The drained test thus yields a much higher shear strength than does the CU test (66 versus 28 psi).

OVERCONSOLIDATED CLAY. On the other hand, a clay may be consolidated first under an all-round pressure $\bar{p}_c$ and then allowed to rebound under a smaller pressure $\bar{\sigma}_3$ before it is loaded to failure. Such a specimen at failure yields a Mohr's circle A (Fig. 8.12). For clays sheared under a pressure σ_3 less than its preconsolidation pressure, the failure envelope is represented by a curve ab (Fig. 8.12) which intersects the y axis at a distance above the origin. This portion of the envelope can be approximated by a straight line as

$$s = \bar{c} + \bar{\sigma} \tan \bar{\phi} \qquad (8.3a)$$

for drained tests or effective stresses in the consolidated-undrained test [Fig. 8.12(*a*)] and

$$s = c_{cu} + \sigma \tan \phi_{cu} \qquad (8.6)$$

for total stresses in the consolidated-undrained test [Fig. 8.12(*b*)].

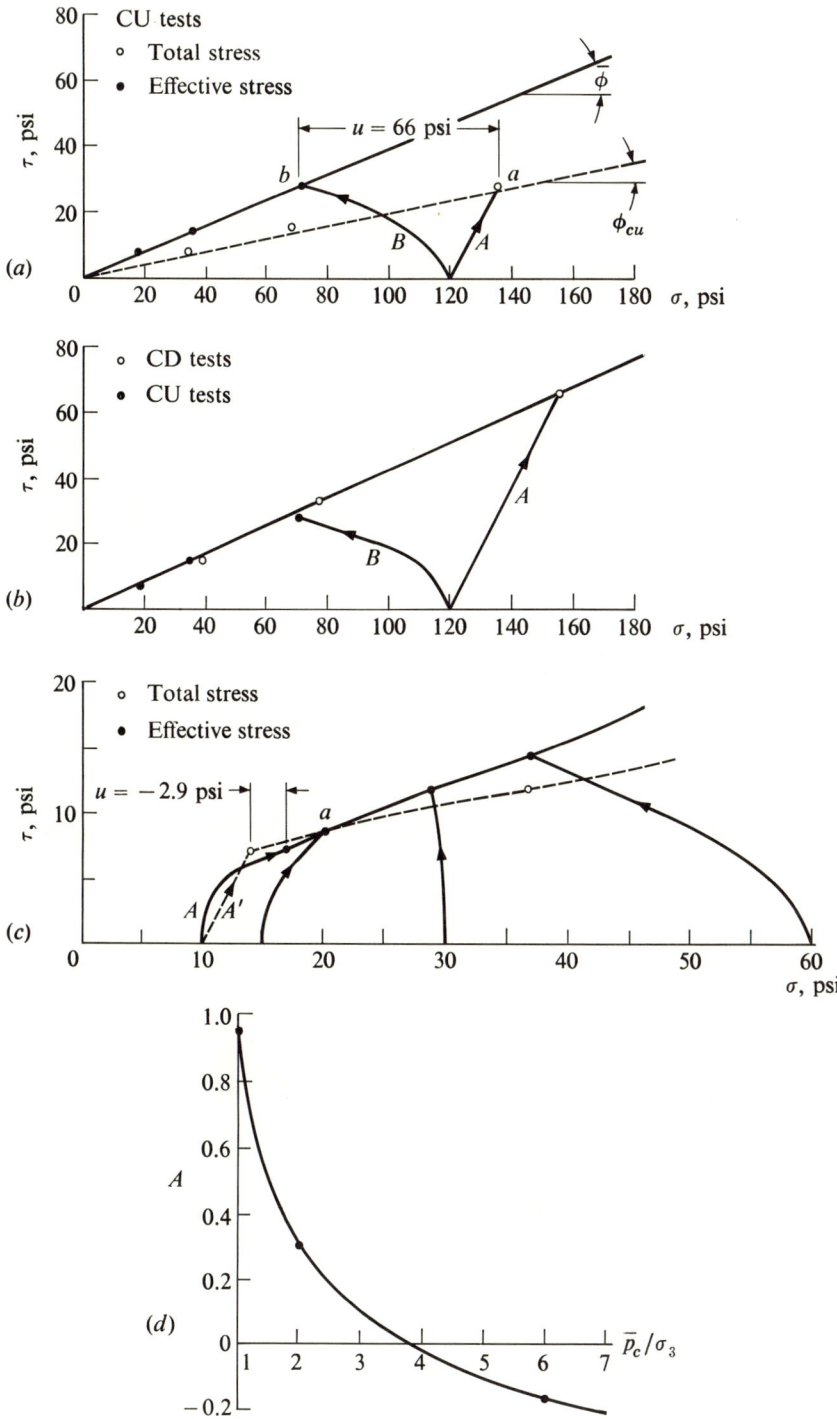

FIGURE 8.11 *Stress paths and failure envelopes.*

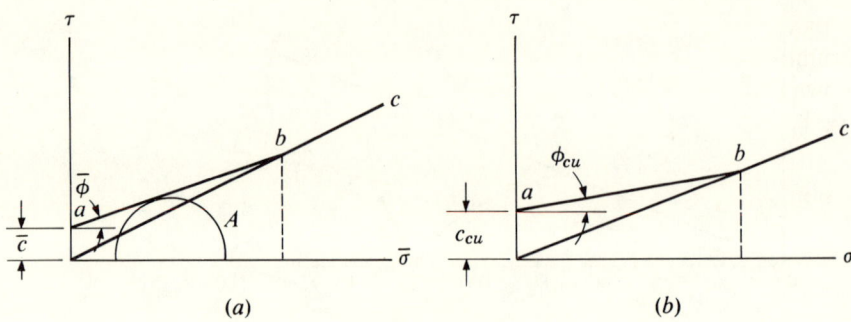

FIGURE 8.12 *Failure envelopes of cohesive soils.*

Triaxial test data on overconsolidated Weald clay [Henkel (1956, 1959)] are given in Table 8.3. The specimens were initially all consolidated under 60-psi pressure. After this they were allowed to rebound under pressures of 10, 15, and 30 psi and then loaded to failure. Specimen No. 1 is normally consolidated. Thus the overconsolidation ratio $\bar{p}_c/\sigma_3$ is 1, 2, 4, and 6 for the four specimens. A ratio of $\bar{p}_c/\sigma_3$ of 1 means a normally consolidated specimen.

TABLE 8.3 *Triaxial Test Results on Overconsolidated Weald Clay*[a]

Spec. No.	Type	σ_3, psi	$\sigma_1 - \sigma_3$, psi	u, psi	$\bar{\sigma}_3$, psi	$\frac{1}{2}(\bar{\sigma}_1 + \bar{\sigma}_3)$, psi	A	$\bar{\sigma}$,[b] psi	τ,[b] psi	p_c/σ_3
1	CU	60.0	33.0	31.4	28.6	45.1	0.95	36.8	14.3	1.0
2	CU	30.0	27.0	8.1	21.9	35.4	0.30	28.6	11.7	2.0
3	CU	15.0	19.5	0	15.0	24.8	0	20.1	8.5	4.0
4	CU	10.0	17.0	−2.9	12.9	21.4	−0.17	17.1	7.4	6.0

[a] Preconsolidation pressure p_c equals 60 psi.
[b] Calculated for $\alpha = 60°$.

The test results are plotted in Fig. 8.11(c). The solid lines represent effective stresses and the dashed lines show total stresses. With a high overconsolidation ratio of 6, the porewater pressure is negative at failure. Then the effective stress path A falls to the right of the total stress path A' at failure. The effective stress envelope lies below the total stress envelope. At an overconsolidation ratio of 4 the porewater pressure at failure is 0, and the two envelopes intersect (point *a*). The shape of the stress path in a CU test is seen to be dependent on the overconsolidation ratio. The measured porewater pressure also enables us to plot the relationship between the porewater-pressure coefficient A and the overconsolidation ratio. This is shown in Fig. 8.11(d).

Since many natural soils have a preconsolidation pressure $\bar{p}_c$, their shear strengths follow the broken line *abc* in Fig. 8.12. The strength of the material is given by line *ab* or *bc*, depending on whether the material is consolidated in the laboratory to a stress $\bar{\sigma}_3$ smaller or greater than the preconsolidation pressure.

SEC. 8.8 Shear strength of saturated clays

UNDRAINED SHEAR STRENGTH. If an undrained triaxial test is performed on a saturated soil with preconsolidation pressure $\bar{p}_c$, the stresses may be analyzed as shown in Fig. 8.13. The soil element in situ is subjected to the stresses shown in Fig. 8.13(a). For simplicity, we use the case of $K_0 = 1$ as an example. If this

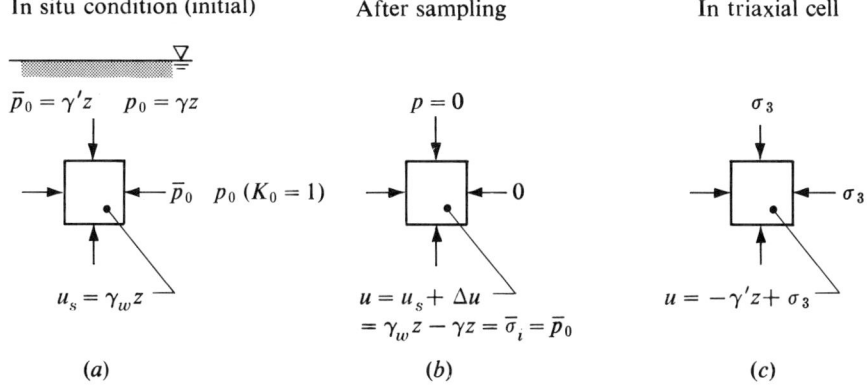

FIGURE 8.13 *Stresses in an undrained test.*

soil is extracted by a perfect sampling technique which introduces no disturbance, the soil sample prior to placement in the triaxial cell would be subjected to the stresses shown in Fig. 8.13(b). The total stress is zero. The removal of the all-around stress p_0 under the undrained condition induces a pore pressure $\Delta u = -p_0$ in the soil. Hence, the pore pressure and effective stress are equal to

$$u = u_s + \Delta u = \gamma_w z - \gamma z - \gamma' z$$
$$\bar{\sigma} = \sigma - u = 0 + \gamma' z = \gamma' z$$

The effective stress at this state is represented by point a on the Mohr diagram, Fig. 8.14. When the hydrostatic pressure σ_3 is applied, the porewater pressure increases by u_b. Since for a saturated soil the porewater-pressure coefficient B is equal to 1.0, u_b is equal to σ_3. Hence, the effective stress is still designated by point a. The total stress at this stage is represented by point b.

When a soil is loaded to failure under a stress $\sigma_1 - \sigma_3$, the porewater pressure u_a is developed. The total and effective principal stresses at failure give, respectively, the Mohr's circles A and B. In an undrained test on a saturated soil the value of $\bar{\sigma}_3$ immediately after the application of the hydrostatic pressure σ_3 is always equal to $0a$ irrespective of the pressure σ_3. Hence, in terms of effective stresses, all undrained tests are the same, and they are all represented by the Mohr's circle B. In terms of total stresses, the tests yield a series of circles with the same diameter and the Mohr envelope is a line parallel to the σ axis. For this condition one obtains

$$\phi_u = 0$$
$$s_u = c_u = \tfrac{1}{2}(\sigma_1 - \sigma_3) \qquad (8.7a)$$

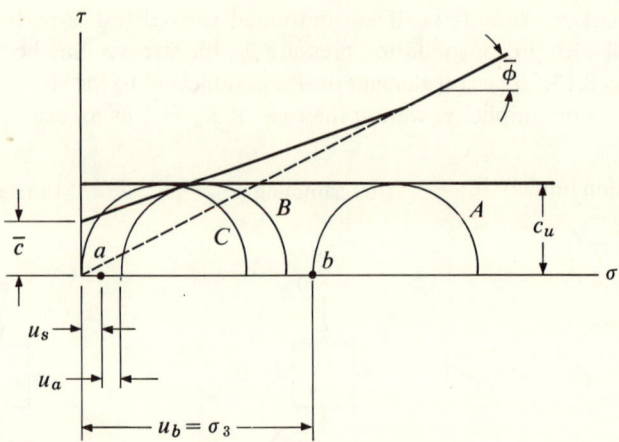

FIGURE 8.14. *Effective-stress and total-stress envelopes for undrained test.*

A special case of undrained triaxial test that is used extensively for cohesive soils is the unconfined compression test. In this test the minor principal stress is equal to 0 and the axial stress at failure is called *unconfined compressive strength*, or σ'_u. Mohr's circle for the unconfined compression test is shown as C. The undrained shear strength is

$$s = \tfrac{1}{2}\sigma'_u \tag{8.7b}$$

Example 8.1. A consolidated-undrained triaxial test is performed on a specimen of saturated clay. The value of σ_3 is 2.0 kg/cm². At failure we have $\sigma_1 - \sigma_3 = 2.8$ kg/cm², $u = 1.8$ kg/cm². If the failure plane in this test makes an angle of 57 deg with the horizontal, calculate the normal and shear stresses on the failure surface and the maximum shear stress in the specimen.

Solution. From the measured stresses at failure, we have

$$\sigma_1 = 2.8 + 2.0 = 4.8 \qquad \sigma_3 = 2.0$$

On the plane $\theta = 57$ deg, Eq. (4.4) gives

$$\sigma = \left(\frac{\sigma_1 + \sigma_3}{2}\right) + \left(\frac{\sigma_1 - \sigma_3}{2}\right)\cos 2\theta$$

$$= \left(\frac{4.8 + 2.0}{2}\right) + \left(\frac{4.8 - 2.0}{2}\right)\cos 114° = 2.83 \text{ kg/cm}^2$$

$$\tau = \left(\frac{\sigma_1 - \sigma_3}{2}\right)\sin 2\theta = \left(\frac{4.8 - 2.0}{2}\right)\sin 114° = 1.27 \text{ kg/cm}^2$$

The effective normal stress on the failure plane is

$$\bar{\sigma} = \sigma - u = 2.83 - 1.80 = 1.03 \text{ kg/cm}^2$$

The maximum shear stress occurs on a plane at 45 deg with the horizontal

$$\tau_{max} = \frac{\sigma_1 - \sigma_3}{2} = \frac{4.8 - 2.0}{2} = 1.40 \text{ kg/cm}^2$$

SEC. 8.8   Shear strength of saturated clays   205

Example 8.2. The clay in Example 8.1 has a $\bar{\phi}$ of 24 deg and $\bar{c}$ of 0.80 kg/cm². Show why failure occurs on the plane $\theta = 57$ deg instead of on the plane of maximum shear stress.

Solution. On the $\theta = 57$-deg plane, we have

$$\bar{\sigma} = 1.03 \text{ kg/cm}^2$$

According to Eq. (8.2) the shear strength is

$$s = \bar{c} + \bar{\sigma} \tan \bar{\phi} = 0.80 + 1.03 \tan 24° = 1.27 \text{ kg/cm}^2$$

Thus we see the shear strength is equal to the shear stress and failure occurs. On the plane of maximum shear stress, $\theta = 45$ deg, we have

$$\sigma = \left(\frac{4.8 + 2.0}{2}\right) + \left(\frac{4.8 - 2.0}{2}\right) \cos 90° = 3.4 \text{ kg/cm}^2$$

$$\bar{\sigma} = 3.4 - 1.8 = 1.6 \text{ kg/cm}^2$$

$$s = 0.80 + 1.60 \tan 24° = 1.51 \text{ kg/cm}^2$$

Although the shear stress is larger when $\theta = 45$ deg, the shear strength is even greater and no failure occurs.

Example 8.3. If the specimen in Example 8.2 is loaded slowly to failure in a drained test with $\sigma_3 = 2.0$ kg/cm², what will be the major principal stress at failure?

Solution. In the drained test the excess porewater pressure is 0. Thus at failure

$$\bar{\sigma}_3 = \sigma_3 = 2.0 \text{ kg/cm}^2$$

On the failure plane, $\theta = 57$ deg,

$$s = \bar{c} + \bar{\sigma} \tan \bar{\phi} = 0.80 + \left(\frac{\bar{\sigma}_1 + 2.0}{2} + \frac{\bar{\sigma}_1 - 2.0}{2} \cos 114°\right) \tan 24°$$

and

$$\tau = \frac{\bar{\sigma}_1 - 2.0}{2} \sin 114°$$

Since the shear stress is equal to shear strength at failure, $s = \tau$, or

$$\frac{\bar{\sigma}_1 - 2.0}{2} \sin 114° = 0.80 + \left(\frac{\bar{\sigma}_1 + 2.0}{2} + \frac{\bar{\sigma}_1 - 2.0)}{2} \cos 114°\right) \tan 24°$$

Solving this for $\bar{\sigma}_1$ we get

$$\bar{\sigma}_1 = 13.7 \text{ kg/cm}^2$$

We can obtain this graphically by means of the Mohr circle. A circle is drawn with $\bar{\sigma}_3$ equal to 2.00 kg/cm² and tangent to the Mohr envelope for $\bar{c} = 0.80$ and $\bar{\phi} = 24$ deg. It gives $\bar{\sigma}_1 = 13.7$ kg/cm².

Example 8.4. A normally consolidated clay deposit has an undrained shear strength equal to 1.00 kg/cm². In the laboratory the clay is found to have a $\bar{\phi}$ equal to 30 deg and a $\bar{c}$ equal to 0. If it fails in the undrained state, what are the effective principal stresses at failure?

Solution. We know that the undrained shear strength is 1.00 kg/cm². Hence

$$c_u = \frac{\bar{\sigma}_1 - \bar{\sigma}_3}{2} = 1.00$$

or

$$\bar{\sigma}_1 = 2.00 + \bar{\sigma}_3$$

Failure will occur on the plane $45° + (\phi/2)$. Hence, using the effective stress envelope, we have on the failure plane,

$$\bar{\sigma} = \frac{\bar{\sigma}_1 + \bar{\sigma}_3}{2} + \frac{\bar{\sigma}_1 - \bar{\sigma}_3}{2} \cos(90° + \phi)$$

$$= \frac{2.00 + 2\bar{\sigma}_3}{2} + \frac{2.00}{2} \cos(120°)$$

$$= \frac{\bar{\sigma}_1 - \bar{\sigma}_3}{2} \sin(90° + \phi) = \frac{2.00}{2} \sin 120°$$

$$s = \bar{\sigma} \tan \bar{\phi} = \left(\frac{2.00 + 2\bar{\sigma}_3}{2} + \frac{2.00}{2} \cos 120°\right) \tan 30°$$

Putting $s = \tau$,

$$\frac{2.00}{2} \sin 120° = \left(\frac{2.00 + 2\bar{\sigma}_3}{2} + \frac{2.00}{2} \cos 120°\right) \tan 30°$$

Solving this, we get

$$\bar{\sigma}_3 = 1.00 \text{ kg/cm}^2$$

and

$$\bar{\sigma}_1 = 2.00 + \bar{\sigma}_3 = 3.00 \text{ kg/cm}^3$$

Graphically, we can solve this problem by drawing the stress circle tangent to the Mohr envelope with $\bar{\phi} = 30$ deg. The diameter of the circle should be 2.00 kg/cm².

Example 8.5. Apply the soil properties given in Example 8.4 to the calculation of the stability of a vertical cut in the clay with height H (Fig. 8.15). The unit weight of the soil is 125 pcf.

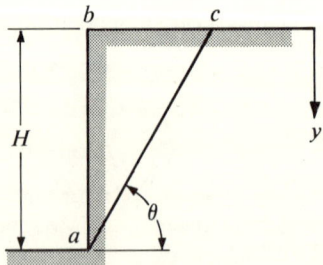

FIGURE 8.15 *Vertical cut in clay.*

Solution. We consider the wedge *abc* sliding on a failure surface *ac*. We see that the vertical stress is equal to the unit weight times the depth, or γy. The horizontal stress is 0 at the face of the cut. We may simplify the problem by assuming that the horizontal stress is 0 everywhere in the wedge. Then the vertical and horizontal stresses are principal stresses and

$$\sigma_1 = \gamma y \qquad \sigma_3 = 0$$

Shear strength of saturated clays

For the average condition, we may write

$$\sigma_1 = \frac{\gamma H}{2} \qquad \sigma_3 = 0$$

Thus the conditions are the same as those in a triaxial test. Under the undrained condition,

$$\theta = 45° + \frac{\phi_u}{2} = 45°$$

$$c_u = \tfrac{1}{2}(\sigma_1 - \sigma_3)$$

or

$$2000* = \frac{1}{2}\frac{\gamma H}{2} = \frac{1}{2}\frac{125 H}{2} \qquad H = 64 \text{ ft}$$

This means the soil mass will fail along ac when H reaches 64 ft.

If we adopt the effective stress analysis, $\theta = 45° + (\bar{\phi}/2) = 60°$. From Example 8.4 we know that for a clay with $c_u = 1.00$ kg/cm², $\bar{\phi} = 30$ deg, and $\bar{c} = 0$, the principal effective stresses at failure are

$$\bar{\phi}_1 = 3.00 \text{ kg/cm}^2 \qquad \bar{\sigma}_3 = 1.00 \text{ kg/cm}^2$$

For the wedge we have

$$\sigma_1 = \frac{\gamma H}{2} = \frac{126 \times 64}{2} = 4000 \text{ psf} = 2 \text{ kg/cm}^2$$

and

$$u = \sigma_1 - \bar{\sigma}_1 = 2.00 - 3.00 = -1.00 \text{ kg/cm}^2$$

This is the average porewater pressure that would exist in the wedge due to the excavation of the cut. We arrive at this answer by means of the condition that the clay has an undrained shear strength of 1.00 kg/cm² and at the same time has a $\bar{\phi}$ of 30 deg and a $\bar{c}$ of 0.

To show the consistency of this approach we proceed to check the stability of the cut by means of effective stress, with the knowledge that the average porewater pressure is -1.00 kg/cm². On the failure surface $\theta = 60$ deg,

$$\sigma = \frac{\sigma_1 + \sigma_3}{2} + \frac{\sigma_1 - \sigma_3}{2} \cos 2\theta = \frac{1}{2}\left(\frac{\gamma H}{2}\right) + \frac{1}{2}\left(\frac{\gamma H}{2}\right) \cos 120°$$

$$= \frac{125 \times 64}{4} + \frac{125 \times 64}{4} \cos 120° = 1000 \text{ psf} = 0.50 \text{ kg/cm}^2$$

$$\bar{\sigma} = \sigma - u = 0.50 - (-1.00) = 1.50 \text{ kg/cm}^2$$

$$\tau = \frac{\sigma_1 - \sigma_3}{2} \sin 2\theta = \frac{1}{2}\left(\frac{\gamma H}{2}\right) \sin 120°$$

$$= \frac{125 \times 64}{4} \sin 120° = 1732 \text{ psf} = 0.87 \text{ kg/cm}^2$$

$$s = \bar{\sigma} \tan \bar{\phi} = 1.50 \tan 30° = 0.87 \text{ kg/cm}^2$$

Therefore on the plane $\theta = 60$ deg, the shear stress equals the shear strength when H is 64 ft. Hence, according to the effective-stress principle, failure would also occur.

---

* 1 kg/cm² equals approximately 1 tsf.

The two analyses yield the same answer with respect to stability* but predict failure along different surfaces; $\theta = 45$ deg for the undrained analysis and $\theta = 60$ deg for the effective-stress analysis. This is because the $\phi_u = 0$ condition does not give us the relationship between the shear strength and the normal effective stress. (In Sec. 8.8, it is shown that all undrained tests on a saturated soil yield only one stress circle.) The soil actually behaves as a frictional material with respect to effective stress and has a value of $\bar{\phi}$ greater than 0. This was noted by Terzaghi as early as 1936, when he observed that the inclination of failure planes in undrained tests was always greater than 45 deg, although the measured shear strengths gave $\phi_u = 0$ [Terzaghi (1936c)]. It is important to note that if failure is defined in terms of the general stress conditions (principal stresses, maximum shear stress, etc.) either analysis gives the correct answer with respect to the failure condition.

8.9 Shear strength of compacted unsaturated clays

The process of compaction usually utilizes very high pressures, so that the compacted clay resembles in some respects a heavily preconsolidated clay. The strength characteristics of compacted clays are, therefore, similar to those represented by part *ab* of the Mohr's envelopes in Fig. 8.12. It is important to note that compacted clays are not fully saturated, and some air is present in the soil. The compression of these air spaces during loading brings about a volume change and a rise in the pressure of the air.

Figure 8.16(*a*) illustrates the behavior of the air and water pressures during an undrained triaxial test. In an unsaturated soil there exist menisci at the air-water interfaces and the water is under capillary tension. Hence, the initial porewater pressure has a negative value equal to u_0. After the all-around pressure σ_3 is applied, the air and water pressures both rise because of the compression of the void spaces. They are shown as points *a* and *b* in Fig. 8.16(*a*). The difference between the poreair pressure u_c and the porewater pressure u_w is equal to the capillary tension. As σ_1 is increased, both u_a and u_w change with strain.

In a system with both air and water under pressure, the pore pressure is given by Eq. (2.5). The effective stress may be written as

$$\bar{\sigma} = \sigma - u_a + \chi(u_a - u_w) \tag{8.8}$$

in which χ is a coefficient that depends upon the degree of saturation. This relationship is evaluated experimentally.

To determine the quantity χ in an unsaturated soil, it is necessary to know both the pressure of air and water. Furthermore, we assume that $\bar{c}$ and $\bar{\phi}$ are the same as those of the saturated soil. At high cell pressures the air is dissolved in the water and the specimen becomes saturated. At low cell pressures saturation is accomplished by slowly passing water through the specimen to displace the air.

---

* This is true only for the failure condition. For stresses below failure, the factor of safety calculated by the two methods are not the same [see Hansen (1962)].

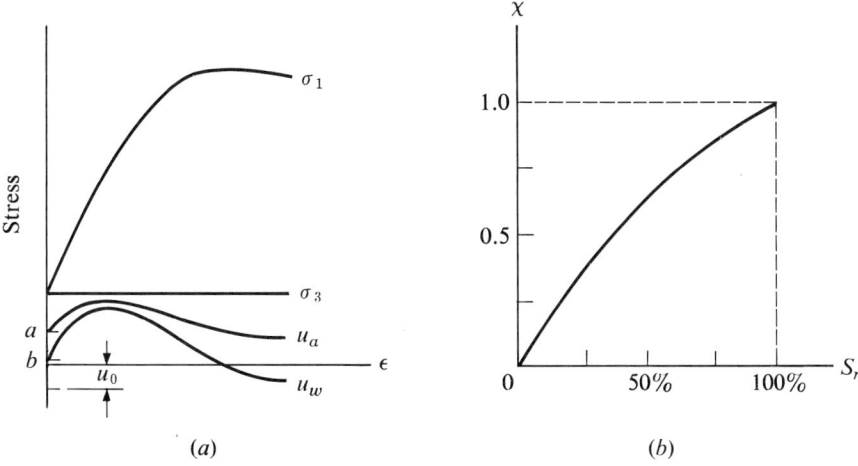

FIGURE 8.16 *Behavior of an unsaturated clay in the triaxial test.*

Thus $\bar{c}$ and $\bar{\phi}$ may be measured for the saturated soil. For the unsaturated soil with given air and water pressures, the value of χ must be such that it satisfies

$$s = \bar{c} + [\sigma - u_a + \chi(u_a - u_w)] \tan \bar{\phi} \qquad (8.9a)$$

Equation (8.9a) may be rewritten as

$$\chi = \frac{s - \bar{c} - (\sigma - u_a) \tan \bar{\phi}}{(u_a - u_w) \tan \bar{\phi}} \qquad (8.9b)$$

In this equation, $\bar{c}$ and $\bar{\phi}$ are known for the soil from tests on saturated specimens, and s, σ, u_w, and u_a are measured for the particular specimen in the triaxial test. The appropriate value of χ can be calculated directly. The objective is to establish the relationship between χ and degree of saturation S_r. If a series of specimens are tested under different consolidation pressures, and therefore at different degrees of saturation, a curve like the one in Fig. 8.16(b) can be obtained. This relationship between χ and S_r is an indispensable soil property if the values of $\bar{c}$ and $\bar{\phi}$ are to be applied to the unsaturated soil.

As an example, we consider an unsaturated soil in an embankment or earth dam at given air and water pressures and degree of saturation. Its strength is s, as determined by laboratory tests. But to calculate the stability it is convenient to use the parameters $\bar{c}$, $\bar{\phi}$. Then we have to use Eq. (8.8) and the appropriate value of χ to define the effective stress. In essence, we have defined first the parameters $\bar{c}$ and $\bar{\phi}$. The quantity χ is that value which, when used in conjunction with the parameters $\bar{c}$ and $\bar{\phi}$, gives the observed shear strength of the soil under the total stress σ and the poreair and porewater pressures u_a and u_w. The reader is referred to the paper by Blight (1967) for a detailed discussion of the evaluation and use of effective stresses in shear strength measurements and the limitations of this approach.

In an undrained test the value of B is smaller than 1.0 and the effective stresses are no longer the same in all tests, as is the case with the saturated soil. The total stress envelope is not a line parallel to the σ axis. Therefore, the unconfined compression test does not give an accurate measure of the undrained shear strength, although the error may not be large in many cases. As the soil becomes saturated at high pressures its behavior becomes the same as that described in Sec. 8.8.

Example 8.6. A compacted clay is found to have $\bar{\phi} = 24$ deg and $\bar{c} = 0.15$ kg/cm² when saturated. An unsaturated specimen with a water content of 16.2 percent is subjected to the consolidated-undrained test with measurement of poreair and porewater pressures. The cell pressure is 1.35 kg/cm², and at failure the deviator stress is 3.45 kg/cm², the poreair pressure is -0.36 kg/cm², and the porewater pressure is -1.00 kg/cm². Calculate the value of χ.

Solution. The stresses at failure are as follows:

$$\sigma_1 = 3.45 + 1.35 = 4.80 \text{ kg/cm}^2 \qquad \sigma_3 = 1.35 \qquad u_a = -0.36 \qquad u_w = -1.00$$

Putting these values into Eq. (8.9),

$$s = \bar{c} + [\sigma - u_a + \chi(u_a - u_w)] \tan \bar{\phi}$$
$$= 0.15 + [\sigma + 0.36 + \chi(-0.36 + 1.00)] \tan 24°$$
$$= 0.31 + 0.445\sigma + 0.285\chi$$

Also,

$$\sigma = \frac{\sigma_1 + \sigma_3}{2} + \frac{\sigma_1 - \sigma_3}{2} \cos 2\theta = \frac{4.80 + 1.35}{2} + \frac{4.80 - 1.35}{2} \cos(90° + 24°)$$
$$= 2.37 \text{ kg/cm}^2$$

$$\tau = \frac{\sigma_1 - \sigma_3}{2} \sin 2\theta = \frac{4.80 - 1.38}{2} \sin 114° = 1.58 \text{ kg/cm}^2$$

Thus
$$s = 0.31 + (0.445)(2.37) + 0.285\chi$$
$$= 1.36 + 0.285\chi$$

Equating s and τ we may solve for χ:

$$1.58 = 1.36 + 0.285\chi \qquad \chi = 0.77$$

The calculations show that the soil, when compacted to this particular degree of saturation and moisture content, has a shear strength expressed by the values of $\bar{c}$, $\bar{\phi}$, and χ. These values mean that when the stresses on the soil are such that

$$\sigma_1 = 4.80 \text{ kg/cm}^2 \qquad \sigma_3 = 1.35 \qquad u_a = -0.36 \qquad u_w = -1.00$$

failure results. Unfortunately, the present state of knowledge does not permit us to predict the value of u_a under a given stress condition. Therefore it is always necessary to determine the poreair pressure by measurement in order to calculate the shear strength.

THEORETICAL CONSIDERATIONS

Despite the widespread use and general success of the Mohr-Coulomb criterion in solving practical problems, we must note that the theory and the laboratory procedures such as the direct shear test and the triaxial test involve many simplifications. In some cases their indiscriminate use may lead to significant errors. While it is not possible to describe all individual cases, the generalized theories outlined in the following sections should alert the reader to shortcomings in the Mohr-Coulomb criterion and some of the practical consequences of these shortcomings.

8.10 The three-dimensional yield surface

A general failure theory should describe failure as some function of the three principal stresses. Such a function may be considered as a surface in the three-dimensional stress space, which is often called a yield surface.

We first construct the yield surface on the assumption that the Mohr-Coulomb criterion holds. Consider the intersection of the yield surface with the octahedral plane, which is the plane perpendicular to the octahedral normal ON [shaded plane in Fig. 8.17(a)]. When viewed along the normal NO axis, the σ_1, σ_2, and σ_3 axes appear as shown in Fig. 8.17(b), and the intersection of the yield surface with the plane is shown as $abcdef$. In this section σ_1, σ_2, and σ_3 denote only the principal stresses without the implication that $\sigma_1 > \sigma_2 > \sigma_3$. The distance $a0$ is the value of σ_1 at failure if $\sigma_2 = \sigma_3$. Using Eq. (8.2) together with Eq. (4.4) and $\theta = (\pi/4) + (\phi/2)$ gives

$$\frac{\sigma_1 + \sigma_3}{2} \sin\left(\frac{\pi}{2} + \phi\right) = c + \left[\frac{\sigma_1 + \sigma_3}{2} + \frac{\sigma_1 - \sigma_3}{2} \cos\left(\frac{\pi}{2} + \phi\right)\right] \tan\phi \quad (8.10)$$

If we consider failure under the condition $\sigma_2 = \sigma_3 = 0$, then Eq. (8.10) becomes

$$\sigma_1 = \frac{2c \cos\phi}{1 - \sin\phi} \quad (8.11a)$$

Similarly $e0$ and $c0$ may be found by setting $\sigma_1 = \sigma_3 = 0$ and $\sigma_1 = \sigma_2 = 0$ respectively. The result is the same as given by Eq. (8.11).

To determine $f0$, we consider the case in which $\sigma_1 = \sigma_2 = 0$ and failure is induced by application of $-\sigma_3$ (tension), as shown in Fig. 8.17(c). Substitution into Eq. (8.10) yields

$$\sigma_3 = \frac{-2c \cos\phi}{1 + \sin\phi} \quad (8.11b)$$

Simplifications have been introduced in applying some generalizations of the Mohr-Coulomb criterion to three-dimensional space. The most common versions include the Drucker-Prager (1952) hypothesis, which assumes that the yield surface is a cone. Its intersection with the octahedral plane is a circle [Fig.

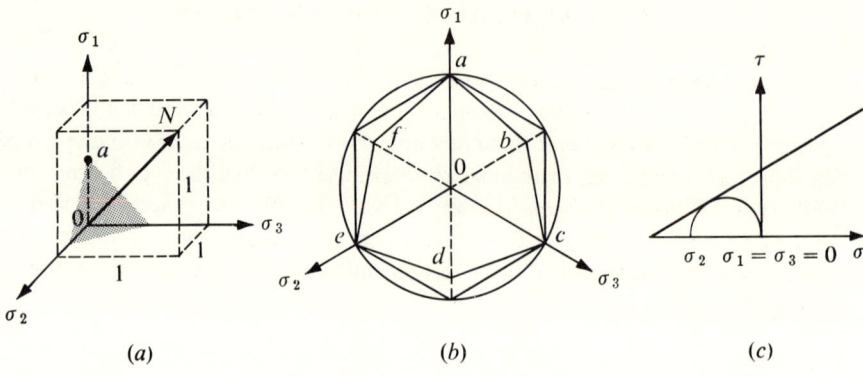

FIGURE 8.17 Yield surface.

8.17(b)] with a radius a0 as given by Eq. (8.11). The other common generalization assumes the yield surface to be a hexagonal pyramid. Its intersection with the octahedral plane is a hexagon [Fig. 8.17(b)].

Experimental studies have been conducted to determine the yield surface of soils. While the results of the various experiments differ somewhat, there is general agreement that under the condition $\sigma_1 = \sigma_2 > \sigma_3$ the strength [$f0$ in Fig. 8.17(b)] is close to that predicted by the Mohr-Coulomb criterion [Eq. (8.11b)]. A condition commonly encountered in practice is plane strain or $\sigma_1 > \sigma_2 > \sigma_3$. Laboratory plane-strain tests have shown that the strength is somewhat higher than that defined by the yield surface $abcdef$. For cohesionless soils, the difference is of the order of a few degrees in the value of ϕ [see Bishop (1966)].

8.11 The void-ratio criterion

Section 8.5 describes the volume changes that follow the application of stresses. The relationship between void ratio and stress for saturated Weald clay [Henkel (1959, 1960)] is presented in Fig. 8.18 as an illustration.* The stresses in a conventional triaxial test are shown in Fig. 8.18(a). In the $\bar{\sigma}_1$, $\bar{\sigma}_2$, $\bar{\sigma}_3$ coordinates we can conveniently plot the stresses in the shaded plane $0lmn$. Because $\bar{\sigma}_2$ and $\bar{\sigma}_3$ are equal, $0n$ represents their resultant. $0n$ makes angles of 45 deg with the $\bar{\sigma}_2$ and $\bar{\sigma}_3$ axes and is equal to $\sqrt{2}\bar{\sigma}_2$ or $\sqrt{2}\bar{\sigma}_3$. The relationship between $\sqrt{2}\bar{\sigma}_3$ and $\bar{\sigma}_1$ is shown in Fig. 8.18(b). When a specimen is consolidated under all-around pressure, the stress state is represented by points along the line $0A$. Thus two specimens consolidated under different pressures are indicated by points a and b. In consolidated-undrained test, the deviator stress is then increased to failure

---

* This method of plotting was first done by Rendulic (1937). Such plots are commonly called Rendulic diagrams.

without drainage. The effective stress changes during this stage are plotted as curves aa' and bb' and represent the stress paths leading to failure at a' and b'. Since the water content does not change during this stage, aa' and bb' are also curves of equal water content in the stress plane $0lmn$. Any stress combination that falls on the curve bb', for example, also indicates a water content of 18.3 percent. Thus the stress paths of the consolidated-undrained test are also contours of equal water content. Other contours obtained in a similar way are shown as dashed lines.

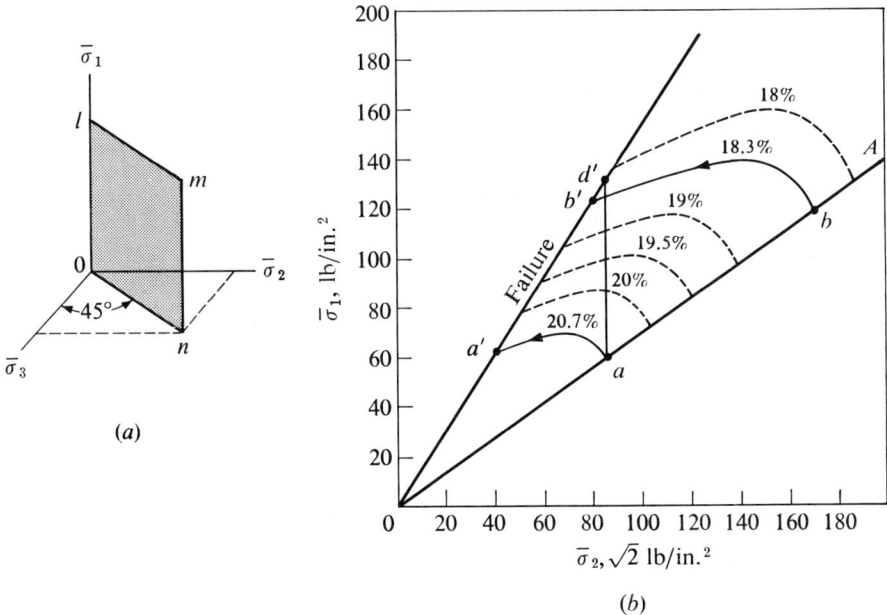

FIGURE 8.18 *Stress path and water content contours.*

The relationship between water content and effective stress is demonstrated by the drained test. If the specimen is failed by increasing $\bar{\sigma}_1$ with $\bar{\sigma}_2$ and $\bar{\sigma}_3$ constant, the stress path is given by ad'. During the loading process the specimen drains and the water content is steadily reduced. This is reflected by the stress path ad' as it begins at a with a water content of 20.7 percent and ends at d' with a water content of 18.2 percent. This relationship makes it possible to predict the water content at any stage of a drained test from the results of consolidated-undrained tests. Conversely, the results of drained tests can be used to predict the effective stress and porewater pressure in the consolidated-undrained test. Thus, for normally consolidated samples there is a unique relationship between the water content at failure and the stresses at failure.

In the theory of plasticity, the soil may be considered as a work-hardening material. When consolidated under a given pressure [point a, Fig. 8.18(b)] the yield surface consists of the Mohr envelope $0a'$ and the water-content contour aa'. Plastic deformation occurs when the stress point is on this surface. If the stress changes move the stress point inside of this surface, only elastic deformation is produced. Hydrostatic compression [ab, Fig. 8.18(b)] or axial loading ad' produce plastic (or irreversible) strains and expand the yield surface to $bb'0$. This is called work hardening. For detailed treatments of this topic see Drucker et al. (1955) and Roscoe et al. (1963a). The use of the yield surface in the solution of boundary value problems has been illustrated by Wroth and Simpson (1972).

8.12 Cohesion and internal friction

The use of the terms *cohesion* and *internal friction* to designate the shear-strength parameters c and ϕ implies a physical mechanism. We therefore examine the relationship between the shear-strength parameters and the forces at the contact points between the soil grains.

The internal friction of a granular, cohesionless soil is equal to the sliding friction plus the additional resistance provided by the interlocking of the particles. Figure 8.19(a) shows an idealized picture of particle movement during shear.

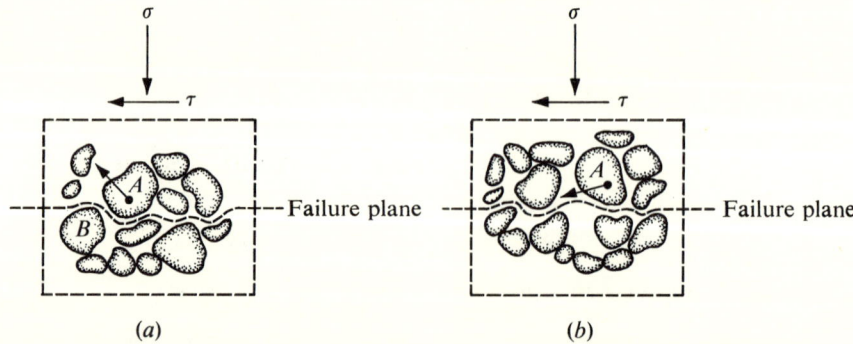

FIGURE 8.19 *Particle displacement during shear.*

When a soil with dense structure slides along the failure plane, a soil particle A must rise over its neighbor particle B, and work must be done against the normal stress σ. The mechanism may be appreciated if we consider the case of a number of uniform rods arranged in a dense packing as shown in Fig. 8.20(a). To produce shear failure rod A must slip over rod B [Fig. 8.20(b)]. If the forces on the rod are P_1 and P_3 along the principal axes 1 and 3, respectively, their relationship when sliding occurs is

$$P_1 = P_3 \tan(\phi_u + \beta) \tag{8.12}$$

in which β is the angle between the direction of slip and the 1 axis and ϕ_u is the angle of friction of the material. If l_1 and l_3 are the distances between alternate rows of rods in the 1 and 3 directions, respectively, then

$$\sigma_1 = \frac{P_1}{l_1} \qquad \sigma_3 = \frac{P_3}{l_3}$$

Substituting this into Eq. (8.12) we have

$$\frac{\sigma_1}{\sigma_3} = \frac{P_1 l_3}{P_3 l_1} = \frac{l_3}{l_1} \tan(\phi_\mu + \beta) = \tan \alpha \tan(\phi_\mu + \beta) \qquad (8.13)$$

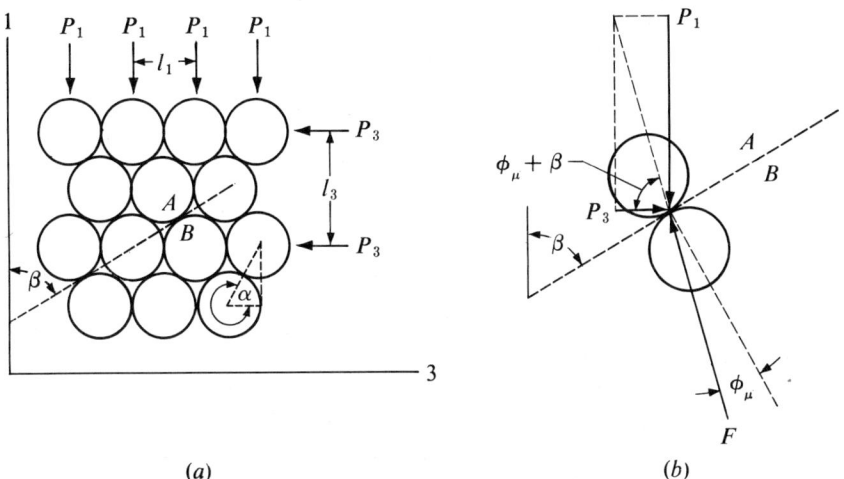

FIGURE 8.20 *Relationship between internal friction and friction.* [After Rowe (1962).]

For a cohesionless soil the ratio σ_1/σ_3 is also equal to $\tan^2[45° + (\phi/2)]$ at failure. Comparing this with Eq. (8.13) we see that the angle of internal friction is considerably greater than the angle of friction ϕ_μ between the solid particles and is dependent on the packing of the particles [Bishop (1954), Rowe (1962)].

A soil with very loose structure is shown in Fig. 8.19(b). As the soil particle A slips over its neighbor B it falls into a void and work is released by the normal stress σ. Thus the angle of internal friction ϕ may be less than the angle of friction ϕ_μ. For very loose sands Bjerrum et al. (1961) have shown that the angle of internal friction may be less than 10 deg. It can also be seen from Fig. 8.19 that soils with dense structure undergo a volume expansion when particles must ride over others during shear. In a soil with a loose structure, the material is compressed. At very large strains the initial soil structure in the slip zone is completely destroyed and no further volume changes take place with additional strain.

Similar behavior has been observed for cohesive soils. At very large strains the clay particles in the slip zone are strongly oriented in the direction of shearing. When the soil is parted along the slip zone the surface appears polished.

Such a surface is called a slickenside. The shearing resistance developed at large strains is called the residual strength, denoted by $\bar{c}_r$ and $\bar{\phi}_r$, and may be quite different from the strength at small strains [Skempton (1964)]. Kenny (1967) has observed that the residual strength of clays is dependent primarily on the mineral composition and is not related to plasticity or particle size distribution. The general range in values of $\bar{\phi}_r$ is given in Table 8.4. Hvorslev (1937) considered

TABLE 8.4 Values of $\bar{\phi}_r$ for Various Clay Minerals [after Kenny (1967)]

Quartz, feldspar, calcite	$\bar{\phi}_r > 30°$
Micaceous minerals (mica, illite)	$\bar{\phi}_r > 17°$
Montmorillonitic minerals	$\bar{\phi}_r < 11°$

cohesion to be the result of physical-chemical bond forces between the soil particles and dependent on the average spacing between the particles, which is represented by the void ratio. Thus we have

$$s = c_e(e) + \bar{\sigma} \tan \phi_e \qquad (8.14)$$

in which c_e and ϕ_e are called *true cohesion* and *true angle of internal friction*. The difference between this equation and Eq. (8.6) is that in this case c_e is not a constant but varies with the void ratio. This concept provides an explanation for the difference between the failure envelopes of normally consolidated and over-consolidated clays. Consider a normally consolidated and an overconsolidated sample of the same clay. They are represented by points a and c respectively on the pressure vs. void ratio curve in Fig. 8.21(a). Both are loaded to failure in the undrained condition, and the effective stresses at failure are shown as Mohr's circles a and c in Fig. 8.21(b). Lines A and B are the effective stress envelopes of the normally consolidated and overconsolidated clays. Since the two samples have the same void ratio (or water content) at failure, the value of c_e should be the same. The difference in strength is attributed to the difference in $\bar{\sigma} \tan \phi_e$. The values of c_e and ϕ_e are as shown in Fig. 8.21(b). Measurements of c_e and ϕ_e have been described by Bishop and Henkel (1962), Bjerrum (1954), and Olson (1962).

8.13 Rate of loading

The presentation of the principles of shear strength in the preceding sections makes no mention of the rate and manner of loading. In practice, these can vary over a tremendous range. For example, in the construction of a large structure, the gradual increase in load may extend over a period of several months. In contrast, the load from a blast or explosion reaches its peak value in less than a second. The problems of very rapid loading and vibrations fall in the realm of soil dynamics and beyond the scope of this book. However, even in static

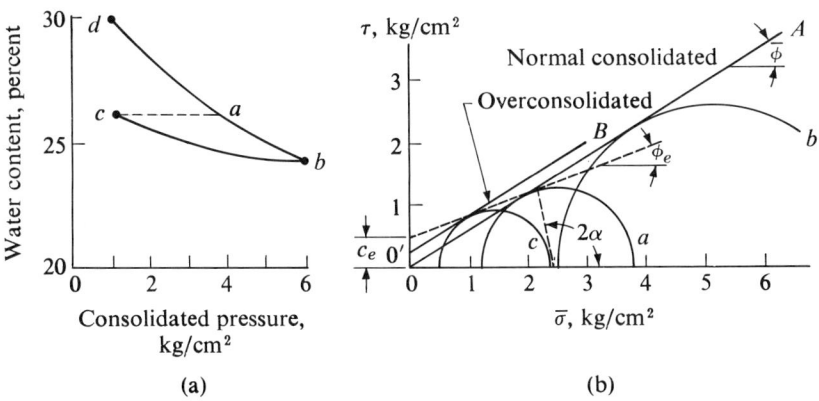

FIGURE 8.21 *Hvorslev's true cohesion and true angle of internal friction.*

problems, the influence of the rate of loading should be recognized. The primary effect is on the strength and deformation. Strength generally increases, whereas porewater pressure decreases, with the rate of loading. The importance of the loading rate varies with the type of soil.

Casagrande and Wilson (1951) studied this problem with long-time unconfined compression tests, in which the stress was increased at various rates. The time of loading ranged from 1 to 10^4 min. Using a time of loading of 1 min as reference, it was found that the strength decreased about 20 percent for some clays at very slow loading. A few soils showed some increase in strength with longer time. Another kind of loading used was the creep test. The stress was applied instantaneously, and the elapsed time to failure was measured. Under this condition it was found that the stress necessary to produce failure decreased with increased elapsed time. For elapsed time up to 10^4 min, this decrease was of the order of 20 percent for five clays but was as large as 80 percent for Cucaracha clay-shale. The exact relationship between strength and rate of loading is very complicated and still only partially understood. For additional information on soil behavior under different rates of loading the reader may refer to the works by Crawford (1959) and Whitman and Healey (1962).

Our knowledge of the effects of rate of loading is not complete enough for the formulation of theories that can be applied to design. Nevertheless, it should be recognized that this problem exists and may be expected to affect the reliability of test results when applied to real problems.

PROBLEMS

8.1 A specimen of cohesionless sand was subjected to the direct shear test under a normal stress of 1.0 kg/cm². It failed when the shear stress reached 0.60 kg/

cm². (a) Plot the stresses on the Mohr diagram and calculate the value of ϕ. (b) At what shear stress would the soil fail if the normal stress is 2.5 kg/cm²?
Answer: 31°; 1.5 kg/cm²

8.2 In the consolidated-undrained triaxial test, two specimens were loaded to failure after consolidation under all-around pressures of 2 and 4 kg/cm². The results are (in kg/cm²):

Spec. no.	σ_3	σ_1 at failure	u_a at failure
1	2	3.5	1.4
2	4	7.0	2.8

Calculate: (a) the values c and ϕ for total stress, (b) the values $\bar{c}$ and $\bar{\phi}$ for effective stress, and (c) the shear and normal stress on the failure plane in specimen no. 2.
Answer: $c = 0$, $\phi = 16°$; $\bar{c} = 0$, $\bar{\phi} = 33°$; $\bar{\sigma} = 1.85$ kg/cm², $\tau = 1.30$ kg/cm²

8.3 The saturated clay in Problem 5.2 is found to have the following shear strength:

Undrained: unconf. comp. strength = 3.50 kg/cm²
Drained: $\bar{c} = 1.0$ kg/cm² $\bar{\phi} = 25°$

If the normal stress on a plane in the clay layer is increased suddenly to 3.0 kg/cm²: (a) What is the shearing resistance along this plane immediately after the normal stress is increased? (b) What is the shearing resistance along this plane a long time afterwards?
Answer: 1.75 kg/cm²; 2.40 kg/cm²

8.4 A sandy soil has $\phi = 30$ deg and $c = 0$. (a) If a specimen of the soil is subjected to major and minor principal stresses of 2.0 and 1.5 kg/cm², respectively, will the specimen fail? Why? (b) A minor principal stress of 1.0 kg/cm² is imposed on the specimen and the major principal stress is increased steadily. Can the specimen support a major principal stress of 4.0 kg/cm²? Why?
Answer: no; no

8.5 The following results (in tsf) are obtained from direct shear tests on a silty sand:

Normal stress	Shear stress at failure
1	0.40
2	0.70
3	1.22

(a) Determine the values of c and ϕ. (b) Show which of the following stresses when applied on a plane will produce shear failure in this material. Why?

$\tau = 0.20$ tsf $\sigma = 0.25$ tsf
$\tau = 1.00$ $\sigma = 4.00$
$\tau = 0.20$ $\sigma = 1.00$

Answer: $c = 0$, $\phi = 22°$; yes, no, no

8.6 Consolidated-undrained triaxial tests were performed on two specimens of a clay. The porewater pressures were measured during the tests. The stresses at failure of the two specimens were as follows (in kg/cm²):

Spec. no.	σ_1	$\bar{\sigma}_1$	σ_3	$\bar{\sigma}_3$	u
1	1.4	1.1	1.0	0.7	0.3
2	2.8	2.2	2.0	1.4	0.6

If the normal effective stress on a plane through the clay deposit is 1.2 kg/cm², what is the shear strength along that plane?
Answer: 0.3 kg/cm²

8.7 Two undisturbed specimens were obtained from a uniform deposit of soft clay as shown in Fig. 8.22. Specimen No. 1 was subjected to the unconfined compression test and failed under an axial stress of 1000 psf. An undrained triaxial test was performed on specimen No. 2 with an all-around pressure of 750 psf. The specimen failed when the axial stress reached 1750 psf. The clay also has the following physical properties:

$$w = 26\% \qquad \text{sp. gr.} = 2.70$$
$$\text{LL} = 30\% \qquad \text{PL} = 15\%$$
$$\gamma_{sat} = 2.00 \text{ g/cc}$$

What is the shearing resistance per square foot of the soil along plane *A-A* if the shear force is applied rapidly so that no change in water content occurs during the process?
Answer: 500 psf

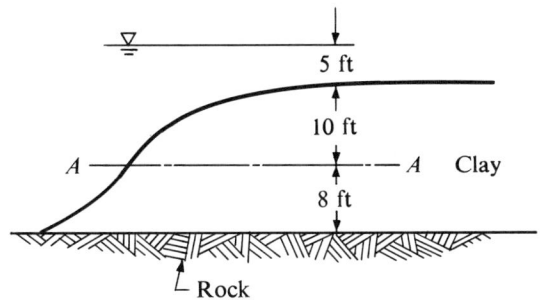

FIGURE 8.22

8.8 Using the shear-test data in Fig. 8.9 ($e = 0.423$) calculate the value of $\bar{\phi}$: (*a*) if the maximum shear stress is taken as the failure criterion, and (*b*) if the ultimate shear stress (at a displacement of 0.15 in.) is taken as the failure criterion.
Answer: 44°; 35.5°

8.9 Consider the vertical cut shown in Fig. 8.15. The depth *H* is 7 m. Drained shear tests have shown that $\bar{c} = 0.2$ kg/cm² and $\bar{\phi} = 28°$; $\gamma = 2.0$ g/cc. Will

the cut fail if the cut is excavated slowly (drained) and the porewater pressure is zero along the failure surface *ac*? Assume that $\theta = 45° + \phi/2$; $\sigma_x = 0$ and $\bar{\sigma}_y = \gamma y$ along *ac*.
Answer: No, fails at $H = 6.65$ m

8.10 For the cut in Problem 8.9 consider the case in which the water table is at the ground surface and the excavation is made slowly under water. Will failure take place? Assume that $\sigma_x = \gamma_w y$ and $\sigma_y = \gamma y$.
Answer: Yes, fails at $H = 13.30$ m

8.11 After the excavation described in Problem 8.10 is completed, the cut is rapidly pumped dry. Will failure take place? Assume that the pumping occurs so rapidly that the porewater pressure along *ac* remains equal to that for Problem 8.10 immediately after the cut has been pumped dry. *Hint:* $\sigma_x = 0$.
Answer: Yes

8.12 Undrained shear tests on the soil described in Problem 8.9 have shown that $\bar{c}_u = 0.4$ kg/cm². (*a*) What is the maximum depth of excavation at failure if the excavation is rapid? (*b*) What is the maximum depth of excavation if the excavation is slow and the pore pressure along *ac* is 0? (*c*) If an excavation is made rapidly to a depth of 8 m and allowed to stand, what will happen?
Answer: 8.50 m, 6.65 m, fails after excess hydrostatic pore pressure dissipates

9

Plastic Equilibrium

9.1 Introduction

Plastic equilibrium deals with the stresses in soil masses at failure. The soil is then in the plastic range shown in Fig. 6.1. The ideally plastic material at failure yields and undergoes deformation at a constant rate without change in stress. The stress-strain characteristics of real soils may be considerably different (see Figs. 8.7 and 8.9). In some problems, the shear strength may not be fully developed along the entire failure surface. The primary value of plasticity theory is that it provides a theoretical basis with which the actual behavior of soil masses may be compared.

This chapter presents solutions to a number of problems commonly encountered in soil mechanics. Even though we are dealing with problems that differ widely in practice, their solutions by plasticity theory are so similar that the underlying principles are best illustrated when they are taken up in one chapter.

RIGOROUS SOLUTIONS

The basic equations in plastic equilibrium consist of the equations of equilibrium and the condition of yield or failure. The solution of these equations gives the stresses at every point in a soil mass. Such solutions are called limiting equilibrium solutions. Because of mathematical complications, it is possible to obtain limiting equilibrium solutions for only a few simple problems. Several common problems are described in the following sections.

9.2 Slip plane

If we consider a two-dimensional element of soil stressed to failure under principal stresses σ_1 and σ_3, failure occurs along a "slip plane" inclined at an angle α

with the major principal plane. To determine α, use is made of the failure criterion,

$$s = c + \sigma \tan \phi \tag{8.2}$$

The normal and shear stresses (σ and τ) on the failure surface are given by Eq. 4.4. Letting $\theta = \alpha$

$$\sigma = \frac{\sigma_1 + \sigma_3}{2} + \frac{\sigma_1 - \sigma_3}{2} \cos 2\alpha$$
$$\tau = \frac{\sigma_1 - \sigma_3}{2} \sin 2\alpha \tag{4.4}$$

At failure the shear stress τ on the failure surface must be equal to the shear strength s. Substituting Eq. (4.4) in (8.2) and setting τ equal to s,

$$\tfrac{1}{2}(\sigma_1 - \sigma_3) \sin 2\alpha = c + \left(\frac{\sigma_1 + \sigma_3}{2} + \frac{\sigma_1 - \sigma_3}{2} \cos 2\alpha\right) \tan \phi$$

By application of the trigonometric relations,

$$\sin 2\alpha = 2 \sin \alpha \cos \alpha \qquad \cos 2\alpha = 2 \cos^2 \alpha - 1$$

this may be transformed into

$$\sigma_1 \sin \alpha \cos \alpha - \sigma_3 \sin \alpha \cos \alpha = c + \tan \phi [\sigma_3 + (\sigma_1 - \sigma_3) \cos^2 \alpha]$$

or

$$\sigma_1 = \sigma_3 + \frac{c + \sigma_3 \tan \phi}{\sin \alpha \cos \alpha - \cos^2 \alpha \tan \phi} \tag{9.1}$$

The quantity α is determined by the condition that the stresses on the failure plane are the critical ones and control the value of σ_1 for a given value of σ_3. Once failure develops along this plane, σ_1 cannot be increased any further, even though failure has not taken place on all the other possible planes. In other words, the failure plane is the one with angle α that gives a minimum value of σ_1 for a given value of σ_3. We see that in Eq. (9.1), σ_1 is a minimum if the denominator ($\sin \alpha \cos \alpha - \cos^2 \alpha \tan \phi$) is a maximum. Hence, the critical value of α may be determined by setting

$$\frac{d}{d\alpha}(\sin \alpha \cos \alpha - \cos^2 \alpha \tan \phi) = 0$$

This leads to

$$\cos^2 \alpha - \sin^2 \alpha + 2 \tan \phi \sin \alpha \cos \alpha = 0$$

from which we obtain

$$\alpha = 45° + \frac{\phi}{2}$$

Since the stresses are symmetrical with respect to the 1 and 3 axes, we get two sets of failure planes making angles of $\pm \alpha$ with the major principal plane (Fig. 9.1).

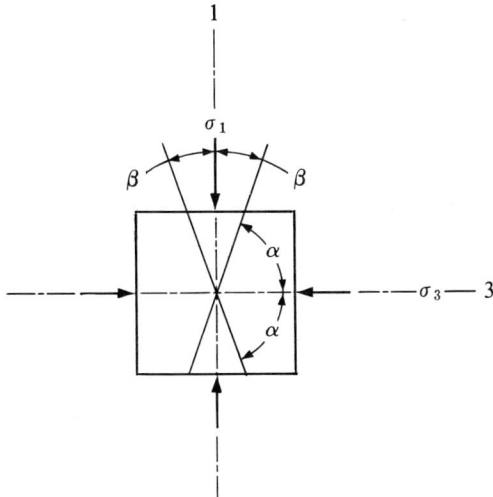

FIGURE 9.1 *Failure planes.*

9.3 Rankine's theory of earth pressure

We begin the study of failure stresses in soil masses with the simple problem in which the surfaces of failure are planes. The major and minor principal planes at all points lie in the same directions. The solution of this problem is *Rankine's theory*.

Figure 9.2 shows a mass of soil bounded by a frictionless wall *a-a* that extends to infinite depth. An element of soil at a depth y is subjected to a vertical

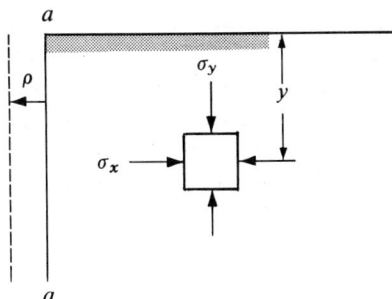

FIGURE 9.2 *Stresses in a soil mass behind a frictionless wall.*

stress σ_y and a horizontal stress σ_x. There exist no shear stresses on the vertical and horizontal planes, and these are the principal planes. Therefore, σ_x and σ_y are also the principal stresses. If the frictionless wall *a-a* is allowed to move away from the soil mass, the value of σ_x decreases. If this deformation continues, σ_x

soon reaches a value such that failure of the soil occurs and plastic equilibrium is attained in the soil. Since failure is obtained by reducing σ_x, it must be the minor principal stress σ_3, while σ_y is the major principal stress σ_1. The relationship between σ_1 and σ_3 at failure is given by Eq. (9.1). Substituting $\alpha = 45° + (\phi/2)$ into Eq. (9.1) one obtains

$$\sigma_3 = \sigma_1 \tan^2\left(45° - \frac{\phi}{2}\right) - 2c \tan\left(45° - \frac{\phi}{2}\right) \quad (9.2)$$

or

$$\sigma_x = \sigma_y \tan^2\left(45° - \frac{\phi}{2}\right) - 2c \tan\left(45° - \frac{\phi}{2}\right) \quad (9.3a)$$

If the wall a-a is moved against the soil so that the soil is compressed laterally, the horizontal pressure σ_x is the major principal stress σ_1 and the vertical pressure σ_y is the minor principal stress σ_3. If these are substituted into Eq. (9.2), we get

$$\sigma_x = \sigma_y \tan^2\left(45° + \frac{\phi}{2}\right) + 2c \tan\left(45° + \frac{\phi}{2}\right) \quad (9.3b)$$

Since in this problem, the vertical stress is equal to the weight of the soil,

$$\sigma_y = \gamma y$$

Eqs. (9.3a) and (9.3b) become, respectively,

$$p_a = \sigma_x = \gamma y \tan^2\left(45° - \frac{\phi}{2}\right) - 2c \tan\left(45° - \frac{\phi}{2}\right) \quad (9.4a)$$

and

$$p_p = \sigma_x = \gamma y \tan^2\left(45° + \frac{\phi}{2}\right) + 2c \tan\left(45° + \frac{\phi}{2}\right) \quad (9.4b)$$

The distributions of the stress σ_x and the slip planes are as shown in Figs. 9.3 and 9.4. The two conditions of stress are often called *Rankine's active and passive states of stress* and the pressures p_a and p_p, the *active and passive earth pressures*. They are, respectively, the minimum and maximum pressures that the soil can exert on a smooth wall extending to infinite depth.

The ratio of the horizontal to vertical pressure σ_x/σ_y is called the coefficient of earth pressure and is denoted by K_a and K_p for the active and passive states, respectively. For a cohesionless soil ($c = 0$), this ratio is constant and is given by

$$K_a = \tan^2\left(45° - \frac{\phi}{2}\right) \quad (9.5a)$$

$$K_p = \tan^2\left(45° + \frac{\phi}{2}\right) \quad (9.5b)$$

If the wall does not move at all, the pressure is called the *at-rest earth pressure*. The earth pressure coefficient for this state is denoted by K_0. Since the at-rest state does not require failure in the soil, K_0 cannot be calculated from plastic theory. Experimental measurement of K_0 is discussed in Sec. 10.3.

SEC. 9.3   Rankine's theory of earth pressure   225

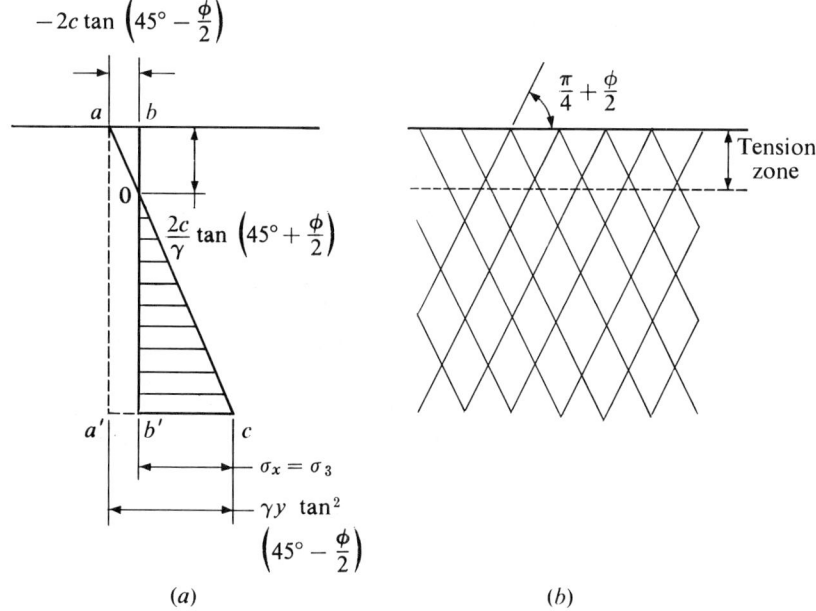

FIGURE 9.3   *Rankine's active state.*

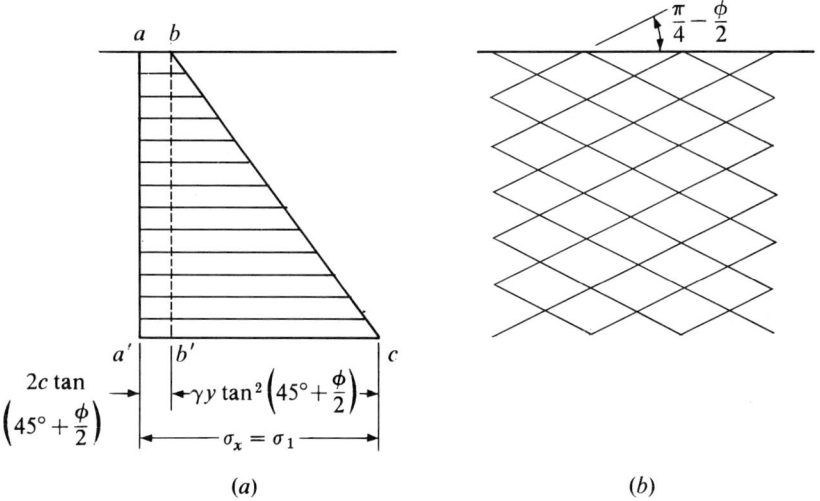

FIGURE 9.4   *Rankine's passive state.*

The resultant horizontal forces in the active and passive states are designated as P_a and P_p, respectively, and are equal to

$$P_a = \tfrac{1}{2}\gamma y^2 \tan^2\left(45° - \frac{\phi}{2}\right) - 2cy \tan\left(45° - \frac{\phi}{2}\right) \qquad (9.6a)$$

$$P_p = \tfrac{1}{2}\gamma y^2 \tan^2\left(45° + \frac{\phi}{2}\right) + 2cy \tan\left(45° + \frac{\phi}{2}\right) \qquad (9.6b)$$

The point of application of the forces P_a and P_p can be determined by considering the pressure distribution as a composite of the rectangular area $abb'a'$ and the triangular area $aa'c$. The resultant of the first part acts at the midpoint and that of the second part acts at the lower third point.

If a uniform surcharge q is placed on the ground surface, its effect is to increase the vertical stress at every point by q. Hence, instead of γy in the first term of Eqs. (9.4a) and (9.4b), we have $\gamma y + q$. The active and passive earth pressures are increased by amounts equal to $q \tan^2[45° - (\phi/2)]$ and $q \tan^2[45° + (\phi/2)]$, respectively, and we may write

$$p_a = \gamma y \tan^2\left(45° - \frac{\phi}{2}\right) - 2c \tan\left(45° - \frac{\phi}{2}\right) + q \tan^2\left(45° - \frac{\phi}{2}\right) \qquad (9.7a)$$

$$p_p = \gamma y \tan^2\left(45° + \frac{\phi}{2}\right) + 2c \tan\left(45° + \frac{\phi}{2}\right) + q \tan^2\left(45° + \frac{\phi}{2}\right) \qquad (9.7b)$$

The increase in earth pressure due to q is independent of y; it is uniform along the height of the wall. It should be noted that when c is greater than 0, the active earth pressure is tensile near the ground surface. Since most soils cannot resist tensile stresses over extended periods, cracks often develop in the tensile zone. For this reason, the tensile stress represented by the triangle $ab0$ is often neglected when computing the earth pressure.

The active and passive states of stress can also be determined by the use of Mohr's circle. In Fig. 9.5 the vertical stress σ_y is represented by point 1. Under expansion the horizontal stress σ_x decreases until the Mohr's circle becomes

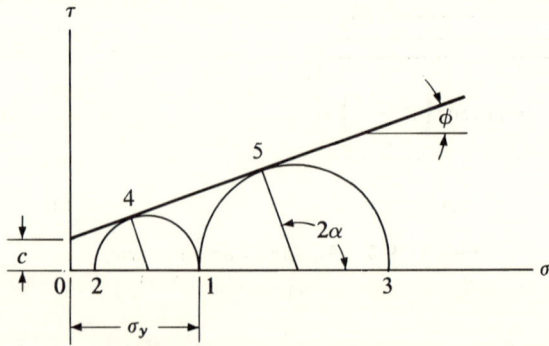

FIGURE 9.5 Mohr's circle of stress for Rankine's active and passive states.

tangent to the envelope at point 4. At this point failure occurs, and the value of σ_x is given by the distance 02. When the soil is compressed laterally, σ_x increases and at failure reaches a value equal to 03. Points 4 and 5 represent the state of stress on the planes of failure. Equations (9.4) and (9.5) can also be obtained from the geometry of the Mohr's circles.

The active and passive states of stress can also be obtained for a ground surface sloped at an angle β to the horizontal (see Problem 9.10). For solutions of other geomechanics problems with polar coordinates, see Nadai (1963, pp. 446–54, 463–65).

9.4 Bearing capacity

In this section we study the problem of a strip foundation first solved by Prandtl (1920). A strip foundation with a smooth base is located on the ground surface as shown in Fig. 9.6(a). As the load Q is increased, the penetration of the strip

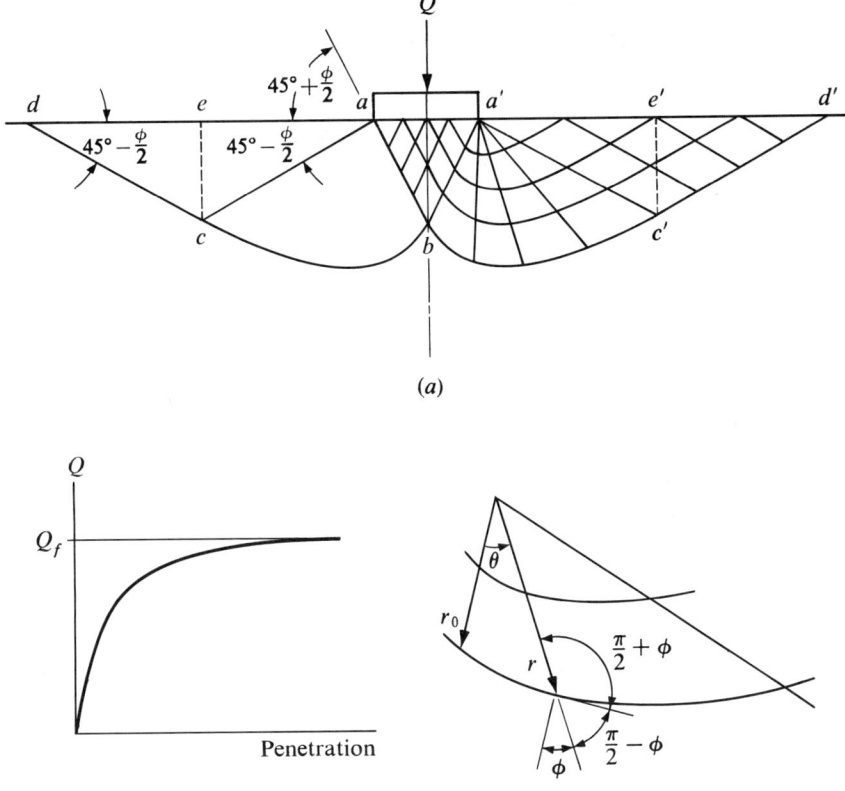

FIGURE 9.6 *Failure of a loaded strip.*

increases as shown in Fig. 9.6(b). When the load Q_f is reached the penetration increases indefinitely. At this point a bearing capacity failure is said to occur. We wish to calculate the force Q_f. The unit pressure q_f at failure is called the bearing capacity.

In this problem the direction of the principal stresses varies from point to point, and the slip surfaces are curves as shown in Fig. 9.6(a). Since the plate is frictionless, immediately beneath the loaded plate, the major principal stress is in the vertical direction and the failure surface in $a'ba$ makes an angle $45° + (\phi/2)$ with the horizontal. Also, away from the loaded plate, in zone acd, the movement is predominantly horizontal. Along the free surfaces ad and $a'd'$, the major and minor principal stresses are in the horizontal and vertical directions, respectively. Therefore, the failure surfaces in adc intersect the free surface ad at an angle of $45° - (\phi/2)$ with the horizontal. The slip surfaces in abc connect those in aba' with those in adc. Thus, we postulate a series of straight lines that radiate from a (or a') and a set of curves. We note that the two sets of slip surfaces must intersect each other at an angle of $90° + \phi$ [Fig. 9.7(b)]. From the geometric relationship between two slip surfaces shown in Fig. 9.6(c) we can deduce that the normal to the curve makes an angle ϕ with the radius. This requirement is met if the curve has the shape given by

$$r = r_0 e^{\theta \tan \phi} \tag{9.8}$$

where r_0 is the reference radius, θ is the angle between r_0 and the radius r. Such a curve is called a logarithmic spiral.

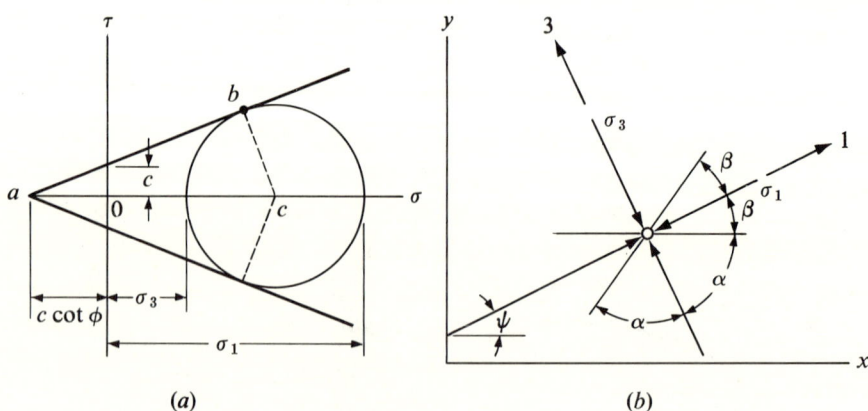

FIGURE 9.7 (a) *Mohr's circle of stress at failure.* (b) *Slip planes and principal axes in the xy plane.*

Experimental determination of the shape of the slip surface beneath model footings has yielded results in close agreement with the calculated slip surfaces. The failure surface beneath a model footing is shown in Fig. 9.8.

To solve this problem we again make use of the failure condition [Eq. (8.2)]

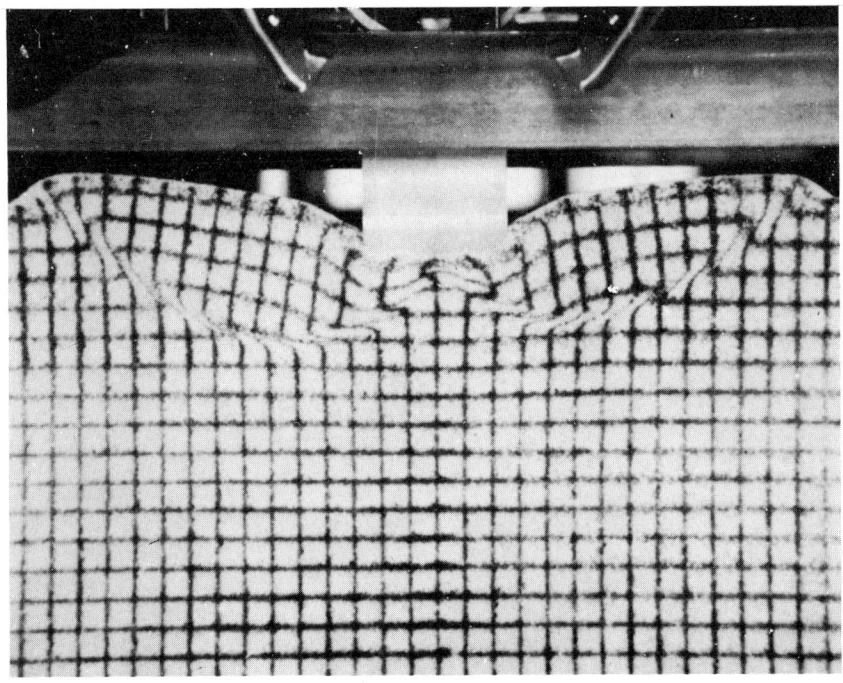

FIGURE 9.8 *A photograph of the slip surface beneath a model footing.* (Courtesy of E. T. Selig.)

and the equilibrium condition, which must now be written in differential form. In two dimensions the differential equations of equilibrium are [see Eq. (6.4)]

$$\frac{\partial \sigma_x}{\partial x} + \frac{\partial \tau_{xy}}{y} = X \tag{9.9a}$$

or

$$\frac{\partial \sigma_y}{\partial y} + \frac{\partial \tau_{xy}}{\partial x} = Y \tag{9.9b}$$

The failure condition is defined by Eq. (8.2). In order to combine it with the equilibrium equation (9.9), we transform Eq. (8.2) by the following operations. The stresses at failure are described by the Mohr circle shown in Fig. 9.7(a). The distance ac is $\frac{1}{2}(\sigma_1 + \sigma_3) + c \cdot \cot \phi$ and bc is $\frac{1}{2}(\sigma_1 - \sigma_3)$. From the geometric relationship, we may write

$$\begin{aligned}\tfrac{1}{2}(\sigma_1 + \sigma_3) &= [\tfrac{1}{2}(\sigma_1 + \sigma_3) + c \cdot \cot \phi] - c \cdot \cot \phi \\ \tfrac{1}{2}(\sigma_1 - \sigma_3) &= [\tfrac{1}{2}(\sigma_1 + \sigma_3) + c \cdot \cot \phi] \sin \phi \end{aligned} \tag{9.10}$$

If the principal axis is inclined at an angle ψ from the x axis ($\theta = \psi$), we also have from Eq. (4.4),

$$\left.\begin{array}{c}\sigma_x \\ \sigma_y\end{array}\right\} = \tfrac{1}{2}(\sigma_1 + \sigma_3) \pm \tfrac{1}{2}(\sigma_1 - \sigma_3) \cos 2\psi \tag{9.11}$$

and

$$\tau_{xy} = \tfrac{1}{2}(\sigma_1 - \sigma_3) \sin 2\psi \tag{9.11}$$

Substituting Eqs. (9.10) in Eq. (9.11) we obtain the expression for σ_x, σ_y, and τ_{xy} at failure,

$$\left.\begin{array}{c}\sigma_x\\\sigma_y\end{array}\right\} = [\tfrac{1}{2}(\sigma_1 + \sigma_3) + c \cot \phi](1 \pm \sin \phi \cos 2\psi) - c \cot \phi$$

$$\tau_{xy} = [\tfrac{1}{2}(\sigma_1 + \sigma_3) + c \cot \phi] \sin \phi \sin 2\psi \tag{9.12}$$

Equation (9.12) is the failure condition expressed in terms of σ_x, σ_y, τ_{xy}, σ_1 and σ_3 instead of σ and τ. The directions of the slip lines and the principal axes are shown in Fig. 9.7(b).

In this section we solve the simple problem of a long frictionless plate penetrating a purely cohesive material ($\phi = 0$) that is weightless ($X = 0$, $Y = 0$). The general problem of $\phi > 0$, $X > 0$, $Y > 0$ may be solved numerically by the method of characteristics. This is described in Chapter 12.

For the case of $\phi = 0$, the logarithmic spiral reduces to a circular arc, and the failure surfaces are as shown in Fig. 9.9(a). The shear strength of the material is given by Fig. 9.9(b)

$$\tfrac{1}{2}(\sigma_1 - \sigma_3) = c$$

Equation (9.11) then becomes

$$\left.\begin{array}{c}\sigma_x\\\sigma_y\end{array}\right\} = \tfrac{1}{2}(\sigma_1 + \sigma_3) \pm c \cos 2\psi$$

$$\tau_{xy} = c \sin 2\psi$$

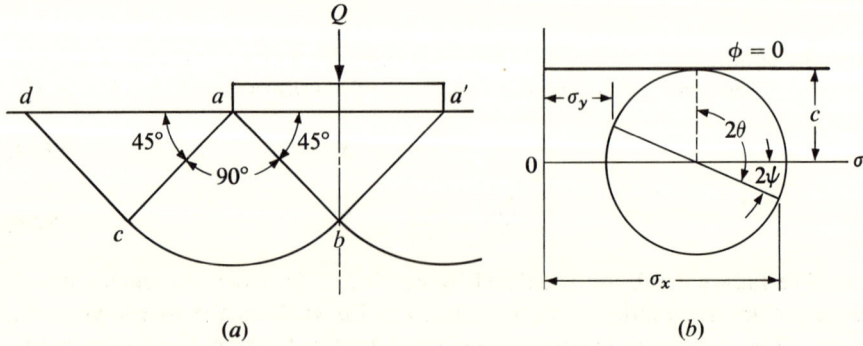

FIGURE 9.9 Failure of a frictionless, weightless soil under a strip load.

Since the directions of the slip lines have been established, we can calculate the stress changes along the slip lines. We let θ denote the angle between the x axis and the first slip line [Fig. 9.9(b)], then $2\theta = 2\psi + (\pi/2)$, and the above equations become

$$\left.\begin{array}{c}\sigma_x\\\sigma_y\end{array}\right\} = \tfrac{1}{2}(\sigma_1 + \sigma_3) \pm c \cdot \sin 2\theta$$

$$\tau_{xy} = c \cdot \cos 2\theta$$

Substituting this into Eq. (9.9), we have

$$\frac{\partial}{\partial x}\left(\frac{\sigma_1 + \sigma_3}{2}\right) + 2c\left(\cos 2\theta \frac{\partial x}{\partial \theta} - \sin 2\theta \frac{\partial \theta}{\partial y}\right) = 0$$

$$\frac{\partial}{\partial y}\left(\frac{\sigma_1 + \sigma_3}{2}\right) - 2c\left(\cos 2\theta \frac{\partial \theta}{\partial y} + \sin 2\theta \frac{\partial \theta}{\partial x}\right) = 0$$

If we let the x and y axes coincide with the slip lines, then $\theta = 0$, and

$$ds_a = dx, \quad ds_b = dy$$

in which s_a and s_b are the lengths along the a and b slip lines, respectively. The above equations then become

$$\frac{\partial}{\partial s_a}\left(\frac{\sigma_1 + \sigma_3}{2}\right) + 2c\frac{\partial \theta}{\partial s_a} = 0 \qquad (9.13a)$$

$$\frac{\partial}{\partial s_b}\left(\frac{\sigma_1 + \sigma_3}{2}\right) - 2c\frac{\partial \theta}{\partial s_b} = 0 \qquad (9.13b)$$

Equations (9.13a) and (9.13b) describe the stress changes along the failure surfaces and are often called *Kötter's equations* [Kötter (1903)]. For a derivation of Kötter's equations for $\phi > 0$, see Nadai (1963, pp. 454–56). The equations are independent of our choice of the x and y axes and therefore are not restricted to the special case of $\theta = 0$ assumed in the proof.

If the slip lines are straight lines, then

$$\frac{\partial \theta}{\partial s_a} = 0 \qquad \frac{\partial \theta}{\partial s_b} = 0$$

Equation (9.13) then leads to

$$\frac{\partial}{\partial s_a}\left(\frac{\sigma_1 + \sigma_3}{2}\right) = 0 \qquad \frac{\partial}{\partial s_b}\left(\frac{\sigma_1 + \sigma_3}{2}\right) = 0 \qquad (9.14)$$

Thus the average normal stress $[(\sigma_1 + \sigma_3)/2]$ remains constant. If the slip lines consist of sets of concurrent straight lines and concentric circles as in zone *abc* [Fig. 9.9(a)], then along the straight slip lines

$$\frac{\partial \theta}{\partial s_a} = 0 \qquad \frac{\partial}{\partial s_a}\left(\frac{\sigma_1 + \sigma_3}{2}\right) = 0 \qquad (9.15)$$

Along the circular slip lines,

$$s_b = r\theta \qquad \frac{\partial \theta}{\partial s_b} = \frac{1}{r}$$

in which r is the radius of the circular slip line. Substituting in Eq. (9.13b)

$$\frac{1}{r}\frac{\partial}{\partial \theta}\left(\frac{\sigma_1 + \sigma_3}{2}\right) - \frac{2c}{r} = 0 \quad \text{or} \quad \frac{\partial}{\partial \theta}\left(\frac{\sigma_1 + \sigma_3}{2}\right) = 2c$$

Upon integration, this yields

$$\frac{\sigma_1 + \sigma_3}{2} = A + 2c\theta \tag{9.16}$$

in which A is the constant of integration. It is given by

$$A = \left(\frac{\sigma_1 + \sigma_3}{2}\right)_{\theta=0}$$

Stating at ad [Fig. 9.9(a)], we have $\sigma_3 = 0$, $\sigma_1 = 2c$. Thus,

$$\tfrac{1}{2}(\sigma_1 + \sigma_3) = c$$

This holds for the zone acd by virtue of Eq. (9.14). In zone abc Eq. (9.15) applies along the circular arc cb. On ac, $\theta = \pi/4$ and on ab, $\theta = 3\pi/4$. Thus,

$$\left(\frac{\sigma_1 + \sigma_3}{2}\right)_{\theta=\frac{3\pi}{4}} = \left(\frac{\sigma_1 + \sigma_3}{2}\right)_{\theta=\frac{\pi}{4}} + 2c\left(\frac{3\pi}{4} - \frac{\pi}{4}\right) = c(1 + \pi)$$

In zone aba', the term $[(\sigma_1 + \sigma_3)/2]$ is again constant and

$$\sigma_1 = \left(\frac{\sigma_1 + \sigma_3}{2}\right) + \left(\frac{\sigma_1 - \sigma_3}{2}\right) = \left(\frac{\sigma_1 + \sigma_3}{2}\right) + c = c(1 + \pi) + c = 5.14c \tag{9.17}$$

Since aa' is the major principal plane, σ_1 is the bearing capacity.

APPROXIMATE SOLUTIONS

The rigorous solutions described in the preceding sections can be obtained only for a relatively small number of problems in which the boundary conditions are simple. In most engineering problems the boundary conditions are more complex. For example, the frictionless plate in Fig. 9.6 is certainly an idealization. A real plate of some material like steel or concrete would have considerable friction. Under these nonideal conditions it becomes necessary to make reasonable assumptions to simplify the solution. The formulation of these assumptions requires intuition and ingenuity. They should simplify the problem, but should be compatible with the actual behavior of the material; otherwise, the resulting solution becomes irrelevant. The approximate solutions to a number of common soil mechanics problems are illustrated in the following sections.

9.5 Terzaghi bearing-capacity theory

The calculation of the load Q_f on a strip footing at failure is often based on the simplifying assumption that the failure surface is composed of a straight line ce and a logarithmic spiral bc as found for the weightless material (Fig. 9.10). The effect of the weight of the soil on the shape of the failure surface is ignored. Since the base of the footing is not smooth, it is assumed that the friction between the footing and soil is large enough so that no shear displacements occur along *ad*. Then Rankine's active state cannot exist in the zone *aba*, and this wedge must move with the plate. Terzaghi's solution assumes that *ab* makes an angle ϕ with the horizontal. Then the spiral portion of the failure surface *bc* must be vertical at point *b*, because *ab* is also a failure surface, and failure surfaces intersect each other at angles of $90° + \phi$. The load at failure may be calculated by considering the forces acting on the soil mass *abcd* (Fig. 9.10) [Terzaghi (1943b)].

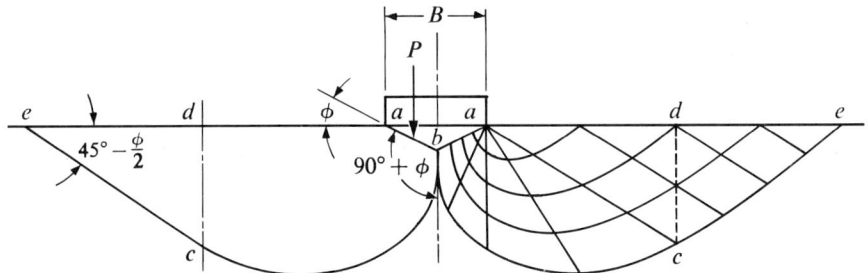

FIGURE 9.10 *Terzaghi bearing-capacity theory, showing the slip surface beneath a strip footing with a rough base.*

At failure the shear stress along the failure surface *bce* equals the shear strength of the soil as given by Eq. (8.2). We consider separately the resistance developed by the two components of the shear strength, c and $\sigma \tan \phi$. The forces that act on the mass *abcd* for these two cases are shown in Fig. 9.11.

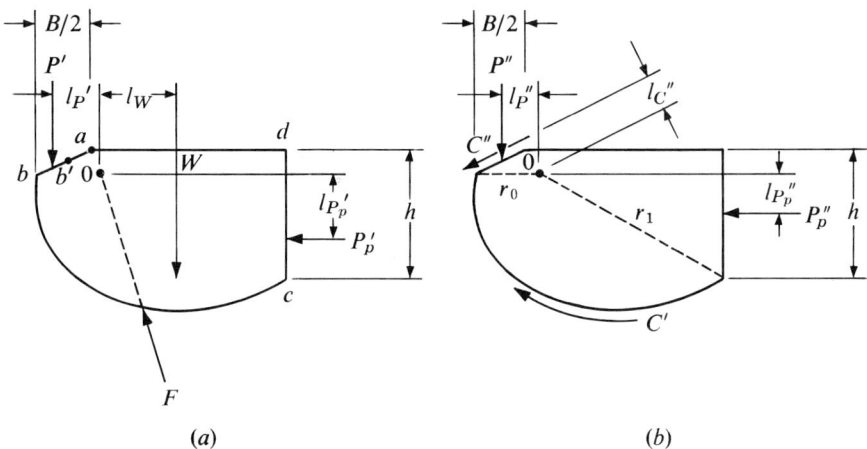

FIGURE 9.11 *Forces on soil mass abcd in the Terzaghi bearing-capacity theory.*

Figure 9.11(a) shows the forces that act on the mass if the shear strength is equal to $\sigma \tan \phi$ only (or $c = 0$). Since the failure surfaces are planes in acd, Rankine's passive state of stress exists in this zone. The passive earth pressure P'_p on cd is, therefore,

$$P'_p = \tfrac{1}{2} h^2 \gamma \tan^2\left(45° + \frac{\phi}{2}\right) \qquad (9.18)$$

P'_p acts at a distance of $\tfrac{2}{3}h$ from the surface. The weight of the mass $abcd$ is equal to W and it acts through the centroid of $abcd$. On the failure surface bc there exist a normal stress σ and a shear stress equal to the shear strength of the soil, $\sigma \tan \phi$. Therefore, the resultant of σ and $\sigma \tan \phi$ makes an angle ϕ with the normal to the failure surface at every point. The resultant force F on the spiral part of the failure surface also makes an angle ϕ with the slip surface. F therefore passes through the center of spiral, 0. The force that acts on the failure plane ab is designated by P', and it also acts at the lower third point.* This system of forces may be solved by summation of moments about point 0. The force F passes through point 0 and produces no moment. Hence,

$$M_0 = P'_p l'_{P_p} + W l_W - P' l'_p = 0$$

or
$$P' = \frac{1}{l'_p} (P'_p l'_{P_p} + W l_W) \qquad (9.19)$$

We next consider the resistance to failure developed by c, and the forces are shown in Fig. 9.11(b). The passive pressure P''_p is

$$P''_p = 2ch \tan\left(45° + \frac{\phi}{2}\right) \qquad (9.20)$$

and acts at the midpoint of cd. The shear strength is c everywhere along curve bd, and its moment about 0 may be found by integrating the unit stress c along the spiral (Fig. 9.12):

$$dM_c = c\, ds \cos \phi \, r = rc \frac{r\, d\theta}{\cos \phi} \cos \phi = cr^2\, d\theta$$

$$M_c = \int_0^{\theta_1} dM_c = \frac{c}{2 \tan \phi} (r_1^2 - r_0^2) \qquad (9.21)$$

The force C'' on plane ab is equal to the unit stress c times the distance $\overline{ab}$. P'' acts at the midpoint of $b*$ and summation of moment about 0 brings

$$\sum M_0 = P''_p l'''_{P_p} + M_c - C'' l''_c - P'' l''_p = 0$$

or
$$P'' = \frac{1}{l''_p} (P''_p l'''_{P_p} + M_c - C'' l''_c) \qquad (9.22)$$

If P' and P'' are known, the load Q may be determined by considering the triangular mass aba (Fig. 9.10) as the free body. The forces are P', P'', C'', Q,

---

* See Problem 9.8.

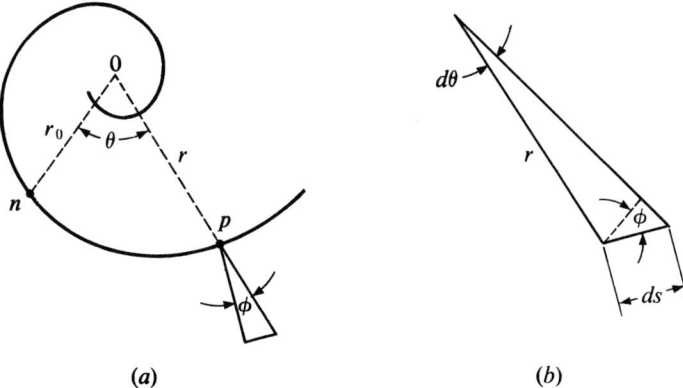

FIGURE 9.12 *Properties of the logarithmic spiral.*

and the weight of the mass aba'. Taking the summation of forces in the vertical direction, one finds that

$$Q_f = 2(P' + P'') + 2C'' \sin \phi - \left(\frac{\gamma B^2}{4}\right) \tan \phi \qquad (9.23)$$

The preceding discussion assumes that the failure surface bce is known. Actually, the failure surface is not accurately established because of the approximations mentioned at the beginning of this section. Therefore, the critical surface must be determined by trial. Computations of Q should be made for a number of trial surfaces, using different locations of the center of spiral. The trial surface that results in the minimum value of Q is the critical one.

9.6 Bearing capacity of shallow foundations

The analysis presented in Sec. 9.5 is used to determine the bearing capacity of shallow foundations. Figure 9.13 shows a section beneath a continuous footing located at a depth D_f beneath the ground surface. A shallow foundation is defined as one in which D_f is equal to or less than B [Terzaghi (1943b)]. To simplify the calculations, the part of the soil mass above ad is treated as a surcharge exerting a pressure q on the surface ad. The surface of failure is assumed to terminate at d. This ignores the shearing resistance of the soil located above ad, and therefore tends to underestimate the bearing capacity.

It can be seen from the solution described in Sec. 9.5 that the value of bearing pressure at failure $q_f = Q_f/B$ represents the sum of the following three components. The weight of the soil mass $abcd$ and the passive earth pressure on plane dc constitute the first part of the bearing capacity. This is P' in Sec. 9.5 [Eq. (9.19)]. The weight W increases with the square of $B/2$. Equation (9.18) shows that the passive pressure P'_p is proportional to h^2 and hence proportional

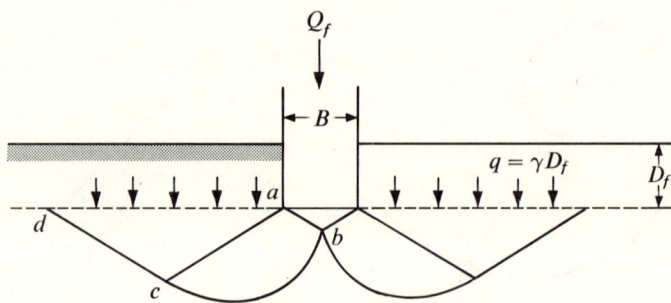

FIGURE 9.13 *Bearing capacity of a shallow foundation.*

to $(B/2)^2$. Thus the contribution to Q_f by W and P'_p increases with $(B/2)^2$ and their contribution to q_f increases with $(B/2)$. Furthermore, both W and P'_p are proportional to the unit weight γ. If this portion of the bearing capacity is denoted as q_γ, we can write for simplicity

$$q_\gamma = \tfrac{1}{2}\gamma B N_\gamma$$

in which N_γ is the proportionality factor, called a *bearing-capacity factor*. Its value can be computed, after P' is calculated, as outlined in Sec. 9.5.

By similar examination we see that P'' and C'' are proportional to $B/2$ and c. Therefore the contribution of the cohesion q_c to the bearing capacity is independent of B and we have

$$q_c = c N_c$$

in which N_c is the bearing-capacity factor for cohesion. The effect of the surcharge is also independent of B and is proportional to D_f and γ. Thus

$$q_q = \gamma D_f N_q$$

in which N_q is the bearing-capacity factor for surcharge. The total bearing capacity can be expressed as

$$q_f = \tfrac{1}{2} B \gamma N_\gamma + c N_c + \gamma D_f N_q \tag{9.24}$$

N_γ, N_c, and N_q are dimensionless coefficients that are governed only by the value of ϕ. This is because the shape of the failure surface depends on ϕ. It is, therefore, possible to compute these coefficients for a set of values of ϕ, and these values may be used for all bearing-capacity calculations. Values of N_γ, N_c, and N_q as given by Terzaghi (1943b) are presented graphically in Fig. 9.14.

Many other solutions using different assumed failure surfaces have been proposed. The value of N_γ is particularly sensitive to the shape of the failure surface. The largest value computed is of the order of five times the smallest value.

SHAPE FACTORS. Footings with circular and rectangular shapes present very

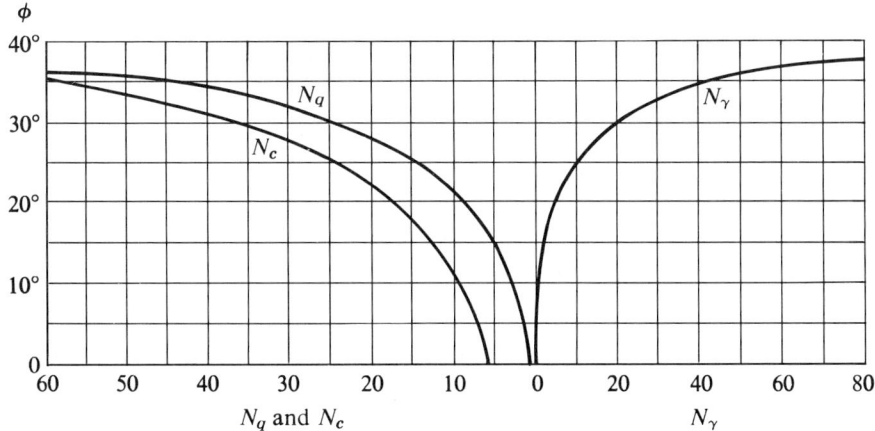

FIGURE 9.14 *Bearing capacity factors.* [After Terzaghi and Peck (1967).]

difficult mathematical problems. Semiempirical equations have been proposed by various individuals. The bearing-capacity equation may be written as

$$q_f = \frac{\gamma}{2} \lambda_\gamma B N_\gamma + \lambda_c c N_c + \gamma D_f N_q \tag{9.25a}$$

where λ_γ and λ_c are shape factors. Values of shape factors proposed include

$$\lambda_\gamma = 0.6 \quad \text{and} \quad \lambda_c = 1.3 \tag{9.25b}$$

for circular footings [Terzaghi (1943b)] and

$$\lambda_\gamma = 1 - 0.2\frac{B}{L} \quad \text{and} \quad \lambda_c = 1 + 0.2\frac{B}{L} \tag{9.25c}$$

for rectangular foundations [Skempton (1951)], where B and L are the width and length of the foundation.

ECCENTRIC LOADS. A foundation supporting an eccentric load can be analyzed with the following additional simplification. The pressure on the base of the foundation is assumed to act over a reduced area. Figure 9.15 shows a strip foundation carrying a load with an eccentricity l. The reaction is assumed to be distributed over a width equal to $B - 2l$ [Meyerhoff (1953)]. The bearing capacity is then equal to that of a concentrically loaded foundation with a width equal to $B - 2l$.

9.7 Bearing capacity of deep foundations

The bearing capacity of foundations located at some depth below the ground surface has been derived by Meyerhoff (1951), Berezentzev et al. (1961), Hu

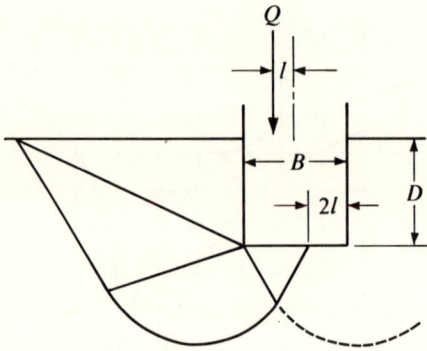

FIGURE 9.15 *Bearing capacity of foundations subjected to an eccentric load.* [After Meyerhoff (1953).]

(1965) and Durgunoglu and Mitchell (1973), among others. The solution by Durgunoglu and Mitchell is described here. Figure 9.16 shows the load applied through a strip foundation buried deep inside the soil mass. The failure zone is bounded by ABC. The failure surface consists of a straight line AB and a section of the logarithmic spiral BC. At C the spiral is tangent to the vertical line CD. It is assumed that no shearing resistance is developed along CD. If the depth of foundation is relatively small BC may intersect the ground surface at an angle. For this case the derivation is more complicated and the reader should consult the paper by Durgunoglu and Mitchell.

COHESION. To simplify the calculations we determine the bearing capacity due to cohesion separately from that due to $\sigma \tan \phi$ and weight. For the cohesion part, we consider the soil as weightless, and the failure surface can be obtained as follows.

Line $0B$ is located first. The stresses in $0AB$ are as shown in Fig. 9.16(c). Since $0B$ is a slip line, its stress state is represented by point b on the failure envelope. $0A$ is not a slip line, so the stresses on that plane are represented by some other point on the Mohr's circle. If the angle of friction and adhesion between soil and foundation are δ and c_a, then the stresses on $0A$ must fall on line $0a$. On this line the shear stress is

$$\tau = c_a + \sigma \tan \delta$$

Thus point p, which is the intersection of the Mohr circle with $0a$, satisfies this condition and is the stress state on $0A$. With ξ known the spiral may be drawn tangent to AB, ending at C with a vertical tangent. Thus θ_0 is known.

The stresses on $0B$ can be calculated from the geometry of Mohr's circle. For a weightless material, there are no stresses on $0C$. The stresses σ_b, τ_b as

SEC. 9.7 Bearing capacity of deep foundations

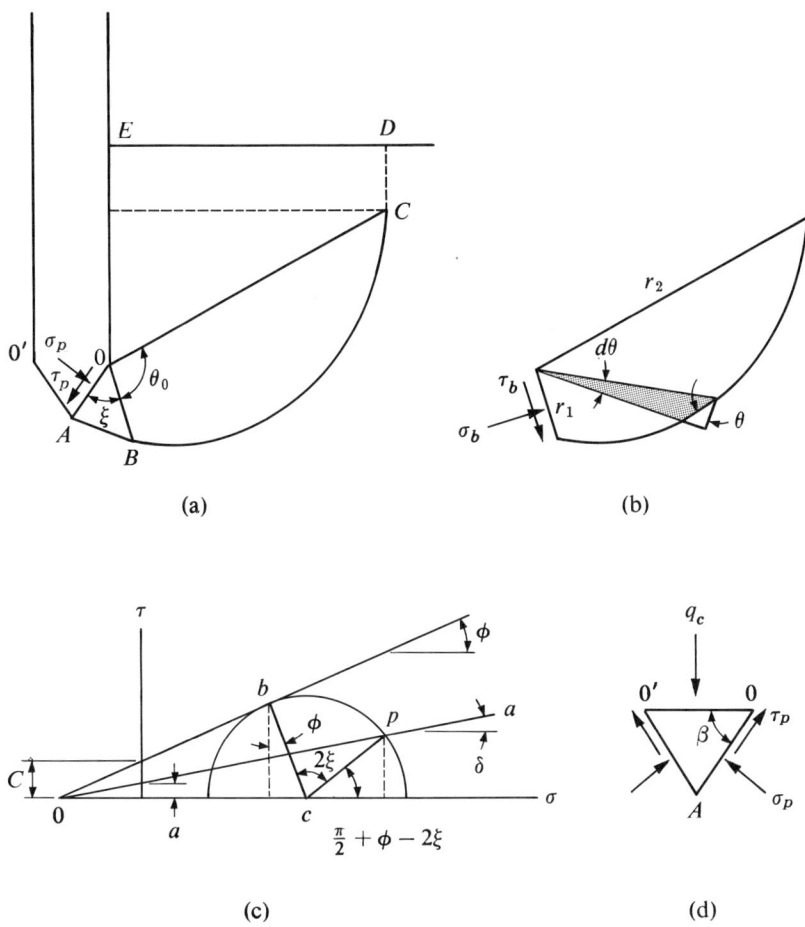

FIGURE 9.16 *Bearing capacity of a deep foundation, cohesion only.* [After Durgunoglu and Mitchell (1973).]

indicated by point b on the Mohr's circle may be obtained by summation of moments about 0 [Fig. 9.16(b)]:

$$\sum M_0 = -\sigma_b \frac{r_1^2}{2} + \int_0^{\theta_0} cr^2 \, d\theta = 0$$

or
$$\sigma_b = \frac{c}{\tan \phi} [e^{2\theta_0 \tan \phi} - 1] \quad (9.26a)$$

Since point b lies on the failure envelope,

$$\tau_b = c + \sigma_b \tan \phi = ce^{2\theta_0 \tan \phi} \quad (9.26b)$$

The stresses σ_p, τ_p may be obtained from the geometry of Mohr's circle [Fig. 9.16(c)].

$$\sigma_p = \frac{\tau_b}{\tan \phi} - \frac{c}{\tan \phi} + \frac{\tau_b}{\cos \phi} \sin \phi + \frac{\tau_b}{\cos \phi} \sin (2\xi - \phi)$$

$$= \frac{\tau_b}{\cos \phi} \left[\frac{1}{\sin \phi} + \sin (2\xi - \phi) \right] - \frac{c}{\tan \phi}$$

and

$$\tau_p = \frac{\tau_b}{\cos \phi} \cos (2\xi - \phi)$$

Substitution of Eq. (9.26) into the above yields

$$\sigma_p = \frac{c e^{2\theta_0 \tan \phi}}{\cos \phi} \left[\frac{1}{\sin \phi} + \sin (2\xi - \phi) \right] - \frac{c}{\tan \phi} \quad (9.27a)$$

$$\tau_p = \frac{c e^{2\theta_0 \tan \phi}}{\cos \phi} \cos (2\xi - \phi) \quad (9.27b)$$

We now calculate the bearing capacity q_c by considering the stresses on the wedge $0A0'$ [Fig. 9.16(d)]. Summation of forces in the vertical direction gives

$$\Sigma F_y = q_c \cos \beta - \sigma_p \cos \beta - \tau_p \sin \beta = 0$$

or

$$q_c = \sigma_p + \tau_p \tan \beta$$

Substitution of Eq. (9.27) into this yields

$$q_c = \left\{ \frac{e^{2\theta_0 \tan \phi}}{\cos \phi} \left[\frac{1}{\sin \phi} + \sin (2\xi - \phi) \right] - \frac{1}{\tan \phi} \right. \\ \left. + \frac{e^{2\theta_0 \tan \phi}}{\cos \phi} \cos (2\xi - \phi) \tan \beta \right\} \quad (9.28a)$$

We may express the bearing capacity due to cohesion as

$$q_c = c N_{pc} \quad (9.28b)$$

Then the bearing capacity factor is the quantity in { } in Eq. (9.28a).

INTERNAL FRICTION. The contribution of internal friction and weight to the bearing capacity is derived by consideration of the forces shown in Fig. 9.17(a). Summation of moments about 0 gives

$$\Sigma M_0 = -P_b l_b - P_1 l_1 + P_2 l_2 + W_1 l_{w1} + W_2 l_{w2} + M_w = 0 \quad (9.29)$$

where M_w is the moment produced by $0BC$. M_w may be computed by expressing the moment of the shaded triangular element as

$$dM = \tfrac{1}{2} \gamma r^2 l \, d\theta$$

From the geometry of the spiral,

$$r = r_1 e^{\theta \tan \phi}$$

$$l = \tfrac{2}{3} r \cos (\theta_0 - \alpha - \theta)$$

SEC. 9.7　　　　　Bearing capacity of deep foundations　　　　　　　241

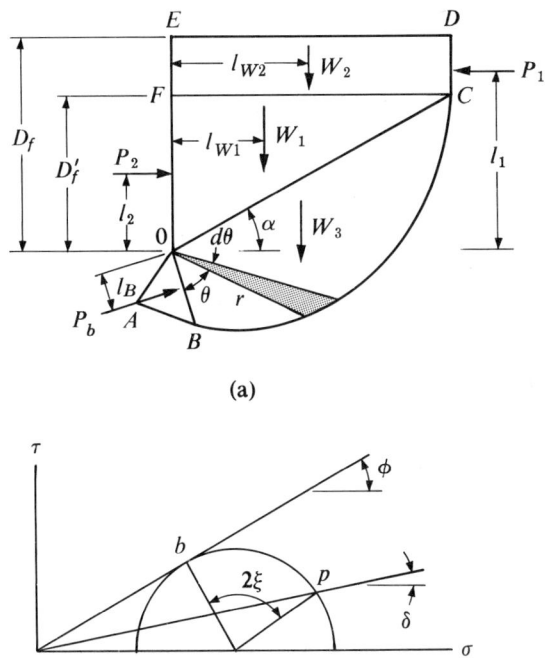

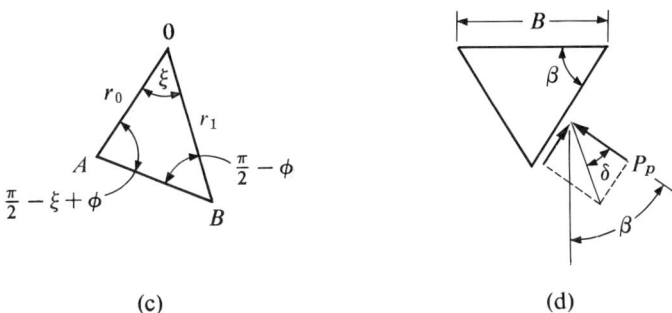

FIGURE 9.17 *Bearing capacity of a deep foundation, weight only.* [After Durgunoglu and Mitchell (1973).]

Substituting these into the above and integrating, we have

$$M_w = \int_0^{\theta_0} dM_w = \frac{\gamma r_1^3}{3} \{3 \tan \phi [e^{3\theta_0 \tan \phi} \cos \alpha - \cos(\theta_0 - \alpha)]$$
$$+ e^{3\theta_0 \tan \phi} \sin \alpha + \sin(\theta_0 - \alpha)\} \frac{1}{1 + 9 \tan^2 \phi} \quad (9.30a)$$

For simplicity we write

$$M_w = \frac{\gamma r_1^3}{3} N_\theta \qquad (9.30b)$$

where N_θ is the influence number equal to the quantity in { } in Eq. (9.30a).

The other forces in Eq. (9.29) and the distances are

$W_1 = \frac{1}{2}\gamma(OF)(CF)$ $\qquad l_{w1} = \frac{1}{3}(CF)$
$W_2 = \gamma(EF)(CF)$ $\qquad l_{w2} = \frac{1}{2}(CF)$
$P_1 = \frac{1}{2}K_0\gamma[D_f - (OF)]^2$ $\qquad l_1 = (OF) + \frac{1}{3}(EF)$
$P_2 = \frac{1}{2}K_0\gamma D_f^2$ $\qquad l_2 = \frac{1}{3}D_f \quad l_b = \frac{2}{3}r_1$
$(OF) = r_1 e^{\theta_0 \tan \phi} \sin \alpha$ $\qquad (CF) = r_1 e^{\theta_0 \tan \phi} \cos \alpha$
$(EF) = D_f - (OF) = D_f - r_1 e^{\theta_0 \tan \phi} \sin \alpha$

Substituting these and Eq. (9.30a) into Eq. (9.29) and solving for P_b, we obtain

$$P_b = \tfrac{1}{2}\gamma r_1^2 N_\theta + \tfrac{1}{4}\gamma r_1^2 \sin\alpha \cos^2\alpha\, e^{3\theta_0 \tan\phi} + \tfrac{3}{4}\gamma r_1 \cos^2\alpha\, e^{2\theta_0 \tan\phi}$$

$$(mB - r_1 \sin\alpha\, e^{\theta_0 \tan\phi}) + \tfrac{1}{4}K_0\gamma \left(\frac{B^3}{r_1}\right) m^3$$

$$- \tfrac{1}{4}K_0\gamma \left(\frac{B^3}{r_1}\right)(m - m')(m + 2m') \qquad (9.31)$$

where $m = D_f/B$ and $m' = D_f'/B$.

With the force P_b given by Eq. (9.31), $\sigma_b = F_b/r_1$ is known. The stress σ_p may be calculated from the geometric relationship shown in Fig. 9.17(b) and we get

$$\sigma_p = \frac{\tau_b}{\cos\phi}\left[\frac{1}{\sin\phi} + \sin(2\xi - \phi)\right]$$

From Fig. 9.16(c) and (d) we see that

$$P_p = \sigma_p r_0 \quad \text{and} \quad r_1 = r_0 \frac{\cos(\xi - \phi)}{\cos\phi} = \frac{B\cos(\xi - \phi)}{2\cos\beta\cos\phi}$$

Then, we obtain

$$P_p = P_b \frac{1 + \sin\phi \sin(2\xi - \phi)}{\cos\phi \cos(\xi - \phi)} \qquad (9.32)$$

where P_b is as given by Eq. (9.31).

To obtain the bearing capacity q_γ we resolve the forces shown in Fig. 9.17(d) as follows

$$\Sigma F_y = -q_\gamma \frac{B}{2} + \frac{P_p}{\cos\delta}\cos(\beta - \delta) = 0 \qquad (9.33)$$

Combining Eqs. (9.31), (9.32), and (9.33) gives

$$q_\gamma = \gamma B \left\{ \frac{\cos(\beta - \delta)}{\cos\delta}\left[\frac{1 + \sin\phi \sin(2\xi - \phi)}{\cos\phi \cos(\xi - \phi)}\right]\left[\frac{\cos^2(\xi - \phi)}{4\cos^2\beta \cos^2\phi} N_\theta \right.\right.$$

$$+ \frac{3\cos(\xi - \phi)}{4\cos\beta \cos\phi}\cos^2\alpha\, e^{2\theta_0 \tan\phi}(m - \tfrac{2}{3}m') + K_0 \frac{\cos\beta \cos\phi}{\cos(\xi - \phi)} m^3$$

$$\left.\left. - K_0 \frac{\cos\beta \cos\phi}{\cos(\xi - \phi)}(m - m')^2(m + 2m')\right]\right\} \qquad (9.34a)$$

The bearing capacity due to internal friction and weight is expressed as

$$q_\gamma = \gamma B N_{p\gamma} \tag{9.34b}$$

The bearing capacity factor N_γ is the quantity inside { } in Eq. (9.34a).

Values of N_c and N_γ have been published by Durgunoglu and Mitchell (1973). The results for a flat end ($\beta = 0$) are applicable to strip foundations, and are reproduced in Fig. (9.18). Two cases are shown. The conditions $\delta/\phi = 0$ and $\delta/\phi = 1$ represent respectively the perfectly rough and perfectly smooth foundations. The failure surfaces assumed for the two cases are shown in Fig. 9.18.

For buried foundations, Eq. 9.32 gives the bearing capacity of the end $0A0'$ [Fig. 9.15 (a)]. To this we must add the shearing resistance along the side of the foundation $0E$ produced by friction f between soil and foundation. Thus the total load at failure is given by

$$Q_f = q_f B + 2fD_f \tag{9.35}$$

SHAPE FACTORS. For piles and piers with circular or rectangular cross-sections, Durgunoglu and Mitchell have recommended the shape factors proposed by Hansen (1961) since they agree with experimental results. The shape factors for large values of m are

$$\lambda_c = \lambda_\gamma = 1 + (0.2 + \tan^6 \phi)\frac{B}{L} \tag{9.36a}$$

and

$$\lambda_c = \lambda_\gamma = 1 + (0.2 + \tan^6 \phi) \tag{9.36b}$$

for rectangular and circular foundations respectively.

SIMPLIFIED SOLUTION. Terzaghi and Peck (1967) adopted the following simplification for the calculation of the bearing capacity. The failure surface is taken to be bc, as shown in the left half of Fig. 9.19, instead of the curve bd' in the right half. The portion bc is assumed to be the same as bcd in the case of shallow footings (Fig. 9.13). The shear strength of the soil along cd is neglected. The bearing capacity of the end aa' of the foundation thus becomes the same as that expressed by Eqs. (9.24) and (9.25). In addition to the end-bearing capacity, there exists the shearing resistance f between the soil and the foundation. Therefor, the total bearing capacities of square and circular foundations are, respectively,

$$Q_f = B^2(0.4\gamma B N_\gamma + 1.3cN_c + \gamma D_f N_q) + 4fBD_f \tag{9.37a}$$

$$Q_f = \pi R^2(0.6\gamma R N_\gamma + 1.3cN_c + \gamma D_f N_q) + 2f\pi R D_f \tag{9.37b}$$

It can be seen that errors of considerable magnitudes may be introduced by these simplifications. However, at our present state of knowledge, this inaccuracy is overshadowed by the uncertainty of the soil behavior around deep foundations (see Sec. 11.7).

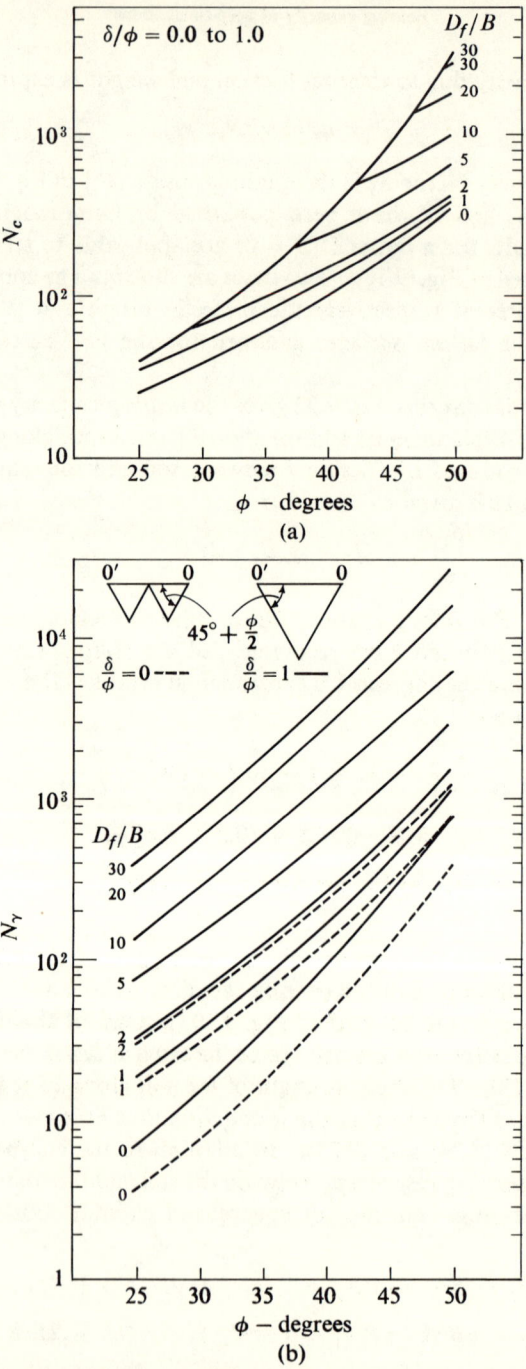

FIGURE 9.18 *Bearing-capacity numbers for deep foundations.* [After Durgunoglu and Mitchell (1973).]

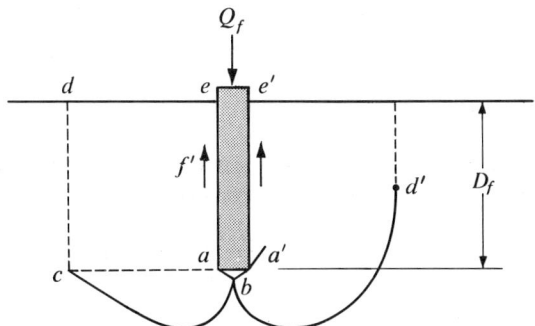

FIGURE 9.19 *Comparison of slip surfaces for deep foundations.*

Example 9.1. A continuous strip foundation 8 ft wide is located 10 ft below the ground surface. The subsoil consists of a thick deposit of cohesionless sand with ϕ equal to 35 deg, γ equal to 130 pcf, and δ equal to 25 deg. The water table is at very great depth and its effect may be ignored. Calculate the bearing capacity of the foundation.

Solution. According to the definition, this is to be considered a deep foundation. We first calculate the bearing capacity using Durgunoglu and Mitchell's solution. For $D_f/B = 1.25$, Fig. 9.18(b) gives $N_{p\gamma} = 75$. Then

$$q_f = \tfrac{1}{2} \times 8 \times 130 \times 75 = 39{,}000 \text{ psf}$$

Assuming that $K_0 = 0.5$,

$$f = \tfrac{1}{2} \times 10 \times 130 \times 0.5 \tan 25° = 150 \text{ psf}$$

Therefore,

$$Q_f = 39{,}000 \times 8 + 2 \times 150 \times 10 = 315{,}000 \text{ lb/ft}$$

We next calculate the bearing capacity from the simplified Terzaghi theory. From Fig. 9.14, $\phi = 35°$; $N_\gamma = 43$, $N_q = 43$:

$$Q_f = (8)(\tfrac{1}{2} \times 130 \times 8 \times 43 + 130 \times 10 \times 43) + 2 \times 137 \times 10$$
$$= 628{,}000 \text{ lb/ft}$$

If we assumed a shallow foundation, the Terzaghi theory gives

$$Q_f = (8)(\tfrac{1}{2} \times 130 \times 8 \times 43 + 130 \times 10 \times 43)$$
$$= 625{,}000 \text{ lb/ft}$$

Example 9.2. A uniform soil deposit has the following properties: $\gamma = 135$ pcf, $\phi = 28$ deg, $c = 800$ psf. What is the allowable soil pressure for a square foundation located at 4 ft below ground surface and supporting a load of 920 kips? A safety factor of 2.5 is required against failure.

Solution. For a shallow foundation Fig. 9.14 gives

$$\phi = 28°: \quad N_\gamma = 15, \quad N_c = 32, \quad N_q = 18$$

However, Eq. (9.25) shows that the bearing capacity depends on the width B, which is unknown. We assume B to be 10 ft.

$$q_f = 0.4 \times 135 \times 10 \times 15 + 1.3 \times 800 \times 32 + 135 \times 4 \times 18$$
$$= 51{,}000 \text{ psf}$$

The bearing capacity is the pressure that produces failure in the soil mass. For a safety factor of 2.5, the allowable pressure q_a is

$$q_a = \frac{q_f}{F_s} = \frac{51{,}000}{2.5} = 20{,}400 \text{ psf}$$

The required foundation size is

$$\frac{920{,}000}{20{,}400} = 46 \text{ ft}^2 \qquad B = 7.0 \text{ ft (approx.)}$$

If the foundation is made 7 ft square, our initial estimate of B is in error. The revised value of q_f would be

$$q_f = 0.4 \times 135 \times 7 \times 15 + 1.3 \times 800 \times 32 + 135 \times 4 \times 18$$
$$= 48{,}560 \text{ psf}$$
$$q_a = 19{,}400 \text{ psf}$$

The change in the soil pressure is unimportant in this case.

9.8 Modifications of bearing capacity equations

Experimental evidence [Vesic (1963)] suggests that the slip surfaces used in the derivations of the bearing capacity equations in Sec. 9.5 through 9.7 are not always realized. For footings resting on the surface of cohesionless soils a slip surface of the type illustrated in Fig. 9.8 is realized only if the sand is dense. With sands of medium density considerable volumetric compression takes place in the soil and the slip surface is only partly developed [Fig. 9.20(a)] even at penetrations as large as $0.5B$. No visible slip surface is developed in loose sands even at very large penetrations [Fig. 9.20(b)]. The three cases have been called general shear, local shear, and punching shear [Terzaghi (1943), Vesic (1963)]. Modifications of the bearing capacity equations for shallow foundations to account for soil compressibility have been suggested by Vesic (1973).

As the depth of the foundation increases, the local shear and punching shear become more important. For very deep foundations in sand, general shear is not developed even if the sand is dense. Experimental data indicate also that the stress state in the soil is influenced by the method of installation. The general observation is that the bearing capacity due to end bearing and side friction do not increase indefinitely with depth as suggested by Eq. (9.35) or (9.37). Beyond a depth of about $20B$ both components are essentially constant with increasing depth (Vesic, 1967). A more reliable prediction of bearing capacity requires consideration of the stress conditions in the soil around the pile, and solutions using numerical techniques have been suggested [i.e. Desai (1974)].

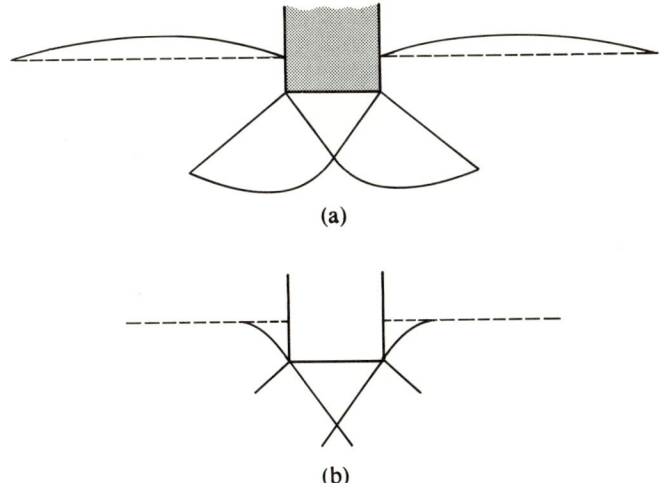

FIGURE 9.20 *Slip surface for (a) local shear; (b) punching shear.*

9.9 Slope analysis by circular-arc method

Consider a slope of finite height [Fig. 9.21(*a*)]. The failure surface may be approximated by a logarithmic spiral or a circular arc. Taylor (1948) showed that calculations made with logarithmic spiral and circular arc as failure surfaces give very close results. Extensive studies by Swedish engineers of slope failures in clay during the early part of this century also revealed that the failure surface closely approaches the circular arc. Figure 9.22 shows the displacement of the soil mass produced by a slope failure. The analysis of slope stability using a circular slip surface was developed by Fellenius (1936) and others and is often referred to as the *Swedish circle method.*

CIRCULAR ARC METHOD. Figure 9.21(*a*) shows a soil mass bounded by the slip surface *abcd.* Since the circular arc is an assumed slip surface, its position is not determinate. The arc *abcd* is therefore an arbitrarily assumed arc. The forces acting on the soil mass *abcd* are the weight of the soil mass W and the force on the failure surface. The force on any element *bc* of the slip surface is composed of a normal component dF_n and shear or tangential component dF_t. At failure, dF_t must be equal to the shear strength of the soil as expressed by Eq. (8.2), which in this case can be written as

$$dF_t = c\, dl + dF_n \tan \phi$$

The shear force dF_t is therefore composed of a term $c\, dl$, independent of the normal stress and a term $dF_n \tan \phi$, which is proportional to the normal stress. To simplify calculations, we combine the part $dF_n \tan \phi$ with the normal force

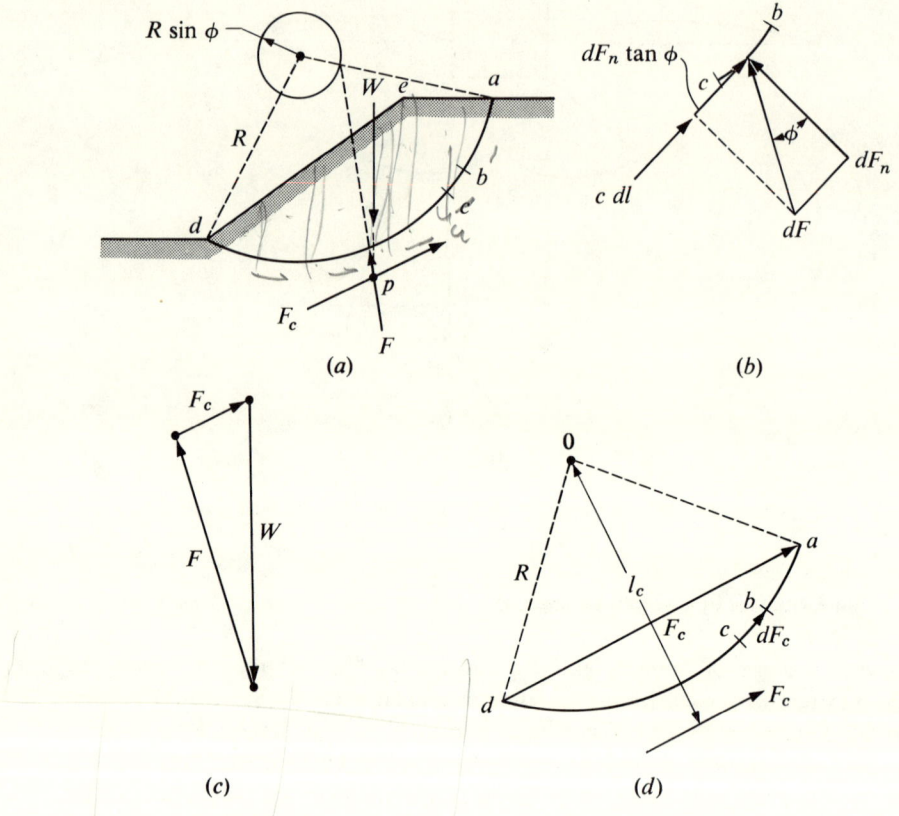

FIGURE 9.21   *Circular-arc method of slope analysis.*

as shown in Fig. 9.21(*b*). This force is designated as dF and makes an angle ϕ with the normal. The resultant of $c\,dl$ and dF over the entire failure surface is represented by F_c and F, respectively.

The resultant F_c of the cohesion distributed uniformly along the arc *ad* is equal to the unit cohesion times the chord distance $\overline{ad}$, or

$$F_c = c(\overline{ad})$$

This can be understood from examination of the forces shown in Fig. 9.21(*d*). The cohesion along *ad* is the summation of infinitesimal vectors whose directions follow the arc *ad*. Thus their resultant vector F_c has a magnitude and direction equal to that of the chord. The line of action of F_c must be such that its moment about any point (for convenience, point 0) is the same as that of the cohesion distributed along the failure surface.

The moment of the uniformly distributed cohesion along section *bc* of the

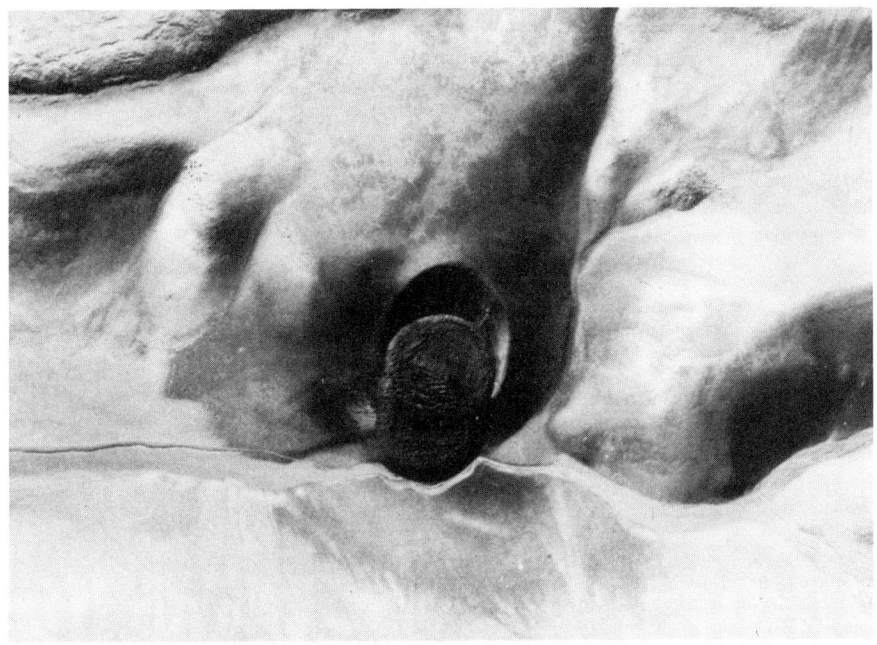

FIGURE 9.22 *Slope failure caused by stream erosion at the toe of the slope.* (Inter-American Geodetic Survey in Chile.)

arc is $c\,dl\,R$. For the entire length of the arc, it is $c\,\widehat{ad}\,R$. The moment of the resultant F_c is $F_c l_c$. Hence, equating the two we have

$$F_c l_c = c(\widehat{ad})\,R$$

or

$$l_c = \frac{c(\widehat{ad})\,R}{F_c} = \frac{c(\widehat{ad})\,R}{c(\overline{ad})} = \frac{(\widehat{ad})}{(\overline{ad})}\,R \tag{9.38}$$

Thus we see that l_c is larger than R since the arc distance $\widehat{ad}$ is always larger than the chord distance $\overline{ad}$.

For moment equilibrium the resultant F along arc ad must pass through point p, the intersection of the vectors F_c and W. Furthermore, as noted in Fig. 9.21(b), dF makes an angle ϕ with the normal to the failure surface. Thus the resultant F must make an angle approximately equal to ϕ with the normal. We may make an approximation and say F should be tangent to a circle with radius $R \sin \phi$ [Fig. 9.21(a)]. In this way, the direction and line of action of the force F is determined. At failure, the three forces W, F_c, and F are in equilibrium, as illustrated by the force polygon in Fig. 9.21(c).

Failure occurs if we imagine that the weight W is progressively increased by increasing the density γ until the equilibrium conditions shown in Fig. 9.21(c) are fulfilled. Since the slip surface is arbitrarily chosen, many slip circles must be tried. The one which satisfies the failure conditions at the smallest value of density γ is the one that will fail first. This is called the *critical circle*.

STABILITY NUMBER. We note that the equilibrium conditions at failure require that F, F_c, and W have the relationship shown in Fig. 9.21(c). Actually five parameters are involved when we consider the three forces F, F_c, and W. They are the shear strength of the material as represented by c and ϕ, the unit weight of the material γ, and the dimensions of the slope β and H (Fig. 9.23). If four of

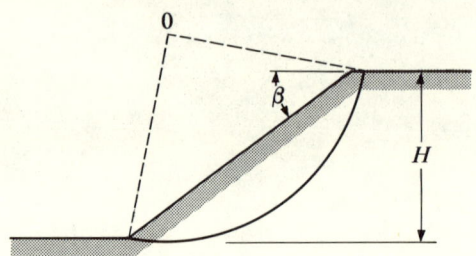

Soil Properties Unit weight: γ
Strength: $s = c + \sigma \tan \phi$

FIGURE 9.23 *Slope parameters included in stability number.*

the five parameters are given, the fifth one can be calculated. Since the five parameters account for all the physical properties of a slope on a homogeneous material, their relationship can be calculated once and for all. Taylor (1937) published the results of these calculations in the form of charts. To simplify the presentation, three of the parameters, c, γ, and H, are combined into a new parameter N_s, called the *stability number*, defined as

$$N_s = \frac{\gamma H}{c} \qquad (9.39)$$

Curves are constructed relating N_s to β for various values of ϕ, as shown in Fig. 9.24.

A very common case of nonhomogeneity is that of a very stiff material located at a depth $n_d H$ below the top of the slope (Fig. 9.25). This poses a restriction on the failure surface, as it cannot penetrate the stiff material. In this connection we should define the different types of failure surface. They are the toe circle, which passes through the toe [Fig. 9.25(a)]; the slope circle, which intersects the slope at some point above the toe [Fig. 9.25(b)]; and the midpoint circle whose center lies on a vertical line that passes through the midpoint of the slope [Fig. 9.25(c)]. Calculations show that for all values of ϕ greater than 3 deg, the critical circles are all toe circles. If the value of ϕ is close to or equal to 0, the position of the critical circle and the stability number is influenced by the dis-

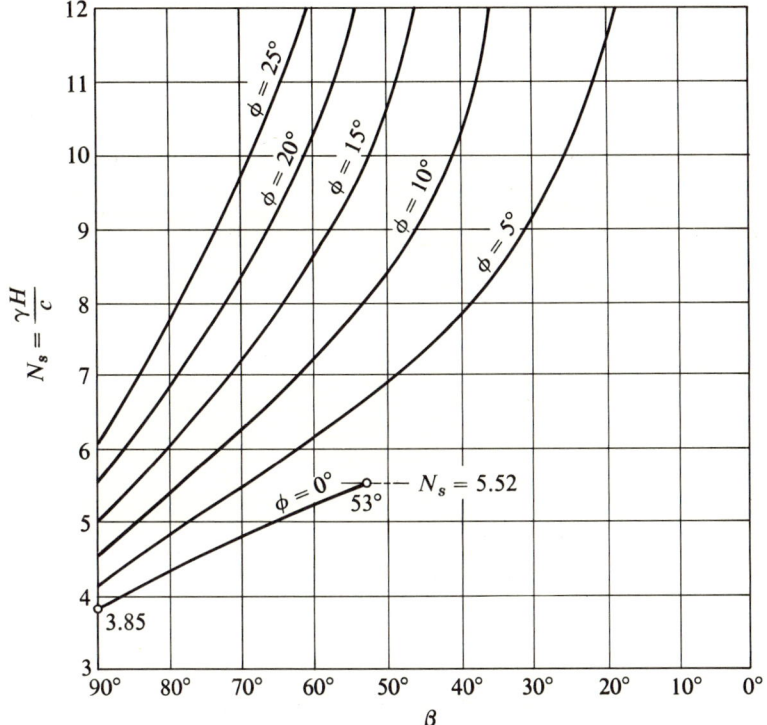

FIGURE 9.24 *Stability numbers.* [After Terzaghi and Peck (1967).]

tance to the stiff bottom. The stability numbers and the types of failures as determined by analysis are shown in Fig. 9.26. For all slopes with angle β larger than 53 deg, failure occurs along the toe circle. If β is less than 53 deg, the type of failure depends on the depth factor n_d. Failure may occur along the slope circle or midpoint circle as well as the toe circle. Failure occurs along the midpoint circle when n_d exceeds 4.

We now introduce a slightly different approach. Instead of asking what is the density at failure, as is done in the bearing-capacity problem, we ask how close is a given slope to failure, the properties of the material being known. It is equivalent to a comparison of the shear stress on the potential failure surface under the given conditions to the shear stresses at failure, which is equal to the strength of the material. The ratio of the shear strength to the shear stress under given conditions, is called the *factor of safety* F_s, or

$$F_s = \frac{s}{\tau} \tag{9.40}$$

It can also be considered as the ratio of the load at failure to the given load. To evaluate the factor of safety for a given slope, we first determine the parameters

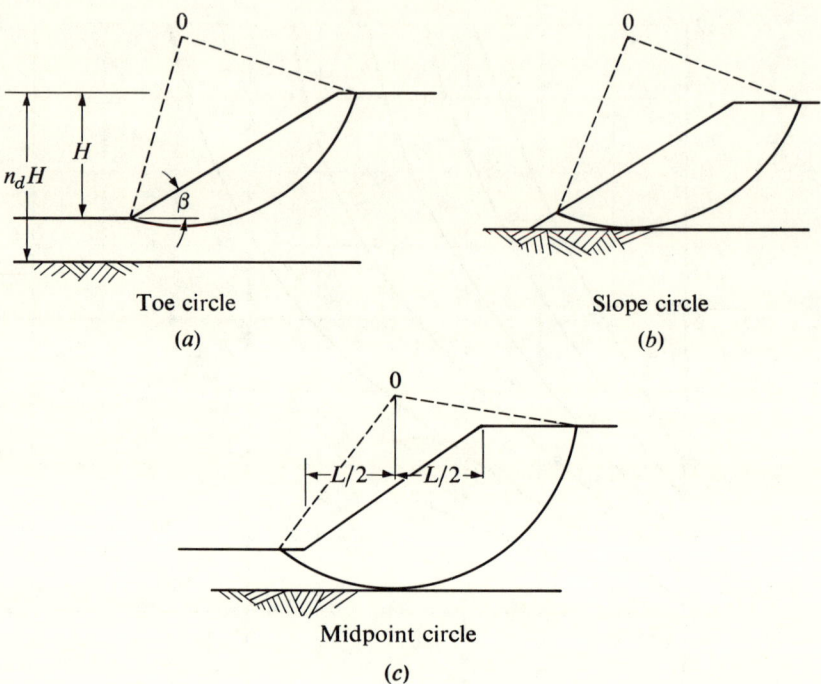

Figure 9.25 *Types of failure surface.*

c, H, and γ at failure. This can be obtained from Figs. 9.24 or 9.26 in the form of the stability number N_s for the given slope angle β and angle of friction ϕ. For the given H and c, we can calculate by Eq. (9.39) what γ should be in order to yield this particular stability number. This is the value that would result in failure of the slope with the given values of c, ϕ, H, and β. The factor of safety is then the ratio of the γ required for failure to the given γ.

Example 9.3. The simple slope shown in Fig. 9.23 has a height of 35 ft and a slope 1 vertical to 1 horizontal. The soil properties are as follows: $\gamma = 125$ pcf, $\phi = 15$ deg, $c = 700$ psf. By means of the stability chart, determine the factor of safety with respect to failure.

Solution. For a slope of 1 vertical to 1 horizontal we have
$$\tan \beta = 1.0 \qquad \beta = 45°$$
The stability number for $\phi = 15$ deg and $\beta = 45$ deg is 11.6 (Fig. 9.24). Thus at failure
$$N_s = \frac{\gamma H}{c} = 11.6$$
For the given conditions we have
$$N_s = \frac{125 \times 35}{700} = 6.25$$

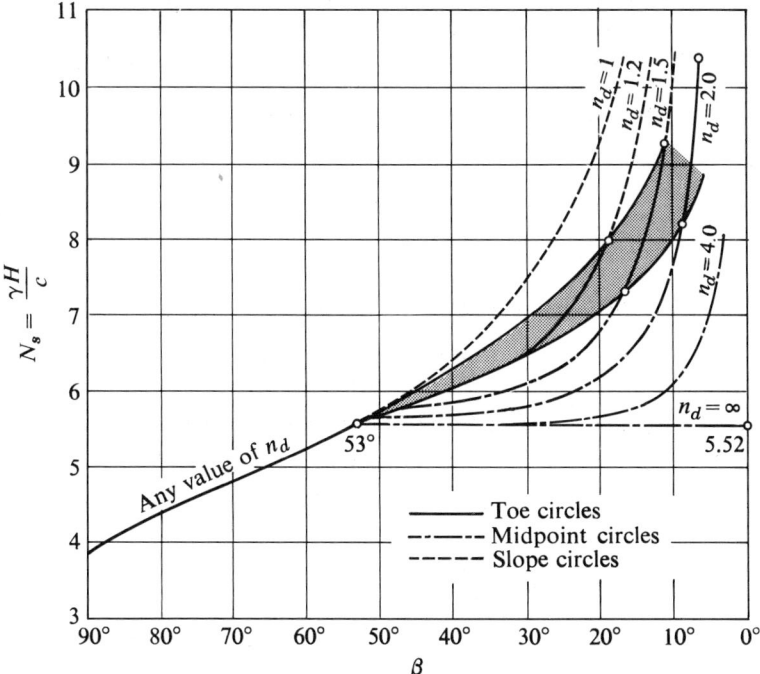

FIGURE 9.26 Stability numbers and types of slope failures for $\phi = 0$. [After Terzaghi and Peck (1967).]

which is much smaller than N_s required for failure. A stability number of 11.6 can be reached if the height or density is increased or if the cohesion is reduced. Failure would occur if

$$\gamma = \frac{N_s c}{H} = \frac{11.6 \times 700}{35} = 232 \text{ pcf}$$

or

$$H = \frac{N_s c}{\gamma} = \frac{11.6 \times 700}{125} = 65 \text{ ft}$$

The factor of safety is then the density (or height) that the soil can support, divided by the given density (or height):

$$F_s = \frac{232}{125} = \frac{65}{35} = 1.86$$

The cohesion required to make N_s equal 11.6 if γ and H are as given is

$$c = \frac{\gamma H}{N_s} = \frac{125 \times 35}{11.6} = 378 \text{ psf}$$

The factor of safety may also be the ratio of the given cohesion to that required for equilibrium, or

$$F_s = \frac{700}{378} = 1.86$$

It should be noted that this safety factor is for cohesion only. The value of ϕ used to enter the chart is 15 deg; hence the safety factor with respect to ϕ is 1.0. Therefore the value of 1.85 is not the factor of safety defined by Eq. (9.40), which considers the total shear strength.

The safety factor that applies equally to both c and ϕ may be expressed as

$$F_s = \frac{c + \sigma \tan \phi}{\tau}$$

or

$$\tau = \frac{c}{F_s} + \sigma \frac{\tan \phi}{F_s}$$

To calculate this safety factor, we first estimate F_s to be 1.60. Then

$$\frac{\tan \phi}{F_s} = \frac{\tan 15°}{1.60} = 0.168$$

This corresponds to a ϕ of 9.5 deg. We enter the chart (Fig. 9.24) with ϕ of 9.5 deg and find

$$N_s = 9.0 \qquad c = \frac{125 \times 35}{9.0} = 488 \text{ psf}$$

$$F_s = \frac{700}{488} = 1.43$$

Thus we can have factors of safety of 1.60 for ϕ and 1.43 for c instead of 1.00 for ϕ and 1.85 for c. If we want the same safety factor for c and ϕ, we must revise our first estimate of 1.60. Subsequent trials give a value of 1.49 for F_s that holds for both c and ϕ.

9.10 Slope analysis by method of slices

The method of slices is an important modification of the circular-arc method, because it enables one to calculate with reasonable accuracy the distribution of stresses along the failure surface. It is particularly useful in solving problems involving nonuniform soil and porewater-pressure conditions.

In the method of slices, the mass of soil *abcd* (Fig. 9.27) is divided into a number of segments. The forces acting on each segment are evaluated from the equilibrium of the segment.

We consider the forces on an individual slice *cdef*. The forces consist of the weight of the slice ΔW, the surface load acting on the slice Q, the normal and shear forces ΔF_n and ΔF_t acting on the failure surface *ef*, and the normal and shear forces P_n, T_n, P_{n+1}, and T_{n+1} on the vertical faces *df* and *ce*. The system is statically indeterminate, and to arrive at a solution it is necessary to make certain assumptions concerning the magnitudes and points of application of the forces P and T.

APPROXIMATE SOLUTION. An approximate solution may be obtained by assuming that the resultants of P_n and T_n are equal to that of P_{n+1} and T_{n+1} and that

SEC. 9.10   Slope analysis by method of slices

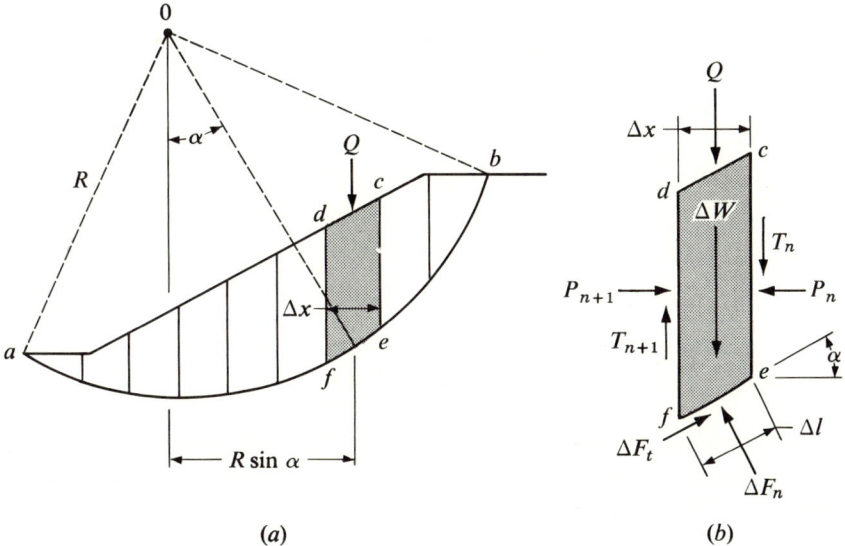

FIGURE 9.27   *Slope analysis by the method of slices.*

their lines of action coincide. This leaves only the forces ΔW, Q, ΔF_n, and ΔF_t. Applying the condition of equilibrium to the segment, we obtain

$$\Delta F_n = (\Delta W + Q) \cos \alpha$$
$$\Delta F_t = (\Delta W + Q) \sin \alpha \qquad (9.41)$$

The unit pressure on *ef* is equal to

$$\sigma_n = \frac{1}{\Delta l}(\Delta W + Q) \cos \alpha$$
$$\tau_n = \frac{1}{\Delta l}(\Delta W + Q) \sin \alpha \qquad (9.42)$$

The shearing resistance on a segment of the arc is

$$s\,\Delta l = (c + \sigma \tan \phi)\,\Delta l = c\,\Delta l + (\Delta W + Q) \cos \alpha \tan \phi$$

For moment equilibrium about 0 at failure,

$$\sum Rs\,\Delta l - \sum R \sin \alpha (\Delta W + Q) = 0$$

The summation is made over all the slices. If the loads are not sufficient to produce failure, then the "resisting moment" $\sum Rs\,\Delta l$ is greater than the "driving moment" $\sum R \sin \alpha (\Delta W + Q)$. We define the ratio of the resisting moment to the driving moment as the factor of safety, or

$$F_s = \frac{\sum Rs\,\Delta l}{\sum R \sin \alpha (\Delta W + Q)} = \frac{\sum [c\,\Delta l + (\Delta W + Q) \cos \alpha \tan \phi]}{\sum (\Delta W + Q) \sin \alpha} \qquad (9.43)$$

Since the failure surface is not predetermined, it is necessary, as before, to locate the critical circle (minimum factor of safety) by trial.

The foregoing analysis does not distinguish between total stresses and effective stresses. Whichever stress is adopted, the appropriate shear-strength parameters should be used. As an example, we analyze the stability by means of effective stresses. In this case, the shearing resistance on a segment of the slip surface becomes

$$s \, \Delta l = [\bar{c} + (\sigma - u) \tan \bar{\phi}] \, \Delta l = \bar{c} \, \Delta l + [(\Delta W + Q) \cos \alpha - u \, \Delta l] \tan \bar{\phi}$$

Consequently, the factor of safety is

$$F_s = \frac{\sum \{\bar{c} \, \Delta l + [(\Delta W + Q) \cos \alpha - u \, \Delta l] \tan \bar{\phi}\}}{\sum (\Delta W + Q) \sin \alpha} \tag{9.44}$$

REFINED SOLUTION. We can improve the accuracy of the analysis by taking forces P and T into consideration. For the element in Fig. 9.27(b), summation of forces in the vertical direction gives

$$\sum F_y = -\Delta F_n \cos \alpha - \Delta F_t \sin \alpha + [(\Delta W + Q) + (T_n - T_{n+1})] = 0$$

or

$$\Delta F_n \cos \alpha = [(\Delta W + Q) + (T_n - T_{n+1})] - \Delta F_t \sin \alpha$$

For effective-stress analysis we must deduct the porewater pressure over ef, which is $u \, \Delta l$, or

$$\Delta \bar{F}_n \cos \alpha = [(\Delta W + Q) + (T_n - T_{n+1})] - u \, \Delta l \cos \alpha - \Delta F_t \sin \alpha \tag{9.45}$$

If the slope is not on the verge of failure ($F_s > 1$) the tangential force ΔF_t is equal to the shearing resistance on ef divided by F_s,

$$\Delta F_t = \frac{\bar{c} \, \Delta l}{F_s} + \Delta \bar{F}_n \frac{\tan \bar{\phi}}{F_s}$$

Substituting this in the Eq. (9.45) and solving for $\Delta \bar{F}_n$, we get

$$\Delta \bar{F}_n = \left[(\Delta W + Q) + (T_n - T_{n+1}) - u \, \Delta l \cos \alpha - \frac{\bar{c}}{F_s} \Delta l \sin \alpha \right] \frac{1}{\cos \alpha + (\tan \bar{\phi} \sin \alpha / F_s)} \tag{9.46}$$

The factor of safety of the entire slope is

$$F_s = \frac{\sum (\bar{c} \, \Delta l + \Delta \bar{F}_n \tan \bar{\phi})}{\sum (\Delta W + Q) \sin \alpha} \tag{9.47}$$

Substituting Eq. (9.46) in (9.47) gives

$$F_s = \frac{\sum \{\bar{c} \, \Delta l \cos \alpha + [(\Delta W + Q - u \, \Delta l \cos \alpha) + (T_n - T_{n+1})] \tan \bar{\phi}\} \dfrac{1}{\cos \alpha + (\tan \bar{\phi} \sin \alpha / F_s)}}{\sum (\Delta W + Q) \sin \alpha} \tag{9.48}$$

To obtain F_s from Eq. (9.48), the quantities $T_n - T_{n+1}$ must be evaluated. This is done by successive approximation. Trial values of P_n and T_n that satisfy the equilibrium of each slice, and the conditions

$$\sum (P_n - P_{n+1}) = 0 \qquad \sum (T_n - T_{n+1}) = 0$$

are used. The calculations can be simplified if the term $\sum (T_n - T_{n+1}) \tan \bar{\phi}$ is assumed to be 0. This avoids the necessity of evaluating P_n and T_n. The value of F_s may then be calculated by first assuming a value for F_s. This assumed value, together with the soil properties and given conditions of the slope, are substituted in Eq. (9.48) and F_s is calculated. If the calculated quantity differs appreciably from the assumed value, a second approximation is made and the calculation is repeated. To aid the calculations the quantity

$$m_\alpha = \cos \alpha + (\tan \bar{\phi} \sin \alpha / F_s)$$

may be plotted for various values of α, $\bar{\phi}$, and F_s, as in Fig. 9.28.

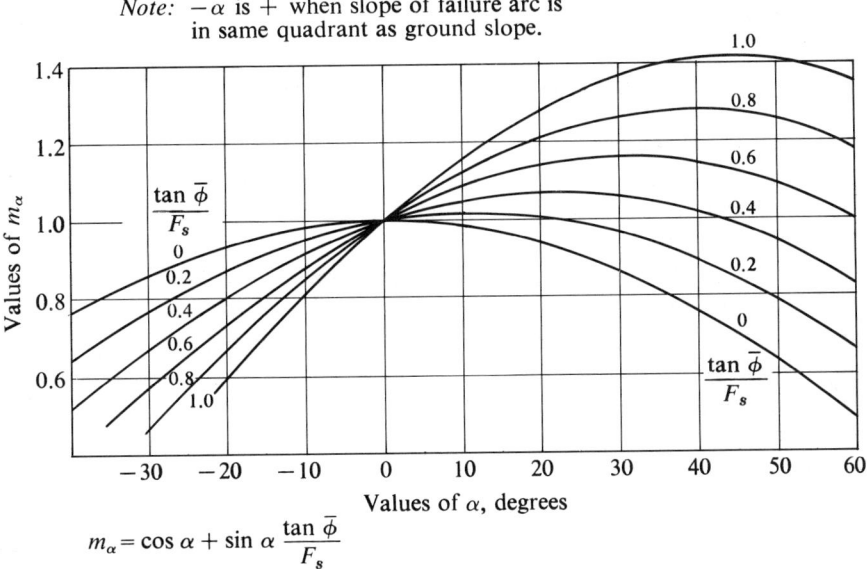

FIGURE 9.28 *Values of m_α.* [After Janbu et al. (1956)].

The method of slices has also been extended to the analysis of noncircular failure surfaces by Janbu (1955) and Morgenstern and Price (1965).

9.11 Effect of seepage and porewater pressure on stability

The role of porewater pressure is very important in all problems involving stability and failure in soils. To take the porewater pressure into account, the calculations should be made in terms of effective stress. This section illustrates the principles of effective-stress analysis. The significance of total stress and effective-stress analyses with respect to the stability of a soil mass is given detailed examination in Chapter 11.

If the effective-stress analysis is made for a stability problem, one should include the porewater pressure acting on the soil mass in our analysis (see Sec. 9.10 for example). Second, the effective-stress shear-strength parameters $\bar{c}$ and $\bar{\phi}$ should be used.

As illustrations, we consider the following three cases. The foundation in Fig. 9.10 rests on sand and the water table is at the ground surface. To calculate the bearing capacity using the effective stress we include the porewater pressure on the surface *abcd* in addition to the forces shown in Fig. 9.11. If we consider the drained state, then the excess hydrostatic porewater pressure is 0 and the porewater pressure on *abcd* is hydrostatic. This means that the resultant water pressure is equal to the buoyancy force on the mass *abcd*. This is equivalent to the replacement of the weight W by W', which is the submerged weight of *abcd*. In this case, where seepage and excess hydrostatic porewater pressure are absent, the effective-stress analysis is made by taking the submerged weight of the soil below the water table.

The second example is the submerged slope shown in Fig. 11.1(*b*). We use the circular-arc method of analysis and consider the entire sliding mass as the free body. If we again consider the drained state the porewater pressure on this mass consists of the hydrostatic pressure on the slope and on the slip surface. Their resultants are shown as P_{W1} and P_{W2}, respectively. As in the first example, we see that the forces P_{W1} and P_{W2} are the same as the buoyancy force. It follows that P_{W1}, P_{W2}, and W may be replaced by the submerged weight of the mass W'.

As a third example we consider a slope subjected to steady seepage, as shown in Fig. 9.29(*a*) and during rapid drawdown, Fig. 9.29(*b*). The flow net is constructed, and this makes it possible to evaluate the porewater pressure at every point (Sec. 3.6). If we analyze the stability by the method of slices, the porewater pressure u on any segment (such as *cd*) of the slip surface can be determined from the flow net. This quantity is used in Eq. (9.44) or (9.48).

UPPER AND LOWER BOUND SOLUTIONS

The approximate solutions demonstrate that it is often necessary to make assumptions regarding the shape of the slip surface in order to arrive at a solution. One should realize that for a given problem there can be more than one reasonable slip surface. Since these solutions all stem from what appear to be reasonable assumptions, there exists no a priori reason to prefer one solution over another. Sometimes it is difficult to choose "reasonable" assumptions regarding the slip surface. When one is faced with a wide variety of possibilities, it becomes helpful to establish limiting values to the answer. These are called the *upper and lower bound solutions*.

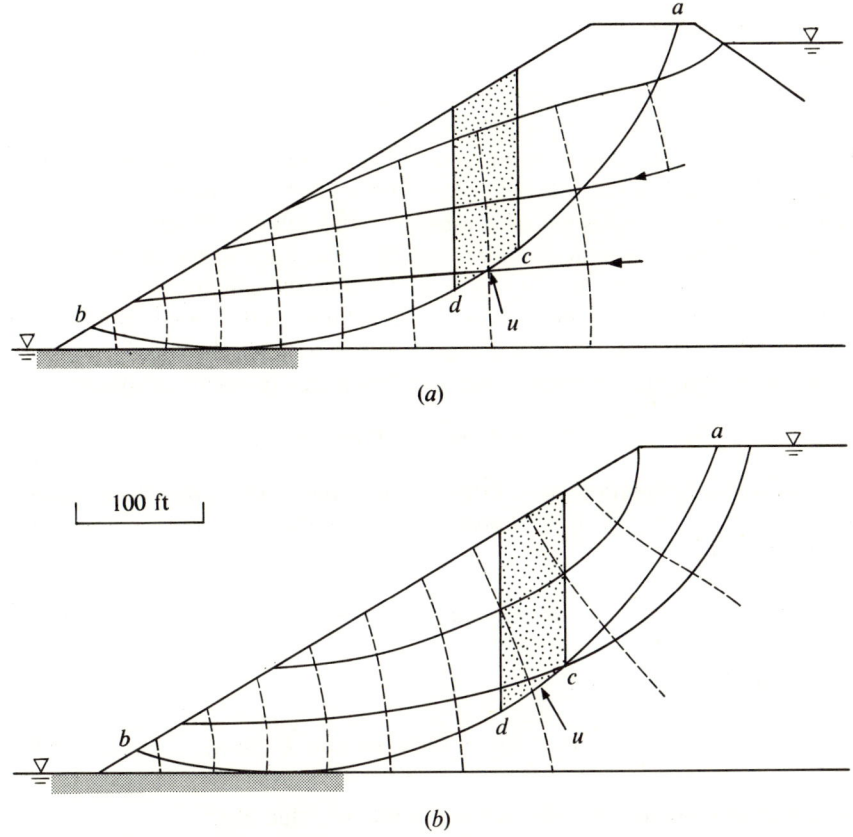

FIGURE 9.29 *Stability analysis by method of slices:* (a) *slope under steady seepage;* (b) *slope after drawdown.*

9.12 Statically admissible and kinematically admissible states

To obtain the upper and lower bound solutions, we make use of two important theorems in the theory of plasticity. They are stated as follows: (a) If a safe, *statically admissible state* of stress exists for a given loading, failure does not occur under this loading; (b) if a *kinematically admissible failure state* can be found at any loading, failure must impend or have taken place already. A safe, statically admissible state of stress is one in which the stress distribution satisfies the equilibrium condition under the given loads, and in which the stresses are less than the yield stress at every point. Thus according to the first theorem, failure does not occur as long as one can find a stress state that can satisfy the condition of equilibrium and not exceed the yield stress at any point. In a kinematically admissible state the displacement velocity is such that the rate of work done by the loads and body forces is equal to the rate of energy dissipation by the stresses. For proof of these theorems, the reader is referred to Drucker et al. (1951).

9.13 Grouser-plate problem

To illustrate the lower and upper bound solutions, we study the grouser plate of a tracked vehicle as solved by Haythornthwaite (1961) (Fig. 9.30). The grouser plate *bcd* is subjected to an inclined force Q. It is required to find the maximum value of Q that the soil can sustain. As a first estimate for the lower bound, we assume that the load Q is entirely resisted by the column of soil within *abcde* which extends to an infinite distance, and that the stress across a section perpendicular to *ab* is uniform [Fig. 9.30(*a*)]. Outside the column *abcde* the stress is everywhere equal to 0. This state obviously is far from the actual state of stress in the soil mass, but it serves as a first estimate. To satisfy the equilibrium condition, the resultant of the stress p acting over the area *fg* must be equal to Q, or

$$Q = p(L \sin \alpha + H \cos \alpha)$$

If the shear-strength parameters of the soil are c and ϕ, the value of p at failure, with 0 stress on *ab* and *de*, is equal to the unconfined compression strength [Fig. 9.30(*b*)], or

$$p = 2c \tan \left(45° + \frac{\phi}{2}\right) \tag{9.49}$$

Thus we have the value of Q as

$$Q_l = 2c \tan \left(45° + \frac{\phi}{2}\right) (L \sin \alpha + H \cos \alpha) \tag{9.50}$$

The subscript *l* denotes that this is a lower bound value of Q.

A first approximation of the upper bound is shown in Fig. 9.30(*c*). Failure or slip is assumed to take place along a zone *ab* with thickness equal to h. To analyze the kinematically admissible state, we first calculate the work done by the load and the body force. From the theory of plasticity it is known that within the slip zone *ab* the material undergoes an expansion of such a magnitude that the displacement velocity at any point *p* in the slip zone lies in the direction $\dot{V}$ which makes an angle ϕ with the slip plane (see Sec. 9.14). The energy dissipation by the stress occurs only within the slip zone because plastic strain exists only in this zone. The stresses on the slip plane consist of the normal stress σ and the shear stress τ, which is equal to $c + \sigma \tan \phi$ at failure. The displacement velocity is denoted by $\dot{V}$.

If we consider the slip zone of thickness h [Fig. 9.30(*c*); see also Fig. 9.33(*b*) and (*c*)], the velocity $\dot{V}$ represents the relative displacement velocity between the top and the bottom of the slip zone. Therefore the strain rates are

$$\dot{\epsilon}_x = 0^* \qquad \dot{\epsilon}_y = \frac{\dot{V} \sin \phi}{h} \qquad \dot{\gamma} = \frac{\dot{V} \cos \phi}{h}$$

---

* See Sec. 9.14, Eq. (9.62).

SEC. 9.13 Grouser-plate problem

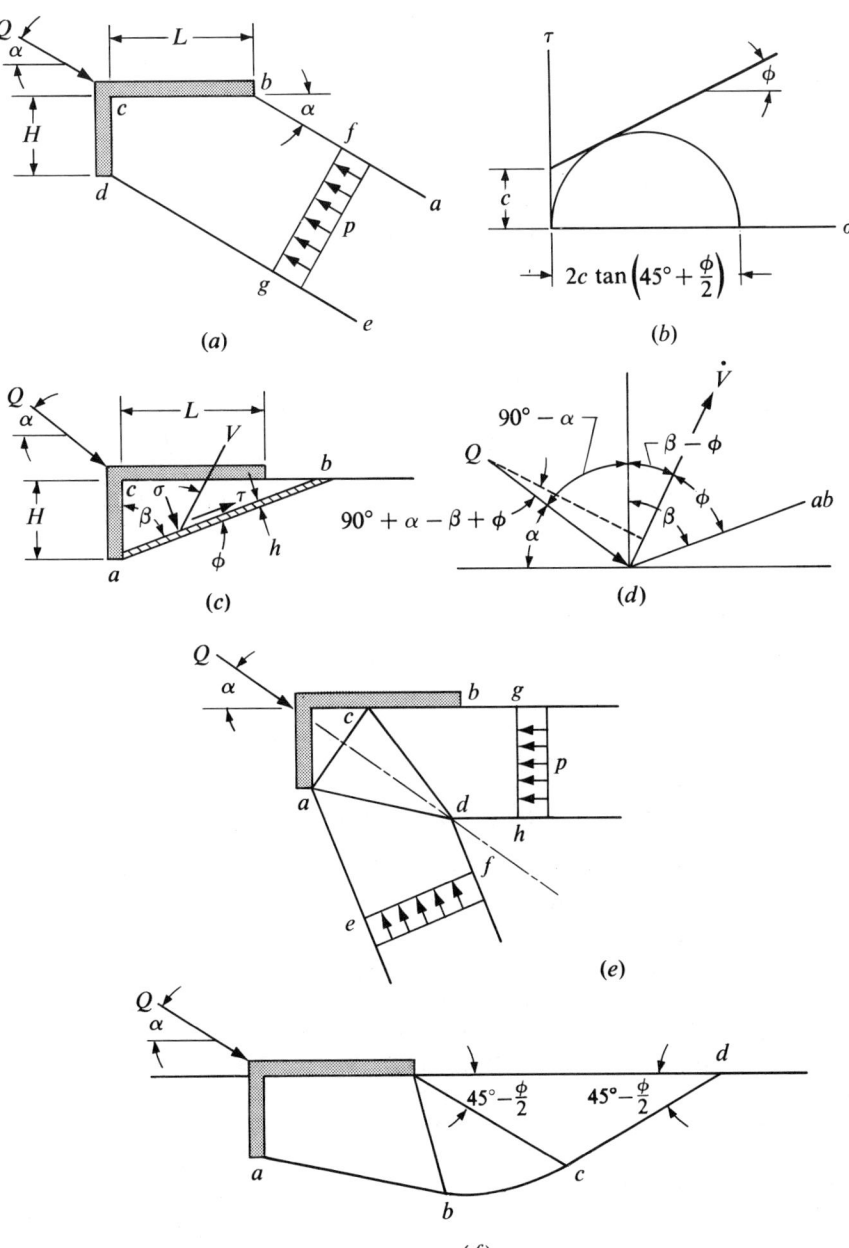

FIGURE 9.30 *The grouser-plate problem.* [After Haythornthwaite (1961).]

The dissipated energy per unit length of the slip zone is

$$dE_d = (-\sigma \dot{\epsilon}_y + \tau \dot{\gamma})h$$
$$= -\sigma \sin\phi \, \dot{V} + \tau \cos \dot{V}$$
$$= -\sigma \sin\phi \, \dot{V} + (c + \sigma \tan\phi) \cos\phi \, \dot{V}$$
$$= c \cos\phi \, \dot{V}$$

The total rate of energy dissipation along ab is thus

$$E_d = c \cos\phi \, \dot{V} \, \overline{ab}$$
$$= c \cos\phi \, \dot{V} H \sec\beta$$

The rate of work done by the load Q is equal to the product of the displacement rate of the grouser plate and the component of Q in the direction of the displacement. The displacement of the grouser plate is in the direction $\dot{V}$, since this is the direction of movement along the slip zone. The component of Q in the $\dot{V}$ direction is $Q \cos(\alpha + \phi + 90° - \beta)$ according to Fig. 9.30(d). Hence the rate of work done is

$$E_w = Q \cos(\alpha + \phi + 90° - \beta)\dot{V}$$

The body force or weight of the soil mass abc is neglected because it is small compared to Q. For the kinematically admissible state, E_w must be at least equal to E_d, or

$$c \cos\phi \dot{V} H \sec\beta = Q \cos(\alpha + \phi + 90° - \beta)\dot{V}$$

Hence
$$Q_u = \frac{Hc \cos\phi}{\cos\beta \cos(\alpha + \phi + 90° - \beta)} \qquad (9.51)$$

The subscript u denotes the upper bound value of Q. In Fig. 9.31(a) the values of Q_l and Q_u are plotted as a function of α. The correct answer should lie between these two limits.

It is to be noted that the range between Q_l and Q_u as found above is rather large for large values of α. This means that we are rather uncertain about the correct value of Q. This large range exists because of the grossly simplifying assumptions we have introduced to obtain the lower and upper bound solutions. To narrow the range we can revise our assumptions to agree better with what is known about the stress and displacement. These are illustrated in Figs. 9.30(e) and (f). The second lower bound assumes that the load Q is resisted by uniform compressive stresses in the two zones $adef$ and $cdgh$ [Fig. 9.30(e)]. The compressive stress p in the two zones at failure is the same as those given in Eq. (9.49). A second choice of the displacement is shown in Fig. 9.30(f). The slip surface $abcd$ is assumed to consist of a logarithmic spiral bc connecting two straight lines. It resembles very much the failure surface in the bearing-capacity problem. The solutions of the second lower and upper bounds are given in Fig. 9.31(b). It is seen that the difference between the second lower and upper bound solutions is much less than that between the first lower and upper bound solutions. This has

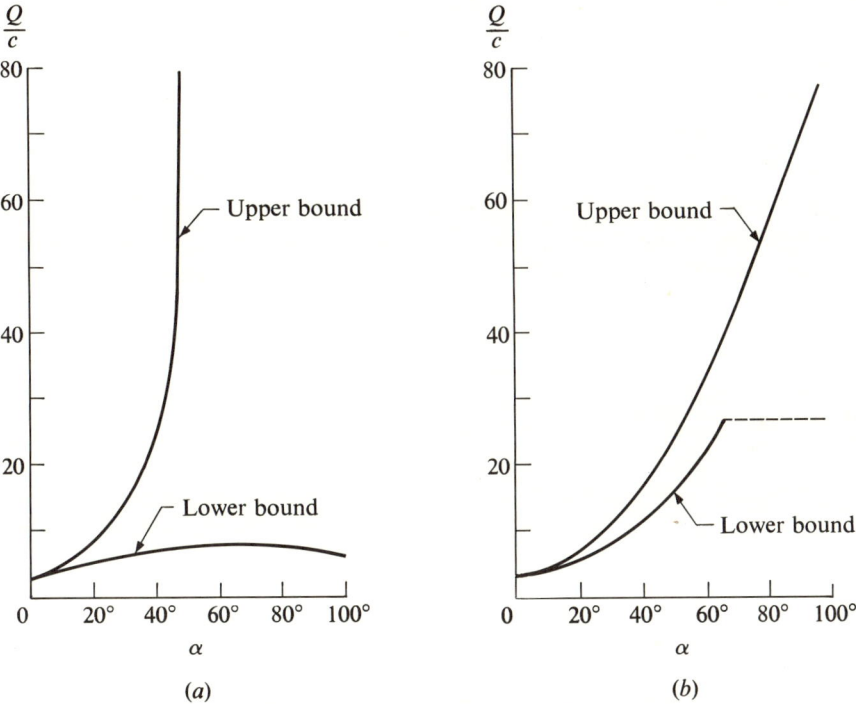

FIGURE 9.31 *Upper and lower bound solutions to the grouser-plate problem.* $L = 2$, $H = 1$, $\phi = 30°$. [After Haythornthwaite (1961).]

brought about a considerable reduction in the uncertainty concerning the correct value of Q. For other examples of upper and lower bound solutions, see Drucker and Prager (1952) and Shield (1955).

VELOCITY SOLUTION

9.14 Velocity field

The theory of plasticity may also be applied to find the velocity distribution in a soil mass undergoing plastic flow (shear failure). This is called the *velocity field*. The velocity field for a material that obeys the Mohr-Coulomb failure theory has been worked out by Shield (1953), and his analysis is presented as follows.

If we consider only the plastic deformations (elastic deformations are neglected), then the plastic strain rates $\dot{\epsilon}_x$, $\dot{\epsilon}_y$, and $\dot{\gamma}_{xy}$ are

$$\dot{\epsilon}_x = \frac{\partial \dot{u}}{\partial x} \qquad \dot{\epsilon}_y = \frac{\partial \dot{v}}{\partial y} \qquad \dot{\gamma}_{xy} = \frac{\partial \dot{u}}{\partial y} + \frac{\partial \dot{v}}{\partial x} \qquad (9.52)$$

in which $\dot{u}$ and $\dot{v}$ are the velocity of displacement in the x and y directions. The stress-strain relationship which is necessary for the velocity solution is obtained from the concept of the plastic potential (Drucker et al., 1951). According to this concept, the plastic strain rates are defined as

$$\dot{\epsilon}_x = -\lambda \frac{\partial \Gamma}{\partial \sigma_x} \qquad \dot{\epsilon}_y = -\lambda \frac{\partial \Gamma}{\partial \sigma_y} \qquad \dot{\gamma}_{xy} = -\lambda \frac{\partial \Gamma}{\partial \tau_{xy}} \qquad (9.53)$$

in which λ is a constant and Γ is the yield function. Γ is defined as the stress conditions that result in plastic yielding of the material. In our case, this is expressed by any one of the Eqs. (8.2), (9.2), (9.10), or (9.12). If we take Eq. (9.2), we have

$$\Gamma = \sigma_1 \tan^2\left(45° - \frac{\phi}{2}\right) - \sigma_3 - 2c \tan\left(45° - \frac{\phi}{2}\right) = 0$$

This can also be written as

$$\Gamma = \frac{\sigma_x + \sigma_y}{2} \sin \phi - [\tfrac{1}{4}(\sigma_y - \sigma_x)^2 + \tau_{xy}^2]^{1/2} + c \cos \phi = 0 \qquad (9.54)$$

by means of Eq. (4.3). Differentiating Eq. (9.54), we obtain the plastic strain rates as

$$\dot{\epsilon}_x = -\frac{\lambda}{2}\left(\sin \phi + \frac{\tfrac{1}{2}(\sigma_y - \sigma_x)}{[\tfrac{1}{4}(\sigma_x - \sigma_y)^2 + \tau_{xy}^2]^{1/2}}\right)$$

$$\dot{\epsilon}_y = -\frac{\lambda}{2}\left(\sin \phi - \frac{\tfrac{1}{2}(\sigma_y - \sigma_x)}{[\tfrac{1}{4}(\sigma_x - \sigma_y)^2 + \tau_{xy}^2]^{1/2}}\right) \qquad (9.55)$$

$$\dot{\gamma}_{xy} = \lambda \frac{\tau_{xy}}{[\tfrac{1}{4}(\sigma_x - \sigma_y)^2 + \tau_{xy}^2]^{1/2}}$$

From the geometrical relationship in the Mohr's circle (Fig. 9.32) the right-hand sides of Eq. (9.55) may be simplified to

$$\dot{\epsilon}_x = -\frac{\lambda}{2}(\sin \phi - \cos 2\psi)$$

$$\dot{\epsilon}_y = -\frac{\lambda}{2}(\sin \phi + \cos 2\psi) \qquad (9.56)$$

$$\dot{\gamma}_{xy} = \lambda \sin 2\psi$$

in which ψ is the angle between the x axis and the major principal axis [Fig. 9.7(b)]. If we denote the angle between the first slip line a and the x axis as θ, then

$$\psi = \theta + \left(45° - \frac{\phi}{2}\right)$$

and Eq. (9.56) becomes

$$\dot{\epsilon}_x = -\frac{\lambda}{2}[\sin \phi + \sin(2\theta - \phi)]$$

$$\dot{\epsilon}_y = -\frac{\lambda}{2}[\sin \phi - \sin(2\theta - \phi)]$$

$$\dot{\gamma}_{xy} = \lambda \cos(2\theta - \phi)$$

Velocity field

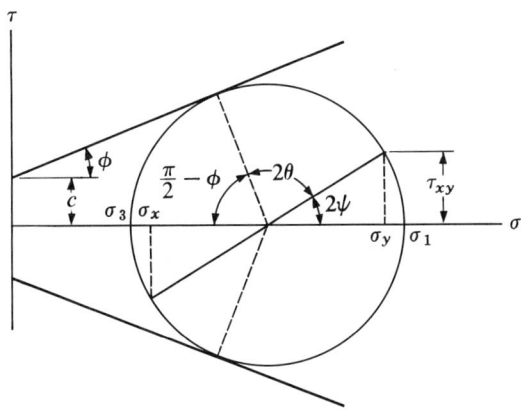

FIGURE 9.32 *Mohr's circle of stress.*

If we let $\theta = 0$ or $\theta = -[(\pi/2) - \phi]$, this means that the x axis coincides with one of the two slip lines. This leads to

$$\dot{\epsilon}_x = \left(\frac{\partial \dot{u}}{\partial x}\right)_{\theta=0} - \left(\frac{\partial \dot{u}}{\partial x}\right)_{\theta=-(\pi/2-\phi)} = 0 \tag{9.58}$$

which means that the strain rate along the slip line is zero.

Equation (9.58) allows us to calculate the velocity along the slip lines. We let $\dot{v}_a$ and $\dot{v}_b$ be the velocities along the slip lines a and b in Fig. 9.33(a). They may be expressed in terms of the components $\dot{u}$ and $\dot{v}$ as

$$\begin{aligned}\dot{v}_a &= \dot{u} \cos\theta + \dot{v} \sin\theta \\ \dot{v}_b &= -\dot{u} \sin(\theta + \phi) + \dot{v} \cos(\theta + \phi)\end{aligned} \tag{9.59}$$

Calculating $\dot{u}$ and $\dot{v}$ in terms of $\dot{v}_a$ and $\dot{v}_b$ from Eq. (9.59), we get

$$\dot{u} = \frac{\dot{v}_a \cos(\theta + \phi) - \dot{v}_b \sin\theta}{\cos\phi}$$

$$\dot{v} = \frac{\dot{v}_a \sin(\theta + \phi) + \dot{v}_b \cos\theta}{\cos\phi}$$

Differentiating $\dot{u}$ with respect to x, we have

$$\frac{\partial \dot{u}}{\partial x} = \frac{\partial \dot{v}_a}{\partial x} \frac{\cos(\theta+\phi)}{\cos\phi} - \dot{v}_a \frac{\sin(\theta+\phi)}{\cos\phi} \frac{\partial\theta}{\partial x} - \frac{\partial \dot{v}_b}{\partial x} \frac{\sin\theta}{\cos\phi} - \dot{v}_b \frac{\cos\theta}{\cos\phi} \frac{\partial\theta}{\partial x}$$

Combining this with Eq. (9.58) we have

$$\left(\frac{\partial \dot{u}}{\partial x}\right)_{\theta=0} = \left(\frac{\partial \dot{v}_a}{\partial x}\right)_{\theta=0} - \dot{v}_a \tan\phi \left(\frac{\partial\theta}{\partial x}\right)_{\theta=0} - \dot{v}_b \sec\phi \left(\frac{\partial\theta}{\partial x}\right)_{\theta=0} = 0$$

or
$$d\dot{v}_a - (\dot{v}_a \tan\phi + \dot{v}_b \sec\phi)\, d\theta = 0 \tag{9.60a}$$

along the first slip line. By similar means we get from $\dot{v}$,

$$d\dot{v}_b + (\dot{v}_a \sec\phi + \dot{v}_b \tan\phi)\, d\theta = 0 \tag{9.60b}$$

along the second slip line.

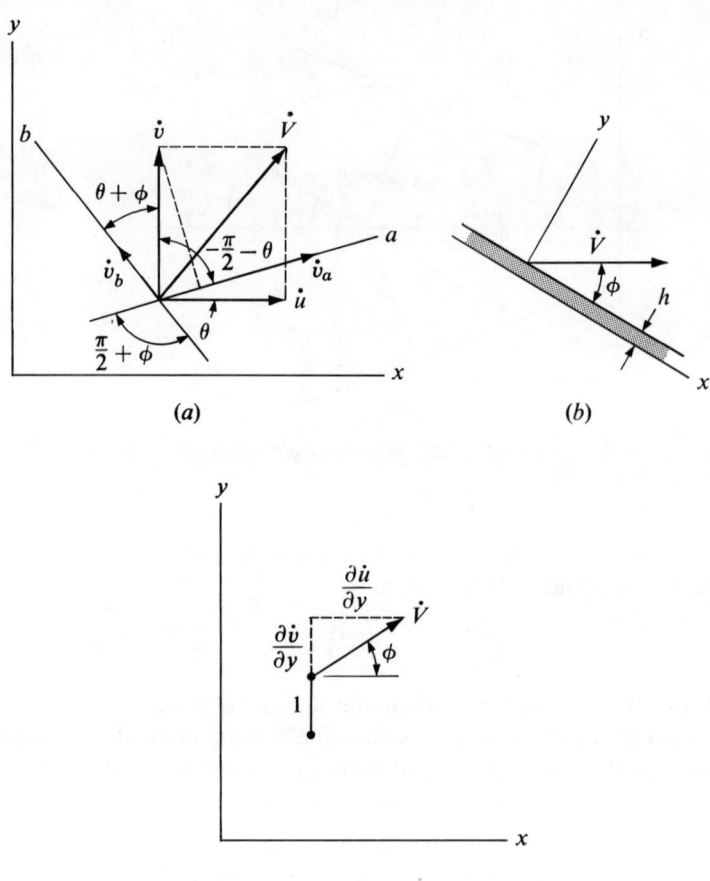

FIGURE 9.33 *Velocity along slip lines.*

Equations (9.60) may be integrated for the velocity boundary conditions to give the velocity field at failure. In addition there is a requirement regarding the volume change or dilation. The dilation rate is $\dot\epsilon_x + \dot\epsilon_y$ and Eq. (9.56) shows that

$$\dot\epsilon_x + \dot\epsilon_y = -\lambda \sin\phi \le 0 \qquad (9.61)$$

Thus plastic deformation must be accompanied by an increase in volume if ϕ is greater than 0.

We consider a thin zone along a slip line as shown in Fig. 9.33(b). Let x be the direction of the slip line as shown in the figure. In this zone $\dot\epsilon_x$ is 0. It must be very small compared with $\dot\gamma_{xy}$. From Eq. (9.57) it is seen that this is possible only if

$$\theta = 0, \quad \pi, \quad -\left(\frac{\pi}{2} - \phi\right), \quad -\left(\frac{3\pi}{2} - \phi\right)$$

For these values of θ, we get

$$\dot{\epsilon}_y = \frac{\partial \dot{v}}{\partial y} = -\lambda \sin \phi$$

$$\dot{\gamma}_{xy} = \frac{\partial \dot{u}}{\partial y} = \pm \lambda \cos \phi \qquad (9.62)$$

As shown in Fig. 9.33(c), the above equations mean that the resultant velocity in this slip zone makes an angle ϕ with the direction of the slip zone.

9.15 Velocity field beneath a strip foundation

We illustrate the calculation of the velocity field with the example of a loaded strip (Fig. 9.6 and Sec. 9.3). If at failure the foundation penetrates the soil mass with a vertical velocity equal to unity, what is the distribution of velocity within the soil mass?

Since the problem is symmetric we need to consider only half of the plastic zone, and Fig. 9.34(a) shows the velocity distribution in the mass $a'bcd$. We first recognize that the failure surface bcd is a discontinuity. The velocity along the slip lines a must be 0 across the discontinuity because otherwise the soil below bcd would also be undergoing plastic flow. The value of $\dot{v}_a$ is 0 to the left of ab and is 0 everywhere in $abcd$. Similarly, $\dot{v}_b$ is 0 to the right of $a'b$. From Eq. (9.60) $\dot{v}_a$ is constant if $d\theta$ is 0 (straight slip lines). In zone $aa'b$ both $\dot{v}_a$ and $\dot{v}_b$ are constant. It is therefore a zone of rigid body movement and the vertical velocity is everywhere equal to 1.

We now consider the velocity along slip line b. The two sides of a slip zone ab move with a relative velocity that makes an angle ϕ with the direction of the slip line [Eq. 9.62]. Hence, on ab, the change in velocity $\Delta \dot{V}$ must make an angle ϕ with ab. We may draw the velocity vector $\Delta \dot{V}$ as shown in Fig. 9.34(b). The velocity $\dot{V}$ is perpendicular to ab because it must make angle ϕ with $a'b$, which is also a slip line. The resultant velocity at b must be equal to 1, which is the rigid body velocity. The magnitude of $\dot{V}$ is then

$$\dot{V} = \tfrac{1}{2} \sec \left(45° + \frac{\phi}{2}\right)$$

In the radial shear zone abc, the velocity along system a is constant along the slip line by virtue of Eq. (9.60a) and so depends only on the direction θ. Hence we let

$$\dot{v}_a = F(\theta)$$

Substituting this in Eq. (9.60b) gives

$$\frac{d\dot{v}_b}{d\theta} + \tan \phi \, \dot{v}_b = -\sec \phi F(\theta)$$

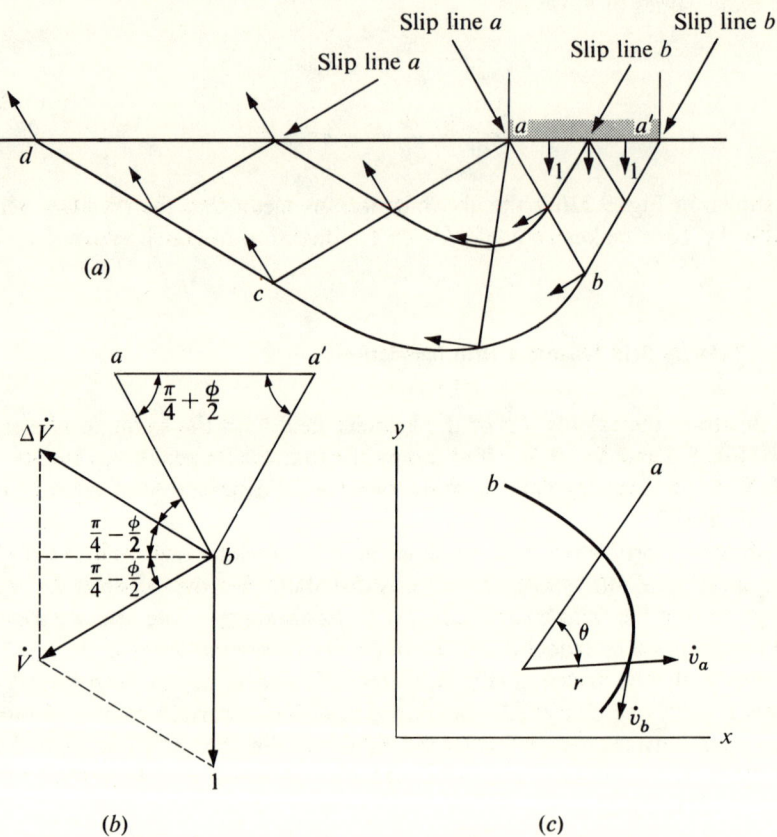

FIGURE 9.34 *Velocity field beneath a strip load.* [After Shield (1953).]

Solving this differential equation yields

$$\dot{v}_b e^{\theta \tan \phi} = -\sec \phi \int_0^{\theta_0} F(\theta) e^{\theta \tan \phi} + A$$

in which A is a constant of integration. Since $\dot{v}_a$ is 0, in abc we have

$$F(\theta) = 0 \qquad \dot{v}_b = A e^{-\theta \tan \phi} \tag{9.63}$$

This means that the velocity along bc varies exponentially with the angle θ. The angle θ is chosen as shown in Fig. 9.34(c) because in order for the volume change to be expansion, the velocity must increase along the slip surface. Along ab we have

$$\theta = 0 \qquad \dot{V} = \tfrac{1}{2} \sec\left(45° + \frac{\phi}{2}\right)$$

Substituting in Eq. (9.63) we find the constant of integration:

$$A = \dot{v}_b = \dot{V} \cos \phi = \tfrac{1}{2} \cos \phi \sec \left(45° + \frac{\phi}{2} \right)$$

Thus $\dot{v}_b$ changes throughout *abc* according to the equation

$$\dot{v}_b = \tfrac{1}{2} \cos \phi \sec \left(45° + \frac{\phi}{2} \right) e^{-\theta \tan \phi} \tag{9.64}$$

Along *ac* the velocity $\dot{v}_b$ is calculated with Eq. (9.61) by letting $\theta = -90$ deg, and

$$\dot{V} = \tfrac{1}{2} \sec \left(45° + \frac{\phi}{2} \right) e^{(\pi/2) \tan \phi}$$

The zone *acd* is again a zone of rigid body movement, as $\dot{v}_a$ is 0 and $\dot{v}_b$ is constant throughout. The velocity vectors are shown as arrows in Fig. 9.34(*a*). The calculated distortion of an initially square grid system is plotted in Fig. 9.35(*a*).*

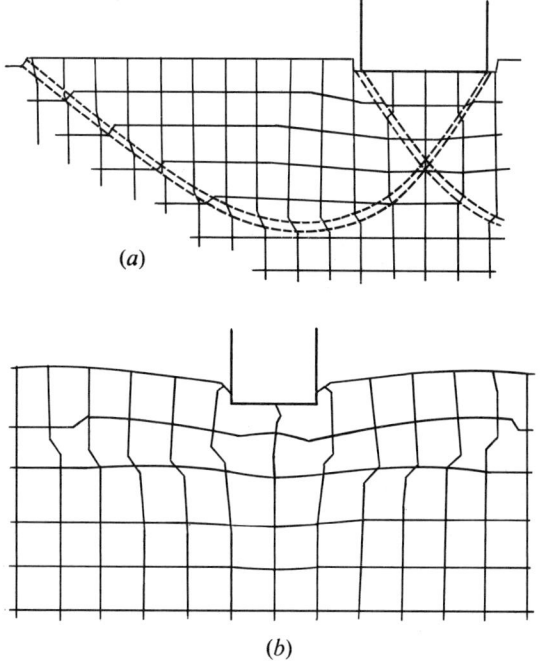

FIGURE 9.35 *Displacements beneath a strip load: (a) calculated* [after Shield (1953)]; (*b*) *measured* [after Sylwestrowicz (1953)].

The displacements in a deposit of sand underneath a strip foundation as measured by Sylwestrowicz (1953) are shown in Fig. 9.35(*b*). It has the same characteristics as the computed velocity field shown in Fig. 9.35(*a*). However, a

---

* An alternative velocity field is also possible (see Shield, 1953).

quantitative comparison cannot be made, since the experiment measured the total displacements for a foundation penetration of 0.4 in., whereas the velocity fields are calculated for a very small penetration. It is important to realize that the velocity solution is probably no more than a solution for initial motion. With continued motion the boundary may deviate from the simple shape at the beginning. It is also questionable whether the volume dilatancy [Eq. (9.61)] holds for any condition other than the instant of failure.

Studies such as those of de Josselyn de Jong (1959) and James and Bransby (1970) indicate that granular soils may show displacements of a nature different from that calculated from the plasticity theory. Little is known about the displacement in clay soils at failure.

Example 9.4. For the case of $\phi = 0$ the velocity solution is shown in Fig. 9.36(a). We may use this to derive the upper bound of the bearing capacity as follows. The rate of work done by Q_f is

$$E_w = q_f B 1$$

The rate of energy dissipation is $(1/\sqrt{2})\,(B/\sqrt{2})c$, $(1/\sqrt{2})\,[(B/\sqrt{2})(\pi/2)]c$, and $(1/\sqrt{2})(B/\sqrt{2})c$ along $a'b$, bc, and cd respectively. Since zones aba' and acd are zones of rigid body movement no energy dissipation occurs inside these zones within abc. The rate of energy dissipation is $(1/\sqrt{2})\,(B/\sqrt{2})\,d\theta\,c$ for the shaded element in Fig. 9.36(b). For the zone abc, the rate of energy dissipation is $(1/\sqrt{2})(B/\sqrt{2})(\pi/2)$ c.

For the kinematically admissible state

$$q_f B = 2\left[\frac{B}{2}c + \frac{\pi B}{4}c + \frac{B}{2}c + \frac{\pi B}{4}c\right]$$

or
$$q_f = c(2 + \pi) = 5.14c$$

We find that this is the same as Prandtl's solution.

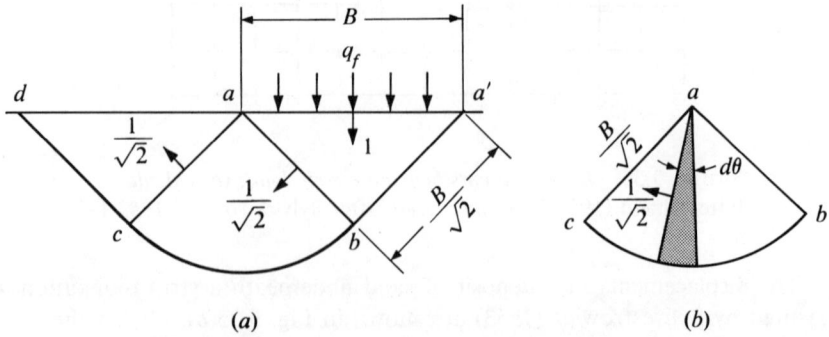

FIGURE 9.36 *Velocity field of a strip footing on clay.*

9.16 Limitations on use of plasticity theory

As is evident from the many examples in this chapter, the theory of plasticity plays a very important role in the solution of soil mechanics problems. It is therefore appropriate to outline the basic assumptions in the theory and examine their validity.

The ideally plastic material assumed in the theory yields plastically at failure as illustrated in Fig. 6.1. It has a known failure condition which is independent of the loading path. This means that the material fails at a given set of stresses, irrespective of the loading sequence that leads up to these stress conditions. Second, at failure, the material in the failure zone undergoes expansion of such a magnitude that the velocity of the displacement is a vector making an angle ϕ with the direction of the slip surface.

Plastic yielding may be considered to be approximately true in many soils at failure. However, there are notable exceptions. Heavily preconsolidated clays with fissures and soft rock often disintegrate at failure (Fig. 13.15), and very loose sands and quick clays may undergo a collapse of the soil structure at failure. In soil mechanics, the Mohr-Coulomb failure criterion has been used almost exclusively. Its validity under the condition of $\sigma_2 = \sigma_3$ has been well established. With an intermediate stress greater than the minor principal stress, the experimental results show some departure from the Mohr-Coulomb criterion. Experiments have also shown that the loading path has relatively little influence on the failure condition (see Secs. 8.10 and 8.11). From a practical viewpoint, the errors that may result from these deviations are probably unimportant when compared with those introduced by the variations in soil property that are likely to be encountered in natural soils.

It should also be noted that the results from the bearing-capacity theory and circular-arc analysis have been compared with observed failures in field and laboratory. Where subsoil conditions are reasonably uniform and the soil properties are sufficiently well known to permit conclusions, the agreement is in general quite satisfactory. Hence this may be accepted as circumstantial evidence of the validity of the plasticity solutions for limiting equilibrium. The comparison between calculated and observed failures is discussed in greater detail in Chapter 11.

By far the greatest uncertainty concerns the plastic strain rate as deduced from the plasticity theory in Sec. 9.14. Experimental studies indicate that the measured expansion may be considerably less than that predicted theoretically under certain conditions. For granular soils, the volume expansion as indicated by Eq. (9.61) may not be fulfilled. Since the upper bound solution and the velocity solution are based on the plastic strain rate, their reliability is also uncertain at present. For details of these studies the reader may consult the publications by Drucker et al. (1955), de Josselyn de Jong (1959), Haythornthwaite (1960), and Poorooshasb et al. (1966, 1967).

PROBLEMS

9.1 For the soil deposit shown in Fig 9.37: (*a*) Calculate the active earth pressure on the imaginary vertical plane *A-A* by Rankine's theory. (*b*) Plot the variation of pressure with depth. (*c*) Draw the direction of the slip lines.

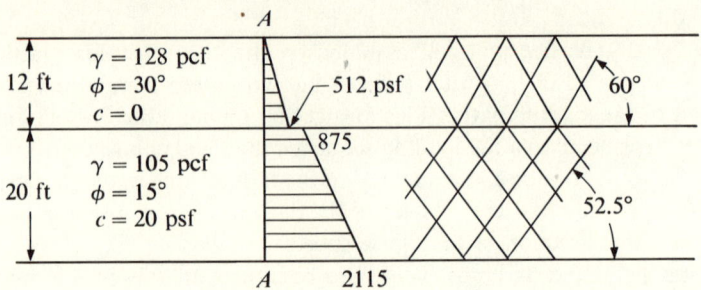

FIGURE 9.37

9.2 (*a*) Calculate the bearing capacity of a strip footing 8 ft wide and located at a depth of 6 ft by the Terzaghi method described in Sec. 9.5. The pertinent soil properties are $\gamma = 120$ pcf, $c = 400$ psf, $\phi = 17$ deg. (*b*) From the minimum value of Q obtained in part (*a*) calculate the bearing-capacity factors N_γ, N_c, and N_q.
Answer: see Fig. 9.38; 4.7, 13.8, 4.9.

9.3 If the water table in Problem 9.2 is at the level of the bottom of the footing, calculate the bearing capacity with effective stresses. The pertinent soil properties are $\gamma = 120$ pcf, $\gamma_{\text{sat}} = 135$ pcf, $\bar{c} = 400$ psf, $\bar{\phi} = 17$ deg.
Answer: 80.9 kips

9.4 A column carries 190 tons and is to rest on the surface of a deposit of moist coarse sand. Drained triaxial tests on specimens of the sand give the following results (in kg/cm²):

Spec. No.	All-around press.	Axial press. at failure
1	0.3	0.98
2	0.6	1.90

The unit weight of the soil is 128 lb/ft³ and the average water content is 12%. Calculate the design soil pressure if it is to have a factor of safety of 2.5.
Answer: 6000 psf

9.5 Consider the loading test in Problem 7.11. (*a*) Calculate the value of ϕ of the sand assuming that bearing-capacity failure occurred at a load of 4800 lb. (*b*) Calculate the bearing capacity of a foundation 8 ft square located 6 ft below the surface. If the information given is inadequate, assume reasonable values for missing data.
Answer: 38° for $\gamma = 120$ pcf; 81 kips/ft²

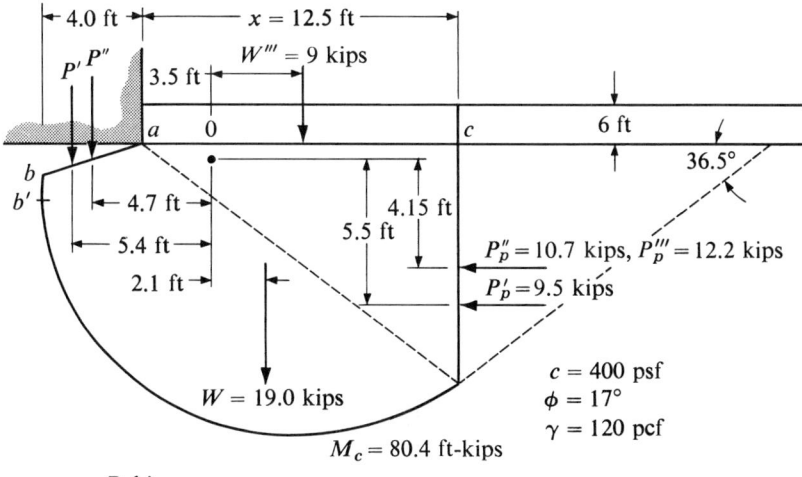

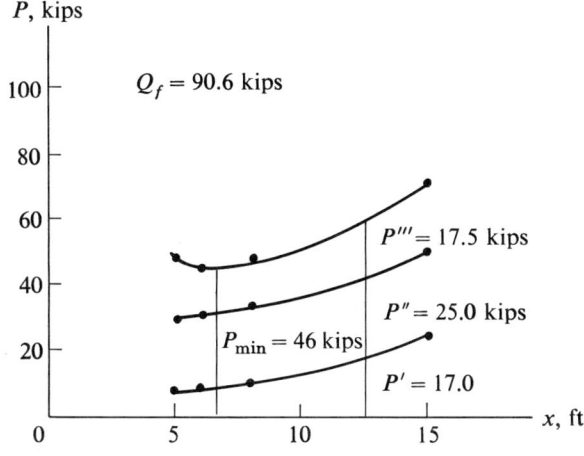

FIGURE 9.38

9.6 Calculate the factor of safety of the slip surfaces in Fig. 9.29 by the method of slices using (a) total stresses, (b) effective stresses. The soil properties are $\gamma_{sat} = 135$ pcf, $w = 14\%$, $c_u = 4000$ psf, $\phi_u = 0$, $\bar{c} = 1000$ psf, $\bar{\phi} = 32$ deg. Assume the material above the phreatic surface to be saturated by capillary action.
Answer: 0.96, 0.96; 1.01, 0.95

9.7 A slope with a slope angle of 30 deg is to be excavated in a homogeneous saturated clay extending to great depth. The excavation will be kept dry by pumping. The soil properties are as follows: $c_u = 1000$ psf, $\phi_u = 0$, $\bar{c} = 300$ psf, $\bar{\phi} = 28$ deg, $\gamma_{sat} = 120$ pcf, $w = 30\%$, LL = 32%, PL = 18%. (a) Calculate by means of stability numbers the depth at which slope failure is expected to occur. (b) Calculate the maximum depth that can be excavated if a safety factor of 2.0 is required against failure.
Answer: 46 ft; 23 ft

9.8 Prove that the resultant and pressures P' and P'' (Fig. 9.12) act at the one-third point and midpoint of ab, respectively.

Answer: The distance l_p between the resultant and point a (or any point) may be written as

$$l_p = \frac{\int_a^b pl\,dl}{\int_a^b p\,dl}$$

If the pressure distribution is uniform, this equation gives the resultant location at the midpoint. If the pressure increases linearly with l, the location is at the third point. We can find out the pressure distribution by considering the resultant pressure on a small segment ab' (Fig. 9.11) and gradually increase ab' until it is equal to ab. From Eq. (9.19) we see that W and P'_p, and consequently P', increase with the square of the distance ab'. Hence the pressure p' increases linearly with ab', and the resultant is at the third point. From Eq. (9.22) we see that M_c, P''_p, C'' increase linearly with ab'. The pressure distribution is therefore uniform, and the resultant of P''_p is at the midpoint

9.9 Calculate an upper and a lower bound value for the height of a vertical slope in a soil with unit weight γ and shear-strength parameters c and ϕ.

Answer: $H_l = (2c/\gamma)\tan[45° + (\phi/2)]$, assuming $\sigma_x = 0$ and $\sigma_y = \gamma y$ in the soil; $H_u = (4c/\gamma)\tan[45° + (\phi/2)]$, assuming slip on a plane making an angle $45° + (\phi/2)$ with the horizontal

9.10 Prove Eq. (9.4) by means of the Mohr's circle of stress.

9.11 For the infinite slope shown in Fig. 9.39, calculate the pressure p for the active and passive Rankine states. Assume the soil to be cohesionless ($c = 0$).

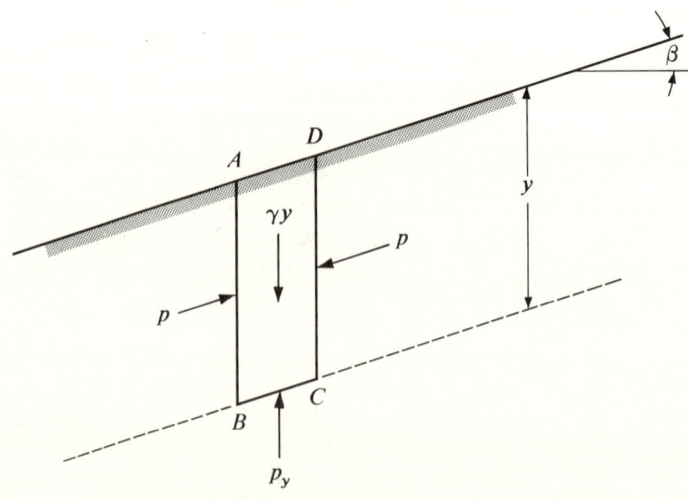

FIGURE 9.39   *Earth pressure in an infinite slope.*

Answer: $p_a = \dfrac{\cos \beta - [\cos^2 \beta - \cos^2 \phi]^{1/2}}{\cos \beta + [\cos^2 \beta - \cos^2 \phi]^{1/2}} \gamma y \cos \beta$

$p_p = \dfrac{\cos \beta + [\cos^2 \beta - \cos^2 \phi]^{1/2}}{\cos \beta - [\cos^2 \beta - \cos^2 \phi]^{1/2}} \gamma y \cos \beta$

Note: Since the soil mass is infinite in extent, the stresses have constant values along any plane at depth y below ground surface (BC in Fig. 9.39). The directions of principal stresses remain invariable throughout the mass. Furthermore, on the vertical planes AB and DC the pressure p acts at an angle β to the horizontal and the pressure p_y on BC acts in the vertical direction. Otherwise forces on $ABCD$ would not meet the condition of $\Sigma M = 0$.

9.12 Prove Eq. (9.28a).

9.13 A pile with a diameter of 18 in. is driven through the subsoil shown in Fig. 9.40. (a) Calculate the bearing capacity of the single pile. (b) Calculate the bearing capacity of nine piles spaced at 2.5 ft in both directions. (c) Do you expect the nine piles to fail as a group or as individual piles?
Answer: 416 kips for $K_0 = 0.5$; 3744 kips; as individual piles

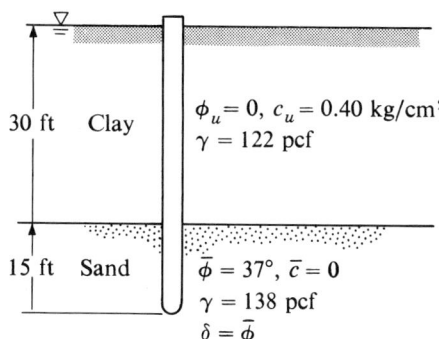

FIGURE 9.40

10

Earth-Pressure Problems

10.1 Introduction

This chapter concerns the application of the principles of soil mechanics to problems involving earth pressure. The objective in these problems is to determine the magnitude and distribution of earth pressure on structures which may be retaining walls of various types, or underground installations such as tunnels or buried silos. The methods of plastic equilibrium are frequently used, but their application is complicated by the behavior of the structure at the soil-structure boundary. These boundary conditions may occur as stresses between structure and soil or as deformation conditions imposed by the structure. The effect of these conditions on the plasticity solution must be evaluated.

Two kinds of earth-pressure problems are studied in this chapter. The first kind concerns walls whose movements are sufficient to develop shear failure in the soil mass. The deformation conditions necessary for shear failure are examined. Even when these requirements are met, many earth-pressure problems contain physical conditions that are rather complex. The quantitative analysis of these problems requires many simplifying assumptions. Thus it is important to study these simplifications and, wherever possible, ascertain their effect on the answer.

Frequently it is necessary to know the earth pressures exerted by a soil mass that is not completely in the plastic state. Shear failure may be localized to a small area or to one or two slip lines. Then it is necessary to consider the elastic behavior of the soil in addition to plasticity. The second kind of earth-pressure problem deals with walls having limited deformations, and the braced excavation is studied as an example. Earth pressure on a yielding base inside a soil mass is a similar problem and is discussed in the last part.

There are, of course, numerous other problems involving earth pressures, but the analyses developed in this chapter should serve to illustrate the methods and considerations required for the solution of these problems. Other topics on earth pressure may be found in the papers by Hansen (1953), Terzaghi (1954), and Tschebotarioff (1962).

10.2 Effect of deformation on earth pressure

The earth pressure exerted by a soil mass on a wall is intimately related to the deformation condition imposed by the wall. It is shown in Sec. 9.2 that the active and passive Rankine states are developed if the soil mass is extended or compressed to induce failure. In the design of earth retaining structures very diverse deformation conditions are encountered.

We begin with the at-rest state, which is the condition of the soil mass before any deformation is introduced. The earth pressure that acts on an imaginary vertical surface is called the earth pressure at rest. Its intensity at any depth, z, below the surface is

$$p = K_0 \gamma z \qquad (10.1)$$

where γ is the unit weight and K_0 is the coefficient of earth pressure at rest. Bishop (1958) has shown experimentally that K_0 is approximately equal to

$$K_0 = 1 - \sin \bar{\phi} \qquad (10.2)$$

for normally consolidated soils.

The deformation conditions may be classified into the following groups. Here we consider only smooth walls. At one extreme we have walls that deform by such magnitudes that the soil mass abc behind the wall is in a state of limiting equilibrium (Fig. 10.1). The deflections ρ_a and ρ_p required to develop the active earth pressure are approximately of the order of $0.001H$ and $0.05H$ for cohesionless soils. The required deflection is not so well defined for the case of passive pressure for cohesive soils. Some of the estimated values can be found in Table 10.1. When these conditions are met, Rankine's state is developed within

TABLE 10.1 *Wall Displacements Required to Develop Active and Passive Earth Pressures*

Soil	State of stress	Type of movement	Necessary displacement
Sand	Active	Parallel to wall	$0.001H$
	Active	Rotation about base	$0.001H$
	Active	Rotation about top	$0.02H$
	Passive	Parallel to wall	$0.05H$
	Passive	Rotation about base	$>0.1H$
	Passive	Rotation about top	$0.05H$
Clay	Active	Parallel to wall	$0.004H$
	Active	Rotation about base	$0.004H$

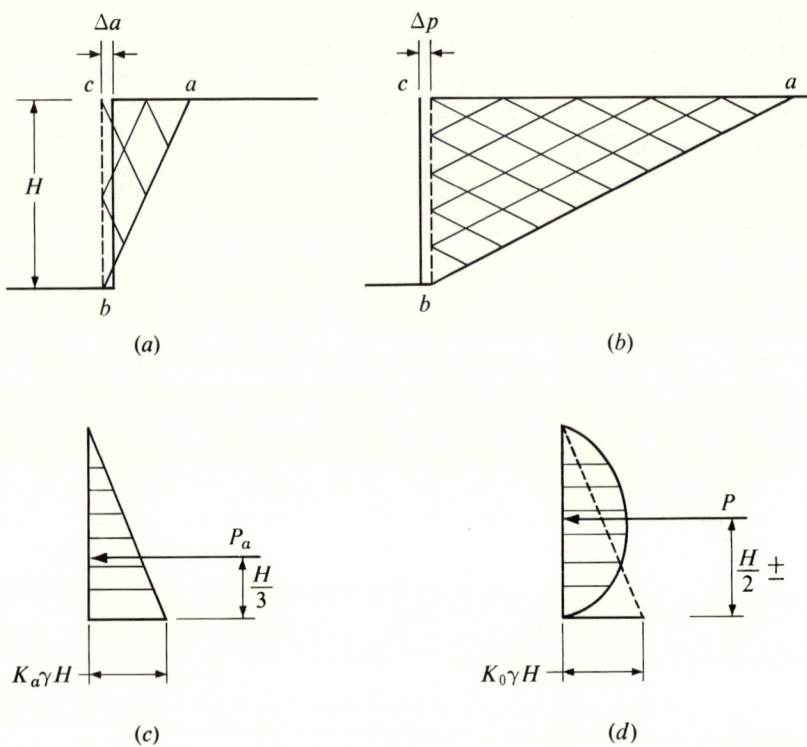

FIGURE 10.1 *Earth pressure on walls.*

the wedge *abc*, while the rest of the soil mass remains in a state of elastic equilibrium. The earth-pressure distribution for a cohesionless soil in the active Rankine state is shown in Fig. 10.1(c). If the movement is less than p_a, the pressure distribution is shown in Fig. 10.1(d). This is similar to that given by the elastic theory.

A condition frequently encountered in retaining wall design is one in which the wall rotates about the bottom, as shown in Figure 10.2. This case was studied by Terzaghi (1934) in his classic experiments. Some of Terzaghi's findings are illustrated in Fig. 10.3, in which the measured earth-pressure coefficient $K = \sigma_x/\sigma_y$ is plotted against the wall movement. The movement is plotted as positive if it is away from the soil mass. The experiment shows that without any movement the value of K is about 0.40. This is the coefficient of earth pressure at rest. As the wall is induced to move away from the soil, shear stresses developed in the soil reduce the earth pressure, and K decreases. This continues until the full shearing resistance of the soil is mobilized. The shear stresses cannot increase any further, and shear failure occurs in the soil mass. This state is Rankine's active state, and K attains the value K_a. The earth-pressure distribution at these three stages is shown in Fig. 10.2(c) for a cohesionless soil.

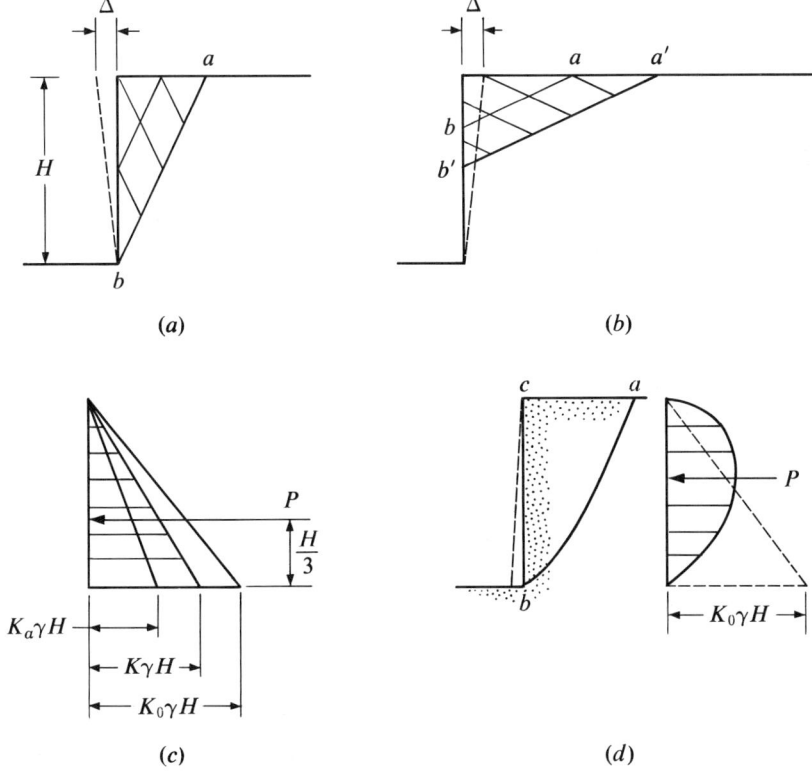

FIGURE 10.2 *Earth pressure on rotating walls.*

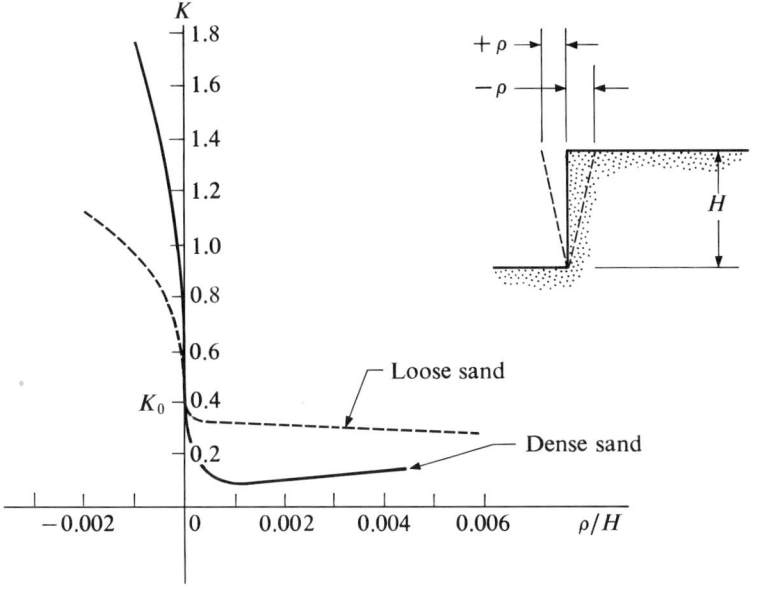

FIGURE 10.3 *Relationship between earth pressure and wall deflection.* [After Terzaghi (1954).]

Similarly, if the wall is moved against the soil mass, the earth pressure steadily increases. However, very large displacements are required to develop the passive state in the soil. Slip surfaces develop first near the top [*ab* in Fig. 10.2(*b*)]. With further movement slip surfaces appear at greater depths [*a'b'* in Fig. 10.2(*b*)]. For dense sand, the experiments of James and Bransby (1970) showed that shear failure in the soil extended to about the midpoint of the wall at a displacement of $0.1H$. For loose sand slip surfaces were not observed even at $0.1H$.

The last case is shown in Fig. 10.2(*d*), in which the wall rotates about the top. At large deflections the slip surface *ab* is formed. However the limiting equilibrium state is not reached throughout the wedge *abc*. The pressure distribution is as shown in Figure 10.2(*e*).

WALLS WITH ADEQUATE DEFLECTIONS

10.3 Earth pressures on smooth walls

Rankine's theory of active and passive earth pressures was used originally to describe the states of stress in a semi-infinite medium. If a smooth wall meets the deformation conditions specified in Figs. 10.1 and 10.2, then the soil in the wedge *abc* is in the plastic state and the earth pressure may be calculated by Rankine's theory.

A situation common to many retaining walls is shown in Fig. 10.4. The force on the back is several times greater than that on the front, so the wall moves forward. If the distance *ad* is sufficiently large, the pattern of the slip lines in the soil behind the wall would not differ significantly from that in Rankine's

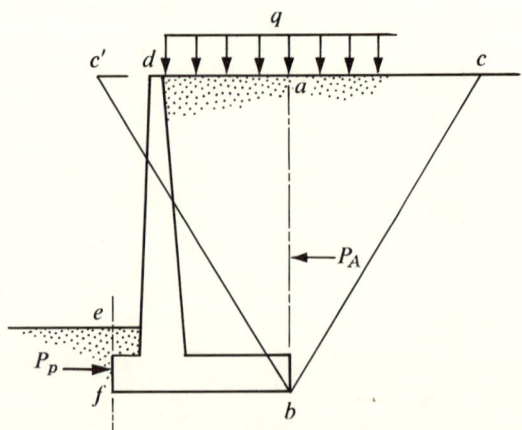

FIGURE 10.4 *Earth pressures on a retaining wall.*

active state. Hence, the earth pressure on *ab* and *ef* may be estimated by Rankine's theory:

$$p_a = \gamma y \tan^2\left(45° - \frac{\phi}{2}\right) - 2c \tan\left(45° - \frac{\phi}{2}\right) \quad (9.4a)$$

$$p_p = \gamma y \tan^2\left(45° + \frac{\phi}{2}\right) + 2c \tan\left(45° + \frac{\phi}{2}\right) \quad (9.4b)$$

If a uniform surcharge q is placed on the ground surface (Fig. 10.4), its effect is to increase the vertical stress at every point by q. Hence we have

$$p_a = (\gamma y) \tan^2\left(45° - \frac{\phi}{2}\right) - 2c \tan\left(45° - \frac{\phi}{2}\right) + q \tan^2\left(45° - \frac{\phi}{2}\right) \quad (9.7a)$$

The analogous case for passive pressure, not shown in Fig. 10.4 is

$$p_p = (\gamma y) \tan^2\left(45° + \frac{\phi}{2}\right) + 2c \tan\left(45° + \frac{\phi}{2}\right) + q \tan^2\left(45° + \frac{\phi}{2}\right) \quad (9.7b)$$

10.4 Influence of wall friction

In Rankine's theory the major and minor principal axes are in the x and y directions. This state is not changed if a perfectly smooth vertical wall is introduced as a boundary. In reality, most retaining walls are far from frictionless. If the wall is vertical, it can be seen that in the immediate vicinity of the wall, the vertical plane is no longer a principal plane. Therefore the stresses in the soil mass may be expected to differ from that in the Rankine state.

With shear stresses between the back of a wall and the soil the directions of the principal stresses differ from those in Rankine's theory. Figure 10.5(a) illustrates the case in which the shear stress acts upward on the soil mass. This is the case for the active state in which the wall moves away from the soil. The wedge of soil slides downward with respect to the wall. The angle between the vertical and the failure plane is equal to ψ instead of $45° - (\phi/2)$. The angle ψ may be determined by Mohr's circle, as shown in Fig. 10.5(b). If the wall friction is equal to $+\delta$, the state of stress at the vertical boundary *ab* is represented by point N according to the sign convention for Mohr's circle. The stress on the slip surface *bc* is represented by point M. Therefore, the angle between *ab* and the failure plane is equal to one-half the angle MCN. The value of ψ is greater than $45° - (\phi/2)$. If the friction acts downward $(-\delta)$, as is the case if the wall moves against the soil, then the angle is equal to $M'CN'$. The shape of the failure surface is illustrated in Fig. 10.5(c). The earth pressures in both cases may be computed by assuming the surface of failure to be composed of a logarithmic spiral *bc* and a straight line *cd* [Fig. 10.5(a) and (c)]. The method outlined in Sec. 9.5 may be used for the calculations. A more elaborate solution, using the velocity field and its results has been described by Roscoe (1970).

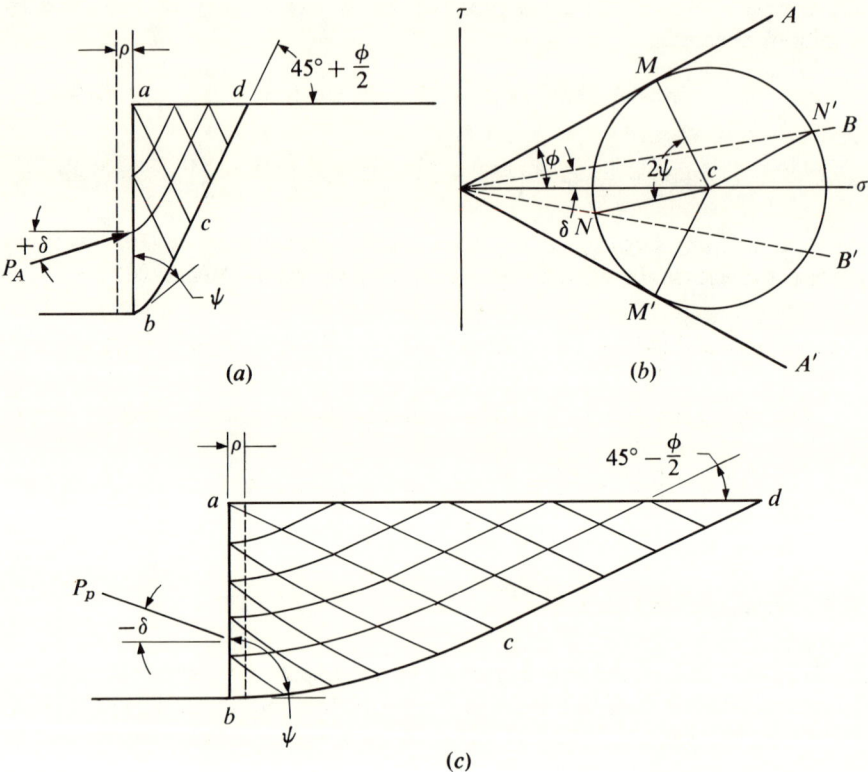

FIGURE 10.5 *Effect of wall friction on a failure surface.*

10.5 Coulomb's theory for cohesionless soils

Coulomb's theory is a simplified solution of the problem presented in the preceding section. The surface of failure is replaced by a plane. The theory assumes that as the wall tilts forward a wedge of soil *abc* tends to slide over a plane of failure as shown in Fig. 10.6(*a*). The downward and forward movement of the mass is resisted by the force *P* between the wall and soil and by stresses along plane *bc*. The limiting value of the force *P* is reached when the shear stress equals the shearing resistance along planes *ac* and *bc*; this is the active earth pressure P_a.

We first consider the case of a cohesionless sand. The shearing resistance of the soil and that between the soil and wall are, respectively,

$$s = \sigma \tan \phi \qquad (10.3)$$

$$s_0 = \sigma \tan \delta \qquad (10.4)$$

In Eq. (10.4), δ represents the angle of friction between the soil and the material that constitutes the wall. The resultant of the shear and normal stresses on the

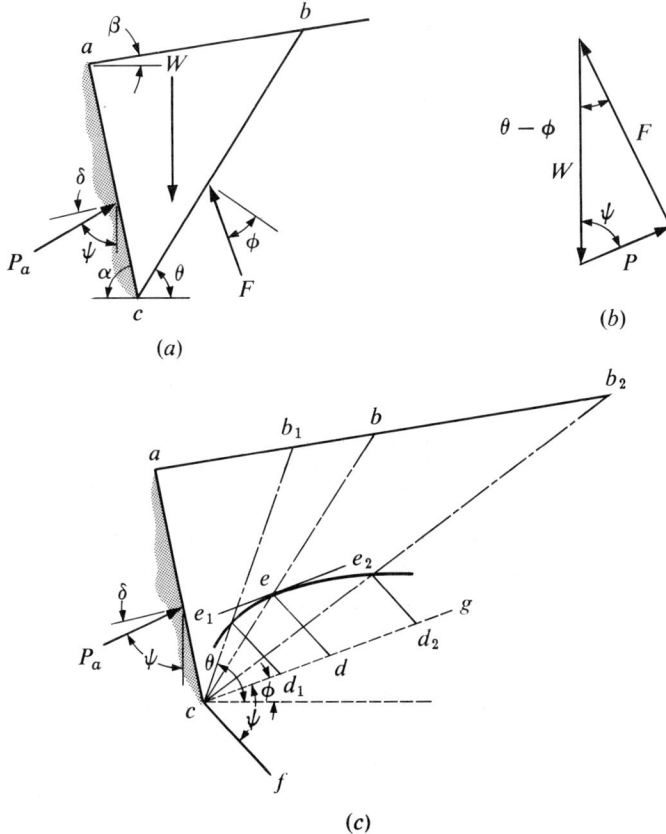

FIGURE 10.6 *Coulomb's theory of earth pressure and Culmann's graphical construction for cohesionless soils.*

surface of failure is F and at failure it makes an angle ϕ with the normal to the surface. Its magnitude is unknown. The force on ac is P, and P makes an angle δ with the normal to the wall. Since the force W is known in magnitude and direction and the forces F and P are known in direction but unknown in magnitude, the system of forces can be solved by means of the force polygon, as shown in Fig. 10.6(*b*). Since there are unlimited numbers of potential failure planes, the critical one that gives the maximum value of P must be determined by trial. This maximum value is the active earth pressure P_a. If a surcharge acts on the ground surface, its presence increases the weight W and can be added to the weight of the soil mass.

If the wall is pushed against the soil to develop passive earth pressure, the directions of the shear stresses are reversed. The forces can be solved in the same manner.

The forces P_a and P_p can also be obtained analytically by the resolution of forces. The result, which was obtained by Coulomb (1776), is

$$P_a = \frac{1}{2} K_a \gamma H^2 \frac{1}{\sin \alpha \cos \delta} \tag{10.5}$$

in which

$$K_a = \frac{\sin^2 (\alpha + \phi) \cos \delta}{\sin \alpha \sin (\alpha - \delta) \left[1 + \sqrt{\frac{\sin (\phi + \delta) \sin (\phi - \beta)}{\sin (\alpha - \delta) \sin (\alpha + \beta)}}\right]^2}$$

and

$$P_p = \frac{1}{2} K_p \gamma H^2 \frac{1}{\sin \alpha \cos \delta} \tag{10.5}$$

in which

$$K_p = \frac{\sin^2 (\alpha + \phi) \cos \delta}{\sin \alpha \sin (\alpha - \delta) \left[1 - \sqrt{\frac{\sin (\phi + \delta) \sin (\phi - \beta)}{\sin (\alpha - \delta) \sin (\alpha + \beta)}}\right]^2}$$

CULMANN'S METHOD. An expedient method for the graphical solution of Coulomb's theory was devised by Culmann. It is illustrated in Fig. 10.6(c) for the case of active earth pressure. For the wall ac, a line cg is drawn, making an angle ϕ with the horizontal. Next, line cf is drawn, making an angle ψ with cg; ψ is the angle between the vertical and the resultant earth pressure P. For a given trial failure plane cb_1, the weight of the wedge W_1 is laid off along cg such that the distance cd_1 is equal to W_1 using an appropriate scale. From d_1 a line $d_1 e_1$ is drawn parallel to cf, intersecting the failure surface at e_1. The magnitude of the resultant earth pressure for this particular failure surface is equal to $d_1 e_1$.

It can be seen that the triangle $cd_1 e_1$ is the same as the force polygon in Fig. 10.6(b) rotated through an angle. The angle $e_1 d_1 c$ is ψ, and the angle $d_1 c e_1$ is $\theta - \phi$. These angles are equal to the angle between W and P and between W and F, respectively, in the force polygon.

The calculations are repeated for a number of failure surfaces (such as cb_2). Finally a curve can be drawn connecting the points $e_1, e_2, \ldots$. This curve expresses the relationship between the magnitude of P and the angle θ (or the direction of the failure plane) and is called the *Culmann line*. The maximum value of P (or P_a) is then determined by tracing a tangent to the curve that is parallel to line cg. The point of tangency is e and it can be seen that de is the maximum value of P. This also locates the critical plane (line cb). Since the actual failure surface does not differ very much from a plane in the case of active earth pressure (Fig. 10.5), the location of the failure plane as determined by the Culmann method may be considered fairly reliable.

Compared with the case of the frictionless wall the net effect of the friction between soil and wall is to reduce the active earth pressure and increase the passive earth pressure. In Fig. 10.7 the earth-pressure coefficients K_a and K_p calculated by Coulomb's theory for a vertical wall with horizontal backfill are plotted against the value of δ and are shown as solid lines. As noted in the first paragraph of this section, Coulomb's theory assumes the failure surface to be a plane, whereas in reality it is a curved surface. If the earth pressure is calculated on the basis of a curved failure surface, a somewhat different result is obtained. The effect is insignificant in the case of active pressure but considerable for passive pressure. The corrected passive earth pressure is represented in the figure by dashed lines. Since Rankine's theory assumes no shear stress on a vertical plane, the earth pressure is equal to that for δ equal to 0.

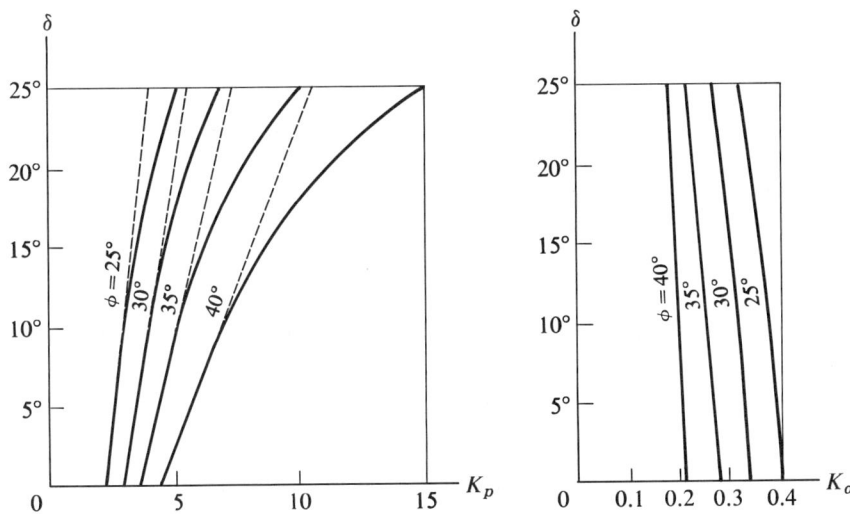

FIGURE 10.7 *Effect of wall friction on earth pressure.* [After Terzaghi (1943b), (1954).]

PRESSURE DISTRIBUTION. To determine the point of application of the resultant earth pressure P_a, it is necessary to calculate the distribution of earth pressure along the back of the wall. This can be accomplished numerically by computing the resultant earth pressure to various depths along the wall. This is illustrated by points d and e in Fig. 10.8. The active earth pressures for heights ad and ae as calculated by Coulomb's theory are represented by P_{ad} and P_{ae}, respectively. The unit pressure p_a is equal to the differential of the resultant pressure with respect to distance. If we consider the increment between d and e, then $\Delta P_a = P_{ae} - P_{ad}$ and $\Delta y = ae - ad$. Hence

$$p_a = \frac{\Delta P_a}{\Delta y} = \frac{P_{ae} - P_{ad}}{ae - ad} \tag{10.6}$$

If the back of the wall is divided into a number of increments, it is possible to calculate the average unit pressure in each of the increments.

If the ground surface behind the wall is level or rises at a constant slope without any surcharge, the pressure distribution is linear with depth. This is

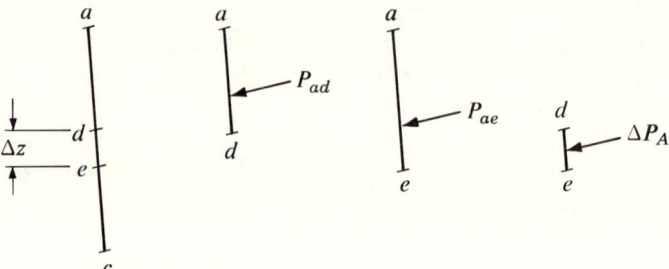

FIGURE 10.8 *Calculation of pressure distribution.*

because the weight of the soil wedge increases with the square of the depth y. Thus the resultant earth pressure increases with the square of y and the unit pressure increases linearly with y.

10.6 Critical height

Figure 9.3 shows that without surcharge ($q = 0$) the pressure for a cohesive soil is negative (indicating tension) to a depth y_0 equal to

$$y_0 = \frac{2c}{\gamma} \tan\left(45° + \frac{\phi}{2}\right) \tag{10.7}$$

If the soil can withstand this tensile stress, then down to a depth $2y_0$ the tensile and compressive stresses cancel each other and the resultant earth pressure is 0.

However, if a vertical cut is made, the tensile and compressive stresses are removed from the free vertical face where all stresses are 0. The stress condition then becomes quite different from that in Rankine's theory. Furthermore most soils cannot withstand even small tensile stresses without cracking. After cracks develop in the soil, the calculated tensile stress can no longer be maintained. To find the height H_c at which a vertical cut can stand unsupported (also called *critical height*) we use Terzaghi's solution for the wedge *abcd* in Fig. 10.9 (Terzaghi, 1943b). The failure surface intersects the crack at c. Hence, the tendency of this mass to slide is resisted by the shearing resistance on the failure plane *bc*. This consists of the shearing resistance due to cohesion, which is

$$C = c(\overline{bc}) = c\,\frac{H_c - y}{\cos[45° - (\phi/2)]} \tag{10.8}$$

and the friction F_t, which is $F_n \tan \phi$. The resultant of F_t and F_n makes an angle ϕ with the normal to *bc*. The weight of the soil mass is

$$W = \tfrac{1}{2}\gamma(\overline{ab} + \overline{cd})(\overline{ad}) = \tfrac{1}{2}\gamma(H_c + y)(H_c - y)\tan\left(45° - \frac{\phi}{2}\right)$$

$$= \tfrac{1}{2}\gamma(H^2 - y^2)\tan\left(45° - \frac{\phi}{2}\right) \tag{10.9}$$

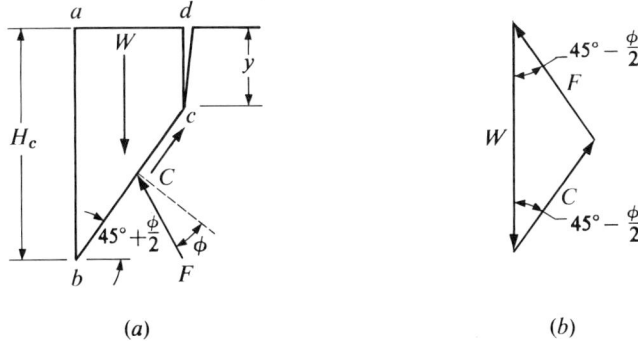

FIGURE 10.9 *Critical height of an unsupported vertical cliff.* [After Terzaghi (1943b).]

Since the three forces W, F, and C are in equilibrium, the force polygon in Fig. 10.9(b) may be constructed. From this we can write

$$W = 2C \cos\left(45° - \frac{\phi}{2}\right)$$

Substituting Eq. (10.9) in this and solving for H_c, we get

$$H_c = \frac{4c}{\gamma} \tan\left(45° + \frac{\phi}{2}\right) - y$$

If we take y as equal to the depth of the tension zone y_0 [Eq. (10.7)] we get

$$H_c = \frac{2c}{\gamma} \tan\left(45° + \frac{\phi}{2}\right) \tag{10.10a}$$

Terzaghi (1943b) cited empirical observations that the depth of the cracks usually does not exceed one-half of the height H_c. In this case

$$y = \frac{H_c}{2} \qquad H_c = \frac{2.67c}{\gamma} \tan\left(45° + \frac{\phi}{2}\right) \tag{10.10b}$$

Thus we see that the critical height is not much greater than the depth of the tension zone.

10.7 Coulomb's theory for cohesive soils

The shearing resistance of a cohesive soil and that between soil and wall are, respectively,

$$s = c + \sigma \tan \phi$$
$$s_a = c_a + \sigma \tan \delta$$

in which c_a is the adhesion between soil and the material of the wall. The principles are the same as in the case of cohesionless soils. One important difference in detail must be considered first—the presence of tension cracks in the upper part of the soil mass.

To apply Coulomb's theory to the calculation of earth pressures of cohesive soils we consider the equilibrium of the mass *abcd* in Fig. 10.10(a) [Terzaghi (1943b)]. The tension crack is assumed to extend down to a depth H_c for simplicity. The trial failure surface *bc* intersects the tension crack at *c*. The wedge *abcd* is subjected to the following forces: the weight of the wedge W, the shearing resistance C along *bc* due to cohesion, and the shearing resistance C_a along a_1b due to adhesion between soil and wall. In addition F is the resultant of the normal force F_n on *bc* and the friction component F_t of the shearing resistance on *bc*. F makes an angle ϕ with the normal. On a_1b, P is the resultant of the normal component P_n and the frictional component P_t of the shearing resistance between wall and soil. The forces C and C_a are given by

$$C = c(\overline{bc}) \qquad C_a = c(\overline{a_1b})$$

and are known for a given failure surface. The weight W is known. Thus we have forces W, C, and C_a that are known in magnitude and direction, and forces P and F that are known in direction but unknown in magnitude. The problem can be solved by considering equilibrium of the forces. Figure 10.10(c) shows the force polygon that may be used to find P and F. As before, P is computed for several trial surfaces, and the one that yields the largest value of P is the critical surface. The corresponding pressure is the active earth pressure.

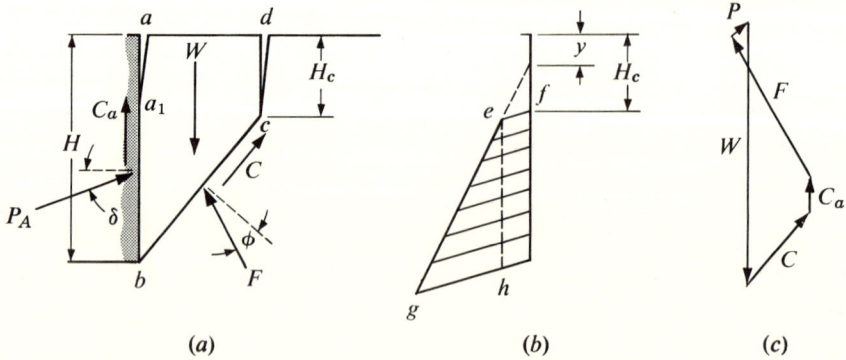

FIGURE 10.10 *Coulomb's theory of earth pressure for cohesive soils.* [After Terzaghi (1943b).]

The distribution of earth pressure is shown in Fig. 10.10(b). The pressure is 0 to depth y_0. However, since the soil can stand without support to a depth H_c, the pressure on the wall may be 0 within this range. If we calculate the pressure distribution according to the method outlined in Sec. 10.5, the earth pressure increases linearly with depth below H_c. The soil mass in the cracked zone affects the earth pressure in the same way as does a surcharge. It offers no shearing resistance and its only effect is to exert a pressure on surface a_1c equal to its own weight. The earth pressure due to this surcharge is ef [Fig. 10.10(b)] and is constant with depth.

10.8 Effect of surcharge loads

The introduction of a distributed surcharge load on the ground surface does not create serious difficulties. If Rankine's active and passive states hold, then the vertical stress at any point inside the soil is increased by q (Sec. 10.3). If Coulomb's theory is used, the surcharge can be included in the weight.

The problem of a concentrated line load Q is more difficult to solve. If the load is small, its influence on the slip surface may be ignored. We may still treat the load Q as part of the weight, if Coulomb's theory is used to calculate the resultant earth pressure. The problem cannot be solved by Rankine's theory. If the load is large and located at some distance from the wall, a composite slip surface develops as illustrated in Fig. 10.11. The solution for a composite slip surface is similar to the analyses described in Chapter 9. The reader may refer to the book by Hansen (1953) for the details as well as tabulated results.

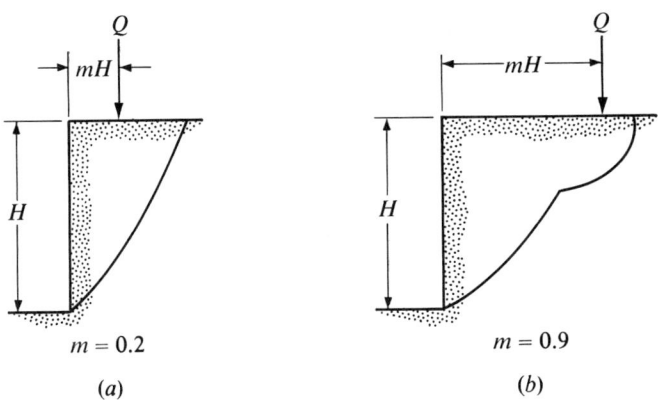

FIGURE 10.11 *Failure surface behind a wall with a surcharge load.*

CULMANN'S METHOD. If Coulomb's theory is used to calculate the resultant earth pressure in a cohesionless soil with a line-load surcharge, Culmann's method is again useful and is illustrated in the following example. Figure 10.12(a) shows a line load Q parallel to the back of the wall. If we carry out Culmann's construction for the earth pressure without the load Q, we get the curve e_3e_1f. When the load Q is added, it does not change the results for trial failure surfaces to the left of cb_1. For the plane cb_3, we see that the load Q does not act on the free body ab_3c. Thus the calculated pressure d_3e_3 is the same as that without the load Q. If we now consider plane cb_1 with b_1 located just to the right of Q, the force Q must be added to the weight of the wedge. If cd_1 represents the weight of the wedge and $d_1 d_2$ represents the load Q, the pressure is equal to $d_2 e_2$. Without Q, the pressure is $d_1 e_1$. Hence the Culmann line is $e_3 e_1 e_2 e_4$ and has a sharp break at e_1. The maximum pressure is again obtained by locating the point of tangency e. The distance ed is the active earth pressure, and bc is the critical plane.

Culmann's construction serves a useful purpose in that it illustrates the effect of the location of Q on the earth pressure. Figure 10.12(a) shows that Q

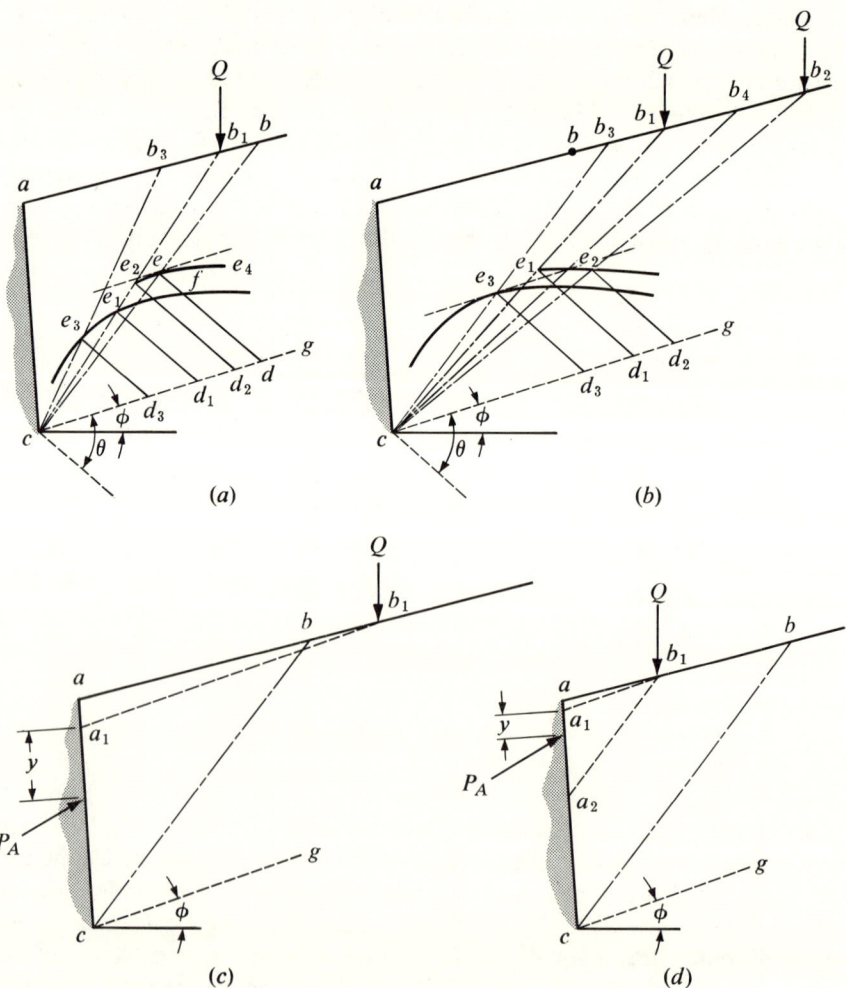

FIGURE 10.12 *Earth pressure due to a line load on the surface.*

increases the resultant earth pressure by an amount equal to *ef*. If Q is placed closer to the wall, say at b_3, the portion of the Culmann line e_2e_4 remains unchanged, and the resultant earth pressure on the wall is still *de*. As long as Q is placed to the left of *b*, its effect on the magnitude of the resultant pressure is the same, irrespective of its position. The critical plane *cb* is usually different from the critical plane for the case of no surcharge. However, the difference is not very large.

If Q is placed to the right of *b*, its effect on the resultant earth pressure decreases as the distance between Q and *b* increases. At a large distance from the wall, it would have no effect on the earth pressure. This is illustrated in Fig. 10.12(*b*); cb_3 is the critical plane obtained with no surcharge and point *b* is the same as point *b* in Fig. 10.12(*a*). If Q is placed at b_2, the earth pressure d_2e_2 is less than d_3e_3, which is the earth pressure with no surcharge. The critical plane in this case is cb_3 and Q has no effect on the earth pressure. If Q is placed at some intermediate point b_1 between *b* and b_4, it increases the pressure on the wall but the amount is less than the amount *ef* shown in Fig. 10.12(*a*).

PRESSURE DISTRIBUTION. No theoretical solution is available for the calculation of the earth pressure distribution along a wall. As an approximation we may assume that the stress distribution is equal to that calculated by elasticity, thus ignoring the plastic behavior of the material. Under this condition we can use Boussinesq's equation [(6.15)]. However, this gives us the horizontal stress at a point inside a homogeneous soil mass. At the boundary between soil and wall, the effect of the wall is to restrict the displacement of the soil under the applied stress. If the wall is considered to be perfectly rigid, Mindlin (1936) has shown that the horizontal stress should be twice that given by Eq. (6.15). Usually, the stiffness of the wall is much greater than that of the soil and the second extreme of a rigid wall is a better approximation. In fact, model tests show good agreement between measured pressure and that calculated by Mindlin's solution when the distance x is greater than $0.5H$ (Fig. 10.13). When the load is closer to the wall, considerable differences are noted (Terzaghi, 1954).

One may still calculate the pressure distribution by determining the change in the resultant earth pressure with depth as outlined in Sec. 10.5. However, the surcharge Q may alter considerably the shape of the failure surface (Fig. 10.11) and thus reduce the accuracy of the results. Thus such refined calculations are often not justified. An empirical method that gives reasonably satisfactory results for cohesionless soils is illustrated in Fig. 10.12(*c*) and (*d*). Point *b* is the location beyond which the effect of Q begins to decrease [same as *b* Fig. 10.12(*a*) and (*b*)]. From point b_1 a line a_1b_1 is drawn parallel to *cg*. If Q is located to the right of the critical surface [Fig. 10.12(*c*)], the center of pressure is located by taking the distance *y* equal to $\frac{1}{3}(a_1c)$. If Q is located to the left of the critical surface [Fig. 10.12(*d*)], line b_1a_2 is drawn parallel to *bc*. The distance *y* is taken as $\frac{1}{3}(a_1a_2)$ [Terzaghi (1943*b*)].

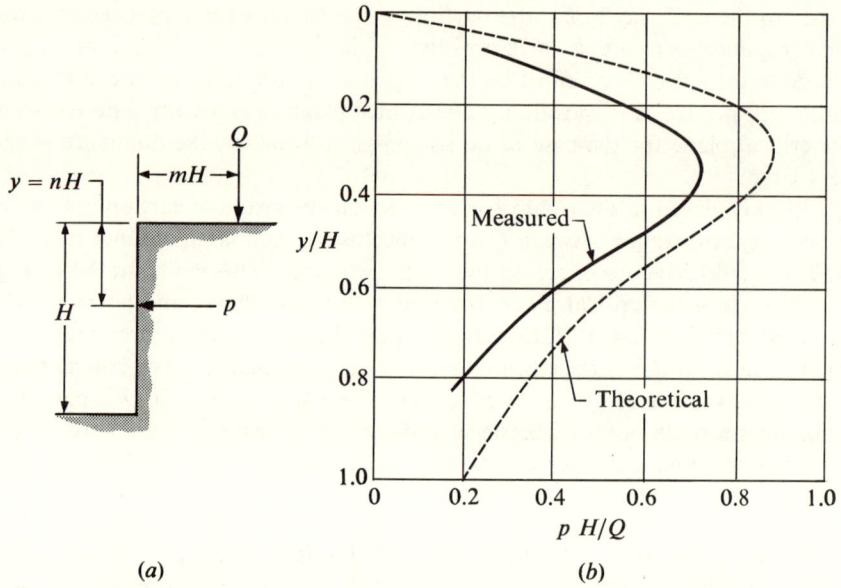

FIGURE 10.13 *Earth pressure due to a line load.* [(b) After Terzaghi (1954).]

10.9 Effect of seepage and porewater pressure

The calculations in the preceding sections do not consider the effect of the porewater pressure or seepage. To do so it is necessary to separate the stress into effective stress and pore pressure. The principles outlined in Secs. 10.5 through 10.8 remain unchanged.

Two examples are shown in Fig. 10.14. In the first one [Fig. 10.14(a)] we calculate Rankine's active earth pressure in a deposit of sand with the water table at a distance H_1 below the ground surface. If we consider the drained state, then the porewater pressure in the soil is the hydrostatic pressure. The earth pressure is calculated with the bulk unit weight γ above the water table and the submerged unit weight γ' below it. Since we are considering the effective stress, the shear strength is $\bar{\phi}$. The calculated pressure is that exerted by the solid particles on the wall. The total pressure on the wall consists of this earth pressure plus the hydrostatic pressure.

Figure 10.14(b) illustrates the calculation of earth pressure by Coulomb's theory with the soil mass subjected to steady seepage. Water enters the ground surface and is drained by a filter along the wall ab. In this case it is necessary to construct the flow net. The porewater pressure along the failure surface bc is determined from the flow net (Sec. 3.7) and is plotted along bc. The resultant water pressure is P_W, and this force is included in the force polygon.

SEC. 10.10   Deformation of braced excavations   293

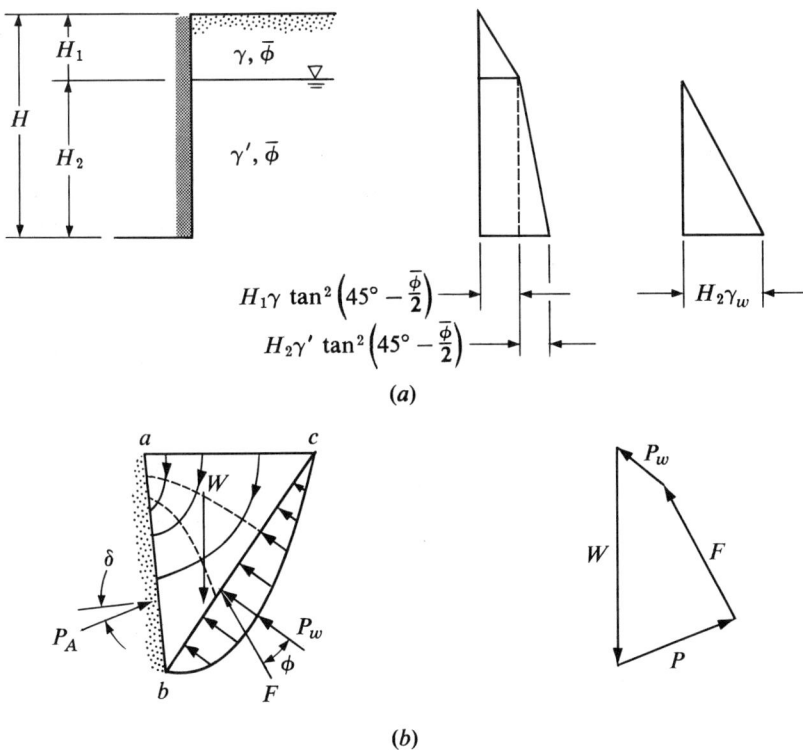

FIGURE 10.14   *Earth-pressure calculations with effective stress.*

The calculations shown above are valid only for the drained condition; the loading of the soil is assumed to correspond to that of the drained shear test. Otherwise, excess hydrostatic porewater pressure is induced in the soil which is not included in these calculations. A comprehensive examination of the drained and undrained loading conditions is made in Chapter 11.

WALLS WITH LIMITED DEFLECTIONS

10.10   Deformation of braced excavations

The deformation of a braced excavation as described by Fig. 10.15 is an excellent illustration of limited deformation. Construction of the unit begins with the driving of the sheeting. Usually a row of struts is placed near the top of the excavation immediately after a small cut [Fig. 10.15(a)]. At this stage there is relatively little movement of the sheeting. With further excavation the sheeting continues to move because the earth pressure is greater on the right side of the wall. However deformation takes place only in the lower part of the excavation, since the top is restrained by the strut already in place [Fig. 10.15(*b*)]. The final condition is illustrated in Fig. 10.15(*c*), when the excavation has reached its full depth. Excavation to the full depth naturally brings about further deformation near the bottom. The final deformation is almost 0 at the top and attains its maximum at the bottom of the excavation. Since this deformation condition bears no resemblance to that assumed in the derivation of Rankine's and Coulomb's theories, the earth pressure distribution is likely to be very different from those predicted by the theories [Terzaghi (1936*a* and *b*)].

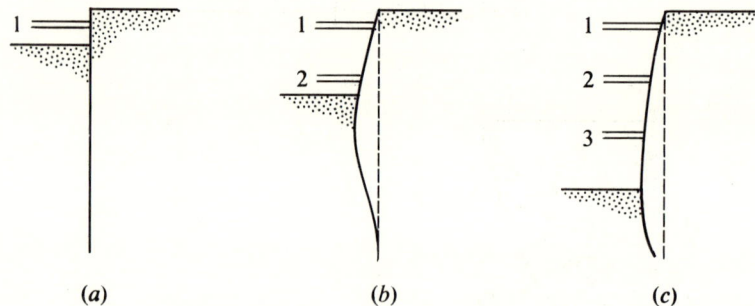

FIGURE 10.15 *Deflection of a braced excavation.*

10.11 Earth pressure on bracing systems

If the deformation at the bottom of the cut is small, the soil mass behind the wall may be in the elastic state. Then the solution given in Sec. 6.9 would be applicable and the pressure distribution would be parabolic, as shown in Fig. 6.18(*c*). For a large deformation at the bottom, shear failure may develop along a surface that would extend to the bottom of the cut (*bc* in Fig. 10.16). In the upper part of the soil mass the deformation is small. Hence, the shear strength may not be completely mobilized along some potential failure surface such as *de*. As a result, the plastic state only extends through a narrow zone along *bc*, as indicated by the slip lines. Therefore it cannot be said that the entire soil mass *abc* is in a state of plastic equilibrium. If the slip surface *bc* passes through the bottom of the cut, the methods of plasticity may be used to calculate the resultant earth pressure P [Terzaghi (1941)*; James and Lord (1972)].

A general description of the displacements in the soil mass and its relationship to the earth pressure has been presented by Bjerrum et al. (1972). Figure 10.17 shows the deflections of initially vertical lines. After excavation, the lines are displaced into positions shown as *ABC*. The displacement is zero below *C*.

---

* Terzaghi's method is called the *general wedge theory*. It is essentially the same as the method described in Sec. 9.5 and Fig. 10.16(b).

SEC. 10.11   Earth pressure on bracing systems

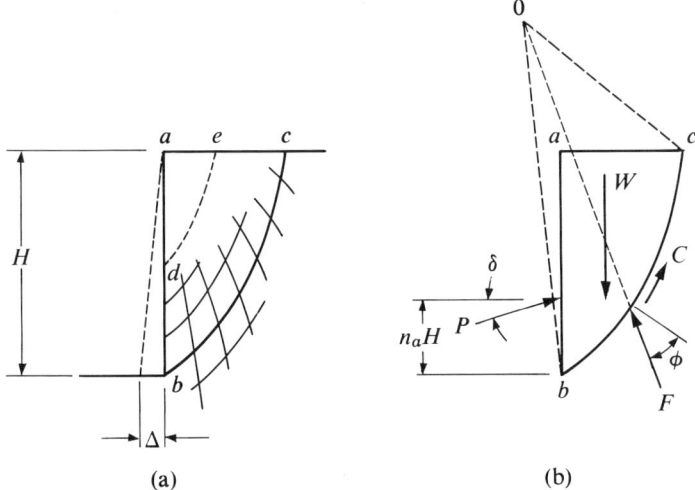

FIGURE 10.16 *General wedge theory of earth pressure.*

Thus, we may expect the pressure exerted by the soil mass on the right side of ABC to be close to the active earth pressure, whereas the pressure on the left side of ABC would be close to the passive earth pressure. Figure 10.17(a) represents an excavation in a stiff soil or in a soft soil that extends only a short distance below the bottom of the cut. Figure 10.17(b) represents an excavation in a soft soil extending to great depth. We note that the displacements are much greater for the case shown in Fig. 10.17(b).

The analysis in Sec. 10.5 shows that the magnitude of the resultant earth pressure is not too strongly influenced by the shape of the failure surface. As an approximation, we may take the resultant active and passive pressures as equal to those given by Rankine's theory. Then we may sum the forces above point C and get

$$P_A = P_P + P \qquad (10.11)$$

in which P is the sum of the strut loads. Bjerrum et al. (1972) have found this relationship to be approximately true for eleven excavations where the depth H_0 could be determined with reasonable certainty. Figure 10.18 shows the measured earth pressures on two sides of the sheeting and compares them with those calculated by Rankine's theory. At present, it is difficult to predict the depth H_0 but empirical data [Bjerrum et al. (1972)] indicate a definite trend in which P/P_A increases with the ratio H_0/H.

The distribution of earth pressure is strongly influenced by the deformation conditions. Qualitatively, its shape would be similar to the shaded areas shown in Fig. 10.17(c) and (d) (see also Fig. 6.18). Depending on the magnitude of the displacement, the resultant earth pressure P acting on the bracing system [shaded areas in Fig. 10.17(c) and (d)] may be greater or less than the resultant active earth pressure to the depth of the cut [shown as abc in Fig. 10.17(c) and (d)].

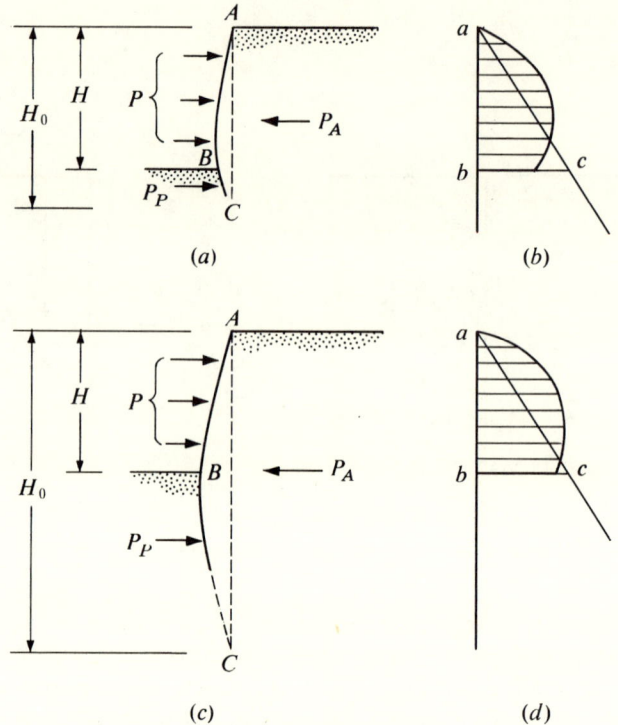

FIGURE 10.17 *Earth pressure on braced excavations.*

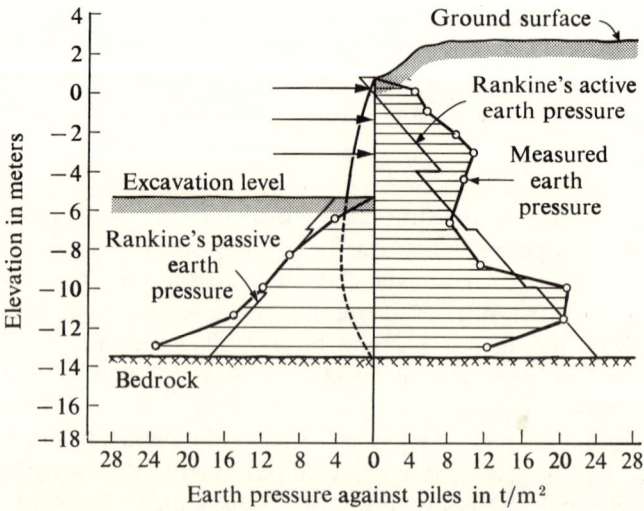

FIGURE 10.18 *Active and passive earth pressure observed by earth-pressure cells in the excavation at Vaterland 1 for Oslo subway.* [After Bjerrum et al. (1972).]

10.12 Field measurement of earth pressure

The preceding section gives the qualitative picture of the magnitude and distribution of earth pressure on braced excavations. Our current design procedures are based largely on the results of measurements of wall deformations and earth pressures under field conditions. Notable among the early measurements are those by Spilker on the Berlin subway [see also Terzaghi (1941)] and by Peck (1943) on the Chicago subway. The Berlin subway was excavated through dense sand above the water table, the Chicago subway in soft saturated clay. Figure 10.19(b) and (c) show representative sheeting deformations and strut loads measured on the bracings of the Chicago subway.

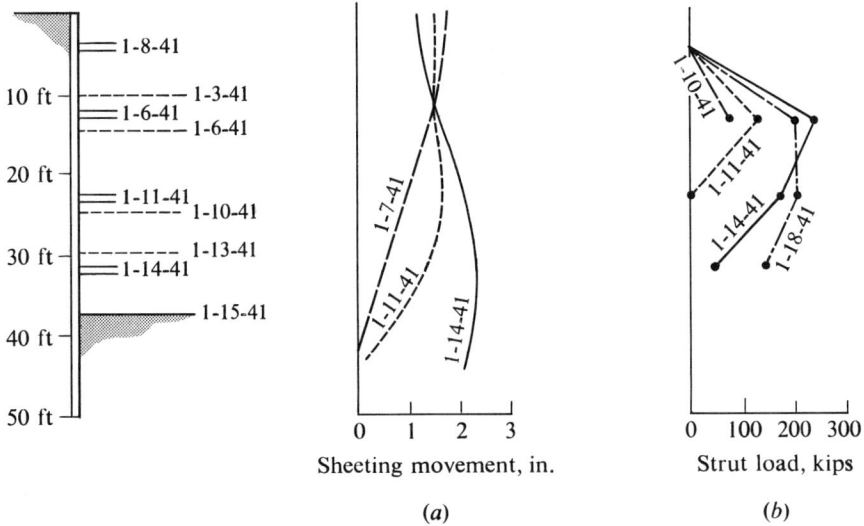

FIGURE 10.19 *Measured deflection and strut loads, Chicago subway.* [After Peck (1943).]

From the measured strut loads it may be seen that the pressure distribution is approximately parabolic with the center-of-pressure well above the lower third point. The value of n_a is about 0.45 for the measured earth pressure. The magnitude of the resultant earth pressure is only slightly larger than the Coulomb or Rankine earth pressures. In the Chicago subway cuts the resultant earth pressure was computed with the undrained shear strength ($\phi_u = 0$), and in the Berlin subway it was computed with the drained shear stength ($\bar{\phi}$) (see Secs. 8.3, 11.4, and 11.6).

One very important feature of the earth pressure was brought out by the measurements. Although along most profiles the earth-pressure distribution was parabolic as predicted by the theory, there was a great deal of scatter in the pressures at different profiles. According to detailed records [Peck (1943)], these variations were traced back to variations in the details of the construction procedure. These variations, which could not be controlled, were responsible for considerable differences in the earth-pressure distribution, although the magnitude of the resultant was consistent for all profiles. For the purpose of design these erratic differences cannot be predicted. Instead the alternative should be to design for the maximum pressure that may be expected at any elevation. This means that the envelope to all the earth-pressure distribution curves should be used in design.

On the basis of results of field measurements made on a number of braced excavations, Terzaghi and Peck (1967) suggested the design earth pressures shown in Fig. 10.20 for sand, soft to medium clay, and stiff-fissured clay. The resultant earth pressures computed by these design rules are substantially larger than those obtained by Rankine's theory, because the design earth pressures represent envelopes to measured earth pressure distributions. The

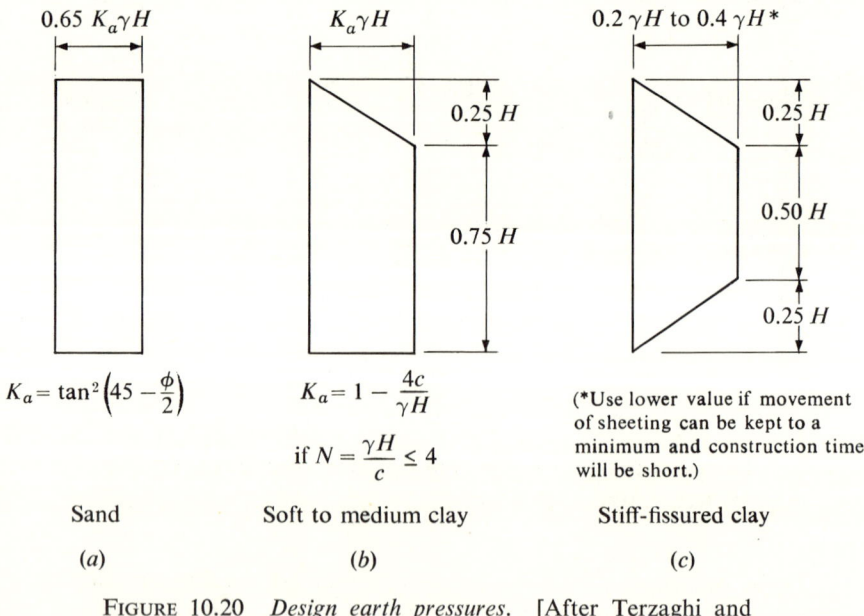

FIGURE 10.20 *Design earth pressures.* [After Terzaghi and Peck (1967).]

design rule in Fig. 10.20(b) may be applied to cases where the depth of the plastic zone H_0 is limited by presence of stiff materials [Fig. 10.17(a)]. Where soft clays extend to great depth [Fig. 10.17(b)], the earth pressure may be considerably greater. Peck (1969b) has analyzed the deformation conditions in detail and has suggested that the dividing line between the two cases should be

$$N = \frac{\gamma H}{c} = 4$$

Below this value, the design rule in Fig. 10.20(b) applies. If N exceeds 4, the value of K_A may be larger. Available data are inadequate for the formulation of design rules for N greater than 4. Limited experiences indicate that for soft, normally consolidated clays, K_A may be given by

$$K_A = 1 - \frac{1.6c}{\gamma H} \tag{10.12}$$

When these pressures are used for the design of bracings, the strut loads are to be computed as shown in Fig. 10.21. The loads on struts 1, 2, and 3 are taken to be, respectively, the earth pressures that act on sections ab, bc, and cd of the sheeting. Points b, c, and d are located midway between struts 1, 2, and 3 respectively.

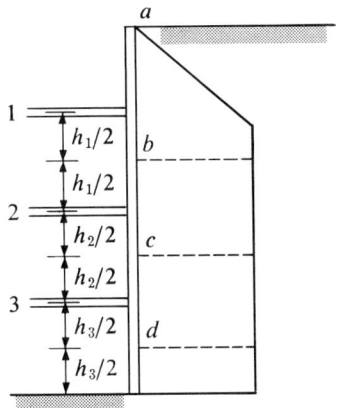

FIGURE 10.21 *Calculation of strut loads.*

YIELDING INSIDE A SOIL MASS

In this section we study the earth pressure inside a soil mass if deformations are introduced within the mass. Such would be the case if an excavation for an underground structure (i.e., a tunnel) were made inside a soil deposit. The deformations induced by this construction alter the at-rest earth pressures that exist prior to excavation.

10.13 Pressure over yielding base

The elasticity solution for pressure on a yielding base is given in Sec. 6.9 and Fig. 6.18. It is noted that at the edge the theoretical stress is infinite and plastic flow would occur in this zone. In Fig. 10.22(a) the zone of plastic flow is delineated by a set of slip lines. Their pattern may be understood as follows. As the strip ab yields, the compressive stress at the edge is steadily reduced. (It changes to tensile stress if elasticity holds.) If the strip ab is assumed to be smooth, then ab is the minor principal plane. The slip lines immediately above ab make angles of $45° - (\phi/2)$ with ab. Over the unyielding part, large compressive stresses exist adjacent to the edge. We may consider this to be a transfer of the overburden pressure from ab to the unyielding base.* In this zone the major principal stress is in the vertical direction. The slip lines make $45° + (\phi/2)$ with the horizontal. Hence we can sketch approximately the pattern of slip lines [Fig. 10.22(b)] with these two zones connected by a zone of radial shear. This holds for a small area immediately surrounding the edge of the yielding trip where plastic flow takes place.

If the yield proceeds far enough and the distance D is small, one or more slip lines may propagate to the ground surface. Such a slip line is shown as ac or bd in Fig. 10.22(a). If D is very large or if the yield of ab is limited, then the slip lines may advance only as far as c_1 and d_1. From this one may visualize the following physical picture. Immediately above the plane ab, the movement of the soil particles is denoted by the arrows A and B on the left side of Fig. 10.22(a). The sand just above ab moves downward and the sand over the unyielding part moves laterally. Above ab the major and minor principal stresses after the displacement are in the horizontal and vertical directions, respectively. Prior to the displacement they are vertical and horizontal, respectively. If K denotes the ratio σ_x/σ_y, one may expect K to increase in the failure zone from an initial value K_0 as the strip ab yields to produce failure. If failure occurs throughout the width of ab, K would approach K_p along the center line as well as at the edge. At some distance above the failure zone, K may increase only slightly or remain at its initial value. This is verified by the model experiments of Terzaghi (1936d). The measured value of K_0 is 0.6. The measured vertical and horizontal at rest pressures above the center of the strip are shown in Fig. 10.23 as solid lines. The pressures after yielding of the strip are shown as dashed lines. Within a distance of $5B$ above the yielding strip, the pressures are greatly reduced. The drop in σ_y is far greater than the drop in σ_x and K increases from 0.60 to somewhere between 1.00 and 1.50. At elevations higher than $5B$ above the strip, the pressures remain unaffected. This suggests that shear failure does not extend beyond a distance of $5B$ above the strip.

For an approximate analysis, it is assumed that the shear strength of the soil is mobilized along ac_1 and bd_1, up to a distance nB above the strip. Furthermore the curves ac_1 and bd_1 are replaced by two vertical lines ae_1 and bf_1. The stresses can be analyzed by considering the equilibrium of the mass abe_1f_1. The weight of the soil above e_1f_1 is replaced by a surcharge q [Fig. 10.22(c)] equal to $\gamma(D - nB)$. Figure 10.22(d) shows the forces that act on an element dy thick. They consist of the unknown vertical stresses σ_y on the upper face and

---

* This phenomenon was called "arching" by Terzaghi; his analysis [Terzaghi (1943b)] is described in this and the next section.

SEC. 10.13   Pressure over yielding base   301

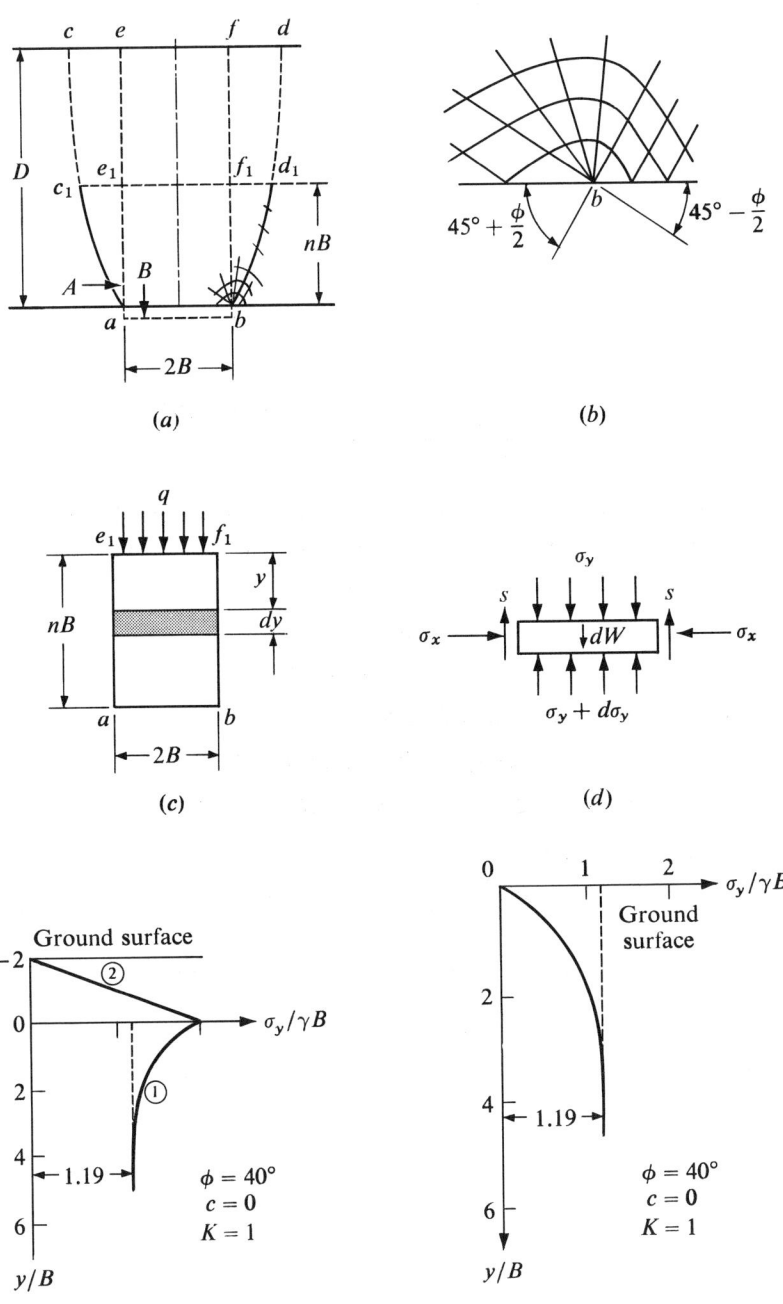

FIGURE 10.22   *Calculation of earth pressure on a yielding base.*
[After Terzaghi (1943b).]

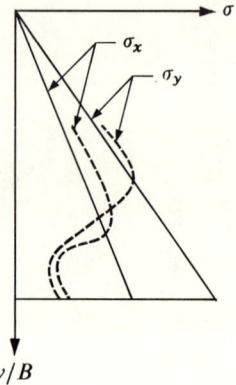

FIGURE 10.23 *Measured earth pressure in a soil mass above a yielding base.* [After Terzaghi (1936d).]

$\sigma_y + d\sigma_y$ on the lower face, the weight of the element dW, and the normal and shear stresses σ_x and s on the side. The values of dW, σ_x, and s are as follows:

$$dW = 2\gamma B\, dy$$
$$\sigma_x = K\sigma_y$$
$$s = c + \sigma_x \tan\phi$$

Summation of forces in the vertical direction gives

$$dW + 2B\sigma_y = 2B(\sigma_y + d\sigma_y) + 2s\, dy$$

or
$$2\gamma B\, dy = 2B\, d\sigma_y + 2c\, dy + 2K\sigma_y \tan\phi\, dy$$

Solving for $d\sigma_y/dy$, we have

$$\frac{d\sigma_y}{dy} = \gamma - \frac{c}{B} - K\sigma_y \frac{\tan\phi}{B} \tag{10.13}$$

The boundary condition is that the pressure of the surcharge acts on $e_1 f_1$,

$$\text{at } y = 0, \quad \sigma_y = q$$

Integrating Eq. 10.13 for the above boundary condition yields

$$\sigma_y = \frac{B(\gamma - c/B)}{K \tan\phi}(1 - e^{-K\tan\phi(y/B)}) + qe^{-K\tan\phi(y/B)} \tag{10.14a}$$

If $c = 0$, the equation reduces to

$$\sigma_y = \frac{\gamma B}{K \tan\phi}(1 - e^{-K\tan\phi(y/B)}) + qe^{-K\tan\phi(y/B)} \tag{10.14b}$$

If $\phi = 0$, we obtain, by l'Hospital's rule,

$$\sigma_y = \left(\gamma - \frac{c}{B}\right)y + q \tag{10.14c}$$

SEC. 10.14   Pressure on tunnels   303

To illustrate the use of Eq. (10.14), we calculate the variation of the vertical pressure σ_y with depth for the case, $c = 0$, $\phi = 40$ deg, and $K = 1$ in the zone of shear [Terzaghi (1943b)]. The depth of the yielding strip is $7B$. In addition, we assume n [Fig. 10.22(a)] to be 5. The calculated values of σ_y at different elevations in the zone $5B$ above the yielding strip are plotted in Fig. 10.22(e) as curve one. The vertical pressure is equal to q at y/B equal to 0, which is the elevation of $e_1 f_1$. The pressure σ_y at elevations between $e_1 f_1$ and the ground surface σ_y is simply the overburden pressure, which increases linearly with depth. This is plotted as curve two in Fig. 10.22(e) for a depth $e_1 e$ equal to $2B$. If the overall depth D is increased to $9B$, for example, and the depth of the shear zone (nB) remains the same, then $D - nB$ increases to $4B$ and q is doubled. However, Eq. (10.14) indicates that the vertical pressure due to q decreases exponentially with y/B. Thus at the elevation of the yielding strip ($y = 5B$), the vertical pressure is not significantly affected by the value of q and remains at approximately $1.20\gamma B$.

For comparison we plot the variation of σ_y with depth for $q = 0$ and the same soil properties. This is the case if the yielding strip is located at a small distance below the ground surface ($D \leq nB$). The calculated pressure is plotted in Fig. 10.22(f). The quantity y/B denotes the depth of the yielding strip. If n is again taken as 5, the maximum depth is $y/B = 5$. Below this elevation we cannot have the full shear strength mobilized along the sides of the column of soil, and the condition of $q = 0$ is no longer valid.

The inclusion of the cohesion term in the shear strength does not alter the major characteristics noted above. It should be noted that these calculations give the average σ_y over the strip ab. The actual pressure distribution is of course not uniform but remains indeterminate. Qualitatively the pressure distribution is of the shape shown in Fig. 6.18.

The above solution can be modified to take into account the stresses on ee_1 and ff_1. Newmark assumed the material above plane $e_1 f_1$ to be elastic. Thus, one may make the approximation that the shear stress along ee_1 and ff_1 is proportional to the displacement. Along $e_1 a$ and $f_1 b$ the shear strength is completely developed as before [see Air Force Special Weapons Center (1962) for details].

10.14 Pressure on tunnels

The theory developed in the preceding section can be readily adapted to the analysis of earth pressure on tunnel linings. The deformations associated with the construction of a tunnel are illustrated in Fig. 10.24(a). Excavation of the tunnel results in the yielding of the roof aa' and the sides ac and $a'c'$. If the movement of ac is large, the zone of shear failure extends through the triangle abc. The wedge abc moves downward and toward the right in very much the same way as the sliding wedge does in Coulomb's theory. Under this condition, downward movement occurs along the strip bb'. Therefore, we may apply the solution in Sec. 10.13, considering bb' as the yielding strip.

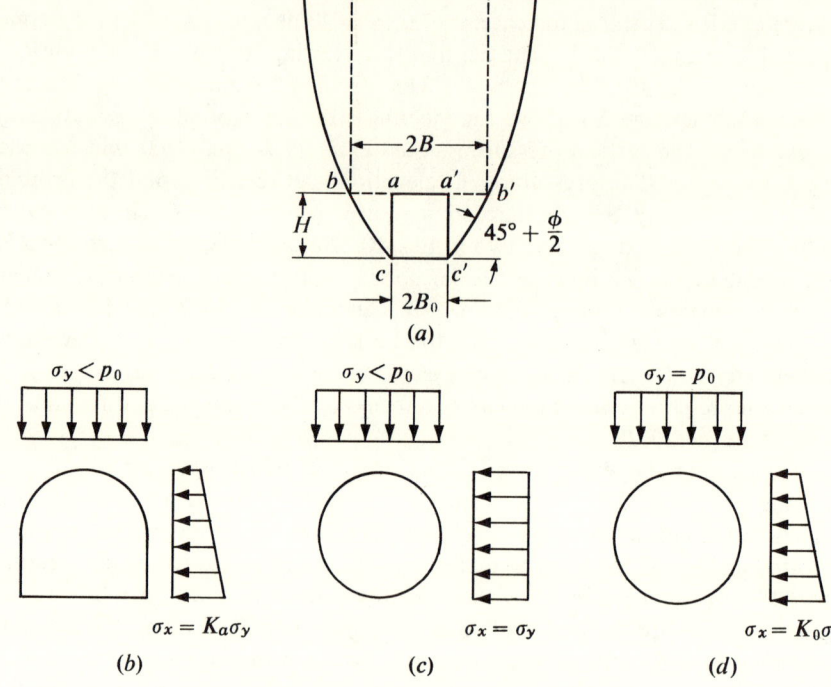

FIGURE 10.24 *Earth pressure on tunnels.*

The width of the strip bb' may be determined approximately if we assume the failure surface cb to be a straight line making an angle of $45° + (\phi/2)$ with the horizontal. This is equivalent to assuming Rankine's active state in zone abc. From this, the width bb' is

$$2B = 2\left[B_0 + H \tan\left(45° - \frac{\phi}{2}\right)\right]$$

in which B_0 is one-half the width of the tunnel. This is the value of B that should be used in Eqs. (10.14) for the calculation of the pressure σ_y. Since the active Rankine state is assumed in zone abc, the horizontal pressure on ac may be calculated from Eq. (9.7a). The elevation of bb' is taken as $y = 0$, and the surcharge q is the pressure σ_y on plane bb'.

It should be emphasized again that the above analysis is based on the assumption that the roof and sides of the tunnel permit adequate yielding. Consequently it is necessary to consider the deformations permitted by the underground structure before the results of the analysis can be applied. As examples, we consider first a temporary tunnel lining that consists of steel ribs supported on footing foundations as shown in Fig. 10.24(b). The footings are free to settle under the load and the sides are free to move in. This satisfies the deformation conditions and Eqs. (10.14) can be expected to give reasonable results. The vertical pressure is considerably less than the overburden pressure p_0, and K is less than K_0 and may approach K_a.

The second example is a flexible steel ring. The flexibility of the ring permits yielding of the roof, but since it is a closed ring the sides must bulge as a result. The vertical yield reduces the value of σ_y, but the lateral bulge increases σ_x and K. A flexible circular ring stops deforming only if it is under hydrostatic pressure. Therefore, after undergoing a certain deformation, the ring stabilizes as σ_x equals σ_y or K equals 1 [Fig. 10.24(c)]. As yielding eventually stops in both cases, the question remains as to whether stress relaxation would take place in the soil. If so, the shear stress along the slip lines decreases. Then the pressure σ_y can be expected to rise with time. It may eventually approach p_0, should the shear stress be completely relaxed.

The third possibility is a rigid structure that does not yield. The excavation of the cavity reduces the pressures to the state shown in Fig. 10.24(b). With stress relaxation, the pressure on the roof may reach the value p_0. However, the rigid structure does not bulge laterally, and the value of K would approach K_0 [Fig. 10.24(d)].

The summary of the soil-structure interaction is derived largely from a number of field measurements of the deflections and earth pressures in tunnels. Tunnels in soft clay were studied by Terzaghi (1943a) and Housel (1943). The vertical pressures were initially about one-half of the overburden, but progressively increased with time. The final vertical pressures were nearly equal to the overburden pressure in both cases. The tunnel of the Chicago subway had a flexible structure and the tube deflections were of the order of 0.5 in. accompanied by hydrostatic pressure distribution ($K = 1$). The Detroit tunnel was rigid, and the final value of K was about 0.5. This indicates that stress relaxation does occur in soft clays. Hence the vertical pressure should be taken as the overburden pressure in the design of the permanent structure. Temporary supports can be designed to support pressures calculated by Eq. (10.14).

Lane (1957) reported on pressure measurements in rigid and flexible tunnel structures at Garrison Dam. The tunnels were located in stiff overconsolidated clay shale. The vertical pressure on the rigid structure very quickly rose to the overburden pressure. In contrast, the flexible structure was subjected to a vertical pressure equal to only 20 percent of the overburden pressure after a period of 2 years. This pressure is of the same order of magnitude as that calculated by Eq. (10.14). Whether this pressure would increase substantially after longer time intervals is not known. Observations by Ward and Chaplin (1957) seem to suggest that the vertical pressure in stiff clays does increase over very long periods of time.

The description in the preceding paragraphs is necessarily a simplification. The earth pressure that acts on real tunnel linings is dependent on many factors. The method of construction is probably one of the most important factors. For detailed discussion of this subject the reader is referred to the comprehensive review by Peck (1969b).

Example 10.1. We study an excavation in a deposit of medium sand above the water table. The requirement is that the bracing system shall consist of the sheeting and one row of struts near the top of the cut. The dimensions of the excavation and soil properties are described in Fig. 10.25(a). The principles outlined in the preceding sections regarding braced excavation are still applicable to this problem. We must, however, note the significance of the use of a single strut near the top. The absence of struts at lower levels may lead to increased deformation of the sheeting at these levels. Short of field or model tests it is not possible to predict accurately the magnitude of the increased deformation and the consequent change in earth-pressure distribution. Hence, we can analyze these factors only on a qualitative basis.

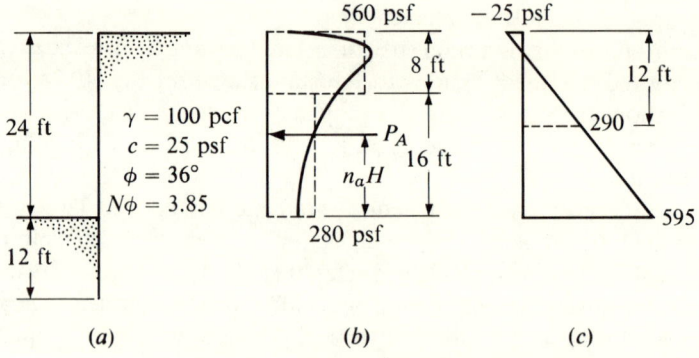

FIGURE 10.25   Example 10.1.

Any increase in sheeting deformation at a particular location serves to reduce the coefficient of earth pressure K at that point. The effect of the larger deflections in the lower portion of the sheeting is to increase the value of n_a over that of a conventional braced cut. The center of pressure moves up and the load on the strut is increased. Thus the strut would be overloaded if it is designed for the pressure distribution shown in Fig. 10.20(b). On the other hand, the moment in the sheeting would be somewhat smaller. A reasonable estimate of the earth-pressure distribution is shown by the curve in Fig. 10.25(b).

If the movement of the sheeting is sufficient to induce the type of failure illustrated in Fig. 10.16, the magnitude of the resultant earth pressure to be used in design should still be close to that given in Fig. 10.20(a), which is 1.3 times the resultant active earth pressure as computed by Rankine's theory. In the present case, there is a small cohesion and the equation for K_a shown in Fig. 10.20(a) cannot be used. We may take the resultant earth pressure as 1.3 times the quantity

$$\frac{1}{2}\gamma H^2 \tan^2\left(45° - \frac{\phi}{2}\right) - 2cH \tan\left(45 - \frac{\phi}{2}\right) = 6890 \text{ lb}$$

or 8950 lb. For convenience, this force is distributed as shown by the dashed lines in Fig. 10.25(b).

A more elaborate calculation of the resultant earth pressure may be made with the assumption that the slip surface is a logarithmic spiral (Terzaghi, 1941). The results of the calculations are shown in Fig. 10.26. For $\delta = 0$, the resultant earth pressure is around 7400 lb. This is close to the value of 6890 lb obtained with Rankine's theory. The resultant pressure of 8960 lb to be used for design is slightly larger because it represents the envelope of the probable earth pressure distributions.

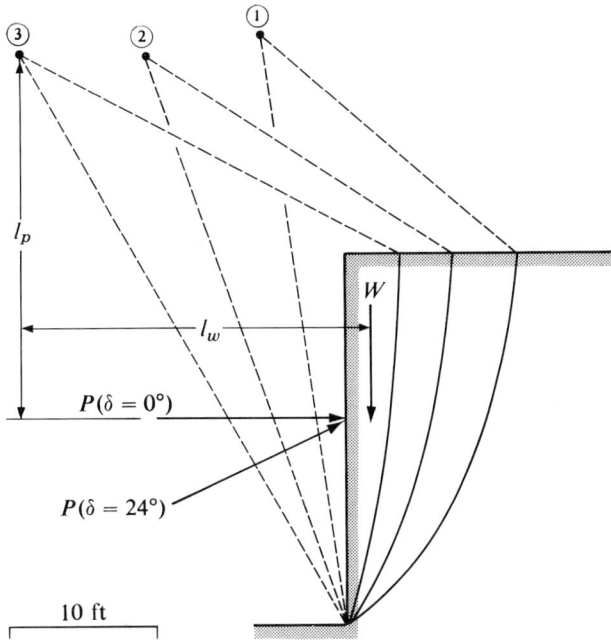

	W, kips	l_w, ft	M_w, ft-kips	r_1, ft	r_2, ft	M_c, ft-kips	$\delta = 24°$		$\delta = 0°$	
							l_p, ft	P, kips	l_p, ft	P, kips
1	14.8	11.5	170	23.0	39.0	17.2	24.5	6.3	24.0	6.4
2	11.2	17.0	190	24.5	40.0	17.2	27.0	6.4	23.5	7.4
3	5.6	24.0	135	29.0	40.5	13.8	30.0	4.1	23.5	5.2

FIGURE 10.26 *Example 10.1, calculation of earth pressure by the general wedge theory.*

We may consider another approach to the problem. If the strut is not installed until the excavation is made to a depth of say 12 ft, the consequence is worth noting. Until the strut is in place, the sheeting acts as a cantilever beam and is free to deflect at the top. The earth pressure may be calculated by Rankine's theory, and the result is shown in Fig. 10.25(c). The magnitude of the sheeting deflection at the top can be calculated from the beam theory and is equal to 0.2 in. If we consider the full depth of the cut, this deflection corresponds to a ratio of ρ/H equal to 0.001, which is adequate to develop the active state of stress in the soil mass [Fig. 10.2(a)]. Then the earth-pressure distribution should approach that calculated by Rankine's theory [Fig. 10.25(c)]. As can be seen, this yields a much smaller strut load but a larger sheeting moment than the pressure given in Fig. 10.25(b).

PROBLEMS

10.1 A retaining structure with a vertical wall 24 ft high supports a deposit of sand. The soil properties are as follows: $e = 0.30$, $\gamma_s = 2.65$ g/cc; $\bar{\phi} = 33$ deg, $\bar{c} = 0$. The water table is at a depth of 10 ft. Assume the soil above the water table to be dry. Calculate the distribution of pressure from the soil and water on the wall: (a) if the wall is rigid and does not move away from the soil, and (b) if the wall moves by an amount sufficient to develop Rankine's state in the soil.

Answer: (a) earth pressure: 579 psf at 10 ft, 1089 psf at 24 ft; water pressure: 0 at 10 ft, 873 psf at 24 ft

(b) earth pressure: 376 psf at 10 ft, 706 psf at 24 ft; water pressure: same as (a)

10.2 A retaining wall 28 ft high is to support a backfill of clayey sand. The samples of the material were subjected to laboratory tests and the following properties were determined: $\gamma_s = 2.70$ g/cc, LL = 11%, PL = 6%; $(\gamma_d)_{max} = 136$ pcf, $w_{opt} = 6\%$. Specimens were then compacted to a dry density of 130 pcf at a moisture content of 5% and subjected to triaxial tests. The shear-strength parameters are $\bar{\phi} = 23$ deg, $\bar{c} = 300$ psf. Steady seepage of groundwater toward the wall is expected as shown in Fig. 10.27. Calculate the resultant earth pressure on the wall by Coulomb's theory.

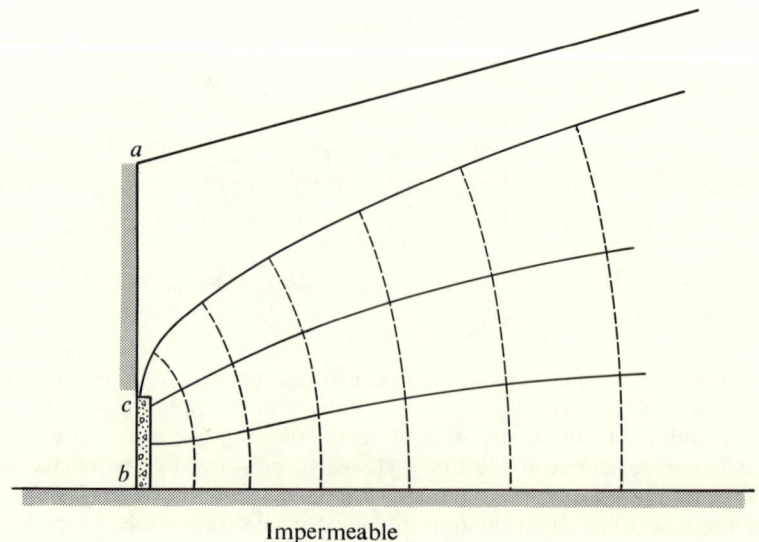

Impermeable

FIGURE 10.27

10.3 What steps and details are necessary in the construction of the wall (Problem 10.2) in order that the conditions for which the pressures are calculated are realized in the real wall?

10.4 Figure 10.28 shows a flexible sheet pile retaining wall driven into the ground and supported at the top by tie rods spaced at regular intervals. The wall is then backfilled with a silty sand. Estimate the earth pressure: (*a*) if the tie-rod anchor and the sheeting are completely rigid; (*b*) if the tie-rod anchor is rigid but the sheeting is flexible; (*c*) if the tie-rod anchor yields under the tie-rod pull.
Answer: see Fig. 10.28

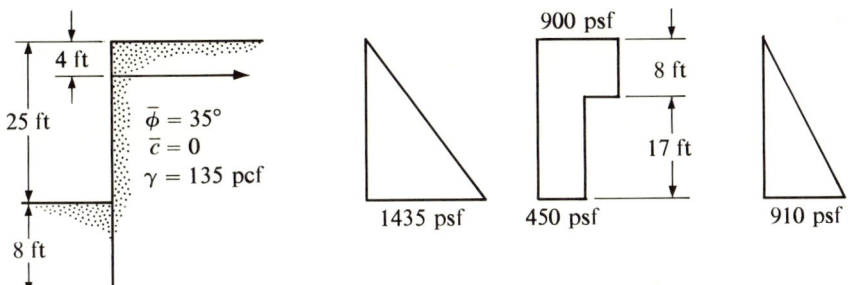

FIGURE 10.28

10.5 Consider the braced excavation in Fig. 10.25. If the sheeting is made very stiff and is driven into a deposit of very dense sand located at a depth of 28 ft, what will be the probable earth pressure on the sheeting?

10.6 The soil behind a retaining wall 30 ft high consists of 12 ft of sand overlying 18 ft of clayey silt. The wall is free to tilt. The water table lies at the top of the clayey silt deposit. The physical properties of the two soils are given as follows:

	Sand	Clayey silt
Unit weight	125 psf	134 psf
w	5%	18%
Shear strength	$\bar{c} = 0; \bar{\phi} = 37°$	$\bar{c} = 80 \text{ psf}; \bar{\phi} = 30°$

(*a*) Calculate the earth-pressure distribution. (*b*) Construct the pattern of slip surface in the soil. (*c*) Do you expect the calculated earth pressure to be the pressure immediately after completion of the construction? Why? (*d*) If not, what additional information is necessary to calculate the earth pressure immediately after completion of construction?
Answer: see Fig. 10.29

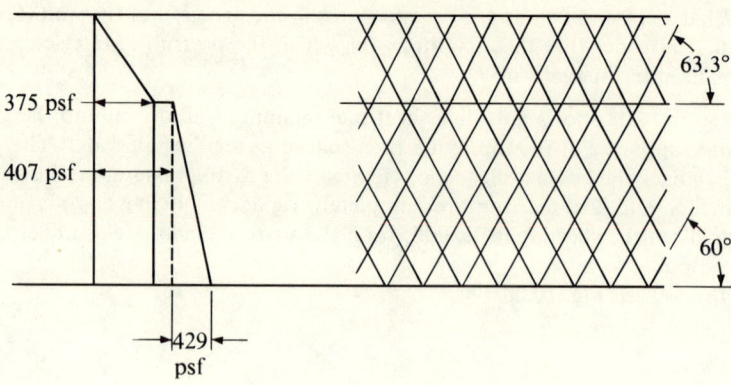

FIGURE 10.29

11

Problems of Stability

11.1 Concepts of design and analysis

This chapter develops the concepts and rationale necessary in the analysis of problems that involve stability or shear failure in a mass of soil. Common examples are bearing-capacity failure and slope failure.

The mechanics of plastic equilibrium and the quantitative evaluation of stresses and forces are treated in Chapters 9 and 10. In the solution of engineering problems, this forms only a part of the analysis. A different kind of analysis is required to decide on the loading conditions for which calculations should be made and, for these selected conditions, one must choose the appropriate soil properties for the calculations.

In this chapter we shall apply our knowledge of soil properties (Chapters 5 and 8) and the principles of mechanics (Chapter 9 and 10) to evaluate soil behavior under various conditions of loading; to examine which of these represent the most critical conditions that should be studied carefully; and to decide what shear-strength parameters best represent these conditions and are to be used in the calculations. It can be seen that this process is just as important as that of making a correct stress analysis, because if the wrong conditions or wrong strength parameters are chosen, no stress calculation, however elaborate, can give relevant results. Finally, the uncertainties and inaccuracies inherent in the analysis are discussed. It is important to realize that all theoretical analyses and laboratory measurements represent simplifications of natural phenomena, which are always much more complicated. To view the analysis in its proper perspective we should have an idea of the order of magnitude of errors that are involved, as well as departures from assumed conditions and their affect on the results of our analysis.

The analysis and critical examination can best be presented by means of examples. Thus a number of common engineering problems are treated in detail in this chapter. These should be considered as examples of the approach and concept in design analysis rather than infallible conclusions. It is obvious that not only will new knowledge of soil behavior and stress analysis alter our concepts and methods, but new problems will arise introducing loading conditions whose effects cannot be anticipated.

11.2 Conditions of loading

Before we consider the effect of loading on strength and stability, we first study several common kinds of loading and the stresses they produce. In addition to loads on the surface or inside a soil mass, we also consider the porewater pressure. Slopes provide excellent examples because of the wide variety of porewater-pressure conditions to which they are subjected. This section illustrates four common types of loading and compares their significance with respect to stability.

CASE 1. The simplest case is that of an unsaturated soil above the water table. If the porewater pressure is small, the case approaches the one analyzed in Sec. 9.9. The forces that act on the soil mass are shown in Fig. 11.1 as W, the weight of the soil; F_n, the normal component of the reaction on the failure surface; and S, the shearing resistance along the failure surface. The three forces are shown as vectors in the force polygon. In the construction of the force polygon to calculate the magnitude of the forces at equilibrium, the shear force along the failure surface F_t is substituted for S. This is because the force F_t required for equilibrium is usually not equal to the resistance S that can be provided by the strength of the soil. The magnitude of F_t is equal to the shearing resistance required for equilibrium; therefore, it is also indicative of the severity of loading.

CASE 2. In the second case we consider a slope that is submerged entirely under water [Fig. 11.1(b)]. If the slope has been under this condition for a long period of time, there is no excess hydrostatic porewater pressure in the soil. Therefore, in addition to the forces W, S, and F_n there exist hydrostatic pressures on the ground surface and on the failure surface. Their resultants are P_{w1} and P_{w2}. The weight W is slightly greater than that in the first case, owing to saturation of the soil. The force polygon shows that the combined effect of P_{w1} and P_{w2} is to reduce W to W', the submerged weight of the soil mass. The magnitude of F_t is much smaller than that for the first case, and therefore this loading is less severe.

CASE 3. If the water above the slope is removed, as in the case of an earth dam subjected to a sudden drawdown in the reservoir, a third condition is created [Fig. 11.1(c)]. The drawdown occurs so rapidly that the soil in the slope remains saturated and the value of W remains the same as that for case 2. The removal of water above the slope reduces P_{w2} to 0. However, P_{w1} is affected differently.

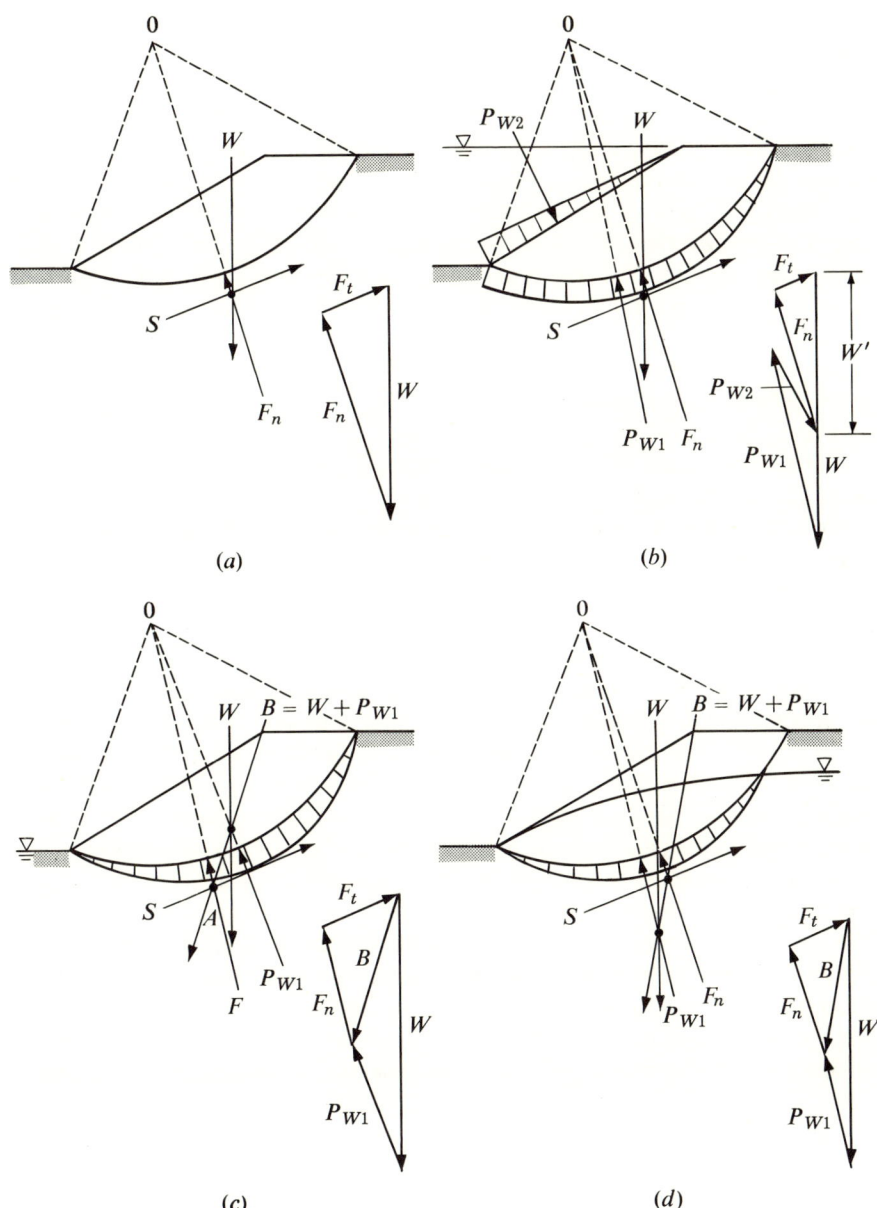

FIGURE 11.1 *Conditions of loading.* [After Taylor (1948).]

The drawdown decreases the total stress on the failure surface, and, therefore, P_{W1} must decrease by an equal amount. Since the change in porewater pressure is not uniform along the failure surface, the direction and position of P_{W1} also changes. In Fig. 11.1(c) the resultant of P_{w1} and W is represented by force B.

The forces B, F_n, and S must be concurrent at point A. The force polygon shows that the shear force F_t is substantially greater than that in case 2 and may be as large as that in case 1.

CASE 4. Finally, Fig. 11.1(d) illustrates the situation a long time after drawdown. There is steady seepage through the soil and the water table is lowered considerably. This reduces the magnitude of P_{W1} to some value less than that in case 3. The direction of P_{W1} is also different. Generally, this case is slightly more stable than case 3. We thus come to the conclusion that out of the four types of loading considered, cases 1 and 3 are more dangerous than cases 2 and 4.

11.3 Conditions of stability

The preceding section illustrates several types of loading to which the slope may be subjected. As we have seen, the safety of a slope may be represented by the factor S/F_t, which is the ratio of strength to stress. Since the strength of a soil is not constant but changes with the effective stress, we must consider the strength changes in an evaluation of the safety factor. The largest value of F_t does not necessarily result in the minimum safety factor. Hence we must consider the shear strength of the soil as influenced by the type of load and by time. The analysis of the relative changes in stress and strength constitutes the first and perhaps most important step in a stability investigation. Through this the engineer obtains a clear picture of the changes in stability throughout the various stages of the project, and is thus able to detect the most critical stages and select these for more careful investigation; other cases may be disregarded. In fact, without such a plan a coherent analysis of the problem is hardly possible.

In the subsequent sections the stability conditions are analyzed for several typical soils. We first study the case of a saturated clay and a cohesionless sand. The two cases can be reasonably well approximated by simplifications that we must necessarily introduce. They represent the two limiting conditions for which the soil behavior is well established in laboratory experiments (Sec. 8.3). The conclusions we draw for these two conditions are so instructive that from them we may, by extrapolation, arrive at general conclusions about less simple soils.

In a mass of soil subjected to load, an excess hydrostatic porewater pressure is created. In the soil at some distance away, the porewater pressure is still equal to the hydrostatic pressure, since the stress changes due to loading are negligible. Hence there is drainage from the soil immediately underneath the loaded area to the adjacent soil. This is a process of consolidation, and at its completion the excess hydrostatic porewater pressure is dissipated. In a clay the permeability is so low that immediately after the application of the load in a normal construction operation, very little drainage has taken place. Therefore the strength of the soil can be measured by the undrained test. The undrained strength of a saturated soil immediately after loading is equal to that before the load is applied. At the other extreme, the permeability of a cohesionless sand is so large that complete drainage occurs during normal loading; consequently no excess hydrostatic porewater pressure can develop even during loading. The stresses induced by the load are effective stresses. Hence at any time the strength of the soil may be measured by the drained test. Intermediate soils such as silts and clayey sands do not yield to such simplifications and their analysis is much more difficult and complicated. It often happens that no definite solution can be obtained for these intermediate soils. Then what we have learned about the saturated clay and cohesionless sand may be used to establish the upper and lower limits for the behavior of the material in question.

As pointed out in the following sections, there are many problems for which the exact answer cannot be determined in advance. Under these circumstances, one can only make certain reasonable assumptions to estimate the stability and arrive at a tentative answer. Field measurements during construction then provide the only source of information with which to check our predictions. Should the measurements reveal that the actual behavior departs significantly from the predicted behavior, then the design should be altered during the early stages of construction. Superficially this is a disadvantage, as one is prevented from arriving directly at a final answer. However, in reality this is the only logical approach, as failure to evaluate the validity of assumptions has on many occasions led to catastrophic results. Some examples of the value of field observation in the evaluation of design assumptions are provided by the papers of Casagrande (1965), Peck (1969a), Terzaghi (1943a, 1960), and Terzaghi and Peck (1957).

11.4 Stability of saturated intact clays

CASE 1. We begin with the problem of an embankment or a shallow foundation constructed over a deposit of clay. For simplicity we consider the embankment material to consist of the same clay (Fig. 11.2). The state of stress at a point a in

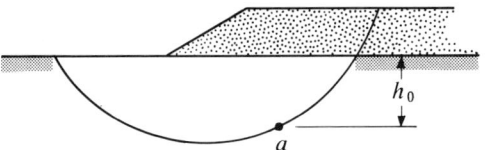

FIGURE 11.2 *Embankment on clay.*

the soil is described in Fig. 11.3(*a*) and (*b*). The shear stress at a increases with the height of the fill and reaches a maximum at the end of construction. The initial porewater pressure is equal to the hydrostatic pressure $h_0\gamma_w$. Since clays have low permeability, the amount of volume change or drainage during the construction period is very small. As a limiting condition, we assume that no drainage and no dissipation of porewater pressure take place during construction. The clay is therefore assumed to be loaded in the undrained condition. The porewater pressure [as given by Eq. (4.18)] increases with increase in the height of fill until the end of construction, as shown in Fig. 11.3(b). Curves for porewater pressure are shown for values of A equal to 1 and -0.5. According to Eq. (4.18), the porewater pressure is a positive quantity unless A has a large negative value. The shear strength of the soil at the end of construction remains equal to the undrained shear strength at the beginning of the construction period [Fig. 11.3(c)].

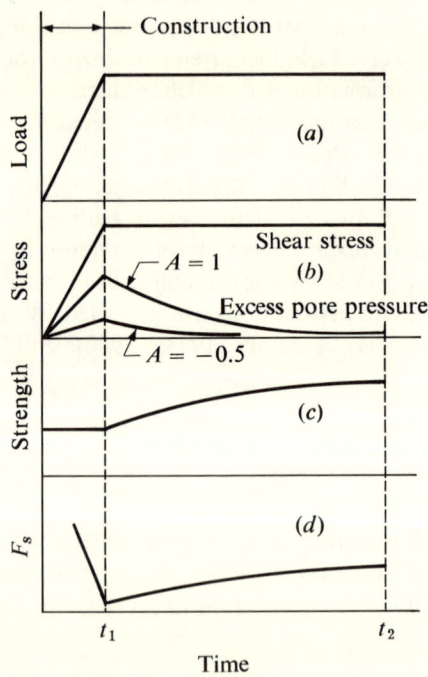

FIGURE 11.3 *Stability conditions of an embankment.* [After Bishop and Bjerrum (1960).]

After the completion of construction, the total stress remains constant, whereas the excess hydrostatic porewater pressure dissipates due to consolidation and becomes equal to 0 at complete consolidation (t_2). The drop in porewater pressure by consolidation is accompanied by a reduction of the void ratio and an increase in the effective stress and the shear strength. The second limiting condition is attained at time t_2, a long time later, when the excess hydrostatic porewater pressure is 0 (drained state). The shear strength at any time can be evaluated from the effective-stress parameters $\bar{c}$ and $\bar{\phi}$ if the porewater pressure (and therefore, the effective stress) is known. Since at time t_2 the excess hydrostatic porewater pressure is 0, effective stresses can be calculated from the load, the weight of the soil mass, and the hydrostatic pressure. The shear strength can be determined from the parameters $\bar{c}$ and $\bar{\phi}$.

Hence, the two limiting conditions involve forces that can be readily calculated. The end-of-construction state is analyzed with total stress and the undrained shear strength and the long-term state with effective stress (excess hydrostatic pressure equals 0) and the effective-stress parameters. From Fig. 11.3(d) it is obvious that the most dangerous condition is encountered at time t_1, immediately at the end of construction. If the embankment survives this state the safety factor increases with time.

CASE 2. A second case is that of a cut made in clay (Fig. 11.4). The excavation of soil reduces the average overburden pressure at point a and brings about a drop in the porewater pressure [negative excess hydrostatic porewater pressure, Fig. 11.5(b)]. This drop depends on the value of the porewater-pressure coefficient

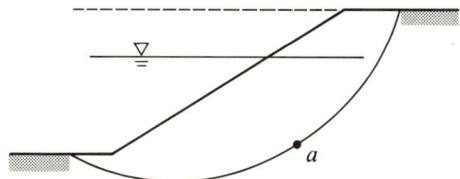

FIGURE 11.4 *Excavation in clay.*

A and the magnitude of the stress change. The porewater-pressure change according to Eq. (4.18) is

$$\Delta u = \Delta \sigma_3 + A(\Delta \sigma_1 - \Delta \sigma_3)$$

if B is equal to 1. In the excavation of a cut the minor principal stress σ_3 decreases more than the major principal stress σ_1. Thus $\Delta \sigma_3$ is negative and $(\Delta \sigma_1 - \Delta \sigma_3)$ is positive in the above equation. In most cases Δu is negative. If A is small, Δu has a large negative value and if A is large Δu has a small negative value.

The shear stress at a increases to a maximum at end of construction. As before, we consider the limiting condition of no drainage during construction. At the end of construction the shear strength remains equal to the undrained shear strength. With time, the negative excess hydrostatic porewater pressure is dissipated, accompanied by a swelling of the clay and a decrease in shear strength. Again the second limiting condition is reached a long time after excavation, with excess hydrostatic porewater pressure equal to 0. As in the previous example, the end of construction stability and long-term stability are represented by the undrained and drained shear strengths, respectively. However, in contrast to

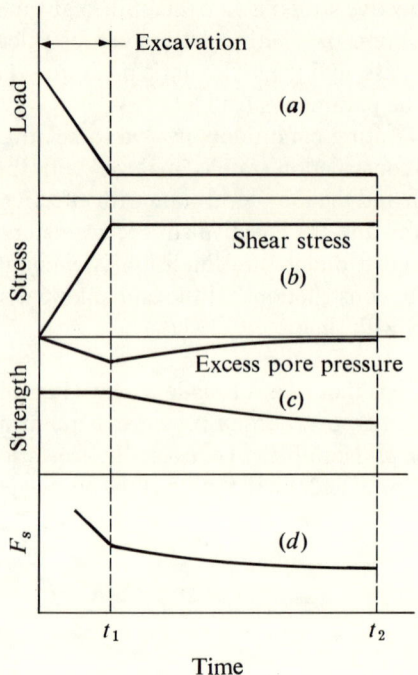

FIGURE 11.5 *Stability conditions of an excavation.* [After Bishop and Bjerrum (1960).]

the previous case, the most unfavorable condition is the long-term stability [Fig. 11.5(*d*)] and is given by the effective-stress analysis.

CASE 3. The long-term stability of a natural slope or hillside on clay presents a problem of another nature. If one is to investigate the stability of an existing slope, it must first be realized that the development of the slope by the forces of erosion took place over a period of many years. The gradual change in stress caused by steepening of the slope has been accompanied by volume change, and no excess hydrostatic porewater pressure exists in the soil. The conditions are, therefore, similar to that of a drained test in the laboratory.

If the porewater pressure is known and is expected to remain as is, the stability analysis can be made with the effective-stress parameters $\bar{c}$ and $\bar{\phi}$. Such an analysis gives the factor of safety of the slope with respect to continued gradual erosion. An interesting example of a slope failure produced by erosion has been described by Hutchinson (1961). The progressive decrease of F_S during the past 90 years was calculated from known profiles and is shown in Fig. 11.6.

CASE 4. As an illustration of the complexity of porewater-pressure behavior, we study a case in which the porewater pressure first increases and then decreases with time. Such conditions arise where large excess hydrostatic porewater pres-

SEC. 11.4 Stability of saturated intact clays 319

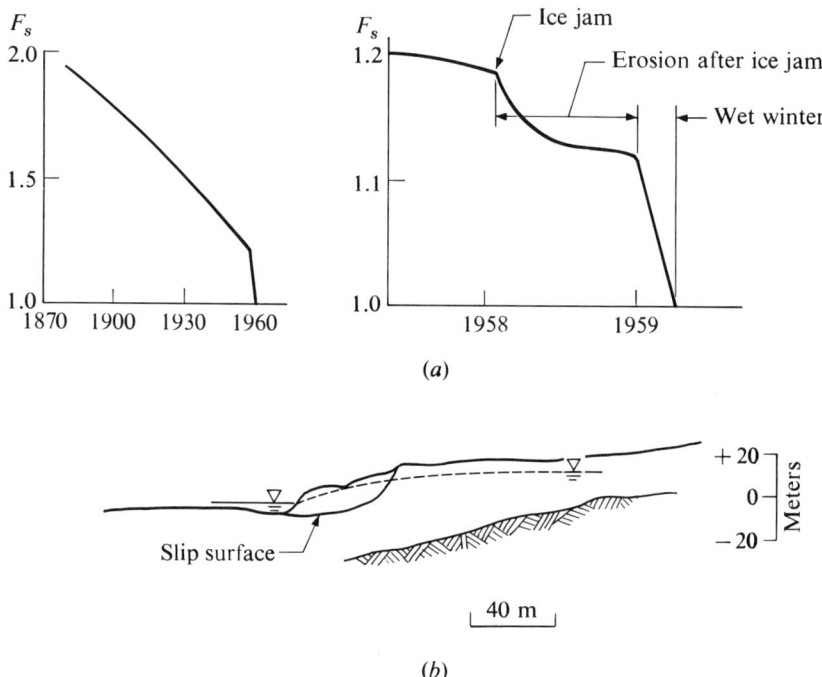

FIGURE 11.6 *The slide at Furre, Norway.* [After Hutchinson (1961).]

sure is induced in an adjacent area due to the construction of a heavy structure or by pile driving (Fig. 11.7). The excess hydrostatic porewater pressure underneath the load q is dissipated by radial drainage, and the flow of water from b to a increases the porewater pressure at a.*

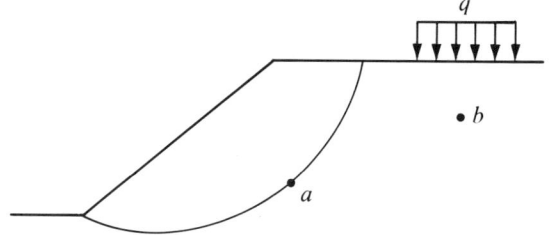

FIGURE 11.7 *Load adjacent to slope.*

The stability condition of the slope is illustrated in Fig. 11.8. We consider a slope that is already in existence. Since the load q is applied at some distance,

---

* The problem of radial consolidation is solved numerically in Sec. 12.6.

it does not change the stresses along the slip circle, and the shear stress remains constant with time [Fig. 11.8(a)]. The application of the load q produces an instantaneous rise in the porewater pressure at b which dissipates through consolidation [curve b, Fig. 11.8(b)]. The porewater pressure at a is increased temporarily by the radial drainage initiating at b, and its behavior is shown as curve a in Fig. 11.8(b). This rise in porewater pressure reduces the shear strength [Fig. 11.8(c)] and the factor of safety [Fig. 11.8(d)]. We see that the minimum factor of safety is reached at some intermediate time t_2. Such problems are potentially very dangerous, because failure may occur in spite of adequate immediate and long-term stabilities.

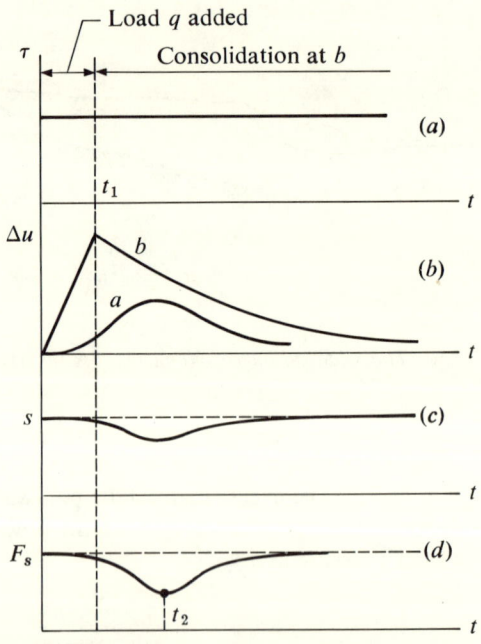

FIGURE 11.8 *Stability conditions of a slope with load.* [After Bishop and Bjerrum (1960).]

STABILITY ANALYSIS. To analyze the stability of a saturated clay, the undrained-shear-strength c_u and effective-stress parameters $\bar{c}$ and $\bar{\phi}$ are determined by laboratory tests on representative undisturbed specimens extracted from the deposit prior to construction or loading. The end-of-construction stability is analyzed with the undrained shear strength $c_u(\phi_u = 0)$ and total stress. For an analysis of long-term stability by effective stress it is necessary to predict the final porewater-pressure condition. Sometimes this may be considerably different from the initial conditions, owing to changes in the groundwater level caused by drainage or seepage subsequent to construction. In most problems, fairly reliable estimates of the condition can be made. A common example is a cut in clay excavated below the water table. If the cut is to be kept dry, the water table must be lowered to the elevation of the bottom of the cut. In the area immediately adjacent to the cut, there will be a steady flow of groundwater toward the cut, and a flow net can be constructed to describe the hydraulic conditions during steady seepage. The porewater pressures can be determined from the flow net. This makes it possible to carry out an effective-stress analysis.

However, if it is desired to analyze the stability at an intermediate stage between t_1 and t_2, the porewater pressure must be evaluated. It is difficult to calculate the rate of dissipation of excess hydrostatic porewater pressure except for very simple cases. Hence, the alternative is to measure the porewater pressure in situ. The field measurement of porewater pressure enables the engineer to check the initial design assumptions regarding porewater pressure and also provides the data which realistically describe the performance and stability of the soil mass. Such data make it possible to evaluate the stability condition at any time during or after construction by effective-stress analysis.

One of the most powerful applications of the effective-stress principle and porewater-pressure measurement is the solution of stability problems of marginal safety. In such problems the safety factor is so small that there is a substantial risk of failure if the loading is carried out in the undrained condition. To increase the safety factor, the rate of loading is controlled so that the porewater pressure may dissipate or be kept below an acceptable value. This allows the shear strength to increase as load and stress increase. For a given load, the porewater pressure that can be tolerated for a required safety factor can be calculated by stability analysis with effective stress.* But before a load increment can be applied, one must ascertain that the porewater pressure in situ does not exceed the required value. Field measurement of porewater pressure provides this vital information and is indispensable to the execution of such a project. Many bold designs have been carried out in this manner [for examples see Casagrande (1949), Skempton and Bishop (1955), Bishop et al. (1960b), Lambe (1962)].

FIELD OBSERVATIONS. The reliability of any set of theories or principles can only be established by observations of actual behavior in the field. The errors involved in the mechanics of stability analysis arise out of simplifications introduced into the solutions. However, by far the greatest concern is about the soil properties that we put into the mathematical solutions. In the application of stability principles, the most important question is whether the strength of a soil in situ behaves according to the laws determined from laboratory experiments on soil specimens.

A good source of information on the reliability of the theories is the failure of real slopes. A number of careful investigations have been made in which the factor of safety of the slope that failed was calculated with measured shear

---

* A more elaborate analysis on the effect of consolidation of a partially saturated soil is given by Bishop (1957).

strengths. If the theory and soil properties used are correct, the safety factor of a slope at failure should be 1. The results of these studies show that for homogeneous clays that are normally consolidated or slightly preconsolidated, the undrained shear-strength analysis and effective-stress analysis for the immediate and long-term stabilities, respectively, are reasonably accurate [Skempton and Golder (1948), Ireland (1954), Sevaldson (1956), Bjerrum and Kjaernsli (1957), among others]. Several studies on bearing-capacity failure also show good agreement.

As examples the results of several investigations are summarized in Tables 11.1 and 11.2. In Table 11.1 are listed seven cases of failure immediately after or during construction. These are failures in the undrained state and were analyzed with the undrained shear strength. The computed factors of safety are all close to 1.0, which shows that the failures should have occurred according to theoretical predictions. Table 11.2 lists five cases of long-term stability; three of them are failures. The loading condition corresponds to the drained state, and effective-stress analysis was used. The calculated factors of safety are close to 1.0 for the failures, and are considerably larger than 1.0 for the stable slopes. These results and others [see Bishop and Bjerrum (1960) for a more comprehensive account) constitute valuable experience for the assessment of the theoretical concepts.

TABLE 11.1 *Immediate Failures in Normally and Slightly Overconsolidated Clay: Undrained Analysis* ($\phi_u = 0$)

Locality	w	LL	PL	F_s	Ref.
Foundations					
1. Kippen	50	70	28	0.95	Skempton (1942)
2. Transcona silo	50	110	30	1.09	Peck and Bryant (1953)
3. Fredrikstad oil tank	45	55	25	1.08	Bjerrum and Øverland (1957)
Embankments					
4. Chingford	90	145	36	1.05	Skempton and Golder (1948)
5. Newport	50	60	26	1.08	Skempton and Golder (1948)
6. Gosport	56	80	30	0.93	Skempton (1948)
Cut slopes					
7. Congress St.	24	33	18	1.10	Ireland (1954)

In Chapter 10 the earth pressure on yielding walls is solved under the assumption of failure in the soil mass. For walls with limited yield (braced excavations), the resultant earth pressure is calculated on the basis of failure along a slip surface. Four representative cases of earth pressure measured in braced excavations in normally consolidated and intact overconsolidated clays are listed in Table 11.3. The agreement between the measured and computed earth pressures also supports the shear-strength theory.

TABLE 11.2 *Long-term Stability in Clay: Effective-stress Analysis* ($\bar{c}, \bar{\phi}$)

Locality	Clay type	w	LL	PL	F_s	Ref.
Slope failures						
1. Lodalen	Overconsolidated, intact	31	36	18	1.05	Sevaldson (1956)
2. Drammen	Normally consol.	31	30	19	1.15	Bjerrum and Kjaernsli (1957)
Stable slopes						
3. Drammen	Normally consol.	31	30	19	1.25	Bjerrum and Kjaernsli (1957)
4. Bakklandet A	Normally consol.	28	25	18	1.85	Bjerrum and Kjaernsli (1957)
5. Borregaard	Normally consol.	18	18	11	1.25	Bjerrum and Kjaernsli (1957)

TABLE 11.3 *Measured and Calculated Earth Pressures*

Locality	w	LL	PL	Ratio: $\dfrac{\text{calc. press.}}{\text{meas. press.}}$	Ref.
Normally consolidated and intact overconsolidated clay, undrained analysis ($\phi_u = 0$)					
1. Chicago S8A & S9C	20	—	—	0.94	Peck (1943)
2. Chicago D8	25	—	—	1.00	Wu and Berman (1953)
3. Shellhaven	90	110	30	1.06	Skempton and Ward (1952)
4. Oslo	40	48	25	1.15 to 0.84	Kjaernsli (1958)
Stiff fissured clay, effective-stress analysis ($\bar{c} = 0$)					
5. Oslo	30	40	21	0.90	Bjerrum and Kirkedam (1958)

The examples of field investigations cited above all involve normally consolidated or slightly overconsolidated (and intact) clays. To date, about 50 slope failures and foundation failures in such clays have been investigated. For about 90 percent of these the discrepancy between calculated and observed safety factors is less than 15 percent. Since most of the clays investigated are fairly uniform deposits, the accuracy of ±15 percent represents the best that can be attained in practice. Some notable exceptions are failures in highly plastic or organic clays. For these cases the use of the undrained shear strength as measured by the unconfined compression or vane shear tests tends to overestimate the safety factor [Bjerrum (1972)].

*Example 11.1.** To illustrate the principles outlined in the preceding section we consider the bearing capacity of a large structure (80 ft × 80 ft) to be erected

---

* I owe the idea of this and the following example to Whitman (1959).

over a deposit of clay. The properties of the soil are described in Fig. 11.9(a). The average undrained shear strength c_u is about 0.40 kg/cm² down to a depth of 80 ft, which would be approximately the depth of the failure zone. To evaluate the effective shear-strength parameters $\bar{c}$ and $\bar{\phi}$, consolidated undrained triaxial tests were performed on undisturbed specimens. The values of $\bar{c}$ and $\bar{\phi}$ are 0 and 23 deg, respectively.

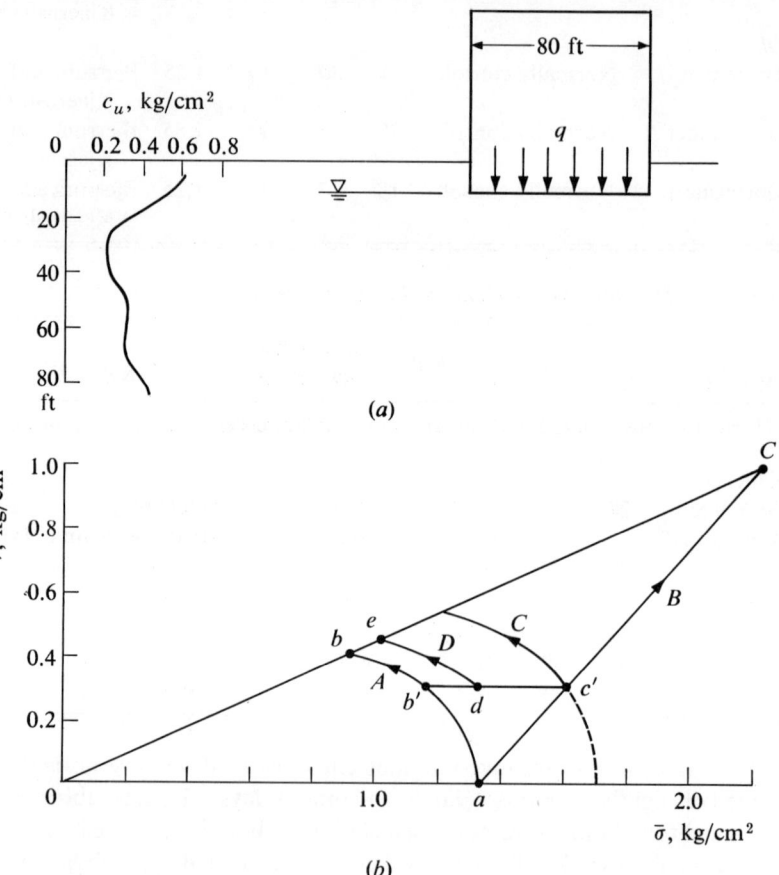

FIGURE 11.9 *Stability of a foundation under undrained and drained loading.*

The structure is subjected to large fluctuations in loads (i.e., a storage tank, a large silo, etc.), and it is assumed that the loading schedule is under our control. Hence we may consider the consequences of rapid and slow loading. If the structure is loaded rapidly in the undrained condition, the bearing capacity may be calculated with c_u. We shall neglect the effect of surcharge, since it is relatively small in this case. Hence

$$q_u = cN_c = 0.40 \times 7.40 = 2.96 \text{ kg/cm}^2$$

SEC. 11.4 Stability of saturated intact clays

Failure would occur if the structure is loaded rapidly to 2.96 kg/cm². If the rate of loading is so slow that no excess hydrostatic porewater pressure is built up in the soil, failure would occur in the drained state. The bearing capacity is then

$$q_d = 0.4\gamma B N_\gamma = (0.4)(128 - 62.5)80 \times 7 = 7.35 \text{ kg/cm}^2$$

Thus in the drained state, the clay can withstand a load more than twice that in the undrained state.

The stress paths of these two cases are shown as A and B, respectively, in Fig. 11.9(b). The initial point a represents the average overburden pressure in the zone of failure:

$$\bar{p}_0 = (40)(128 - 62.4) = 2630 \text{ psf} = 1.32 \text{ kg/cm}^2$$

A shear stress exists at this stage because the horizontal pressure is not p_0 but $K_0 p_0$. However, we shall plot the initial stress as a point and ignore this shear stress for the sake of simplicity. This does not affect the present problem, as the objective is to illustrate the shear strength under various loading conditions. The failure point for undrained loading is b. This is determined by the fact that c_u is 0.40 kg/cm² and that the failure point must fall on the effective-stress envelope. The shape of the curve A is typical of that for a normally consolidated clay.

To locate point c on the drained stress path, it is necessary to know the average normal and shear stresses along the failure surface. This cannot be calculated precisely. For the present purpose we assume that the shear stress is proportional to the load. In the undrained state the shear stress is 0.40 kg/cm² when the foundation is loaded to 2.96 kg/cm². So, if the load is 7.35 kg/cm² in the drained state, the shear stress is estimated to be

$$7.35 \left(\frac{0.40}{2.96}\right) = 1.00 \text{ kg/cm}^2$$

Thus point c is located. The relative positions of b and c explain why the bearing capacity is so much larger in the drained state.

What would happen if the load is rapidly raised to three-quarters of the undrained bearing capacity (2.22 kg/cm²) and maintained at this level for a long period? During the loading, the shape of the stress path is the same as A in Fig. 11.9(b), but the shear stress reaches only 0.3 kg/cm² (point b'). After this the clay consolidates and the effective stress increases while the shear stress remains constant (line $b'c'$). After complete consolidation the stress state comes to point c' on the drained stress path. If the load is again increased rapidly the stress path follows C. We may consider C to be the stress path of a normally consolidated clay that passes through point c'. The stress path has the same shape as A. If we sketch in C we see that the failure shear stress would be approximately 0.53 kg/cm², which corresponds to a bearing capacity of 3.92 kg/cm². Thus loading the clay in stages and allowing for consolidation enables us to increase the bearing capacity considerably.

We may now consider the case where the loads on the structure require that the bearing capacity be at least 3.2 kg/cm². This is clearly untenable if the full load is to be applied at once, because the bearing capacity of the clay at the end of construction is only 2.96 kg/cm². Hence if full load is applied, collapse of the foundation would be inevitable. However, if the loading process can be kept under control, the bearing capacity would ultimately reach 7.35 kg/cm².

The bearing capacity may be substantially increased if the structure is loaded to 2.22 kg/cm² and adequate time is allocated for consolidation so that the clay can follow the stress path $ab'c'$ in Fig. 11.9(b). How can we ascertain that the stress in the clay has actually progressed to point c'? This is a crucial question, because if loading is resumed prematurely, the stress path proceeds along D. If only 25 percent consolidation has taken place after the first loading, the status is denoted by point d. If loading is resumed, a shear strength of only 0.43 kg/cm² is available (point e) instead of 0.53 after 100 percent consolidation. To ensure that we do not overestimate the shear strength (or overload the structure), we must seek reliable information on the magnitude of the excess hydrostatic porewater pressure. This can only be accomplished by measurement of the porewater pressure in situ.

Example 11.2. We next study the stability of a cut slope in a slightly overconsolidated clay [Fig. 11.10(a)]. The cut has a slope of 65 deg and is to be 30 ft deep. The average undrained shear strength of the clay is 0.30 kg/cm² to a depth of 30 ft. The effective-stress parameters $\bar{c}$ and $\bar{\phi}$ are 0.10 kg/cm² and 20 deg, respectively. For simplicity we shall consider a cut excavated under water.

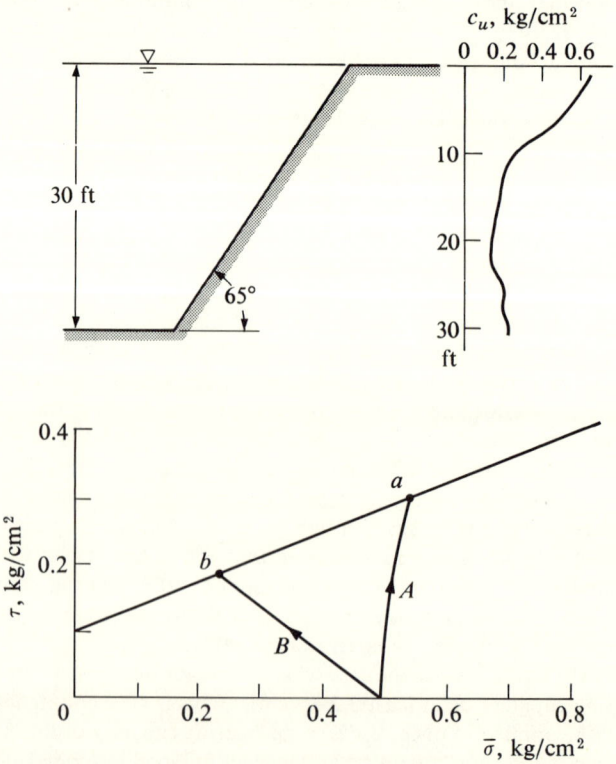

FIGURE 11.10 *Stability of a slope under undrained and drained loading.*

SEC. 11.4 Stability of saturated intact clays

If the excavation is carried out rapidly in an undrained state, the stability can be evaluated with c_u. From Fig. 9.26, for $\phi = 0$, $\beta = 65$ deg, and $n_d = \infty$, we obtain

$$N_c = 5.0$$

This stability number does not include the effect of water pressure on the face of the submerged slope. To do this requires a stability calculation by the circular-arc method. However, for approximate calculations with ϕ_u equal to 0, we may use the submerged weight of $(130 - 62.5)$. For an analysis by total stress, this involves an error which is not large. Then we have $H = 30$ ft, $\gamma = 67.5$ pcf, $c_u = 600$ psf,

$$N_s = \frac{\gamma H}{c} = \frac{67.5 \times 30}{600} = 3.4$$

Since 3.4 is considerably less than 5.0 the cut would be stable, with a margin of safety. The stress path to failure in the undrained state is shown as A in Fig. 11.10(b).

Let us now examine the stability for a very slow rate of excavation and consider the case in which the cut is not dewatered (i.e., a harbor). In the drained state, the stability is analyzed by means of effective stress. For $\bar{\phi} = 20$ deg and $\beta = 65$ deg we obtain from Fig. 9.24,

$$N_s = 9.3$$

For the cut with $H = 30$ ft, $\gamma' = (130 - 62.5)$ pcf, $\bar{c} = 200$ psf,

$$N_s = \frac{67.5 \times 30}{200} = 10.1$$

We see that 10.1 is larger than 9.3; therefore the cut would fail if excavated slowly. Even if the cut is excavated rapidly and it survives in the undrained state, failure would ensue as the excess hydrostatic porewater pressure dissipates and the clay approaches the drained state. For long-term stability the slope cannot withstand a height more than

$$H = \frac{N_s \bar{c}}{\gamma'} = \frac{9.3 \times 200}{67.5} = 27.6 \text{ ft}$$

To compare the drained and undrained stress paths, it is again necessary to estimate the shear strength at failure in the drained state. This can be done approximately as follows. The height for long-term failure is 27.6 ft. If the clay is examined at this stage and subjected to an undrained test, the undrained shear strength should also give a failure height of 27.6 ft. Since $N_s = 5.0$ for $\bar{\phi} = 0$, we have

$$c = \frac{67.5 \times 27.6}{5.0} = 372 \text{ psf} = 0.19 \text{ kg/cm}^2$$

Hence we can locate point b on the failure envelope with a value of τ equal to 0.19 kg/cm². The drained stress path is given by B. In the cut, the effective stress is reduced by the excavation. Therefore the drained shear strength is much lower than the undrained shear strength. Empirical observations have shown that almost all foundation and embankment failures occur immediately after construction or loading, while many slope failures take place months and years after excavation.

11.5 Stability of stiff-fissured clays

The principles of stability analysis with the undrained and effective stress shear strengths presented in Sec. 11.4 apply also to stability of stiff-fissure clays. However, special precaution must be exercised in the evaluation of the shear strength of such clays because of the presence of fissures. The characteristics of some fissures and the measured strengths along the fissures have been described by Skempton and Petley (1967).

The undrained shear strength of a laboratory specimen is strongly dependent, on the number, shape and inclination of fissures in the specimen. The presence of fissures is less likely in small specimens than in large specimens. Hence, there is a tendency for the measured strength to decrease as specimen size increases [Lo (1970), Ward et al. (1965)], and it is difficult to extrapolate from the laboratory shear strength to the in-situ shear strength. Frequently it is necessary to conduct load tests to measure the in-situ strength.

For long-term stability of slopes, calculations with the effective stress shear strength of intact specimens grossly overestimate the safety factor. The shear strength generally decreases with time after construction [Skempton (1948b)] and at failure is somewhere between the peak effective stress shear strength of intact samples and the residual shear strength, cases (1) and (2), Table 11.4. Empirical evidence indicates that the strength may be approximated by ($\bar{c} = 0$, $\bar{\phi}$), where $\bar{c}$ and $\bar{\phi}$ are the conventional effective stress parameters.

This phenomenon has also been observed in earth-pressure studies. Earth pressures on two retaining walls that failed were found to be close to that calculated on the basis of $\bar{c} = 0$ [Henkel (1957)]. Earth pressure on a braced excavation in stiff fissured clay was found to increase with time [Bjerrum and Kirkedam (1958)], which agreed with the progressive reduction of $\bar{c}$. The final ratio of calculated to measured pressure is given in Table 11.3. In general, long-term failures of slopes and retaining walls in stiff clays are common, reflecting the tendency to overestimate the long-term shear strength on the basis of the large value of the undrained shear strength measured during or prior to construction.

If a slope is allowed to undergo continued movement, a slickenside develops along a failure surface. The clay particles along the slickenside are oriented in the direction of the slide. The shear strength on a slickenside surface has been found to be close to the residual shear strength [Skempton (1964), Skempton and Hutchinson (1969)]. An example is given in case (3) Table 11.4. Bjerrum (1967) has cited evidence to indicate that natural slopes in stiff clays and clay shales (without fissures) would also exhibit residual shear strength at failure if sufficient time is allowed for development of "progressive failure".

11.6 Stability of cohesionless soils

Sandy and gravelly soils of high permeability constitute a simpler situation. In these soils volume change occurs rapidly. Under ordinary construction condi-

TABLE 11.4 *Long-term Stability in Stiff-fissured Clay: Effective Stress Analysis* (after Skempton and Hutchinson, 1969)

Locality	Peak shear strength		Reduced shear strength		Residual shear strength[a]	
	Laboratory tests	Computed F_s	Laboratory tests	Computed F_s	Laboratory tests	Computed F_s
First-time slides						
1. Northolt	$\bar{c} = 320$ psf	1.63	$\bar{c} = 0$	0.77	$\bar{c}_r = 20$ psf	0.54
2. Sudbury Hill	$\bar{\phi} = 20°$	2.27	$\bar{\phi} = 20°$	1.05	$\bar{\phi}_r = 13°$	0.74
Movement over old slip surface						
3. Walton's Wood					$\bar{c}_r = 0$ $\bar{\phi}_r = 13°$	1.0

[a] The residual strength denotes the shear strength attained after the sample has been subjected to very large shear displacements.

tions the application of loads induces no excess hydrostatic porewater pressure. Without excess hydrostatic porewater pressure, effective stress can be calculated and stability analysis can be made with the effective-stress shear strength.

The measurement of the drained shear strength or the effective-stress parameters $\bar{c}$ and $\bar{\phi}$ is not difficult in the laboratory, but another kind of obstacle is encountered. This is the difficulty of securing undisturbed specimens of cohesionless soils. Where undisturbed specimens are not available, laboratory tests on disturbed cohesionless soils are not meaningful, because the shearing resistance of such soils is dependent to a very large extent on the density and structure. Hence it is often necessary to determine the density of the soil in situ. A number of field methods are available for this purpose and are described in Sec. 13.13.

Field data on shear failure in cohesionless soils are scarce because relatively few failures have occurred. Our best sources of information are the results of model tests of retaining walls and strip footings. They indicate general agreement between the measured forces at failure and those calculated theoretically. Prediction of bearing capacity failure is complicated by the stress-strain properties of soils. The pattern of failure surface described in Fig. 9.6(*a*) is generally valid for dense sands. Footings on loose sands may undergo large penetrations without developing the failure surface (Sec. 9.8). This is analogous to the problem of passive pressure described in Sec. 10.2. Hence, the full value of the shear strength may not be mobilized at penetrations considered as failure in practice. For the case of local shear (Sec. 9.8) Terzaghi suggested that the design shear strength be taken as two-thirds of that measured in shear tests.

A different stability problem is presented by saturated deposits of very loose sands. Such sands have void ratios well above the critical void ratio and $\bar{\phi}$ may be as small as 10°. A local failure, which may be in an over-steep submerged bank, would produce very large porewater pressures. The porewater pressure parameter A may be as large as 2 [Bjerrum et al. (1961)]. This large porewater pressure reduces the shear strength of the sand to almost zero and "liquefaction" occurs. The sand then flows away and leaves a steep bank behind it, which in turn fails and flows away. Very large areas may be involved in such a failure. Descriptions of such slides have been given by Casagrande (1950), Terzaghi (1957), and Andresen and Bjerrum (1967). Opportunities for comparison of predicted with observed performance are scarce. However, empirically it is known that flow slides occur mostly in saturated deposits of loose, uniform, fine sand and coarse silt.

11.7 Stability of compacted partially saturated clay

The design of partially saturated clay embankments requires consideration of the behavior of air voids in the soil. Since partially saturated clay may undergo volume changes from compression of the air voids, the undrained shear strength is not a constant. One approach is to use the effective-stress analyses to calculate the end-of-construction stability as well as the long-term stability [Bishop and Bjerrum (1960)]. To illustrate the problem, we study the earth dam in Fig. 11.11.

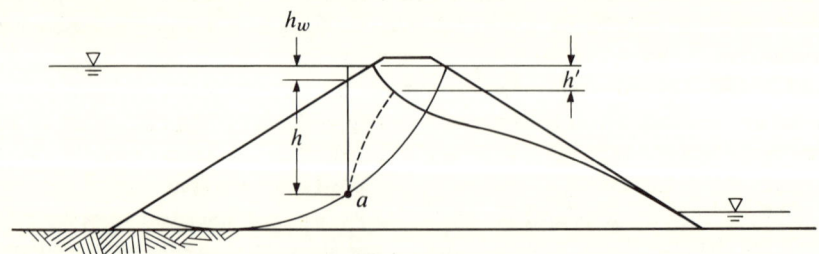

FIGURE 11.11   *The embankment of an earth dam.*

CASE 1.   The embankment material consists of clay and has a low permeability. The end-of-construction condition is characterized by substantial excess hydrostatic porewater pressure in the soil. The porewater pressure at any point consists of an initial porewater pressure u_0, caused by the compaction of the moist soil, and a second part Δu, caused by the weight of the embankment material, so that

$$u_1 = u_0 + \Delta u = u_0 + B[\Delta\sigma_3 + A(\Delta\sigma_1 - \Delta\sigma_3)] \qquad (11.1)$$

in which A and B are the porewater-pressure coefficients and $\Delta\sigma_1$ and $\Delta\sigma_3$ are the major and minor principal stresses induced by the weight of the embankment material.

The values of u_0, A, and B can be measured from laboratory experiments (Secs. 4.6 and 8.9). To calculate the porewater pressure, the stresses $\Delta\sigma_1$ and $\Delta\sigma_3$ must be evaluated. This may be done by the theory of elasticity [Jurgenson (1934)] but the process is rather complicated. A simplification may be accomplished if it is assumed that the soil in the embankment is loaded under the condition of zero lateral strain. This implies the additional assumption that the vertical and horizontal stresses at any point a are the major and minor principal stresses, respectively. For zero lateral strain the minor principal stress is given by

$$\Delta\sigma_x = K_0 \Delta\sigma_y \qquad (11.2)$$

in which K_0 is the coefficient of at-rest earth pressure and $\Delta\sigma_x$ and $\Delta\sigma_y$ are the increments of vertical and horizontal stresses. Putting Eq. (11.2) in (11.1) we get

$$u_1 = u_0 + B[K_0 \Delta\sigma_y + A(\Delta\sigma_y - K_0 \Delta\sigma_y)]$$
$$= u_0 + B(K_0 + A - AK_0) \Delta\sigma_y \qquad (11.3)$$

Since A, B, and K_0 are all soil properties, we can replace them by a single constant B', or

$$u_1 = u_0 + B' \Delta\sigma_y \qquad (11.4)$$

$B' \Delta\sigma_y$ is therefore the rise in porewater pressure in an undrained soil when the major and minor principal stresses are increased by $\Delta\sigma_y$ and $K_0 \Delta\sigma_y$, respectively.

To investigate the stability it is necessary to determine the soil properties. In the case of an embankment, these properties should be known before the embankment is constructed. This can be accomplished by testing samples of the soil compacted in the laboratory to the density and moisture content that are to be attained in the field. To measure B', a soil specimen is subjected to a stress increment $\Delta\sigma_1 = \Delta\sigma_y$ and $\Delta\sigma_3 = \Delta\sigma_x$ in the undrained triaxial test, and the accompanying rise in porewater pressure is measured. By progressively raising σ_1 and σ_3 the relationship between $\Delta\sigma_1$ and Δu can be obtained as shown in Fig. 11.12. The soil has an initial porewater pressure u_0. As the pressures are increased, the porewater pressure rises. The slope of the curve is equal to B'.

As in the case of the A and B coefficients, B', which is the slope of the curve in Fig. 11.12, is not a constant over the entire range of stress. For simplicity in calculations it is taken as a constant representing the average value over the pertinent stress range. The porewater pressure at the end-of-construction stage can also be estimated empirically from the results of field porewater-pressure measurements in earth dams [see Hilf (1948) and Bishop et al. (1960b)].

To evaluate the shear strength with Eq. (8.9a) the air pressure u_a must also be known. If the soil is compacted at a moisture content on the dry side of the optimum, most of air voids are continuous and one may use the approximation $u_a = 0$. When this condition does not hold, the air pressure must be estimated by the same method used to evaluate the porewater pressure.

Alternative methods using the total stress to analyze the immediate stability have been described by Blight (1963) and the Joint ASCE-USCOLD Committee (1967).

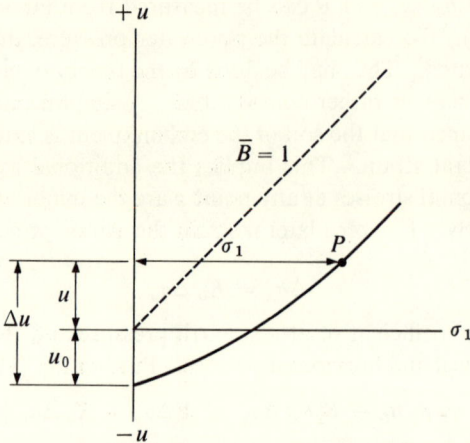

FIGURE 11.12 *Porewater pressure in compacted clay.* [After Bishop and Bjerrum (1960).]

CASE 2. After completion of construction, the excess hydrostatic porewater pressure begins to dissipate due to consolidation (Fig. 11.13). Therefore, both the shear strength and factor of safety increase. The filling of the reservoir again increases the porewater pressure, but at the same time it reduces the shear stresses (Sec. 11.2). As a consequence there is a net increase in the factor of safety. The filling of the reservoir has the effect of reducing the weight of the soil mass. Therefore, this corresponds to an unloading. During the time after filling, the soil tends to swell and there is a slight reduction in the strength. This represents a slight drop in the factor of safety. The porewater pressure at t_2, a long time after filling the reservoir, is equal to the hydrostatic pressure, since there should be no excess hydrostatic porewater pressure and the soil would be saturated. In Fig. 11.11 the porewater pressure at a under steady seepage is equal to

$$u_2 = \gamma_w(h + h_w - h') \tag{11.5}$$

The effective-stress analysis can be made with this porewater pressure and the values of $\bar{c}$ and $\bar{\phi}$ already determined. For an example of "steady seepage" under suction, see Blight (1963).

CASE 3. A sudden drawdown greatly increases the shear stress. The porewater pressure drops by an amount $\gamma_w h_w$, since the hydrostatic head above point a is reduced by h_w. Since the soil is impervious, the embankment remains saturated immediately after drawdown. Therefore at t_3

$$u_3 = \gamma_w(h - h') \tag{11.6}$$

The stability at both t_2 and t_3 can be analyzed by the effective-stress method. The soil is saturated and the porewater pressures are known.

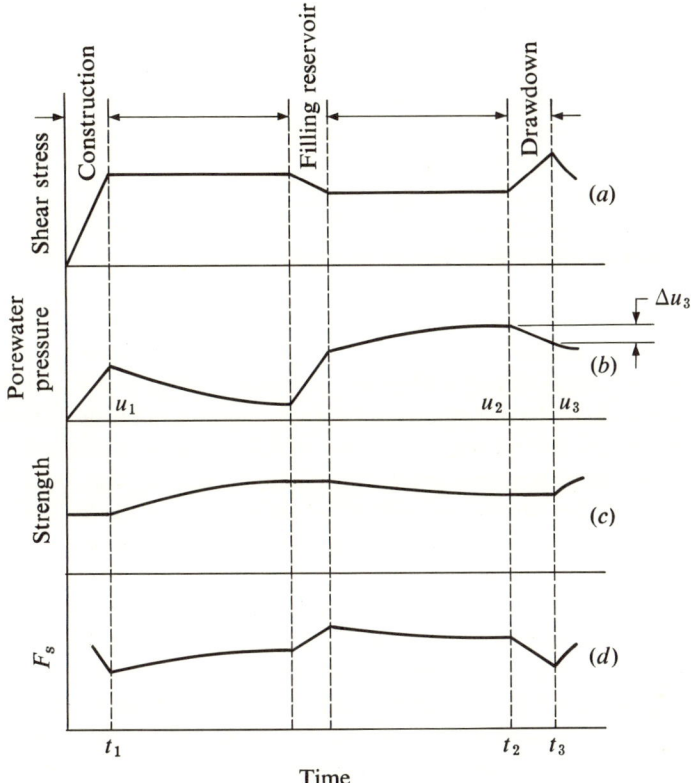

FIGURE 11.13 *Stability conditions of an earth dam.* [After Bishop and Bjerrum (1960).]

The preceding analysis indicates that the first critical stage to be investigated is the end-of-construction stage. If the dam survives, the factor of safety continually increases until the reservoir is filled. There is a gradual drop in the factor of safety, but usually the condition is not very severe. The most dangerous time after t_1 is immediately after drawdown. If the dam survives the sudden drawdown, the factor of safety once more increases with time as the water drains from the upstream slope.

The measured porewater pressure in the Quadradona Dam [Li (1959)] is shown in Fig. 11.14. The dam is composed of a core of compacted silt and shoulders of decomposed rock. The silt is relatively impermeable compared to the rock fill, and the interest as regards the porewater pressure is confined to the silt core. A small section of the core at the toe of the dam [Fig. 11.14(a)] was completed at an earlier date and so contained little excess hydrostatic pressure. Immediately after completion [Fig. 11.14(b)] the porewater pressure in the core increased with the weight of overlying fill (σ_1). The reduced porewater pressure at the bottom and the right side of the core was due to drainage into the foundation and the rock fill. The value of B' is about $\frac{1}{4}$ if we assume the soil behaved as undrained. After the reservoir was filled [Fig. 11.14(c)], the porewater pressure was modified by seepage. Gradually, the excess hydrostatic porewater pressure was dissipated, and the porewater-pressure distribution approached that of steady seepage [Fig. 11.14(d)].

11.8 Stability of soils intermediate between clay and sand

For intermediate soils such as silts and sandy silts, the permeability is large enough so that considerable volume change takes place during construction. The undrained shear strength obtained prior to loading is then no longer an accurate measure of the end-of-construction strength. On the other hand, excess hydrostatic porewater pressure is present, since the permeability is usually not large enough to permit complete consolidation during the loading period. Hence errors of considerable magnitude would result from the application of the two limiting conditions of drained and undrained shear strengths. Furthermore, as pointed out in preceding sections, the rate of dissipation of excess hydrostatic porewater pressure cannot be calculated with a good degree of accuracy. Hence to carry out an accurate stability analysis requires the measurement of the porewater pressure in situ. Since these measurements can be made only after the loads are applied, it follows that accurate stability calculations cannot be made prior to loading. It is therefore necessary to make reasonable estimates of the porewater pressure for the purpose of the design. The measurements during construction are relied upon to provide the final check on these estimates. If they depart significantly from the design conditions, then the design must be altered to meet the conditions as revealed by the measurements.

For a rough estimate of the stability, the limiting states of undrained and drained shear strengths can still be used subject to the condition that the errors are considerable. To approximate the intermediate state of drainage the consolidated-undrained shear strength may be used in the total-stress analysis. However, this is still no more than an estimate, because there is no assurance whatsoever that the porewater pressures in the consolidated-undrained triaxial test are the same as or even close to those in the actual problem.

11.9 Stability of deep foundations

Deep foundations are arbitrarily defined as those whose depth D exceeds its width B. The two types of deep foundations most commonly used are the pile and the pier (or caisson) foundations. The difference between the two lies largely in the method of construction. Usually, piles are driven into the ground and therefore involve the displacement of a mass of soil approximately equal in volume to that of the pile. Piers are customarily constructed inside excavated

SEC. 11.9 Stability of deep foundations

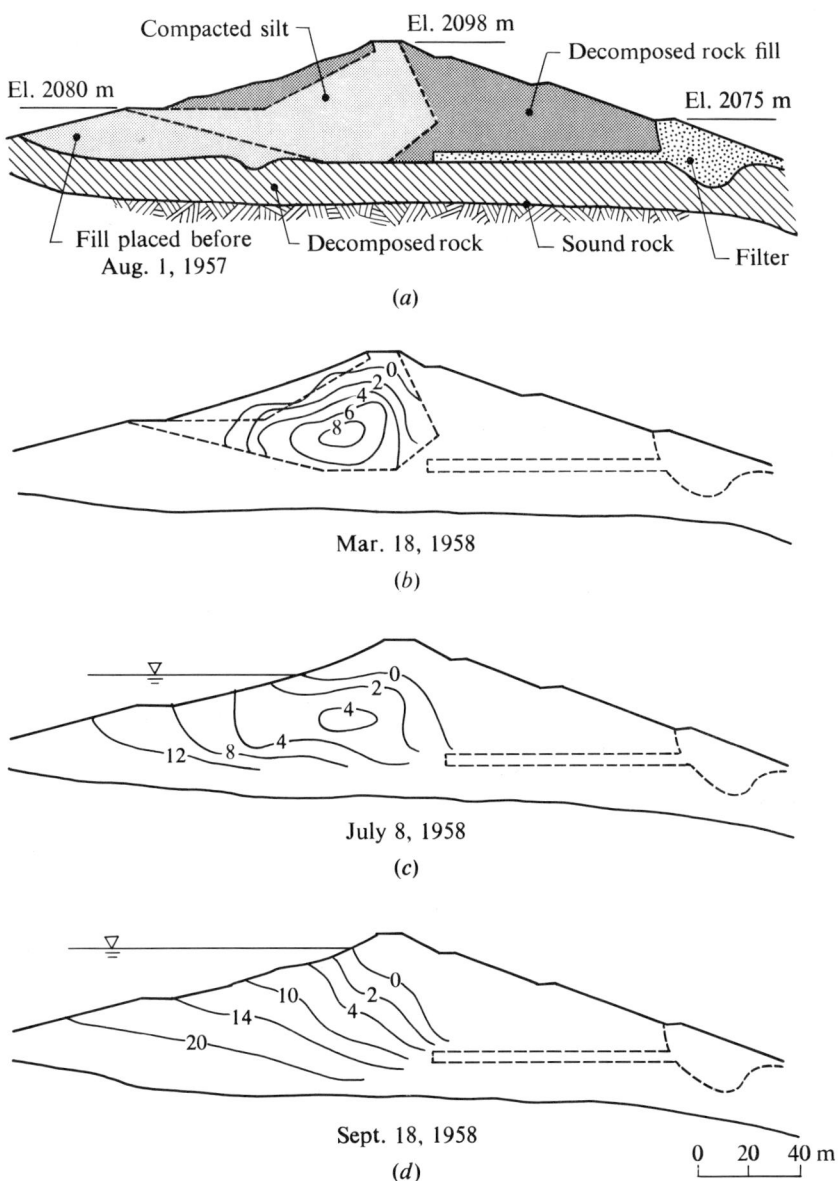

Porewater-pressure contours are head in meters

FIGURE 11.14 *Measured porewater pressure in Quadradona Dam.* [After Li (1959).]

holes. Piles and piers are generally designed to pass through weak soils and derive their support from a firm stratum underneath such as dense sand or hardpan. These are called *point-bearing piles or piers*. Piles may also be driven into a fairly uniform deposit of sand. Their bearing capacity then consists of the end bearing as well as the friction along the side of the pile. These are often called *friction piles*.

The problem of deep foundations is treated as a separate topic because of the important effects of the structure on the soil mass. The basic considerations of drainage and porewater pressure outlined in the preceding sections are still valid, but the installation of any deep foundation produces disturbances and stresses within the soil mass. These exert a profound influence on the strength of the soil.

A common example is a deep foundation passing through soft clay or silt and supported by a deposit of dense sand. In this case, it is convenient to calculate the bearing capacity by the Terzaghi theory. The total bearing capacity of square and circular foundations according to the Terzaghi theory are, respectively,

$$Q_f = B^2(0.4\gamma B N_\gamma + 1.3cN_c + \gamma D N_q) + 4fBD_f \quad \textbf{(9.37a)}$$

$$Q_f = \pi R^2(0.6\gamma R N_\gamma + 1.3cN_c + \gamma D N_q) + 2f\pi R D_f \quad \textbf{(9.37b)}$$

The first term in both equations is the bearing capacity of the end, whereas the second term is the shearing resistance between the sides of the foundation and the adjacent soil. The shear strength (c and ϕ) to be used in the calculation of the end-bearing capacity should be representative of the soil below the end of the foundation and can be evaluated according to the principles outlined in Secs. 11.4 through 11.6.

The evaluation of the end-bearing capacity by Eqs. (9.37) is acceptable if the pile or pier penetrates only a small distance into a deposit of dense sand or stiff clay (see Sec. 9.8). The choice of f requires consideration of the effect of the foundation. For piles driven into clay the shear strength may not be fully developed along the entire length of the pile. The departure becomes particularly important in the case of long piles [McClelland et al. (1967)]. Another complication is introduced by the fact that the soil displacement which accompanies pile driving is very severe and has the effect of remolding the soil. Extensive field measurements on piles [see Cummings et al. (1950), Seed and Reese (1955)] have shown that the strength of the clay is reduced to a value close to that of the remolded clay immediately after driving. However, the driving of the piles also induces stresses and excess hydrostatic porewater pressures in the clay, which leads to consolidation of the clay and an increase in strength with time. The field studies have shown that several weeks after pile driving the water content has decreased by consolidation while the strength of the clay has returned to a value close to the initial undisturbed strength. Hence it is very difficult to predict in advance the strength of the clay at the time the foundation loads are imposed.

Peck, Hanson, and Thornburn (1974) have suggested that the side friction on piles be taken as αs_u where α is a factor that ranges from 0.95 for soft clays ($s_u < 0.5$ tsf) to 0.45 for stiff clays ($s_u = 3.0$ tsf). For concrete piers in clays, a value for the side friction between clay and concrete equal to one-half the undrained shear strength of the undisturbed clay has been recommended by Skempton (1959) on the basis of field observations. This reduction is attributed to the absorption by the clay of moisture from the freshly poured concrete.

If a pile is driven into a deposit of sand the value of f can be estimated if the lateral pressure on the pile is known. At any depth y, this pressure is $K\gamma y$ (Fig. 11.15), and the friction f is $K\gamma y \tan \delta$ where δ is the angle of friction between soil and foundation. Then the average friction along the length of the pile or pier is equal to

$$f = \tfrac{1}{2} K \gamma D_f \tan \delta \tag{11.7}$$

and this value can be substituted in Eq. (9.37). The value of K is around 1.0 for driven piles [Mansur and Hunter (1970)] and may be much smaller for bored and jetted piles. The value of δ for different materials has not been well established. It may range from $\tfrac{2}{3}\phi$ for steel pipe piles to ϕ for concrete piles. However, results of large-scale tests [Vesic (1967)] have shown that the actual behavior of piles departs considerably from the simplified picture described in the preceding paragraph (see Sec. 9.8).

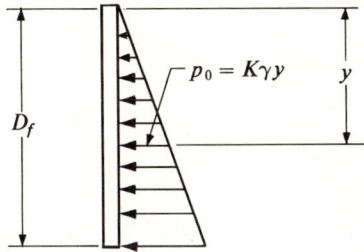

FIGURE 11.15 *Lateral pressure on a pile.*

11.10 Design considerations

The ultimate objective of analyses is to find answers to design problems. This may be in the form of design soil pressures for foundations, or conclusions as to whether a given structure can be expected to perform satisfactorily. In problems of this type the analysis serves the purpose of showing whether a certain design meets the requirements of safety. The requirements usually consist of the following. First, the soil mass under the given loads should have an adequate safety factor with respect to shear failure and, second, the deformation of the soil mass under the given loads should not exceed certain tolerable limits. It is recognized that these requirements are also common in the design of structures or machines and therefore are not special characteristics of soil mechanics. The evaluation of the safety factor is a problem in stability analysis (Chapter 9); the problem of deformation is treated in Chapters 6 and 7.

What constitutes an adequate safety factor and a tolerable deformation is not so simple to establish. First it should be emphasized that the requirements depend on the particular structure and the function it should serve. For example, the failure of a valley slope in a thinly populated area is not very serious, whereas the failure of a building foundation leads to catastrophic consequences (Fig. 11.16). Hence it is reasonable to require a greater factor of safety for the building foundation than for the valley slope. An earth dam may experience a settlement of an inch and still function satisfactorily. On the other hand, if the foundation of a radar tower or precision machinery settles by half the amount, the equipment it supports may become inoperable. It is thus not feasible to put down rigid rules as to what should be the factor of safety or tolerable settlement. (See Sec. 7.5.)

FIGURE 11.16 *Bearing capacity failure of the Transcona grain elevator.* (Courtesy Acker Drill Co., Inc.)

A different phase of design is to devise new solutions when analysis shows that certain proposed schemes are unsatisfactory or unsafe. There is no limit to the possibility of creative designs of this type. However, they represent subtle combinations of intuition and experience which fall outside the scope of this book.

In the following paragraphs are given example problems of design and analysis. They serve to illustrate the complexity of all real problems and the

SEC. 11.10   Design considerations   339

simplifications that must be made in the calculations. This usually leads to a considerable degree of uncertainty, which is the characteristic of all applications of theory to natural materials or natural phenomena. It illustrates in a general way both the power and the limitations of theoretical solutions.

Example 11.3. Figure 11.17(a) shows the subsoil conditions underneath the site of a large power-generating installation. The subsoil consists of a layer of dense sand to a depth of 15 ft. Below this is a thick deposit of soft compressible clay. The major load consists of several hoppers that rest on a rigid reinforced concrete mat 30 by 40 ft in size. The total weight of the full hoppers and the foundation is 1800 tons. It is required to evaluate the type of foundation that is most suitable.

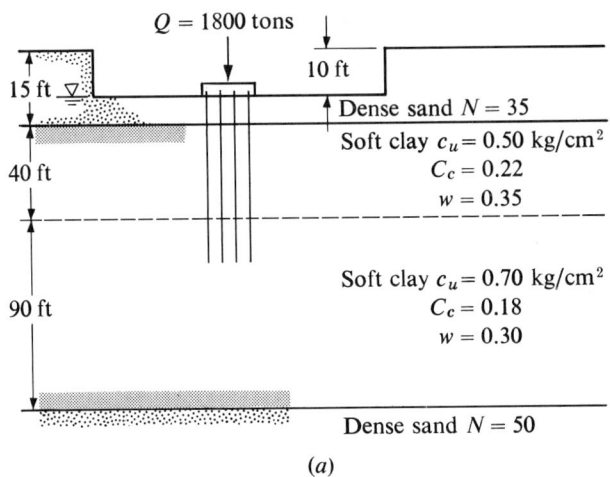

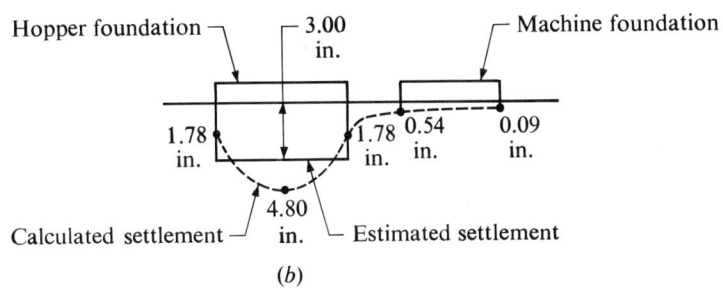

FIGURE 11.17 *Example 11.3.*

Since deep foundations are always expensive, the first possibility is to have the mat supported by the dense sand layer. The sand has an average penetration resistance of 35, and from Fig. 13.18 the value of ϕ should be about 37 deg. We find that the bearing capacity of the sand is more than adequate. Employing the empirical correlation in Fig. 7.14 we find that a design pressure of 1.8 tsf would be acceptable* if a settlement of 1 in. can be tolerated. The actual pressure on the foundation is 1.5 tsf. Hence the sand presents no difficulties.

Despite the capability of the sand to support the foundation loads we should investigate the effect of the loads on the compressible clay layer. A bearing-capacity failure can still occur if the stresses transmitted to the clay exceed its shear strength. We may evaluate the stability of the clay by calculating the increase in vertical stress at the top of the clay due to the foundation loads. By means of the graphical method (Sec. 6.5), this is found to be 2980 psf at the center of the loaded area. The bearing capacity of the clay, according to Eq. (9.25), is

$$q_f = 1.3 \times 5.7 \times 1000 = 7400 \text{ psf}$$

This yields a safety factor of 2.46, which is satisfactory. The settlement of the foundation due to consolidation of the clay, as calculated by Eqs. (7.6) and (7.7) amounts to almost 9 in. While the foundation itself may be able to withstand a settlement of this magnitude, there is likelihood that differential settlements may become too large. For example, located 15 ft from the hopper is a machine foundation. The large settlement of the hopper foundation will cause the machine foundation to tilt; the side close to the hopper foundation will settle more than the side away from it. Such tilting of the foundation will affect the operation of the machine and must be kept to a minimum. It is therefore necessary to investigate the possibility of a deep foundation.

To reach the dense sand below the clay would require piles almost 150 ft long. While this is feasible, it is also expensive. An alternative would be to use friction piles about 50 to 70 ft in length. These would pass through the weakest part of the clay deposit and might be satisfactory. If we consider a pile that is at least 50 ft long and 1 ft in diameter, its bearing capacity may be calculated from Eq. (9.37) as

$$Q_f = \pi \times 1 \times 50 \times 0.50 = 78 \text{ tons}\dagger$$

If the piles are loaded to 30 tons each, the factor of safety is 2.60. The foundation load of 1800 tons would then require about 60 piles. When designing a foundation on a group of piles, one should consider the possibility of the pile group failing as a unit. In this case the soil between the piles simply moves with the piles. To study this we calculate the bearing capacity of the group, which consists of the shearing resistance along the periphery of the group plus the bearing capacity of the base. Taking the dimensions of the pile group as 36 by 25.5 ft, this is found to be 7437 tons, which is ample.

It remains to calculate the settlement of the pile foundation. However, the settlement of a friction-pile foundation cannot be calculated according to the theory outlined in Chapter 7. This is because the equations in Chapter 7 are developed for pressures applied at a given elevation (i.e., the bottom of the foundation). In a friction-pile foundation, a large portion of the foundation load is transmitted to the soil by means of shear stresses along the periphery of the foundation. Thus the soil surrounding the pile group is also subjected to stress.

---

* See Problem 11.8.
† The point resistance is very small, and therefore it is not taken into account.

As an approximation, Terzaghi and Peck (1967) suggested that the foundation loads may be assumed to be transmitted to the soil at a depth equal to two-thirds of the length of the pile (Fig. 11.18). This is only an approximation. Bjerrum et al. (1957) measured the settlement of a bridge pier on friction piles and compared it with that computed according to Terzaghi and Peck's method. The calculated settlement was found to be considerably smaller than the measured value. Hence the results of the calculation represent only an estimate of the magnitude of the settlement. The calculated settlements for a pile foundation that penetrates 60 ft into the clay layer are shown in Fig. 11.17(b).

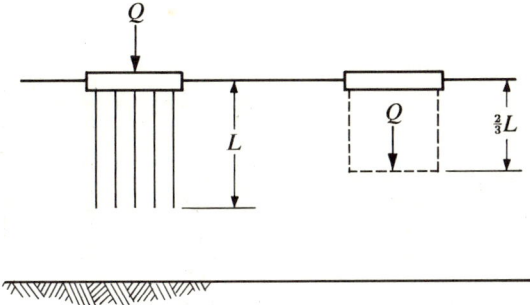

FIGURE 11.18 *Settlement of a friction pile group.*

The calculated settlements are 4.80 in. at the center and 1.78 in. at the corner. It must be remembered that the calculations are based on the assumption that the foundation is perfectly flexible, which is not the case in reality. For a rigid foundation, differential settlement is very limited. The probable settlement of the mat is estimated to be about 3.00 in. The calculated tilt of the machine foundation is about 0.45 in., which is considered acceptable.

The pile foundation using piles 60 ft long is considered satisfactory. In concluding, it is worthwhile to consider a point of a different nature. The driving of 60 piles into a deposit of clay results in the displacement of about 2800 ft^3 of soil. This usually brings about considerable lateral displacement of the soil together with heave of the ground surface. To minimize the undesirable effects, it is good practice to begin pile driving with piles near the middle of the group and proceed outward. Displacements can also be reduced by the use of piles with a small volume displacement, such as steel H piles, or by pre-excavation. This illustrates the important fact that many considerations besides the theory of soil mechanics enter into the design of a project.

Example 11.4. A natural hillside slope 50 ft high and with a slope angle of about 30 deg has undergone continual movement. The movement, up to 3 to 4 in. a year, occurs intermittently and is endangering the installations at the top of the slope. An investigation is made to determine the cause of the displacements and to design remedial measures.

Movements of such continual nature are usually the result of large shear stresses in the soil mass. At the site several large cracks exist at the top of the slope, indicating the top of a slip surface (Fig. 11.19). Detailed examination of the recorded movements and meteorological data reveal the fact that large movements usually occur during years of heavy rainfall. It is concluded that large shear stresses are responsible for the movement but the stresses are not large enough to produce an outright failure. The safety factor must be slightly larger than 1.0. During a wet year the extra moisture increases both the weight of the soil mass and the porewater pressure, reducing the safety factor to about 1.0. The movements occur as a result.

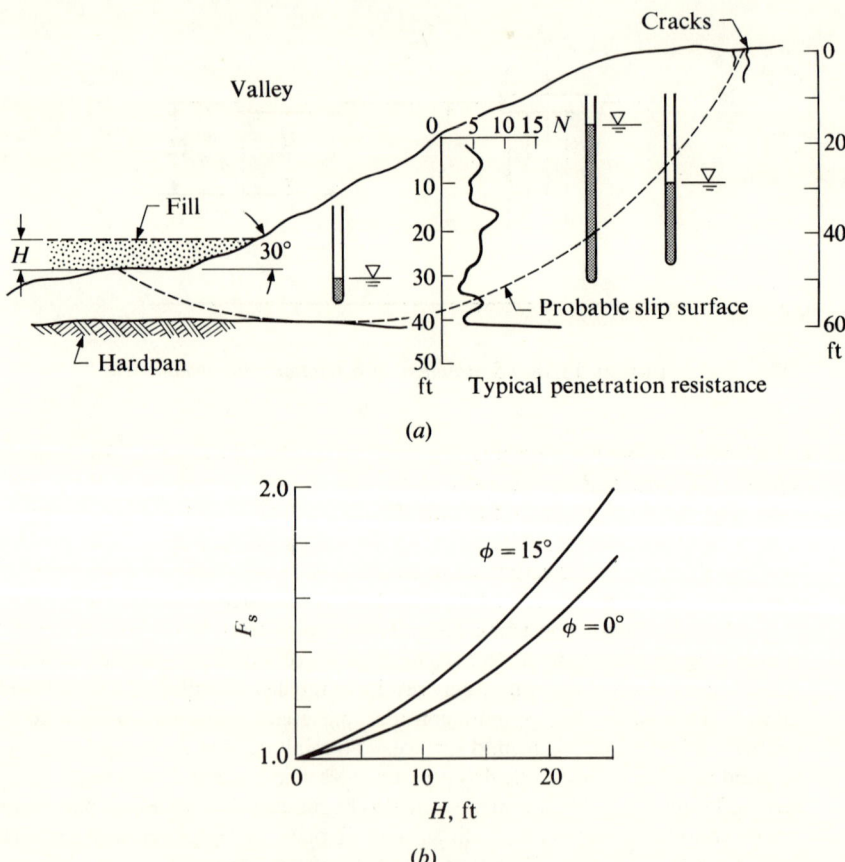

FIGURE 11.19 *Example 11.4.*

In the example the subsoil at the site consists of irregular layers of clayey sand and clay mixed with lenses of medium to coarse sand. No definite stratification can be detected. The penetration resistance averages about 6, indicating a weak material. The results of unconfined compression tests made on some of the clay specimens are erratic. The value of c_u ranges between the wide limits of 0.5 kg/cm² and 3.0 kg/cm². At approximately 10 ft below the valley floor, a layer of hardpan is encountered. Thus the slip surface cannot extend below this depth.

According to the principles developed earlier in this chapter, a natural slope undergoing shear corresponds to a drained test. To analyse the stability with effective stresses, the porewater pressure must be determined. Results of field measurements with piezometers show that the porewater pressure varies from place to place [Fig. 11.19(a)]. The nonuniform conditions of porewater pressure and erratic soil properties rule out the possibility of a stability analysis based on well-defined soil properties and porewater-pressure conditions. We must therefore employ theoretical analysis in a somewhat different manner.

If the movements are the result of shear stress, correction would mean either the reduction of the shear stress or improvement of the shear strength. The former can be achieved by removal of material at the top of the slope or construction of a fill at the toe. To improve the shear strength, drains may be installed to reduce the porewater pressure and water content. Both approaches are logical. However, the irregular nature of the subsoil would require a relatively large number of drains and the cost would be high. The construction of a fill at the toe as shown in the figure is a simpler procedure and is chosen for this problem. The question of how high the fill should be can only be determined from stability analysis.

Even though we cannot evaluate accurately the overall shear strength of the soil from laboratory tests on soil specimens, we can make use of our earlier conclusion that the factor of safety is very close to 1. Since the porewater pressure is not accurately known, we make an approximate analysis with total stresses. What we need is a comparison of the shear stress under the existing conditions with those under the revised conditions. For example, how much would the shear stress be reduced by a 10 ft fill at the toe? The calculations are made for values of ϕ equal to 0 and 15 deg. For these two cases we can calculate the value of c that would yield a factor of safety of 1 on the basis of total-stress analysis. Stability calculations give values of 963 psf and 288 psf for the conditions $\phi = 0$ and $\phi = 15$ deg, respectively.

We next calculate the effect of the fill on the safety factor. Assuming that the existing shear strengths are equal to $\phi = 0$, $c = 963$ psf, and $\phi = 15$ deg, $c = 288$ psf, we can compute the safety factor for the slopes with various heights of the fill. The results are shown graphically in Fig. 11.19(b). It can be seen that insofar as the safety factor is concerned, the value of ϕ used in the calculations does not significantly affect the results. To increase the safety factor to 1.2, the height of the fill should be of the order of 10 ft.

This example illustrates the fact that even though lack of precise soil properties precludes an elaborate stability analysis, the principles of stability are still valid and useful.

PROBLEMS

11.1 A cut is to be excavated in clay below the water table as shown in Fig. 11.20. A year later the excavation is to be dewatered and kept dry. How would the factor of safety change with time?

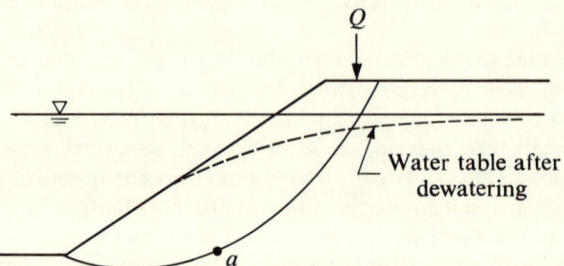

FIGURE 11.20

11.2 The undrained shear strength of a specimen taken at a before dewatering the cut in Problem 11.1 is 1000 psf. The effective-stress parameters are $\bar{c} = 500$ psf and $\bar{\phi} = 23$ deg. Immediately after dewatering, a load is placed near the top of the cut, making the total stress at a equal to $\sigma_3 = 1300$, $\sigma_1 = 3500$. What is the shear strength at a that should be used in analyzing the slope stability under the load?
Answer: 1000 psf assuming dewatering to be a rapid process.

11.3 For the slope in Problem 11.1, consider two clays with values of A equal to 1.0 and -0.3. (*a*) Which clay would have the larger safety factor immediately after excavation? (*b*) Which clay would have the larger long-term safety factor? Assume all other soil properties to be equal.

11.4 For the embankment in Problem 9.6 calculate by effective stress analysis the safety factor of the slip surface immediately after construction. Take the value of B' as equal to 0.3.
Answer: 1.07

11.5 (*a*) If the soil in Problem 11.4 has a porewater-pressure coefficient B' equal to 0.6, is the safety factor greater or smaller than the value calculated in 11.4? Why? (*b*) Since the value of B' increases with the degree of saturation, what conclusion can be drawn regarding the moisture content that should be used during compaction of the embankment?

11.6 Compare the safety factor obtained in Problem 11.4 with that for steady seepage and for rapid drawdown (Fig. 9.29).

11.7 During the steady-seepage state, specimens were taken from the embankment and their average properties are as follows: $w = 16\%$, $c_u = 2000$ psf, $\phi_u = 0$. (*a*) Calculate the safety factor by means of total stresses. (*b*) Is this analysis justified?

11.8 A footing foundation is to be constructed above a thick deposit of moist sand. The water table is located at a very great depth. If sometime after the end of construction, the water table rises to the level of the bottom of the footing: (*a*) What happens to the safety factor? Why? (*b*) What happens to the settlement? Why?

11.9 Calculate the factor of safety at the end of construction for the embankment shown in Fig. 11.21.

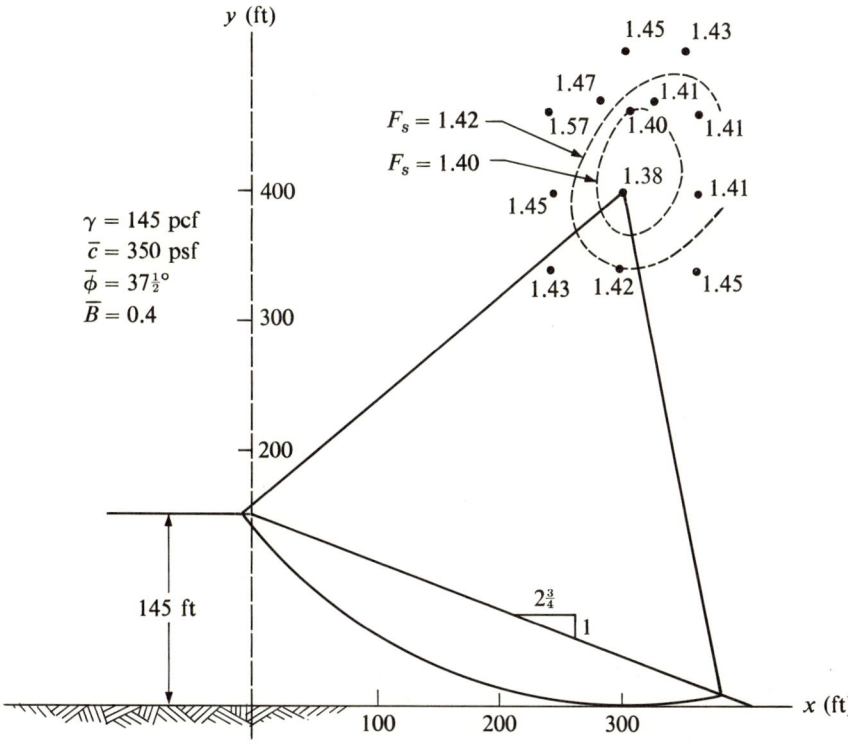

FIGURE 11.21

11.10 A layer of sand lies parallel to the ground surface as shown in Fig. 11.22. The soil properties are as follows:

Silty clay: $\gamma = 128$ pcf, $\bar{\phi} = 20°$, $\bar{c} = 400$ psf
Sand: $\gamma = 132$ pcf, $\bar{\phi} = 35°$, $\bar{c} = 0$

(a) How large must the porewater pressure in the sand be to cause failure of the slope? (b) Will failure occur in the sand or in the clay?
Answer: 765 psf; in the sand

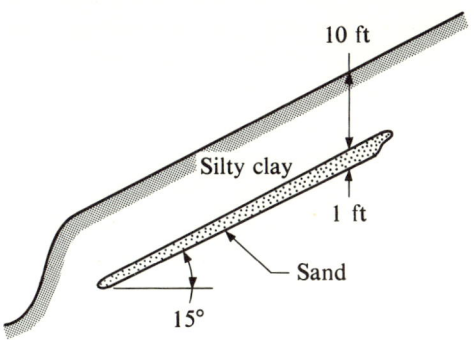

FIGURE 11.22

12

Numerical Solutions

12.1 Introduction

The topic of numerical methods deserves attention because it lends itself very readily to the solution of some typical soil mechanics problems. Among the most severe limitations of theoretical solutions are the idealized conditions, such as homogeneity, that are often assumed to facilitate the mathematical solution. These idealized conditions are rarely realized in natural phenomena, and their departure from the real conditions is often considerable. The great advantage of the numerical methods lies in their ability to solve problems with conditions that are far too complicated for analytical methods. This enables us to introduce into the problem at least some of the complexities and evaluate their effects on the result.

Numerical methods have the additional advantage that they can be handled by computer. Thus voluminous computations that would overwhelm the human calculator can be easily accomplished by machines. This makes it feasible to investigate a large number of variables such as subsoil conditions, loading conditions, etc. The engineer can thus systematically analyze the effect of these possible conditions on the final result such as stability or seepage.

In this chapter the treatment is limited to numerical solution by means of the finite-difference equation. The objective is to introduce the application of numerical methods to the solution of soil mechanics problems rather than to present numerical methods. In recent years the method of finite elements has become a powerful tool in soil engineering. This subject is beyond the scope of this book and the reader should consult the well-known works of Zienkiewiez (1967), and Desai and Abel (1972) for authoritative treatments of the finite element method.

It should be noted that solutions such as the method of slices for stability analysis and the determination of the critical failure surface by trial presented

SEC. 12.2   **Finite-difference equation**   347

in the preceding chapters are also numerical solutions. Many of these operations can be programmed directly for computer calculation.

12.2 Finite-difference equation

For approximate numerical solutions, a differential equation can be put in the form of finite differences. We consider a function $h = f(x)$ (Fig. 12.1) and locate

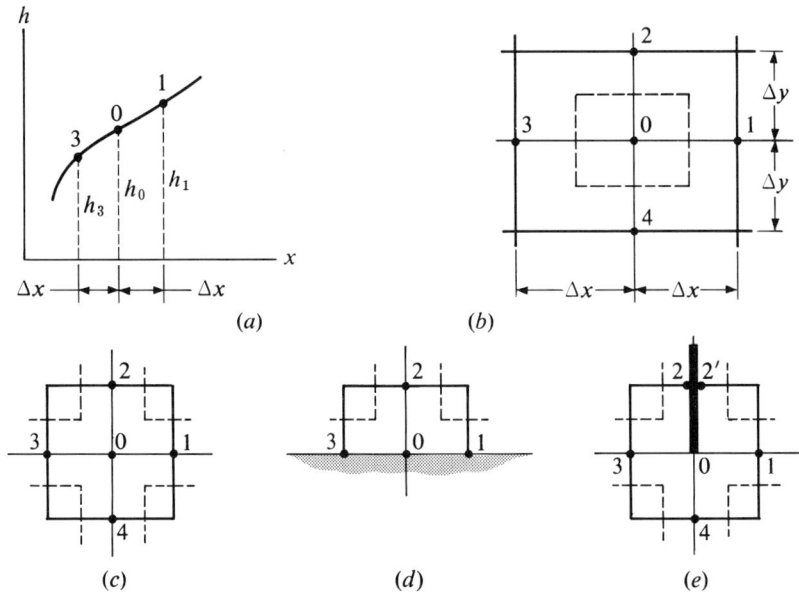

FIGURE 12.1   *Finite differences.*

points 0, 1, and 3 at distances Δx apart. The values of h at points 1 and 3 may be calculated by Taylor's series:

$$\left.\begin{aligned} h_1 &= h_0 + \Delta x \left(\frac{dh}{dx}\right)_0 + \frac{(\Delta x)^2}{2!}\left(\frac{d^2h}{dx^2}\right)_0 + \frac{(\Delta x)^3}{3!}\left(\frac{d^3h}{dx^3}\right)_0 + \frac{(\Delta x)^4}{4!}\left(\frac{d^4h}{dx^4}\right)_0 \cdots \\ h_3 &= h_0 - \Delta x \left(\frac{dh}{dx}\right)_0 + \frac{(\Delta x)^2}{2!}\left(\frac{d^2h}{dx^2}\right)_0 - \frac{(\Delta x)^3}{3!}\left(\frac{d^3h}{dx^3}\right)_0 + \frac{(\Delta x)^4}{4!}\left(\frac{d^4h}{dx^4}\right)_0 \cdots \end{aligned}\right\}$$

(12.1)

Subtracting the two equations, we obtain

$$\left(\frac{dh}{dx}\right)_0 = \frac{1}{2\,\Delta x}(h_1 - h_3) - \frac{(\Delta x)^2}{3!}\left(\frac{d^3h}{dx^3}\right)_0 \cdots$$

If the internal Δx is made sufficiently small, the value of

$$\frac{(\Delta x)^3}{3!}\left(\frac{d^3h}{dx^3}\right)_0$$

and of all succeeding terms are very small. If we neglect these terms we obtain

$$\left(\frac{dh}{dx}\right)_0 = \frac{1}{2\,\Delta x}(h_1 - h_3) \tag{12.2}$$

Similarly, on adding the two Eqs. (12.1) and neglecting the higher-order terms we obtain

$$\left(\frac{d^2h}{dx^2}\right)_0 = \frac{1}{(\Delta x)^2}(h_1 + h_3 - 2h_0) \tag{12.3}$$

Equations (12.2) and (12.3) may also be obtained if we consider the points 3, 0, and 1 to be close enough that h may be taken as linear between these points. Then we have

$$\frac{dh}{dx} = \frac{h_1 - h_0}{\Delta x} \quad \text{from 0 to 1}$$

$$\frac{dh}{dx} = \frac{h_0 - h_3}{\Delta x} \quad \text{from 3 to 0}$$

or

$$\frac{dh}{dx} = \frac{1}{2}\left(\frac{h_1 - h_0}{\Delta x} + \frac{h_0 - h_3}{\Delta x}\right) = \frac{1}{2\,\Delta x}(h_1 - h_3) \quad \text{from 3 to 1}$$

They are called, respectively, the forward, backward, and central differences. The second derivative is

$$\left(\frac{d^2h}{dx^2}\right)_0 = \frac{\left(\frac{h_1 - h_0}{\Delta x}\right) - \left(\frac{h_0 - h_3}{\Delta x}\right)}{\Delta x} = \frac{1}{\Delta x^2}(h_1 + h_3 - 2h_0)$$

If h varies and both x and y, we introduce points 2 and 4 in the y direction and equations similar to (12.2) and (12.3) can be obtained for $\partial h/\partial y$ and $\partial^2 h/\partial y^2$.

The above expressions allow us to write a differential equation in terms of finite differences such as $h_1 - h_3$. We may take the continuity Eq. (3.14) as an example. Using Eq. (12.3), we get for the points in Fig. 12.1(b),

$$k_x \frac{\partial^2 h}{\partial x^2} + k_y \frac{\partial^2 h}{\partial y^2} = \frac{k_x}{(\Delta x)^2}(h_1 + h_3 - 2h_0) + \frac{k_y}{(\Delta y)^2}(h_2 + h_4 - 2h_0) = 0 \tag{12.4}$$

If k_x and k_y are equal and Δx is made equal to Δy, the equation reduces to

$$h_1 + h_2 + h_3 + h_4 - 4h_0 = 0 \tag{12.5a}$$

12.3 Seepage through homogeneous material

We illustrate the use of the difference equation with a simple problem in steady seepage through a homogeneous soil around a sheet pile as shown in Fig. 12.2. A grid or mesh system is drawn, using Δx equal to Δy.

At any of the grid points, the equation of continuity (12.5a) must be satisfied. It is helpful to consider the equation as a direct expression of the flow to

SEC. 12.3 Seepage through homogeneous material

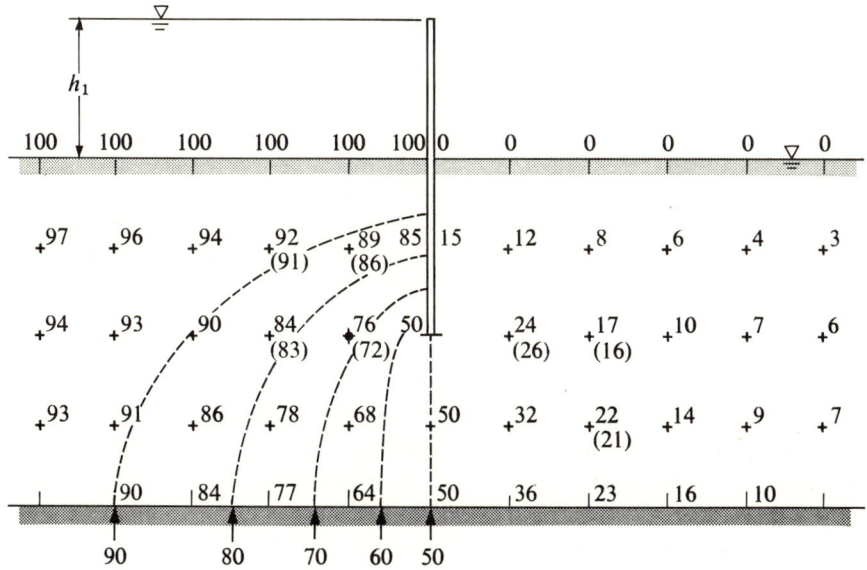

FIGURE 12.2 *Flow around a sheet-pile wall.*

and from a point 0 along the channels illustrated in Fig. 12.1(c) by the dashed lines. Thus from Darcy's law we have

$$(q)_{3-0} = k \frac{h_0 - h_3}{\Delta x} \Delta y$$

$$(q)_{0-1} = k \frac{h_1 - h_0}{\Delta x} \Delta y$$

$$(q)_{2-0} = k \frac{h_0 - h_2}{\Delta y} \Delta x$$

$$(q)_{0-4} = k \frac{h_4 - h_0}{\Delta y} \Delta x$$

Since the total flow to and from point 0 is equal to zero and Δx equals Δy, we have

$$k[(h_1 - h_0) - (h_0 - h_3) + (h_4 - h_0) - (h_0 - h_2)]$$
$$= k(h_1 + h_2 + h_3 + h_4 - 4h_0) = 0$$

If we take a point 0 on the boundary [Fig. 12.1(d)], there is only flow from three directions. The continuity equation then becomes

$$k[\tfrac{1}{2}(h_1 - h_0) - \tfrac{1}{2}(h_0 - h_3) - (h_0 - h_2)] = k(\tfrac{1}{2}h_1 + h_2 + \tfrac{1}{2}h_3 - 2h_0) = 0 \quad (12.5b)$$

Finally at the tip of the sheet pile [Fig. 12.1(e)] the flow channel from 2 to 0 is divided into two parts by the barrier. Thus there are two values of h at point 2, h'_2 at the left side of the sheeting and h''_2 at the right side. We then have

$$k[(h_1 - h_0) - (h_0 - h_3) + (h_4 - h_0) - \tfrac{1}{2}(h_0 - h'_2) - \tfrac{1}{2}(h_0 - h''_2)]$$
$$= k[h_1 + \tfrac{1}{2}(h'_2 + h''_2) + h_3 + h_4 - 4h_0] = 0 \quad (12.5c)$$

The boundary conditions of the problem are as follows. Along the upstream face ab, the head is everywhere equal to h_1 and along the downstream face de it is equal to 0. We shall take the value of h_1 as 100 for convenience. The sheet pile and the impermeable base are flow lines.

To solve the problem numerically, one of the methods is successive approximation. We estimate the value of h at the grid points and this gives us the first approximation. These values are shown on the grid system in Fig. 12.2. This first approximation of course is not the correct solution. Hence it does not satisfy the continuity equation at every grid point. For example, if we take point A, we find that the values of h_0 through h_4, when substituted into Eq. (12.5a), give $50 + 89 + 84 + 68 - 4(76) = 13 \neq 0$. Therefore it is necessary to revise the values, and this leads to our second approximation, which is given in parentheses in Fig. 12.2. This is again checked by Eq. (12.5a). This process is repeated until the values of h satisfy Eq. (12.5a) at every grid point (or until the errors are sufficiently small). Equipotential curves constructed from the values of h at the grid points are shown as dashed curves. Besides iteration, the problem can also be solved by relaxation. This falls beyond the scope of this book and the reader may refer to the well-known book by Southwell (1940).

12.4 Seepage through nonhomogeneous material

The great advantage of numerical methods lies in the ability to handle problems with complex boundaries. We next consider the problem of a nonhomogeneous subsoil [Fig. 12.3(a)]. The subsoil of the sheet-pile problem contains a pocket of highly permeable material whose extent is delineated by the shaded area. The permeability of this material is 10 times that of the surrounding soil. To evaluate the effect of this on the flow and porewater pressure, we simplify the problem by replacing this irregular-shaped pocket with a rectangular block $abcd$. The soil inside the rectangle has a permeability equal to 10 times that of the soil outside.*

The next step in this problem is to consider the grid points 0, 1, 2, 3, 4 on the boundary as shown in Fig. 12.3(b). For flow across the boundary, it is necessary to modify Eq. (12.5a). Because of the different permeability coefficients, the velocity of flow is also different in the two materials. Referring back to Eq. (12.4) the first term on the right side should be rewritten so that k_x represents the average permeability in the x direction. Since the element

---

* For a closer approximation of irregular boundaries see Scott (1963).

SEC. 12.4   Seepage through nonhomogeneous material   351

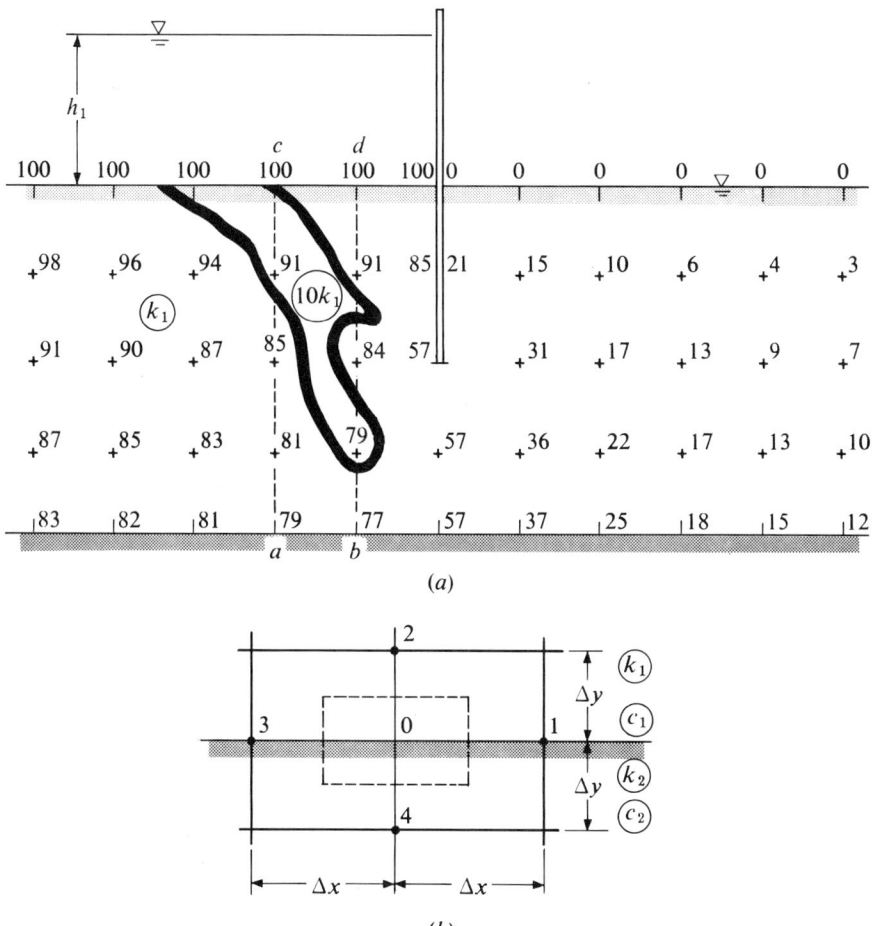

FIGURE 12.3 *Flow through a nonhomogeneous subsoil.*

covered by points 0, 1, 2, 3, 4 lies half in material (1) and half in material (2), we have

$$k_x = \tfrac{1}{2}(k_1 + k_2)$$

To take into account the different permeabilities in the second term of Eq. (12.4) we make use of the transformation of scales (Sec. 3.8). We replace the material (2) by material (1) but with a head of h_4' instead of h_4. In order for the velocity to remain unchanged we have

$$k_1 \frac{h_4' - h_0}{\Delta y} = k_2 \frac{h_4 - h_0}{\Delta y}$$

Thus

$$h_4' = h_0 + \frac{k_2}{k_1}(h_4 - h_0)$$

The finite-difference form of the continuity equation then becomes

$$\frac{k_1 + k_2}{2(\Delta x)^2}(h_1 + h_3 - 2h_0) + \frac{k_1}{(\Delta y)^2}\left[h_2 + \frac{k_2}{k_1}(h_4 - h_0) - h_0\right] = 0 \quad (12.6a)$$

If Δx and Δy are taken as equal, the above equation is reduced to

$$h_1 + \frac{2k_1}{k_1 + k_2}h_2 + h_3 + \frac{2k_2}{k_1 + k_2}h_4 - 4h_0 = 0 \quad (12.6b)$$

The solution of the nonhomogeneous case follows the same procedure as that for the homogeneous case. The only exception is that at the boundary the values of h must satisfy Eq. (12.6) instead of Eq. (12.5a).

12.5 Consolidation by vertical drainage

The differential equation for one-dimensional consolidation is

$$\frac{\partial u}{\partial t} = c_v \frac{\partial^2 u}{\partial z^2} \quad (5.7)$$

Following the derivation by Scott (1963), we transform this equation into dimensionless variables U, T, and Z such that

$$U = \frac{u}{u_m} \quad (12.7a)$$

$$Z = \frac{z}{H_m} \quad (12.7b)$$

$$T = \frac{t}{t_m} \quad (12.7c)$$

Here u_m, H_m, and t_m are arbitrary reference units of porewater pressure, distance, and time. From these we also have

$$\frac{\partial u}{\partial t} = \frac{u_m}{t_m}\frac{\partial U}{\partial T} \qquad \frac{\partial^2 u}{\partial z^2} = \frac{u_m}{H_m^2}\frac{\partial Z^2}{\partial^2 U}$$

Substituting this in Eq. (5.7) we obtain

$$\frac{c_v}{H_m^2}\frac{\partial^2 U}{\partial Z^2} = \frac{1}{t_m}\frac{\partial U}{\partial T} \quad (12.8)$$

For convenience, we can choose a value of t_m so that

$$\frac{1}{t_m} = \frac{c_v}{H_m^2}$$

This simplifies Eq. (12.8) to

$$\frac{\partial^2 U}{\partial Z^2} = \frac{\partial U}{\partial T} \quad (12.9)$$

The problem is to solve for U in terms of Z and T. The final answer can be transformed back to the variables u, z, and t when necessary.

SEC. 12.5   Consolidation by vertical drainage

To solve the problem numerically, we again put Eq. (12.9) in terms of finite differences. Using the grid system in Fig. 12.1(b),

$$\frac{\partial^2 U}{\partial Z^2} = \frac{1}{(\Delta Z)^2}(U_{2,T} + U_{4,T} - 2U_{0,T}) \tag{12.10}$$

We note that the porewater pressure changes with time. Hence the additional subscript T is used to denote the value of U at a particular time T. The derivative $\partial U/\partial T$ can be written as the forward difference

$$\frac{\partial U}{\partial T} = \frac{1}{\Delta T}(U_{0,T+\Delta T} - U_{0,T}) \tag{12.11}$$

Substituting these equations in (12.9) yields

$$U_{0,T+\Delta T} = \frac{\Delta T}{(\Delta Z)^2}(U_{2,T} + U_{4,T} - 2U_{0,T}) + U_{0,T} \tag{12.12}$$

This equation allows us to calculate the porewater pressure at point 0 at a time $T + \Delta T$ from known porewater pressures at time T. As before, we establish a mesh by choosing suitable values of ΔT and ΔZ. However, for the equation to converge, $\Delta T/(\Delta Z)^2$ should be smaller than $\frac{1}{2}$. This is necessary in order that the errors introduced in the calculation of $U_{0,T+\Delta T}$ do not become progressively larger with subsequent steps.

If the subsoil is nonhomogeneous, we have to take the different permeabilities and compressibilities into account at the boundary. As in the preceding section, we make the simplification that the boundary coincides with the grid lines [Fig. 12.3(b)]. The coefficients of consolidation of the two materials are c_1 and c_2. It is convenient to adopt the same time and distance intervals Δt and Δz for both materials.

Beginning with the differential Eq. (5.7) we can write

$$\frac{k}{c_v}\frac{\partial u}{\partial t} = k\frac{\partial^2 u}{\partial z^2} \tag{12.13}$$

As explained in Sec. 5.5, the left-hand side represents the reduction in void volume with time due to changes in the porewater pressure. The right-hand side is the difference between the rate of flow of water into and out of the element. For an element that is on the boundary, the volume change is the sum of the volume changes of the half that consists of material (1) and the other half, which consists of material (2). Hence

$$\frac{k}{c_v}\frac{\partial u}{\partial t} = \frac{1}{2}\left[\left(\frac{k_1}{c_1}\frac{1}{\Delta t}\right) + \left(\frac{k_2}{c_2}\frac{1}{\Delta t}\right)\right](u_{0,t+\Delta t} - u_{0,t})$$

Following the derivation of Eq. (12.6a) the difference in the rate of flow is

$$k\frac{\partial^2 u}{\partial z^2} = \frac{1}{2}\left(\frac{k_1}{\Delta z^2} + \frac{k_2}{\Delta z^2}\right)\left(\frac{2k_1}{k_1+k_2}u_{2,t} + \frac{2k_2}{k_1+k_2}u_{4,t} - 2u_{0,t}\right)$$

Thus Eq. (12.13) becomes

$$\left(\frac{k_1}{c_1} + \frac{k_2}{c_2}\right)(u_{0,t} - u_{0,t+\Delta t})$$

$$= \frac{\Delta t}{(\Delta z)^2}(k_1 + k_2)\left(\frac{2k_1}{k_1 + k_2}u_{2,t} + \frac{2k_2}{k_1 + k_2}u_{4,t} - 2u_{0,t}\right) \quad (12.14)$$

As before, we introduce dimensionless variables U, T, and Z and let

$$U = \frac{u}{u_m} \qquad T = \frac{t}{t_m} \qquad Z + \frac{z}{H_m}$$

and in addition we take $1/t_m = c_2/H_m^2$. Equation (12.14) then becomes

$$\frac{1}{t_m \Delta T}\left(\frac{k_1}{c_1} + \frac{k_2}{c_2}\right)(U_{0,T} - U_{0,T+\Delta T})$$

$$= \frac{1}{H_m^2(\Delta Z)^2}(k_1 + k_2)\left(\frac{2k_1}{k_1 + k_2}U_{2,T} + \frac{2k_2}{k_1 + k_2}U_{4,T} - 2U_{0,T}\right)$$

and we get

$$U_{0,T+\Delta T} =$$

$$\frac{1 + (k_1/k_2)}{1 + [(k_1/k_2)(c_2/c_1)]}\frac{\Delta T}{(\Delta Z)^2}\left(\frac{2k_1}{k_1 + k_2}U_{2,T} + \frac{2k_2}{k_1 + k_2}U_{4,T} - 2U_{0,T}\right) + U_{0,T}$$
$$(12.15)$$

This is the equation that should be used for grid points along the boundary. At all other points Eq. (12.10) holds. It should be noted that the reference time t_m is based on the consolidation coefficient of material (2). It can be taken as H_m^2/c_1 equally well, and the resulting equation is also valid.

Figure 12.4(a) is an example of consolidation in a subsoil that consists of two different clay layers between layers of free-draining sand. The initial porewater pressure U after loading is taken as equal throughout the clay deposit and assigned a value of 200 for convenience. At the top and bottom faces of the clay deposit the initial porewater pressure is 0, because of the free-draining sand. The time interval is chosen to be 10 days and the depth interval 10 ft, and H_m is taken as 40 ft. Thus we get

$$\Delta t_1 = \Delta t_2 = 10 \text{ days} \qquad \Delta T_1 = \frac{c_1 \Delta t_1}{H_{m1}^2} = 1.5 \times 10^{-2}$$

$$\Delta T_2 = \frac{c_2 \Delta t_2}{H_{m2}^2} = 1.1 \times 10^{-3}$$

$$M_1 = \frac{\Delta T_1}{(\Delta Z)^2} = 0.24 \qquad M_2 = \frac{\Delta T_2}{(\Delta Z)^2} = 0.18$$

The increments of time Δt and ΔT and the increments of depth Δz and ΔZ are plotted on the abscissa and ordinates in Fig. 12.4(b). The initial value of the porewater pressure equal to 200 is entered at the grid points below $T = 0$. The values of U at a time ΔT later are calculated by means of Eq. (12.12) when

SEC. 12.6   Consolidation by radial drainage   355

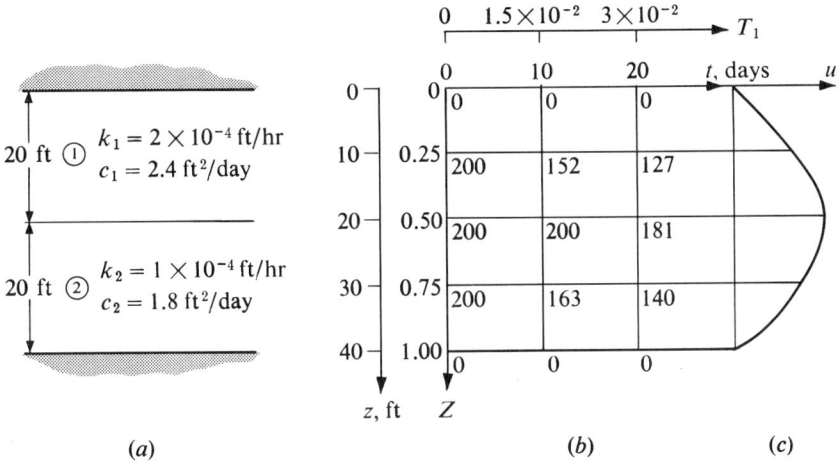

FIGURE 12.4   *Consolidation of two clay layers by vertical drainage.*

the point 0 lies within either layer 1 or 2 and by Eq. (12.15) when the point is on the border. The calculated porewater pressures at 20 days are plotted graphically in Fig. 12.4(c).

The numerical method can also be used to solve the case of consolidation under a gradually increasing load, as shown by the example in Fig. 12.5. The solid curve in Fig. 12.5(b) is the rate at which the load is applied. For the solution, it is necessary to approximate the loading by discreet steps, as shown by the dashed line in the graph. The soil properties are given in Fig. 12.5(a). The calculation of U_T is shown in Fig. 12.5(c). The first load increment of 500 psf produces initial porewater pressures equal to 100 throughout the layer. The porewater pressures at time intervals ΔT after this load increment are calculated by Eq. (12.12) and the values are given in the second and third columns. At the time of 20 days, a second increment of pressure equal to 1000 psf is applied. Therefore the porewater pressure throughout the clay must be increased by 200. This adds to the porewater pressure that is already in the clay as a result of the first load increment. The calculated porewater-pressure changes at a depth of 5 ft ($Z = 0.25$) for the stepwise loading are plotted in Fig. 12.5(d).

12.6 Consolidation by radial drainage

The problem of one-dimensional radial consolidation is solved by writing the consolidation equation in cylindrical coordinates.* It takes the form

$$\frac{\partial u}{\partial t} = c_v \left(\frac{\partial^2 u}{\partial r^2} + \frac{1}{r} \frac{\partial u}{\partial r} \right) \qquad (12.16)$$

---

* For transformation to cylindrical coordinates, see any book on calculus, for example, Hildebrand (1962), pp. 303–305.

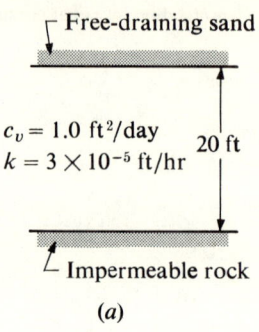

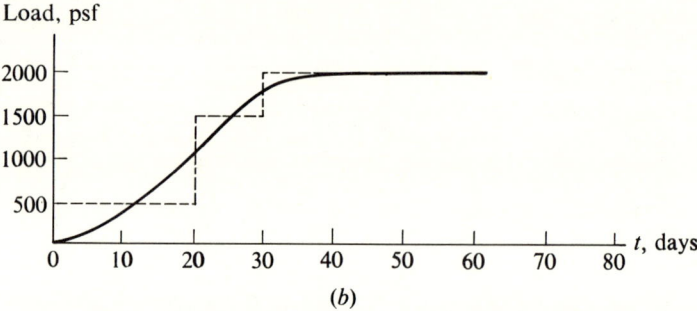

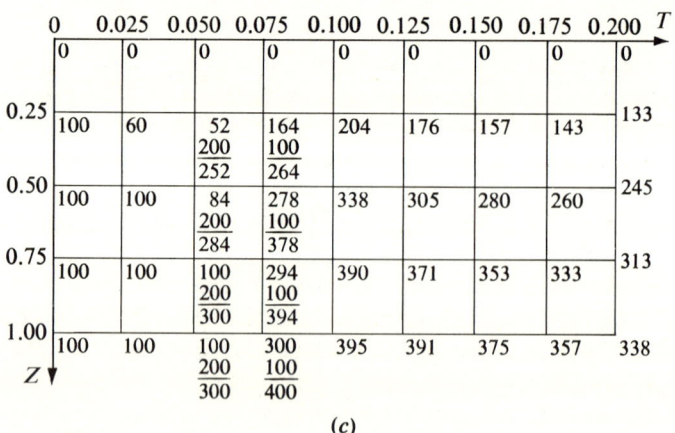

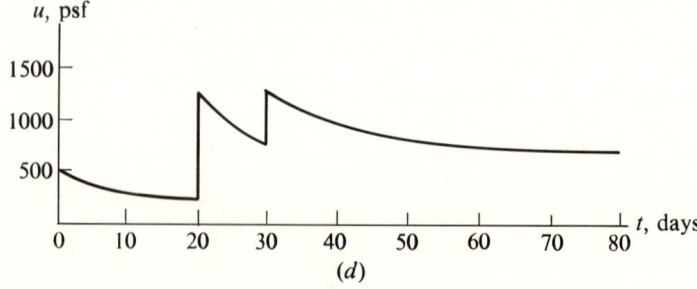

FIGURE 12.5 Consolidation under a gradually increasing load.

SEC. 12.6 Consolidation by radial drainage

in which r is the radial distance. This condition arises if a load is applied to a layer of clay confined between two impermeable layers as shown in Fig. 12.6(a). In this case the excess hydrostatic porewater pressure immediately beneath the load is large, whereas at some distance away it is negligible. Hence drainage occurs radially beginning at the center line of the load, which is the axis of symmetry. If, in addition, the top and bottom of the consolidating layer are bounded by permeable material, the solution for one-dimensional consolidation (Sec. 12.5) can be superimposed on the solution for radial consolidation [see Gibson and Lumb (1953)]. The practical significance of such drainage patterns is pointed out in Sec. 7.6.

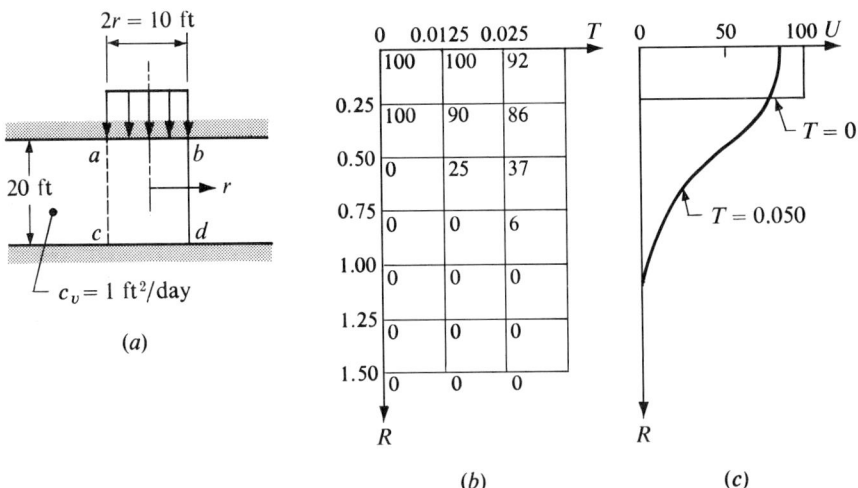

FIGURE 12.6 *Consolidation of a clay layer by radial drainage.*

Using the dimensionless variables U and T [Eq. (12.7a and b)] and letting

$$R = \frac{r}{H_m} \qquad \frac{1}{t_m} = \frac{c_v}{H_m^2} \tag{12.7d}$$

Eq. (12.16) is transformed to

$$\frac{\partial U}{\partial T} = \frac{\partial^2 U}{\partial R^2} + \left(\frac{1}{R}\frac{\partial U}{\partial R}\right) \tag{12.17}$$

This can be put in the form of finite differences by means of the relationships

$$\frac{\partial U}{\partial T} = \frac{1}{\Delta T}(U_{0,T+\Delta T} - U_{0,T})$$

$$\frac{\partial^2 U}{\partial R^2} = \frac{1}{\Delta R^2}(U_{2,T} + U_{4,T} - 2U_{0,T})$$

$$\frac{1}{R}\frac{\partial U}{\partial R} = \frac{1}{R}\left(\frac{U_{2T} - U_{4T}}{2\,\Delta R}\right)$$

Substituting these in Eq. (12.17) yields

$$U_{0,T+\Delta T} = \frac{\Delta T}{(\Delta R)^2}\left[U_{2,T} + U_{4,T} - 2U_{0,T} + \frac{U_{2,T} - U_{4,T}}{2(R/\Delta R)}\right] + U_{0,T} \quad (12.18)$$

At $r = 0$,

$$\frac{1}{r}\frac{\partial u}{\partial r} \to \frac{\partial^2 u}{\partial r^2}$$

Thus the equation becomes,

$$U_{0,T+\Delta T} = \frac{2\,\Delta T}{(\Delta R)^2}(U_{2,T} + U_{4,T} - 2U_{0,T}) + U_{0,T} \quad (12.19)$$

The solution of the radial drainage problem is illustrated in Fig. 12.6. The clay layer is 20 ft thick and lies between two impermeable layers. The circular loaded area has a radius of 5 ft. We assume the initial excess hydrostatic porewater pressure to be 100 everywhere underneath the loaded area (in *abcd*) and 0 outside *abcd*. We take $H_m = 20$ ft and $\Delta R = 0.25$. The initial excess hydrostatic porewater pressure is therefore 100 for R of 0 and 0.25 in Fig. 12.6(*b*). The calculated porewater pressures are plotted against R in Fig. 12.6(*c*) and point out a very important feature of radial consolidation. When the porewater pressure underneath the loaded area decreases, that in the surrounding area ($R > 0.25$) increases. This is because water flows from *abcd* into the adjacent area. Hence as the soil underneath the load consolidates, the soil outside *abcd* first swells and then consolidates. Strictly speaking, we should use the coefficient for swelling instead of c_v in Eq. (12.16) for the swelling part. However, this greatly complicates the computations, and, furthermore, according to Gibson and Lumb, the error introduced by the use of c_v throughout the layer is not serious.

12.7 Bearing-capacity problem

As an example of the use of numerical methods in plasticity problems, we can follow Sokolovski's solution (1956) of the bearing-capacity problem described in Sec. 9.4.

The basic equations are (9.9) and (9.11). They may be combined as follows. Let σ_a denote

$$\sigma_a = \frac{\sigma_1 + \sigma_3}{2} + c \cot \phi \quad (12.20a)$$

Then substituting Eqs. (9.11) and (12.20a) into (9.9a) and (9.9b), we get

$$(1 + \sin\phi \cos 2\psi)\frac{\partial \chi}{\partial x} + \sin\phi \sin 2\psi \frac{\partial \chi}{\partial y}$$

$$- \cos\phi \left(\sin 2\psi \frac{\partial \psi}{\partial x} - \cos 2\psi \frac{\partial \psi}{\partial y}\right) = \frac{\cot\phi}{2\sigma_a} X \quad (12.20b)$$

SEC. 12.7 Bearing-capacity problem

$$\sin\phi \sin 2\psi \frac{\partial \chi}{\partial x} + (1 \sin\phi \cos 2\psi)\frac{\partial \chi}{\partial y}$$
$$+ \cos\phi\left(\cos 2\psi \frac{\partial \psi}{\partial x} + \sin 2\psi \frac{\partial \psi}{\partial y}\right) = \frac{\cot\phi}{2\sigma_a} Y$$

in which

$$\chi = \frac{\cot\phi}{2} \ln \frac{\sigma_a}{c} \tag{12.20c}$$

and c is an arbitrary reference stress.* Equations (12.20b) may be simplified by introducing the new variables ξ and ν defined by

$$\xi = \chi + \psi \qquad \nu = \chi - \psi \tag{12.20d}$$

Equations (12.20b) then become

$$\frac{\partial \xi}{\partial x} + \tan(\psi + \beta)\frac{\partial \xi}{\partial y} = b \qquad \frac{\partial \nu}{\partial x} + \tan(\psi - \beta)\frac{\partial \nu}{\partial y} = a \tag{12.21a}$$

in which $\beta = 45° - \frac{\phi}{2}$

$$\left.\begin{array}{c}a\\b\end{array}\right\} = \pm \frac{X \sin(\psi \pm \beta) - Y \cos(\psi \pm \beta)}{(\sigma_1 + \sigma_3 + 2c \cot\phi)\sin\phi \cos(\psi \mp \beta)} \tag{12.21b}$$

The variables ξ and ν contain a stress term χ plus the term ψ, which denotes the direction of the principal stress. If ξ and ν are determined for a point, then the stresses are readily found by means of Eqs. (12.20a), (12.20c), and (12.20d).

The total derivative $d\xi/dx$ may be written as

$$\frac{d\xi}{dx} = \frac{\partial \xi}{\partial x} + \frac{\partial \xi}{\partial y}\frac{dy}{dx} = \frac{\partial \xi}{\partial y}\frac{dy}{dx} - \frac{\partial \xi}{\partial y}\tan(\psi + \beta) - b$$

This equation is useful because of the following relationship. In the x-y coordinate system, along a curve with a slope

$$\frac{dy}{dx} = \tan(\psi + \beta) \tag{12.22a}$$

the equation reduces to

$$\frac{d\xi}{dx} = -b \tag{12.23b}$$

But we note that $\tan(\psi + \beta)$ is just the slope of the slip line [Fig. 9.7(b)]. Hence Eq. (12.23a) gives us the relationship between the stresses along the slip line. By similar treatment of $d\nu/dx$ we obtain the relationships along the other slip line:

$$\frac{dy}{dx} = \tan(\psi - \beta) \tag{12.22b}$$

and

$$\frac{d\nu}{dx} = a \tag{12.23b}$$

---

* For convenience we have chosen the reference stress equal to the cohesion c.

If the body force is 0 (a weightless material), then a and b are both 0. Thus along one set of slip lines ξ is constant, while along the second set ν is constant [see Sokolovski (1956)].

Equations 12.23(a) and (b) may be integrated along a failure surface for prescribed boundary conditions to give the stresses in a mass under plastic equilibrium. In the theory of partial differential equations these equations are called characteristics. The method followed in this section is sometimes called the method of characteristics.

We return to the problem of a loaded strip shown in Fig. 9.6. Along the free surface ad, the stresses and the direction of the slip lines are known. This means that χ, ψ, a, and b are known. From these the values of ξ and ν along ad can be calculated. This constitutes the boundary condition. If we take two points A and B located on the surface ad (Fig. 12.7) the values of ξ at A (or ξ_A) and ν at B (or ν_B) are known. Also, Eq. (12.23) gives the change in ξ and ν along the two sets of slip lines, which are shown in Fig. 12.7 as AC and BC. This allows us to calculate the values of ξ and ν at C, which is the intersection of the two slip lines, one beginning at A and the other at B. Knowledge of ξ_C and ν_C allows us in turn to calculate χ, ψ, and the stresses at C. By this means the slip lines can be traced throughout the zone of failure, and the stresses that act on surface aa' can be evaluated.

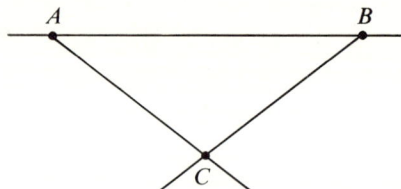

FIGURE 12.7 *Intersection of two slip lines.*

To solve the problem numerically, we first locate the coordinates of the point C (Fig. 12.7), which is the intersection of the two slip surfaces starting at A and B. This can be obtained by means of Eq. (12.22) written in finite differences. For the first set of slip surfaces, we have

$$y_C - y_A = \tan(\psi_A - \beta)(x_C - x_A)$$

and for the second set we have

$$y_C - y_B = \tan(\psi_B + \beta)(x_C - x_B)$$

The subscripts A, B, and C denote the values of x, y, and ψ at the points A, B, and C. Solving the two simultaneous equations, we get

$$x_C = \frac{x_A \tan(\psi_A - \beta) - y_A - x_B \tan(\psi_B + \beta) + y_B}{\tan(\psi_A - \beta) - \tan(\psi_B + \beta)} \quad \text{(12.24a)}$$

$$y_C = y_B + (x_C - x_B)\tan(\psi_B + \beta) \quad \text{(12.24b)}$$

SEC. 12.7   Bearing-capacity problem   361

This allows the coordinates of point C to be computed from known values of x, y, and ψ at points A and B. To calculate the stresses at C we make use of Eqs. (12.23a and b). Written as finite-difference equations, they become

$$v_C - v_A = (x_C - x_A)a_A$$
$$\xi_C - \xi_B = (x_C - x_B)b_B$$

Solving these, we get

$$v_C = v_A + (x_C - x_A)a_A \qquad (12.25a)$$
$$\xi_C = \xi_B + (x_C - x_B)b_B \qquad (12.25b)$$

With known values of v_C and ξ_C, the value of χ_C and ψ_C can be calculated from Eq. (12.20d). This process can be continued to determine the slip surfaces and stresses throughout the failure zone. The numerical example for the simple problem of a weightless material is given in Fig. 12.8 and Table 12.1 [Sokolovski (1956)].

The soil is loaded along the surface $a0$, with a pressure given by $40c + 18cy$. It is required to find the minimum pressure q_f on $0b$ that will prevent a bearing-capacity failure under the load on $a0$. In other words, the soil mass should be on the verge of failure under these pressures.

A series of equally spaced points are chosen along the surface $a0$, and the slip surfaces beginning at these points are numbered 0, 1, 2, 3, 4 in each direction. The point of intersection of the slip lines $m = 0, n = 4$ is designated 0-4. Along $a0$ the interval between points 0-4, 1-3, ... are chosen to be 0.20 and, therefore, the coordinates of these points are known. Also, along $a0$, the stress conditions are known, which means that ξ and v are known. The major principal stress is given by the vertical load which is $40c + 18cy$. This allows us to calculate the quantity σ_a in Eq. (12.21a) The expression for σ_a may be simplified if we note that σ_a is equal to ac in Fig. 12.8(d), while σ_1 is equal to $0b$ and $a0$ is $c \cot \phi$. From the geometry of the Mohr's circle, we get

$$\sigma_1 + c \cot \phi = (ac) + (cb) = (ac)(1 + \sin \phi) = \sigma_a(1 + \sin \phi)$$

Thus

$$\sigma_a = \frac{\sigma_1 + c \cot \phi}{1 + \sin \phi} \qquad (12.26)$$

This relationship allows us to calculate σ_a directly from σ_1. The values of ξ and v are computed from Eq. (12.20d). Since the major principal stress is in the direction of the x axis, the value of ψ is equal to 0 and ξ and v are equal for points along the surface $a0$. Thus for point 0-4, we have

$$\sigma_a = \frac{40c + 2.75c}{1 + 0.342} = 31.8c$$

$$\psi = 0$$

$$\xi = v = \frac{2.75}{2} \ln \frac{31.8c}{c} = 4.75$$

TABLE 12.1 *Bearing Capacity Problem*

n	0	1	2	3	4	5	6	7	8	9	10	11
m												
0												
x					0	0	0	0				
y					0	0	0	0				
ψ					0	-0.525	-1.045	-1.570				
ξ					4.75	3.70	2.66	1.61				
ν					4.75	4.75	4.75	4.75				
1												
x				0	0.143	0.246	0.316	0.281	0			
y				0.20	0.100	0.021	-0.147	-0.398	-0.760			
ψ				0	-0.050	-0.575	-1.095	-1.620	-1.570			
ξ				4.85	4.75	3.70	2.66	1.61	1.71			
ν				4.85	4.85	4.85	4.85	4.85	4.85			
2												
x			0	0.143	0.286	0.470	0.592	0.507	0.214	0		
y			0.40	0.300	0.189	0.030	-0.293	-0.760	-1.100	-1.374		
ψ			0	-0.050	-0.100	-0.625	-1.145	-1.670	-1.625	-1.570		
ξ			4.95	4.85	4.75	3.70	2.66	1.61	1.70	1.81		
ν			4.95	4.95	4.95	4.95	4.95	4.95	4.95	4.95		
3												
x		0	0.143	0.286	0.429	0.677	0.805	0.660	0.366	0.154	0	
y		0.60	0.50	0.395	0.260	0.027	-0.418	-1.034	-1.345	-1.595	-1.796	
ψ		0	-0.045	-0.095	-0.145	-0.670	-1.190	-1.715	-1.670	-1.615	-1.570	
ξ		5.04	4.95	4.85	4.75	3.70	2.66	1.61	1.70	1.81	1.90	
ν		5.04	5.04	5.04	5.04	5.04	5.04	5.04	5.04	5.04	5.04	
4												
x	0	0.143	0.286	0.427	0.570	0.914	1.055	0.822	0.530	0.323	0.173	0
y	0.80	0.70	0.595	0.470	0.333	0.019	-0.530	-1.353	-1.640	-1.866	-2.045	-2.270
ψ	0	-0.055	-0.100	-0.150	-0.200	-0.725	-1.245	-1.770	-1.725	-1.670	-1.625	-1.580
ξ	5.15	5.04	4.95	4.85	4.75	3.70	2.66	1.61	1.70	1.81	1.90	1.99
ν	5.15	5.15	5.15	5.15	5.15	5.15	5.15	5.15	5.15	5.15	5.15	5.15

SEC. 12.7   Bearing-capacity problem   363

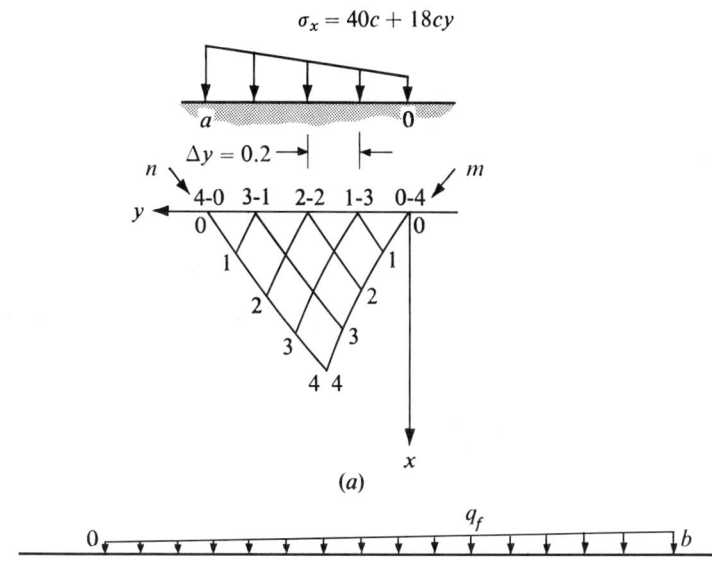

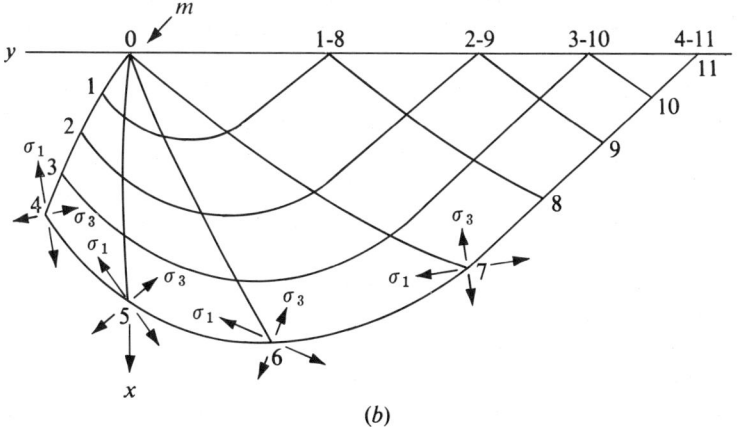

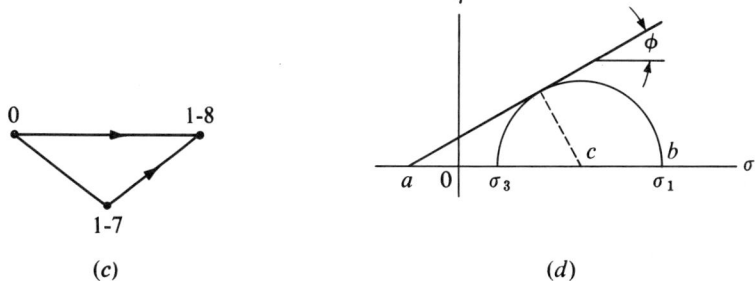

FIGURE 12.8 *Bearing capacity problem.*

The coordinates x, y and the calculated values of ξ and ν are recorded in Table 12.1 along the diagonal 0-4, 1-3, 2-2, ... Equation (12.24) is used to calculate the values of x, y, ξ, ν at the intersection points. For a weightless material, a and b are 0. Hence ξ is the same for each column and ν is the same for each row. To calculate the other terms we use point 2-4 as an example. From Eq. (12.20a) we obtain

$$\psi = \tfrac{1}{2}(\xi - \nu) = \tfrac{1}{2}(4.95 - 4.75) = 0.10$$

Also $\beta = 45° - (20°/2) = 35°$. From Eq. (12.17), we have

$$x = \frac{0.143 \tan(-3° - 35°) - 0.300 - 0.143 \tan(-3° + 35°) + 0.100}{\tan(-3° - 35°) - \tan(-3° + 35°)}$$

$$= \frac{(0.143)(-0.781) - 0.300 - 0.143(0.625) + 0.100}{-0.781 - 0.625} = 0.286$$

From Eq. (12.24b), we have

$$y = 0.100 + (0.286 - 0.143)\tan(-3° + 35°) = 0.189$$

In this way the spaces below the diagonal and to the left of column 4 in Table 12.1 are filled. This gives us all the values within the triangle between the two No. 4 slip surfaces.

To proceed from here we note that point 0-4 must be a point of discontinuity, because the slip surfaces to its right have different directions from those to its left (Fig. 12.8). To the right of 0 the direction of σ_1 is along the y axis and the value of ψ is equal to $-\pi/2$. Actually 0 is an infinitesimally short slip surface and along this slip surface the value of ν is constant [Eq. (12.23b)]. Hence the value of ν is everywhere equal to 4.75 along 0, as shown in Table 12.1, points 0-4 through 0-7. The value of ξ at 0-7 must differ from ν by 2ψ [see Eq. (12.20d)] or π. Therefore ξ is equal to 1.61. For the transition area between slip surfaces 4 and 7 we arbitrarily divide it with two equally spaced slip surfaces, 5 and 6. The value of ξ changes from 4.75 to 1.61 in three equal increments between 4 and 7, with column 4 and row 1 established. The remaining values in columns 5, 6, and 7 can be calculated with Eqs. (12.24) and (12.22).

To compute the values between slip surfaces 7 and 11, we know only the values of x, y, ξ, and ν along 7. In addition we know the values of x and ψ along the surface.

Starting from points 0 and 1-7 we can calculate the necessary quantities at point 1-8. This is done by means of Eqs. (12.24b) and (12.25a). We let A designate point 0 and B designate point 1-7 (Fig. 12.8) and calculate x_c and ξ_c with y_c equal to 0.

After Table 12.1 is complete we can calculate the vertical stress on $0b$, which is the bearing capacity that we are seeking. This can be found from the known values of ξ and ν by means of Eq. (12.20c) and (12.20d). The calculated results are listed in Table 12.2. For point 7-0 the value of q_f is obtained as follows:

$$\xi = 1.61 = x + \psi = \frac{\cot \phi}{2} \ln \frac{\sigma_a}{c} - \frac{\pi}{2}$$

$$1.61 = 1.375 \ln \frac{\sigma_a}{c} - 1.57 \qquad \sigma_a = 10.1c$$

$$q_f = \sigma_3 = \sigma_a(1 - \sin \phi) - c \cot \phi \qquad (12.27)$$

$$= (10.1c)(1 - 0.342) - 2.75c = 3.90c$$

TABLE 12.2 *Bearing Capacity Problem*

Point	σ_a	$q = \sigma_3$
1-8	10.10c	3.89c
2-9	10.85c	4.40c
3-10	12.45c	5.45c
4-11	13.30c	6.01c

PROBLEMS

12.1 If the pressure on the clay deposit in Fig. 12.4 is increased by 2000 psf and the initial overburden pressure is 2800 psf: (a) Calculate the porewater pressure at a depth of 10 ft 30 days after the pressure is applied. (b) Calculate the average percentage of consolidation 30 days after the pressure is applied.
Answer: 1094 psf; 51%

12.2 Calculate the average percentage of consolidation at 30 days after loading the clay deposit in Fig. 12.4, assuming the clay to be homogeneous with k and c_v equal to 2×10^{-4} ft/hr and 2.4 ft²/day, respectively.
Answer: 35%

12.3 Calculate the average percentage of consolidation 30 days after loading the clay deposit in Fig. 12.4 assuming that there is a 6-in. layer of free-draining sand between clay layers one and two.
Answer: 78%

12.4 Compare the errors introduced into the calculated consolidation at 30 days: (a) If the soil-exploration program had not revealed the different properties of the lower clay layer in Fig. 12.4. (b) If the soil-exploration program had not revealed the presence of a 6-in. layer of free-draining sand described in Problem 12.3. (c) Which omission leads to a more serious error in the computed consolidation rate? (d) Which omission leads to a more serious error in the computed final settlement?
Answer: overestimates U; underestimates U; error from (b) is larger; error from (a) is larger

12.5 Consider the consolidation problem described in Fig. 12.5: (a) Making use of the calculated porewater pressures for the step loading [Fig. 12.5(d)], estimate the porewater pressure-time relationship for the actual loading process. (b) If the rate of consolidation is calculated on the basis that the entire load is imposed instantaneously at $t = 0$, what would be the error in the calculated porewater pressure at 30 days and at 60 days? (c) From the answer in (b), what conclusions can be drawn regarding the errors introduced by inaccuracies in estimating the rate of loading?
Answer: about 1100 psf at 30 days; 730 psf at 60 days; underestimates u by 15% at 30 days and 4% at 60 days

12.6 Make a second approximation of the potentials in the problem shown in Fig. 12.3(a).

12.7 Using the data given in Figs. 12.2 and 12.3a: (a) Evaluate the effect of the permeable pocket on the rate of seepage. (b) Evaluate the effect of the permeable pocket on the stability of the soil on the downstream side of the sheeting. (c) Discuss the effect on seepage and stability if the permeable material extends all the way to point A [Fig. 12.3(a)]. Take h_1 as equal to 30 ft, Δx as 10 ft, and the saturated unit weight of the soil as 132 pcf.
Answer: very small effect; reduces F_s from 1.5 to 1.3

12.8 Derive the expression $\sigma_3 = \sigma_a(1 - \sin\phi) - c\cot\phi$ [Eq. (12.27)].

12.9 From the data in Table 12.1 calculate the following for point 4-4: (a) Angle between σ_1 and vertical. (b) σ_x, σ_y, τ_{xy}. (c) σ_1, σ_3.
Answer: 11.45°; 45.45c, 22.35c, 4.88c; 46.30c, 21.31c

13

Properties of Natural Soil Deposits

13.1 Introduction

The preceding chapters deal with various soil properties that are based largely upon examination and measurements made on soil specimens in the laboratory. In the idealized problem, a large mass of soil is considered to be homogeneous, so that its physical properties at any point in the mass are the same as those determined in the laboratory from a few specimens. Unfortunately, soil deposits are often the result of many complicated natural processes and this ideal situation is very seldom, if ever, attained. Most soil deposits are nonhomogeneous. To intelligently evaluate the properties of an extensive soil deposit with the results of a limited number of laboratory experiments it is necessary to understand the processes that are responsible for the formation of the soil deposit and their effects on the soil properties.

In the first part of this chapter, the significant characteristics of some of the most common types of soil are described. Many soil deposits of similar geologic origin possess important characteristics that are common to this type of soil. These characteristics include physical properties such as compressibility or particle-size distribution as well as the nature of variations in the physical properties. An attempt is also made to present a general idea of the distribution and occurrence of the various types of soils. Figure 13.1 is a simplified soil map that describes the distribution of the major soil types in the United States. For detailed accounts of the land forms, geologic history, and soil deposits of different regions of the United States the reader may refer to the works of Fenneman (1931, 1938) and Woods et al. (1962), as well as publications of the geological surveys.

FIGURE 13.1 *Soil deposits of the United States.* [After Peck et al. (1974).]

13.2 Origin of soils

The majority of the soils that cover the earth were formed by the destruction of rocks. The destructive process may be physical, as in the case of erosion by the forces of wind, water, or glacier, and disintegration by alternate freezing and thawing. It may also be chemical decomposition that results in changes in the mineral constituents of the rocks. In the case of erosion, the agencies of wind, water, or ice transport the eroded material to different localities and deposit them in the form of sediments. The property of the transported sediments reflects faithfully the agencies of transportation and deposition. Therefore, they are usually classified as *alluvial* or *fluvial* (waterlaid), *aeolian* (windlaid), or *glacial* soils.

Soils formed by disintegration and chemical decomposition may be subsequently transported by the above agencies. In this case they are classified as alluvial, aeolian, or glacial sediments. However, in many parts of the world, the soils thus formed remain in place. These are called *residual soils*.

In addition to the two major categories of transported and residual soils, there exist a number of soils that are not derived from the destruction of rocks. For example, peat and muck are formed by the decomposition of vegetation in swamps. Marl is the result of precipitation of dissolved calcium carbonate.

ALLUVIAL DEPOSITS

Alluvial sediments are the result of deposition by streams. Two important types of alluvial deposits are braided-stream deposits and meander-belt deposits.

13.3 Braided-stream deposits

Braided streams are streams overloaded with sediment. The water contains far more sediments than it is capable of transporting. As a result, deposition occurs all along the stream channel. The overloading may take place with a sudden decrease in the gradient of the stream. This happens, for example, when a mountain stream emerges on a flat plain. It may also result from excessive contribution of sediment load from a tributary. The braided-stream pattern is marked by the stream being split into large numbers of coalescing channels separated by bars and islands (Fig. 13.2).

The depositional action of a braided stream is reflected by the many shifting channels of the stream. As deposition proceeds at a relatively high rate, a particular channel is soon filled with sediments. The water then must overflow and follow a different channel. Therefore the deposit of a braided stream consists of an agglomeration of fingers and lenses of soils of varying particle sizes. A cross

FIGURE 13.2 *A braided stream in southeastern Alaska, overloaded with sediments released from a glacier, part of which is shown at the top of the photograph.* (Photo by U.S. Navy.)

section through such a deposit is shown in Fig. 13.3(a). The absence of regular stratification is a characteristic of such deposits. The density of the soil is likely to vary to a considerable degree from place to place, depending on the type of soil encountered. The material, however, is usually rather uniform within a particular lens or pocket. Figure 13.3(b) shows the variation in size and density of 12 specimens taken from a lens of medium sand in a braided-stream deposit.

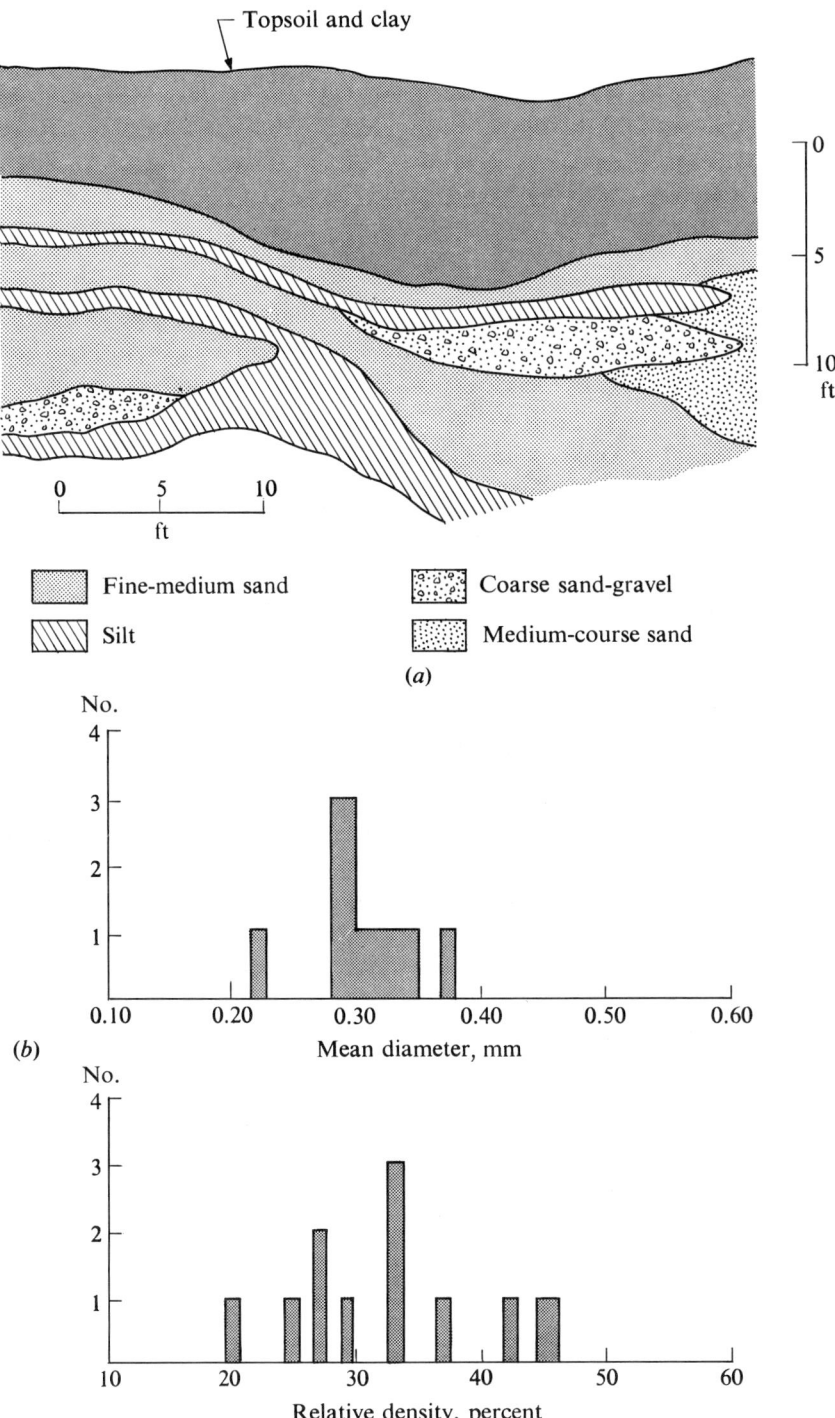

FIGURE 13.3 *Properties of outwash deposits: (a) Cross section of an outwash deposit; (b) Variation in size and density of 12 specimens of outwash sand.*

Braided-stream deposits may contain particles from silt size up to gravel, the size being governed by the velocity of the stream. However, particles of clay size are usually absent.

Deposits of this type may be found locally to cover small areas as a result of sedimentation from braided streams. On a large scale such deposits exist throughout a great part of the western United States. Geologic history indicates that tremendous amounts of soil material were eroded from the Rocky Mountains and deposited on the Great Plains as an extensive mantle of alluvial sediments. Similar events occurred in many other mountain ranges. Hence, many of the valleys are floored with sediments eroded from the mountains. In Fig. 13.1, these are labeled as Great Plains mantle and filled valleys, respectively. Alluvial sand and gravel are also found among coastal plain soils along the Atlantic and Gulf coasts.

In glaciated areas, similar soils, called *outwash*, were formed as sediment-laden glacial meltwaters spread out over plains. Since these deposits were associated with glaciation but were deposited by running water, they are often classified as glacial-fluvial to distinguish them from alluvial soils (see Sec. 13.5).

13.4 Meander-belt deposits

A mature stream meandering in a broad valley does not drop many sediments. Local transportation of sediments takes place as soil eroded from the concave bank (also called *undercut slope*) of the meander is deposited as sand bars along the convex bank. This process is responsible for the development of point-bar deposits along the inside of a bend (Fig. 13.4). Point-bar deposits consist mostly of silt and sand. The fine particles remain in suspension and are carried away by the river.

During a flood a large amount of sediment is deposited and the characteristics of the sediments are very different. Figure 13.5 illustrates the condition in the lower Mississippi River Valley.* When floodwaters overflow the banks of the channel, the sudden reduction in velocity drops the coarser sediments along the river banks. These deposits are called *natural levees*, since over a period of years the sediments accumulate to form ridges along the banks. Particles of clay size remain suspended in the floodwater, which spreads across the entire valley. With the recession of the flood, water is trapped in many low areas of the valley. The fine particles then gradually settle out to form *back-swamp deposits*, which consist of plastic clays. They are subjected to seasonal drying, and as a result their moisture content is considerably lower than the liquid limit.

Meanders may progress to the point where the "neck" between two loops becomes so narrow that water breaks through and a "cutoff" is accomplished (Fig. 13.4). The arc of the meander is then abandoned and becomes an

---

* Fisk (1952) gives an excellent account of the geologic history and soil deposits of the Mississippi Valley.

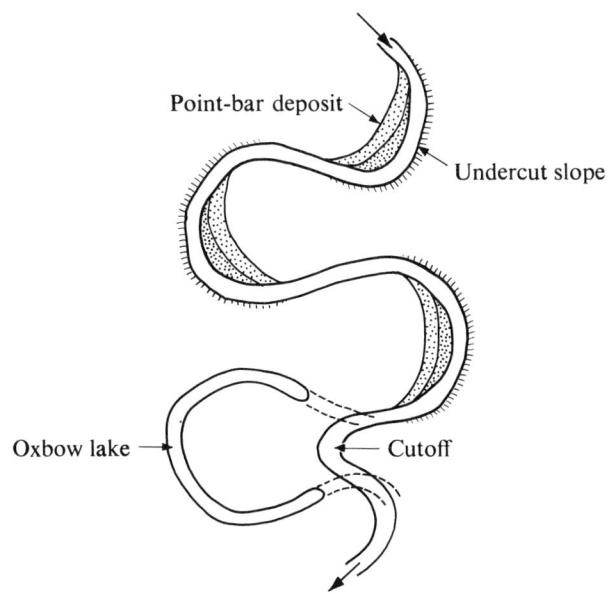

FIGURE 13.4 *Meander development.*

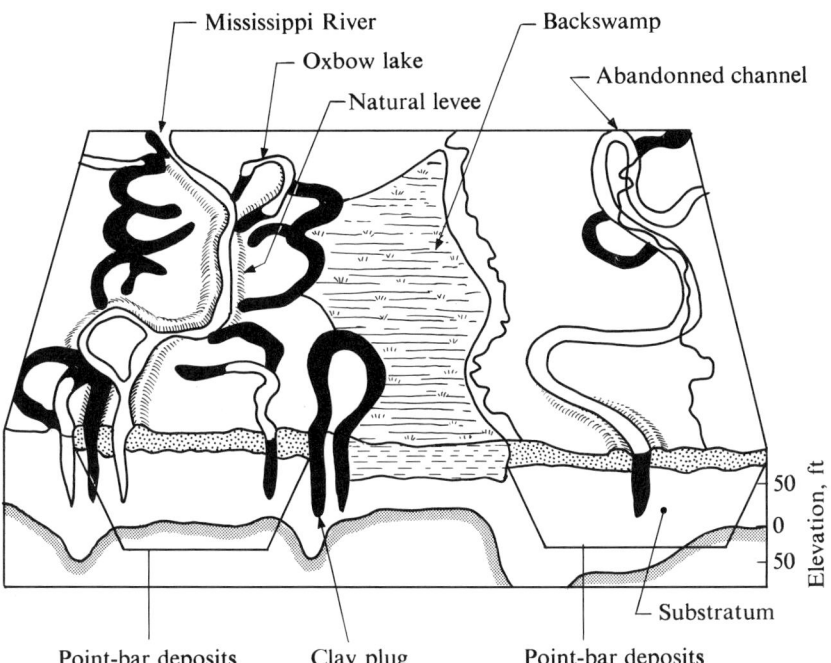

FIGURE 13.5 *Pictorial diagram of the soil deposits of the lower Mississippi Valley.* [After Kolb and Shockley (1957).]

oxbow lake. The oxbow lake gradually becomes blocked from the river channel by sand bars, but water is supplied to it by seepage and by floods. Floodwater brings a continual supply of fine sediments into the oxbow lake and gradually the lake is filled with clay. Such deposits, called *channel fillings* or *clay plugs*, are normally consolidated clays and constitute the most compressible materials that may be encountered in a floodplain.

The preceding paragraphs point out that a number of very different depositional processes are active in a floodplain and that the result is an extremely complex soil formation whose constituents and physical properties are likely to vary over a wide range. One of the most important meander-belt deposits in the United States is found in the lower Mississippi Valley. Other smaller deposits exist in many of the large river valleys. Table 13.1 is a summary of the average physical properties for the different types of sediments in the lower Mississippi Valley.

TABLE 13.1 *Typical Properties of Some Soil Deposits of the Mississippi Valley*[a]

Environment	Particle size and organic content	Water content	LL	I_p	Shear strength
Natural levee	Clay 40%; silt 40%; sand 20%	15–35%	NP–45%	NP–25%	c_{cu} = 180–1200 psf ϕ_{cu} = 0–35°
Point bar	Clay 15%; silt 40%; sand 40%; org. 5%	25–45%	30–55%	10–25%	c_{cu} = 0–850 psf ϕ_{cu} = 25–35°
Abandoned channel	Clay 70%; silt 20%; sand 5%; org. 5%	30–95%	30–100%	10–65%	c_u = 300–1200 psf ϕ_u = 0
Backswamp	Clay 75%; silt 5%; sand 5%; org. 15%	25–70%	40–115%	25–100%	c_u = 400–2500 psf ϕ_u = 0

[a] After Kolb and Shockley (1957).

The changing environment during deposition of alluvium commonly results in stratified soils. Thin layers of weak material may be present and instructive examples include the failures of the Pendleton levee [Fields and Wells (1944) and the discussion by K. Terzaghi], the Thorpe Marsh embankment [Ward et al. (1955)], and the Dartford embankment [Marshland (1961)].

GLACIAL DEPOSITS

Many times during geologic history the earth was covered by continental glaciers. The last period of active glaciation, known as the *Pleistocene Epoch*, left behind many profound features. Among these are the extensive glacial deposits that cover substantial portions of the world, including most of the northern United States. For a detailed description of glacial sediments and glacial geology in North America, see Flint (1957). An excellent account of the formation of moraines and related features as observed in the existing glaciers of Baffin Island is given by Ward (1952).

13.5 Glacial till and outwash

Moving glaciers erode soils and rocks in their paths which are carried along by the glacier and released as the ice melts away. The material deposited directly by the glacier as it melts is called *till*. If the ice front remains more or less stationary for a long period of time, a considerable amount of till would accumulate along the front. After the disappearance of the ice, the material appears as a ridge that traces the position of the former ice front. These ridges are called *moraines*. Tills are also laid down in the form of extensive plains as the glaciers retreat during periods of rapid melting. Tills are characterized by lack of stratification and large range in particle size (Fig. 13.6). Many tills deposited by continental glaciers contain substantial amounts of clay-size particles. They are usually preconsolidated and fairly stiff clays. Even though a deposit of till may be extensive in size and uniform in texture, its strength may vary considerably from point to point.

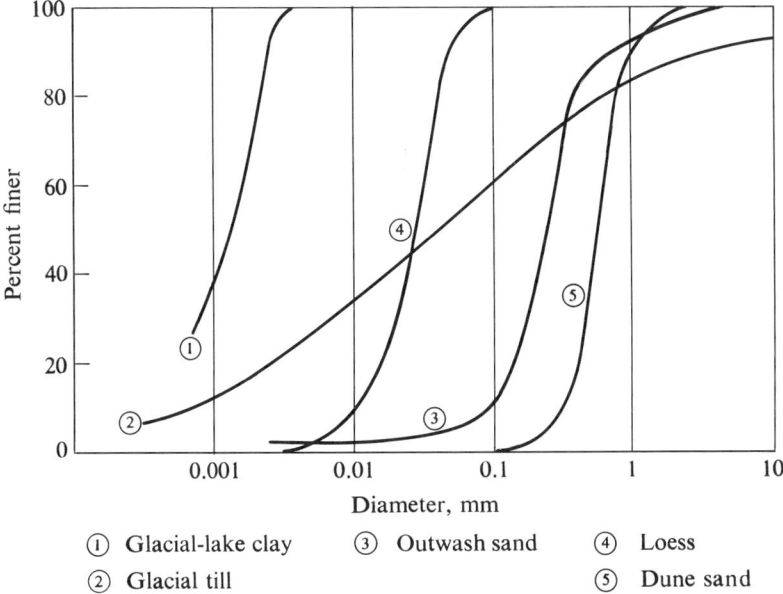

FIGURE 13.6 *Examples of particle-size–distribution curves*

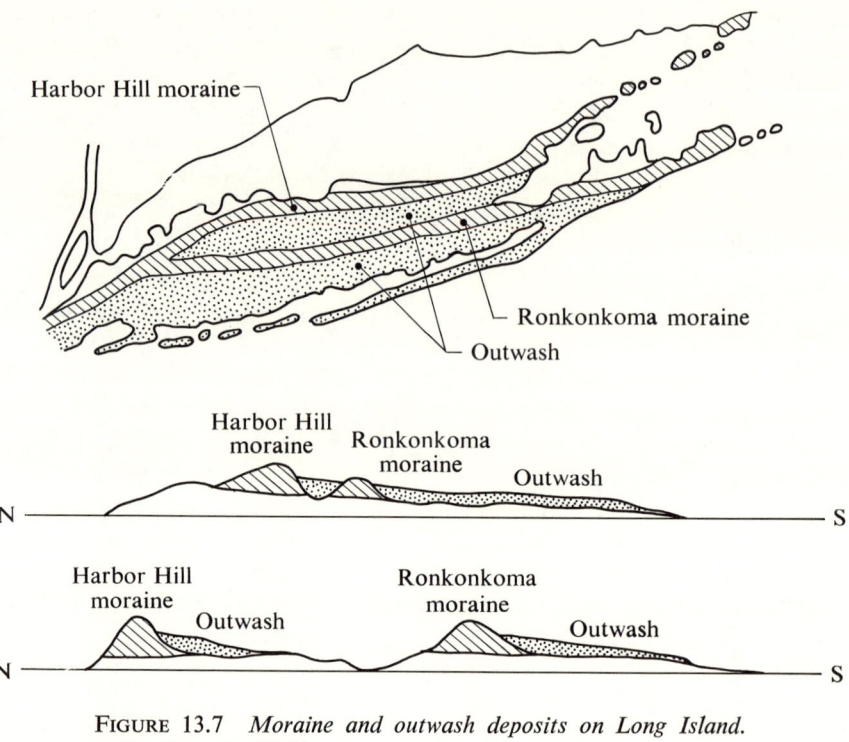

FIGURE 13.7 *Moraine and outwash deposits on Long Island.* [After Lobeck (1939).]

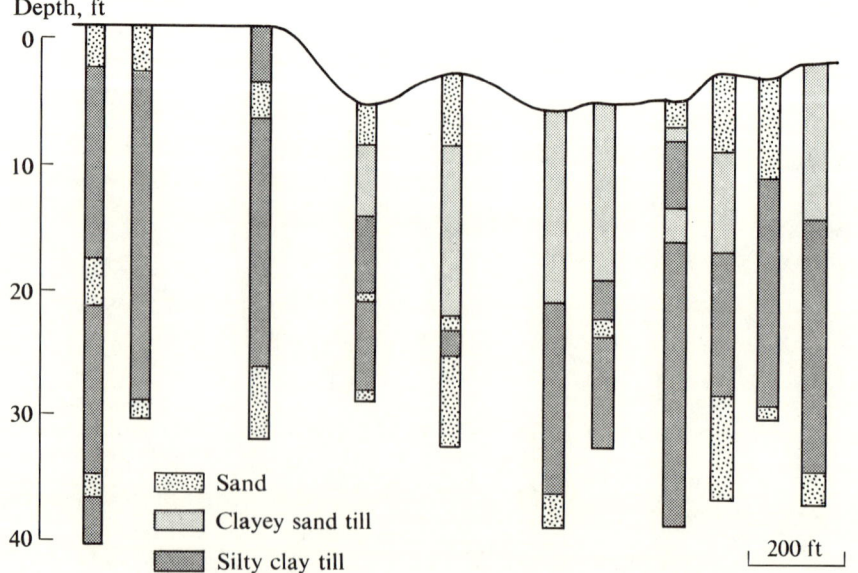

FIGURE 13.8 *Soil profile of a glacial deposit in central Michigan.*

SEC. 13.5   Glacial till and outwash

Along the front of a glacier, water from the melting of the ice gathers to form large torrential streams. These streams are capable of transporting great quantities of sediments. As the streams spread out over the plains, most of the coarse sediment is deposited as alluvium. Thus they have the characteristics of braided-stream deposits and are classified as *glacial-fluvial soils*. They are called *outwash* and consist of granular soils such as gravel, sand, and silt. Outwash usually occurs as a belt in front of a moraine (Fig. 13.7). In some localities extensive areas are covered by outwash up to a hundred feet in thickness. In other areas they exist as thin layers of limited lateral extent included between layers of till. Figure 13.8 shows a soil profile constructed from a series of borings that penetrated a deposit of till with some outwash. Two layers of till can be recognized; a clayey sand on top, and a silty clay below it. The outwash appears

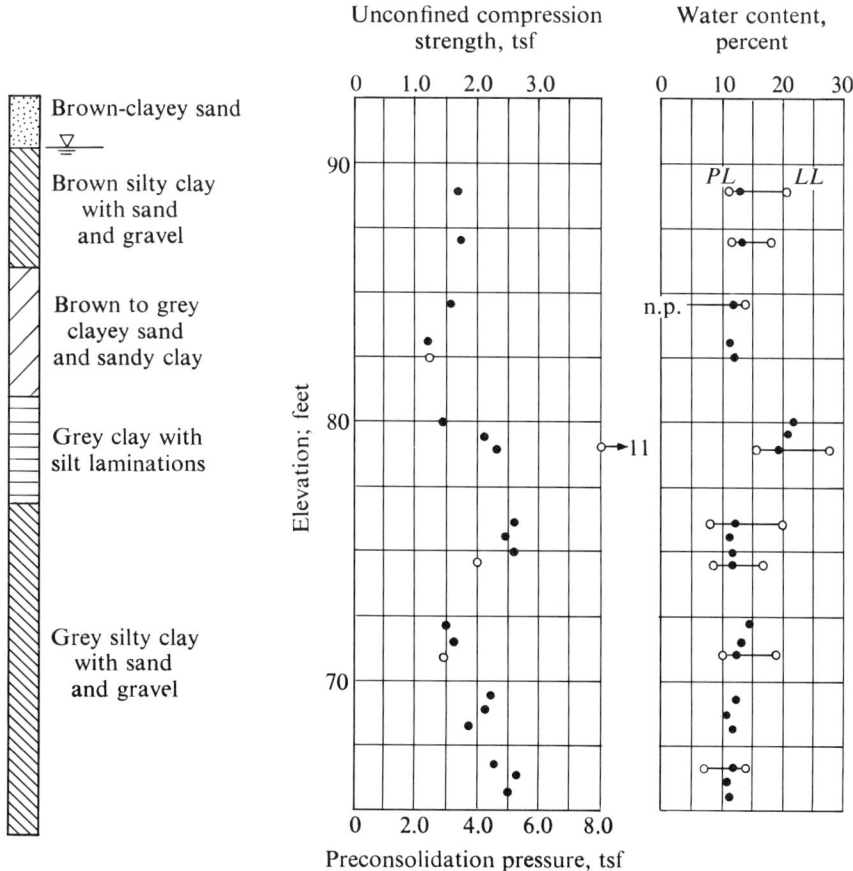

- Unconfined compression strength
- Preconsolidation pressure

FIGURE 13.9 *Physical properties of a glacial deposit in central Michigan.*

as irregular pockets of sand encountered in the borings at various depths. Data on the physical properties of specimens collected from one of these borings are given in Fig. 13.9.

Deposits of glaciofluvial soils may also occur in river valleys (*valley trains*) that once served as drainage outlets for glacial meltwater or locally in the form of ridges (*eskers*) and domes (*kames*) as a result of various complications in the drainage system around glaciers [Lobeck (1939)]. These glaciofluvial soils are composed primarily of silt, sand, and gravel. Some of them are unstratified and others may exhibit irregular stratification.

13.6 Glacial-lake deposits

During Pleistocene time many basins were filled by glacial water to form lakes; many of them, such as the Great Lakes, still exist, although diminished in size. The inflow to a glacial lake consists of glacial meltwater carrying silt and clay. The silt settles first in the quiet water of the lake. In winter, when the lake freezes over and the supply of glacial meltwater and sediment is cut off, the clay settles out. Summer melting brings new supplies of sediment and the cycle is renewed. Glacial-lake deposits therefore exhibit the characteristic stratification of silt and clay called *varved clays*. Figure 13.10 shows a cross section through a varved-clay specimen.

Glacial-lake deposits often exist as normally consolidated clays. Such clays may extend to depths of 100 feet or more. Their low shear strength and high

FIGURE 13.10 *Photograph of a varved clay specimen from Detroit.*

compressibility often create difficult foundation problems. Figure 13.11 shows the boring log and physical properties of a glacial-lake deposit. Since the measured preconsolidation pressure is close to the existing overburden pressure below a depth of about 15 ft, the deposit is normally consolidated. Near the ground surface, desiccation due to seasonal fluctuations in the water table has resulted in overconsolidation of the clay. As many lake levels have been lowered considerably and some were drained completely after the glacial period, many glacial-lake clays became overconsolidated by drying. This substantially increased the strength of the clay.

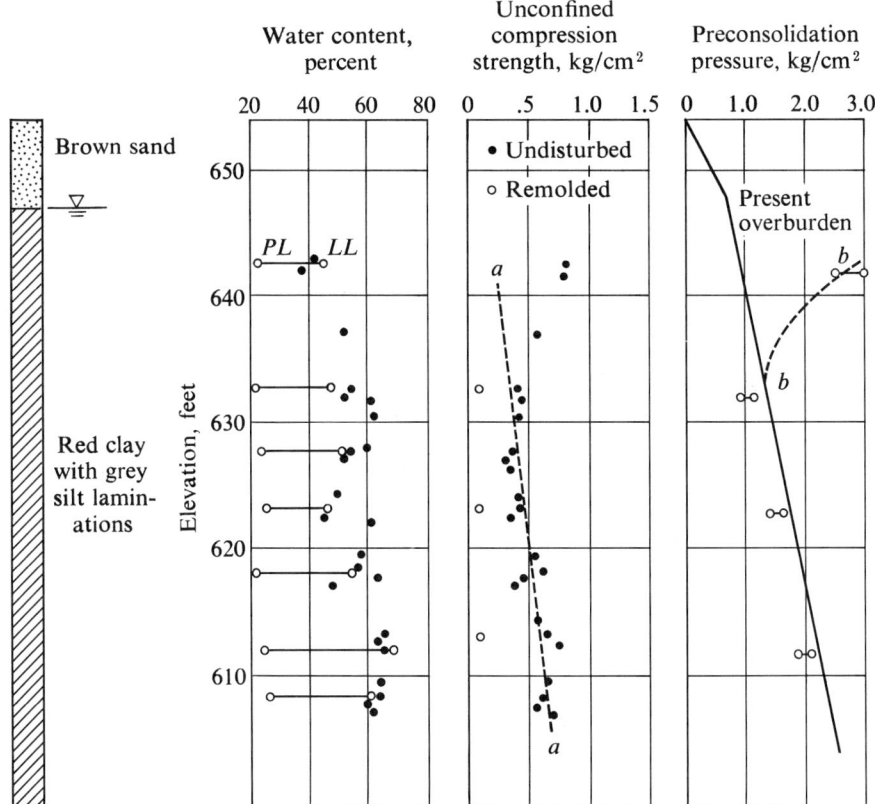

FIGURE 13.11 *Physical properties of a glacial-lake deposit near Sault Ste. Marie, Michigan.*

AEOLIAN DEPOSITS

13.7 Wind-blown sand

Where large areas of sandy soil lie exposed, gusts of wind pick up the sand and drop it at some distance when the wind velocity decreases. The deposits usually take the shape of dunes. The size of the particles transported depends primarily on the wind velocity. Usually the size decreases with the distance from the source; the finest particles are carried to the farthest point. As a result of the sorting action of the wind, dune sands exhibit a high degree of uniformity in particle size (Fig. 13.6). In a typical sand dune, the particles are always being blown over the crest and then rolled down the leeward slope of the dune. As a result, the sand is loose on the leeward side, while on the windward side the force of the wind compacts the sand. The relative density may vary from about 50 percent on the windward slope to almost zero on the leeward slope. Extensive dune sands are found along Atlantic and Gulf coasts and around the Great Lakes. They also occur inland along dry alluvial or rock plains in many parts of the western United States.

13.8 Loess

Loess is a windblown fine silt that occurs in thick layers along the Mississippi Valley and in parts of the western United States. The loess accumulated mostly during interglacial periods when a dry climate prevailed. Strong winds picked up the silt from fresh outwash plains. The broad Mississippi Valley must have been very dry during these periods and was another source of supply. The typical loess in this country is characterized by a system of vertical holes found throughout the soil. These holes, rarely larger than a millimeter in diameter, are believed to be formed by the roots of vegetation that once grew in the soil.

Loess consists primarily of fine-sized particles (Fig. 13.6) cemented together by clay or calcareous material. The cementation binds together the silt particles to form a rigid soil in spite of its unusually low density. Some unusual properties of loess may be attributed to cementation. When the applied stress breaks the cementing bonds, the structure of the soil collapses and the failure resembles that of a brittle material (Fig. 13.12). In regions with a dry climate, the natural

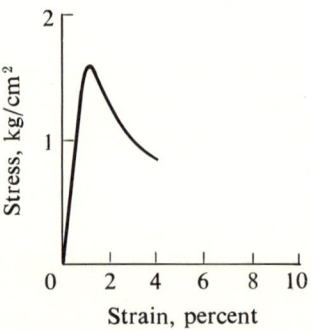

FIGURE 13.12 *Stress-strain curve obtained from an unconfined compression test of a loess specimen.*

water content of the soil is quite low. Any increase in the moisture content usually weakens the cementing bonds. Since the cementing material constitutes a very small fraction of the soil, only a slight increase in moisture content may affect the strength significantly. Under extreme conditions such as saturation or flooding, the structure may break down under the imposed loads and large settlements would result. Figure 13.13 illustrates the drastic change in consolidation behavior brought about by flooding a loess soil. The pressure vs. void ratio curve obtained in a consolidation test on a specimen of loess at its natural water content is represented by the solid curve $0A$. It shows a relatively low compressibility. After loading the specimen to point A, the specimen was completely wetted by submergence in water. The result was a sudden increase in compression as indicated by part AB of the curve. For comparison another specimen was completely wetted prior to loading. The pressure vs. void ratio curve is shown by the dashed curve and indicates a much greater compressibility [Clevenger (1956)].

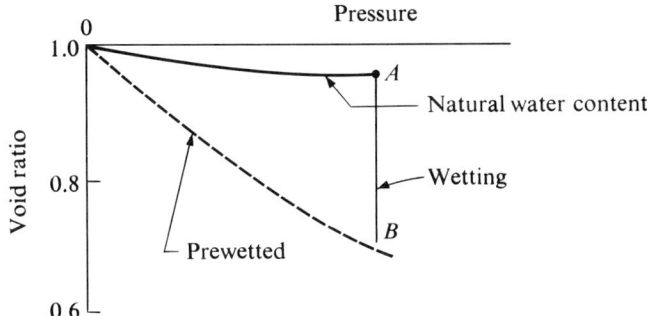

FIGURE 13.13 *Effect of wetting on the consolidation properties of loess.* [After Clevenger (1956).]

ROCKS AND RESIDUAL SOILS

13.9 Residual soils

Chemical weathering of rocks progresses rapidly in areas of warm and humid climate, such as the southeastern United States. If the weathered material remains in place, it forms residual soils. The weathering process may be oxidation. Also, water entering the ground may be acidic or alkaline, and chemical reaction takes place. The properties of a weathered soil depend to a large extent on the parent material and the climate. Figure 13.14 shows a log of a boring in a residual soil derived from weathering of gneiss. The soil profile may be divided into three zones. Above bedrock is the disintegrated gneiss, which represents the incomplete product of weathering. Near the ground surface is the plastic clay resulting from advanced weathering, and in between these one finds the intermediate zone.

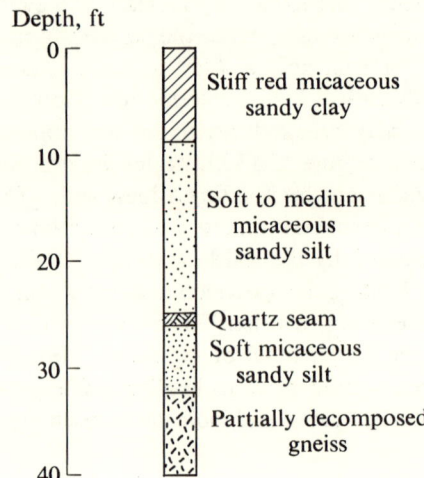

FIGURE 13.14 *A profile of a residual soil deposit.*

Although in a given region weathering of the soil follows a well-established pattern, the depth of weathering varies within extremely wide limits from point to point. It is controlled to a large extent by details in the bedrock, such as the presence of weak and resistant layers, which are most difficult to predict from boring data.

13.10 Decomposed rocks

Sound rocks possess strength that far exceeds the stresses produced by the foundations of structures, and therefore seldom constitute foundation problems. In this section we are primarily concerned with decomposed rocks. These materials have undergone considerable alteration but still retain a great deal of resemblance to the parent material.

Shale, which is a sedimentary rock composed largely of clay-size particles, deteriorates into clay when exposed. The bedding planes in shales are planes of weakness. The shearing resistance along these planes is often much lower than that of intact shale. In addition, shales often contain a secondary structure of fissures. The space between these fissures may vary over a wide range. The fissures may be almost invisible in a confined specimen but may gradually open up after removal of the confining pressure. When such a material fails, it disintegrates along the fissures and behaves as a characteristic brittle material (Fig. 13.15). In western United States and Canada there are extensive shale formations that undergo large volume expansions (swelling) when moisture becomes available. Many of these shales contain appreciable amounts of montmorillonite. Experiences with swelling clays have been described by Holtz and

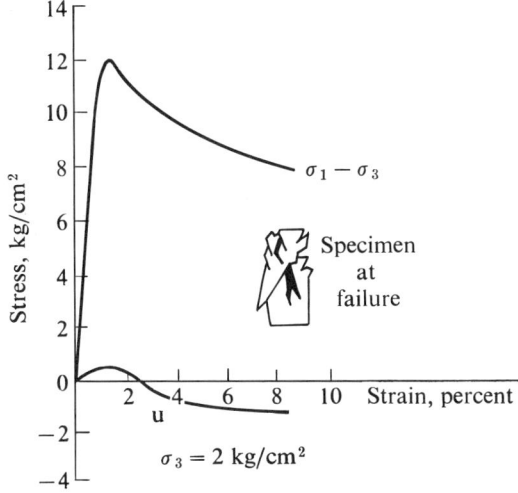

FIGURE 13.15 *Stress-strain curve obtained from consolidated-undrained triaxial test of a clay shale from Garrison, N.D.*

Gibbs (1956), Jennings (1961), and Peterson (1963). Table 13.2 gives the approximate correlation between index properties and swelling for clay shales of western United States.

TABLE 13.2 *Relation between Index Properties and Swelling.* [After Holtz and Gibbs (1954)]

% < 1μ	PI	SL	Vol. exp.(%) from dry to saturation
> 28	> 35	< 11	> 30
20–31	25–41	7–12	20–30
13–23	15–28	10–16	10–20
< 15	< 18	> 15	< 10

Another type of rock that presents characteristic problems of its own is decomposed limestone. Limestone is mostly calcium and magnesium carbonate and is readily soluble in acid. In many localities the groundwater is acidic, and solution of limestone begins with the movement of water through joints and cracks in the rock. A limestone formation in an advanced stage of solution contains many interconnected caverns or underground channels. The strength of such rocks varies from excellent for sound rock to nothing in the cavities. The variation in strength is entirely random, depending upon the location of sound and decomposed rocks. Figure 13.16 shows a boring log which reflects the erratic variations in the physical properties of a limestone formation. The boring went through the clay derived from weathering of the limestone and then encountered a cavity. Sound limestone was found at a depth of 61 ft.

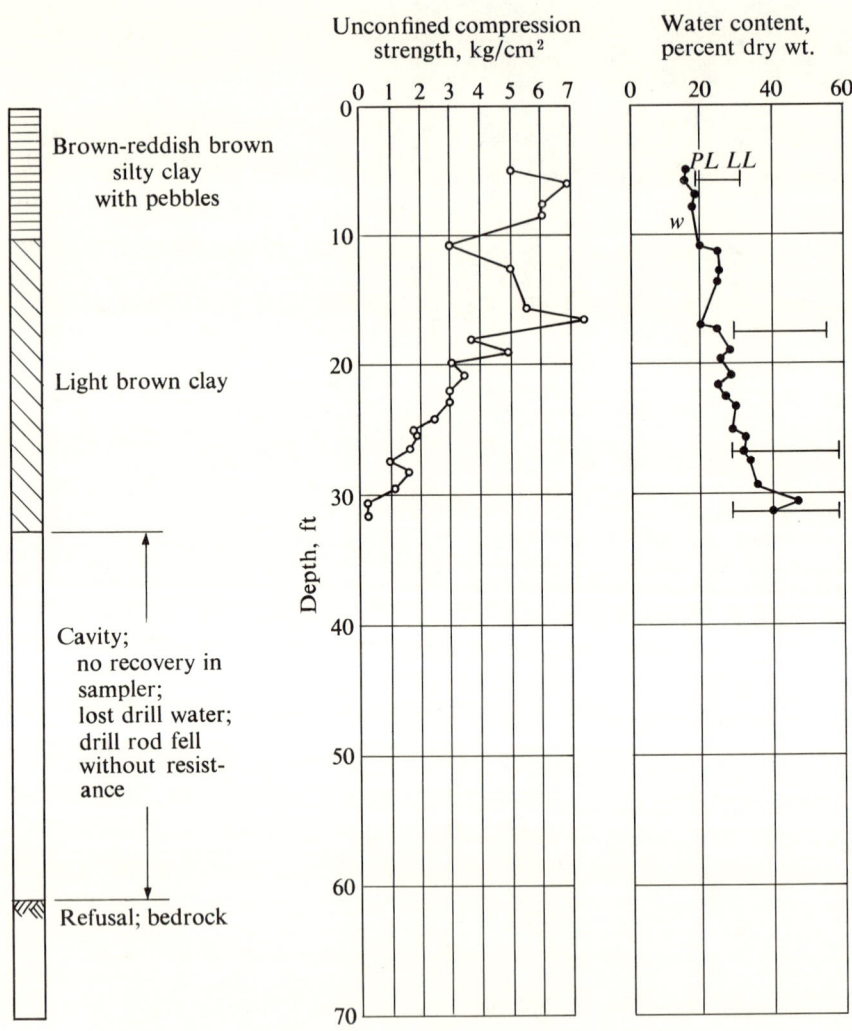

FIGURE 13.16 *Physical properties of a residual soil deposit in Georgia.*

EXPLORATION OF SOIL DEPOSITS

Subsoil exploration usually requires borings from which soil samples may be obtained. The types of specimens may vary from completely disturbed ones that can be used only for classification and identification to carefully taken undisturbed samples for elaborate laboratory tests. The borings may be supplemented by other in situ tests such as sounding and geophysical measurements. The following paragraphs describe the principal features of the different operations.

13.11 Borings

Borings are usually made to permit sampling of the subsoil. There are many types of borings but in general they may be classified as auger borings or wash borings. An auger boring is advanced by means of a screw drill which is rotated by a motor or by hand. The soil is held to the auger and brought to the surface. The samples recovered are badly disturbed but they still contain most of the constituents of the undisturbed soil. They are representative insofar as texture is concerned. In making a wash boring, a casing is first driven into the ground. The soil inside the casing is then broken up by a chopping bit and the material is removed by the circulation of wash water. The soil that is removed from the borehole by the wash water is of course broken up, and loses all the structure and stratification that may be present in the undisturbed material. It can only give a vague indication of what the undisturbed soil may be like. In fact, washings often are not even representative of the undisturbed soil, since coarser particles may temporarily settle to the bottom of the borehole, leaving only the fines to be carried out by the water.

13.12 Sampling

Since the various boring methods yield, at best, disturbed soil samples, it is necessary to employ specially designed sampling devices to retrieve good soil specimens from the bottom of boreholes advanced by one of the methods described in the preceding paragraph. There are a wide variety of sampling devices but they may be classified into two types according to the quality of specimens that they recover. A soil specimen may be somewhat disturbed but still contain all the constituents of the natural soil. It is called a *representative specimen*, as it conveys an accurate picture of the soil. Specimens extracted by means of specially designed samplers may be only very slightly disturbed. The effect of sampling on their strength and compressibility may be quite small; these are usually called *undisturbed samples*. It should be obvious that truly undisturbed specimens do not exist, and the best techniques can yield only slightly disturbed specimens.

A device used extensively to obtain representative specimens is the splitspoon sampler shown in Fig. 13.17(a). The sampler is attached to the end of the drill rod and driven into the intact soil at the bottom of the borehole. The sampler catches a core of soil and is brought up. The spoon consists of a tube split into two halves and held together by couplings at both ends. By unscrewing the couplings the tube is opened and the sample can be removed from the tube. A representative specimen is placed in a jar and shipped to the laboratory. Such specimens may be considered representative specimens, since the original materials and stratifications are usually retained. They are, however, considerably disturbed due to handling and the dimensions of the sampler. It has been well established that the degree of disturbance is closely related to the area ratio of the sampler, which is defined as

$$A_r(\%) = 100 \frac{D_e^2 - D_i^2}{D_i^2} \qquad (13.1)$$

in which D_e and D_i are the external and internal diameters of the sampler [Hvorslev (1949)]. The area ratio of the split-spoon sampler is equal to 110 percent, which results in severe disturbance. For good undisturbed samples, the area ratio should be not greater than 10 percent. The transfer of the specimen from the sampler to jar also causes a slight loss in the moisture content. Nevertheless, in spite of these shortcomings, the samples are useful in providing a good indication of the characteristics of the soil. They may be used for determination of Atterberg limits and particle-size distribution and the approximate values of unconfined compression strength and water content.

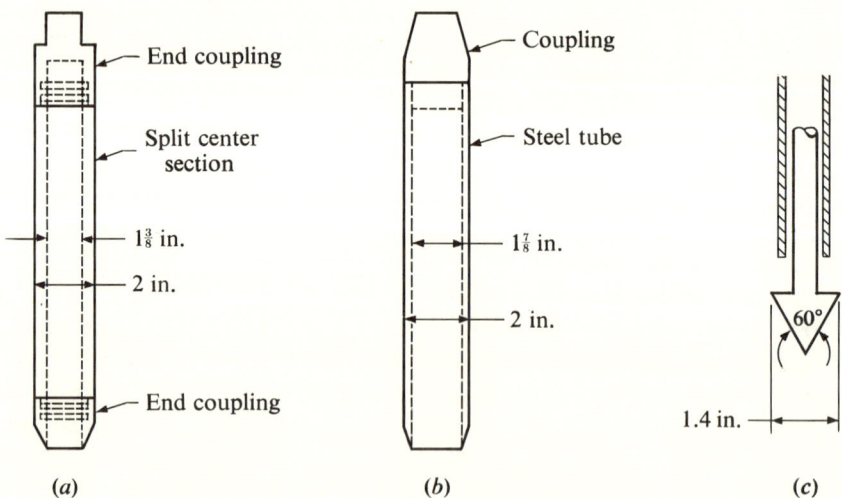

FIGURE 13.17 *Sounding and sampling devices.*

Undisturbed specimens are usually taken in seamless, thin-walled steel tube samplers. The tube's diameter may range between 2 and 6 in. and the wall thickness is usually not more than $\frac{1}{16}$ in. [Fig. 13.17(*b*)]. The tube sampler is attached to the drill rod by means of a coupling and is pushed into the undisturbed soil at the bottom of the borehole. Upon removal from the borehole the tube is sealed at both ends with wax and the entire tube is shipped to the laboratory, where the sample is extruded from the tube for the appropriate tests. Since the specimen is sealed in a tube, it retains its original volume and moisture content. The thin-walled tube sampler gives very satisfactory results when used on cohesive soils. Difficulty is encountered in sampling cohesionless soils, especially when they are located below the water table, as the soil tends to fall out of the tube while it is being removed from the borehole. This difficulty may sometimes be overcome by the use of the piston sampler, which exerts a vacuum on the top of the specimen.

13.13 Soundings

Soundings measure the resistance of a soil in situ against penetration by a standard device. This resistance usually gives some indication of the strength and compressibility of the soil. There are a large variety of sounding devices in use but only two common types are mentioned here.

The most widely used sounding method in this country is the standard penetration test. This method utilizes the split-spoon sampler described in Sec. 13.12. The spoon is driven into the soil at the bottom of a borehole by means of a 140-lb weight dropping a distance of 30 in. The number of blows needed to advance the sampler 1 ft is recorded as the penetration resistance N. The method has the advantage that it can be carried out in a boring operation and also that representative soil specimens can be obtained at the same time. The fact that the sounding is carried out at the bottom of the borehole eliminates the effect of friction on the drive rod.

A sounding method very popular in Europe is the cone-penetration test. In this test, a cone of standard dimensions [Fig. 13.17(c)] is attached to the end of a rod and is driven or forced into the ground. After the cone has reached a given depth the penetration resistance at that depth is measured by forcing the cone into the soil at a steady rate. The static force required to accomplish this is the penetration resistance.

Soundings, besides providing a qualitative and relative description of subsoil properties, can be correlated with significant physical properties such as density and shear strength. Since all such correlations are entirely empirical, their reliability is dependent upon the amount of data that has been collected. Furthermore, these empirical relations are affected by a great many factors, such as the soil type, the moisture content of the soil, and the depth at which the sounding is made. Since only the most important factors can be taken into account, these correlations are always approximate in nature. So far, relatively consistent correlation has been obtained only for cohesionless soils.

The correlation suggested by Peck et al. (1974) is shown in Fig. 13.18 as curve A. The relationship between penetration resistance and relative density has been studied by Gibbs and Holtz (1957). Using their results and data on the relationship between relative density and angle of internal friction [Burmister

(1948), Wu (1957)], a range in ϕ, shown conservatively as B, is obtained. The substantial difference between the two curves illustrates the approximate nature of the correlation.

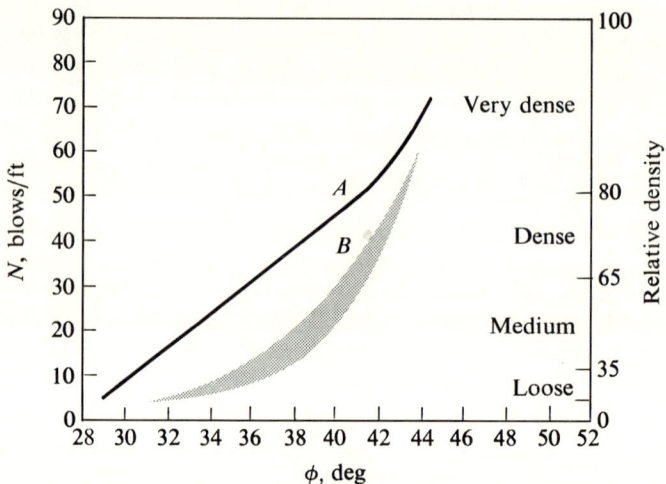

FIGURE 13.18 *Relationship between penetration resistance and angle of internal friction of cohensionless soils.* [After Peck et al. (1974).]

A correlation between the standard penetration and cone penetration is given in Table 13.3.

TABLE 13.3[a] *Correlation between Standard and Cone Penetration Tests*

	Standard penetration resistance, blows/ft	Static cone resistance, tsf
Very loose	Less than 4	Less than 20
Loose	4–10	20–40
Medium	10–30	40–120
Dense	30–50	120–200
Very dense	Over 50	Over 200

[a] After Meyerhoff (1956).

13.14 Geophysical methods

Geophysical methods may be used to measure some in situ property that is related to the significant engineering properties of soils. Common geophysical methods include those that measure the velocity of wave propagation through soils and the electrical resistivity of soils. The wave velocity in a homogeneous or layered medium is theoretically related to the Young's modulus and Poisson's ratio of the material [see Ewing et al. (1957)]. Therefore these values can be calculated if the wave velocity is measured in the field. To measure the wave velocity a shock may be induced at a source point while seismographs located at specific distances from the source record the time required by the wave to travel the given distances.

The electrical resistivity of soils is determined by measurement of the current between two points on the ground under an applied voltage. The electric resistivity of different soils has been investigated empirically and typical values are listed in Table 13.4. The nature of stratification in the subsoil can be explored by recording resistivity between two points while the distance between the points is increased successively. If the subsoil is homogeneous, the resistivity increases uniformly with the distance between the points. The presence of a pocket or layer of a different soil changes the resistivity pattern drastically. Electrical-resistivity methods are therefore very useful in exploring major variations in subsoil stratification [Barnes (1952)].

TABLE 13.4 *Typical Values of Electrical Resistivity of Soils*[a]

Soil type	Specific resistivity, ohm-cm
Clay and satd. silt	0–10,000
Sandy clay and wet silty sand	10,000–25,000
Clayey sand and satd. sand	25,000–50,000
Sand	50,000–150,000
Gravel	150,000–500,000

[a] After Barnes (1952).

13.15 Subsoil exploration program

The determination of the properties of natural soil deposits for engineering projects always requires a detailed subsoil exploration program. The program may range anywhere from soundings from a number of boreholes, to a careful collection of the best undisturbed soil specimens for elaborate laboratory tests. Sometimes these may be followed by large-scale field tests. The nature of the program is governed not only by the project to be undertaken but also, to a very large extent, by the properties of the soil deposit itself.

The first objective of subsoil exploration is to obtain a reliable picture of the type of soil that underlies a given site. This preliminary information can often be secured by borings from which representative but disturbed specimens can be obtained for visual examination and classification. This may be accompanied by geophysical explorations if the conditions so dictate.

The next step depends on the type of information we seek. If the project requires an evaluation of the load-carrying capacity of the soil, the preliminary study may be followed by soundings that reflect the density and stiffness of the soil, sampling for laboratory measurement of strength and compressibility, and field-loading tests to confirm the laboratory data. To determine the permeability of a deposit requires undisturbed specimens for permeability tests. This may be followed by field pumping tests from wells.

The project's size often places an upper limit on the scope of the exploration program. Large and important projects can afford an elaborate program that may yield considerable savings through economic design based on accurate knowledge of subsoil conditions. On the other hand, small undertakings have limited funds for soil investigation, and the incompleteness of information must be counterbalanced by a large factor of safety.

The character of the subsoil often dictates the nature of the subsoil exploration program. For example, a uniform soil deposit needs only a relatively small number of borings to establish a clear picture of subsoil conditions. The remaining funds then may be concentrated on careful and elaborate laboratory tests. The laboratory results can be applied to design with confidence, since they are representative of the soil conditions. Figure 13.19 shows the subsoil profile of a glacial-lake deposit in northern Ohio. The deposit is characterized by its uniform soil properties. Unconfined compression tests on specimens from eight borings at four different sites give remarkably consistent results. With few exceptions, the unconfined compression strength at the four widely scattered sites fall within the narrow range of 0.5 to 0.8 kg/cm^2.

A soil deposit whose properties are extremely variable and erratic cannot justify an extensive laboratory test program on specimens from a few borings, since the specimens may not be at all representative of the soil deposit. In such a case, the available funds should be invested in soundings and borings without undisturbed sampling. Their number should be large, in order to increase the probability of locating weak spots or dangerous areas that may exist at the site. Only limited funds should be expended on laboratory tests, and these should be used to measure the properties of the weakest soils.

A typical example of a subsoil exploration program designed to suit the special conditions of the site is given in Fig. 13.20 [Peck (1953)]. The subsoil consisted of a deposit of dense sand and gravel underneath a loose fill of uncertain extent, and it was considered desirable to place the foundations on the gravel. Preliminary information indicated that the gravel deposit was well capable of supporting the foundation loads, while the loose fill was unsuitable as a foundation material. Hence, the first important question to be answered was the extent of the loose fill. The subsoil exploration program was designed to provide accurate information on the boundary between the fill and the natural soil and the relative density of the sand and gravel deposit, which may be quite erratic in nature. The most suitable means of soil exploration for these purposes would be a type of sounding. Because of the presence of large boulders (up to 3 in. in diameter) in the sand and gravel deposit it was expedient to use a specially designed cone with a diameter of 2.5 in. instead of the split spoon used in

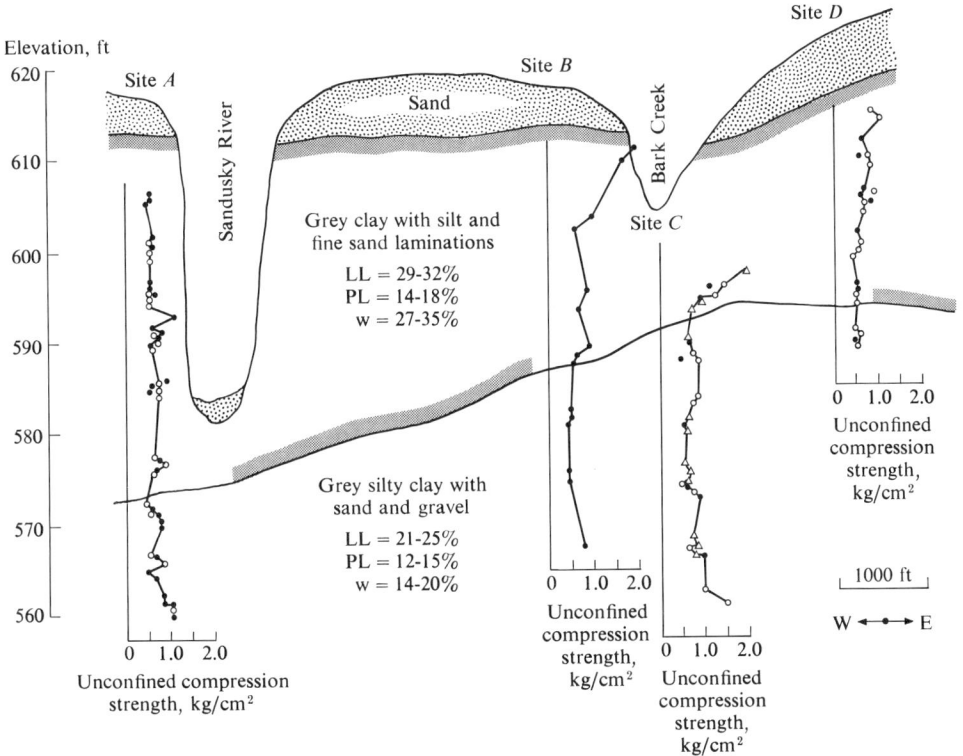

FIGURE 13.19 *A glacial-lake deposit in northern Ohio, showing a high degree of uniformity.*

standard penetration tests. The cone was driven with a weight of 350 lb falling 2 ft. Soundings were made at 50 locations. Two typical logs are plotted in Fig. 13.20. The sounding resistance at footing 20 showed clearly that the boundary between the loose fill and the dense sand and gravel was at elevation -8.0. Below this the penetration resistance varied erratically. At footing 14, the penetration resistance showed that the dense material at elevation -8.0 had been removed, and the fill extended to about elevation -24.0. The dashed curve in Fig. 13.20 represents the boundary as deduced from the soundings. This finding resulted in the decision to locate the structure about 50 ft east of the proposed location. This made it possible to place all footings on the dense material at an elevation of -8.0. Several standard penetration tests were also made so that the cone-penetration resistance could be correlated with the standard penetration resistance.

This case illustrates the important role of a rather simple sounding method in mapping out the major features of the subsoil.

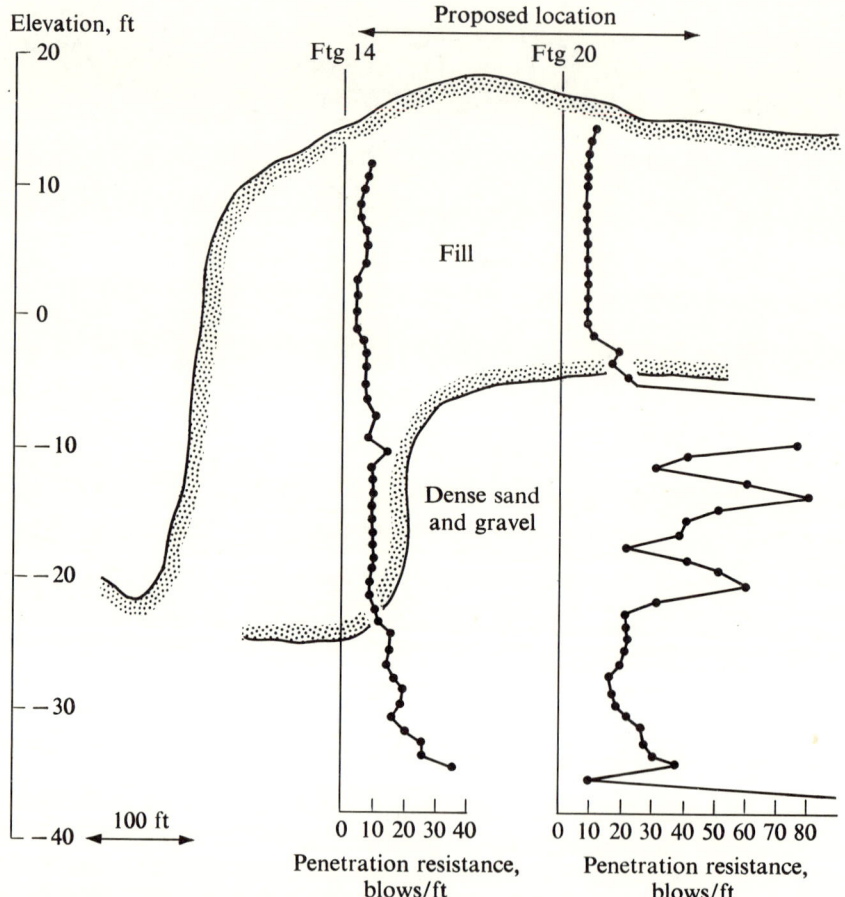

FIGURE 13.20 *Subsoil conditions at the site of the Denver coliseum.*

14

Properties of the Clay Fraction

14.1 Introduction

As defined in Chapter 1, the fraction of the soil that is finer than 2 microns in size is called *clay*.* In this definition, clay denotes soil particles in a certain size range. Often the name *clay fraction* is used for this purpose. The word "clay" is also used to designate a group of minerals that occurs in nature as an "earthy, fine-grained material which develops plasticity when mixed with a limited amount of water" [Grim (1953)]. They consist of very small crystalline particles. Thus the clay fraction of a soil may contain clay minerals as well as nonclay minerals such as finely ground quartz, feldspar, or mica of "clay" size.

Most clay minerals exist as sheets whose thickness is very small relative to the width and length of the sheet. Figure 14.1 shows the shape of kaolinite sheets as photographed through an electron microscope. The shape of the kaolinite particles is hexagonal. For sheet-shaped particles the surface area is so large compared to the volume that their behavior is governed primarily by surface forces. Many soil properties such as plasticity and cohesion are attributed to properties of the clay minerals in the soil. A study of the fundamental laws governing the clay behavior is essential to a better understanding of many properties of cohesive soils.

14.2 Atomic bonds

When we study the behavior of clay minerals we find that the predominant influence is the force that exists between the clay particles. We therefore begin

---

* This line of demarcation is somewhat arbitrary. Many have used 1 micron as the limit, and the U.S. Dept. of Agriculture uses 5 microns.

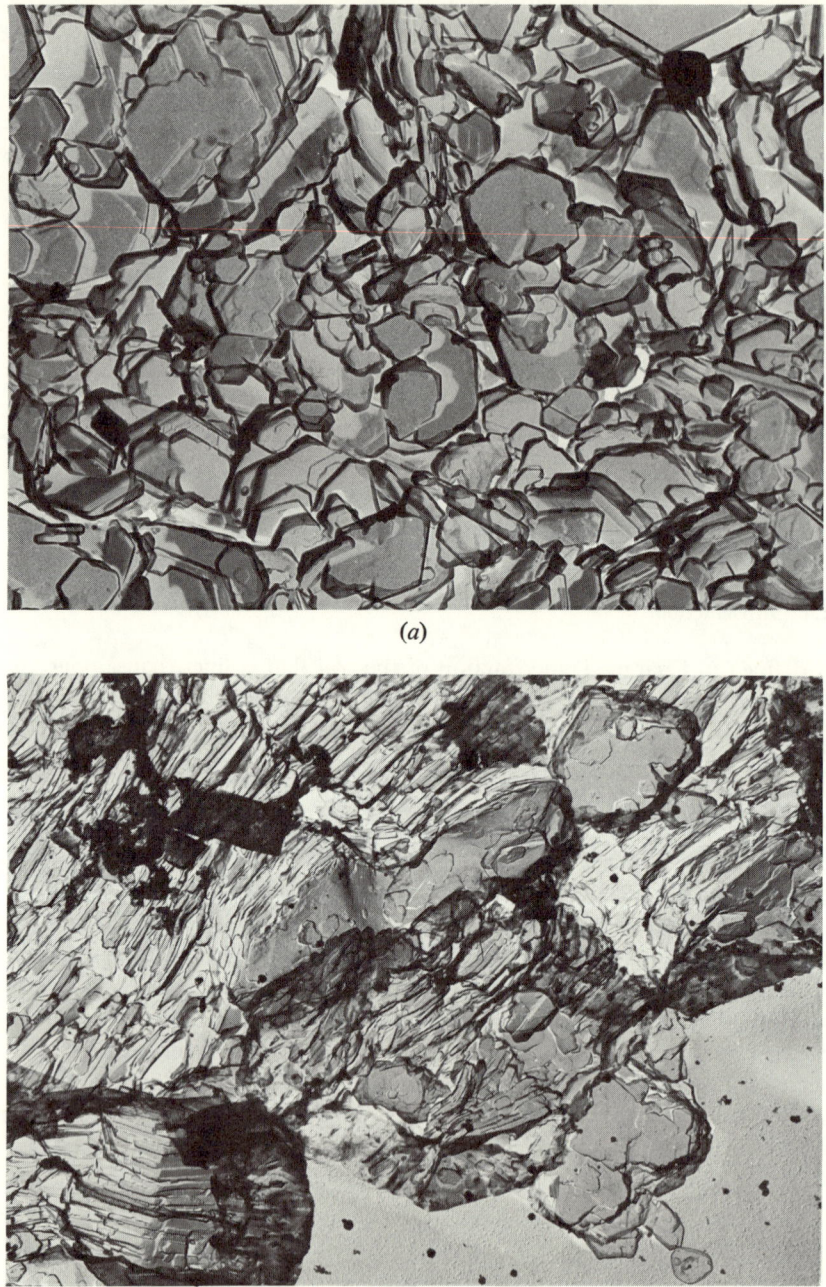

FIGURE 14.1 *Electron micrograph of kaolinite particles: (a) nonaggregated* (15,000× *before 30 percent reduction); (b) aggregated* (32,000× *before 40 percent reduction).* (Courtesy of Georgia Kaolin Co.)

with a review of the types of bonds that hold atoms together. Atomic bonds may be divided into two groups, primary or high-energy bonds and secondary or low-energy bonds. Under the former we may list ionic bonds and covalent bonds, while van der Waals' forces belong to the second group. Hydrogen bonds are considered as intermediate between primary and secondary bonds. Other bonds also exist but we shall describe here only the above four, as they are important to the study of clay-mineral behavior.

Ionic bonds are those that join two elements with incomplete outer-electron shells. The atom of one element loses the electron or electrons in its outermost shell to an atom of the second element. An example is sodium chloride. The sodium atom has only one electron in its outermost shell and the chlorine atom has seven. The sodium atom loses its single electron in the outermost shell and completes the outermost shell of the chlorine atom.

Covalent bonds occur when two atoms each are lacking one or more electrons in their outermost shell. These two atoms then combine by sharing the electrons in the outermost shell. An example is the combination of two oxygen atoms to form the O_2 molecule.

Hydrogen bonds connect the hydrogen cation H^+ to an ion such as oxygen, O^{2-}. A hydrogen bond is formed when the hydrogen of a water molecule is attracted to the oxygen of a neighbor water molecule. This produces a bond between the two water molecules. The hydrogen bond is very weak compared to ionic and covalent bonds. Hydrogen bonds also exist in other molecules such as ammonia, NH_3.

Van der Waals' forces arise when there is attraction between two dipolar molecules. Electrically neutral molecules may have their positive and negative charges distributed unsymmetrically, resulting in a positive and a negative pole. The poles may be attracted to charges of the opposite sign or to the opposite pole of another dipolar molecule.

Van der Waals' forces also exist between induced dipoles, which are formed when a nonpolar molecule is subjected to an electric field of a neighbor molecule. The electric field displaces the positive and negative charges in the nonpolar molecule and it behaves as a dipole.

14.3 Clay minerals

There are many varieties of clay minerals but most of them are made up of two basic units, the octahedral and the silica sheets. The octahedral unit consists of two layers of oxygens or hydroxyls with aluminium atoms embedded in between. The crystal structure of the octahedral sheet $[Al(OH)_3]$ is shown in Fig. 14.2(a). The aluminum may be replaced by magnesium or iron. When magnesium is present to form $Mg_3(OH)_6$, the mineral is called *brucite*. The silica sheet (SiO_2) is made up of tetrahedrons, each consisting of one silicon atom held between four oxygen atoms, as illustrated in Fig. 14.2(b).

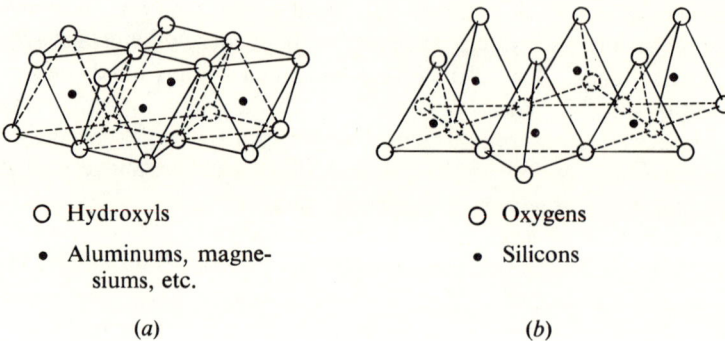

○ Hydroxyls
● Aluminums, magnesiums, etc.

○ Oxygens
● Silicons

(a)   (b)

FIGURE 14.2 *Structure of octahedral and silica sheets.* [After Grim (1953).]

Most clay minerals are the product of the chemical weathering of rock-forming minerals such as feldspar and mica. The three major *clay-mineral groups* are kaolinite, illite, and montmorillonite. The major characteristics of the crystal structure of the groups are shown in Fig. 14.3. Other common clay minerals are halloysite, chlorite, and vermiculite. The atoms in the octahedral and silica crystals are held tightly together by ionic and covalent bonds. Therefore, these bonds are usually not broken under stresses encountered in engineering problems.

In the silica sheet [Fig. 14.2(b)] the silicon Si^{4+} is bonded to 4 oxygen O^{2-} atoms. Each oxygen in the bottom layer is bonded to 2 silicon atoms and has a complete shell of 8 electrons. However, oxygen atoms in the top layer lack one electron to complete the shell. In the kaolinite mineral, the inverted silica sheet is stacked on top of the octahedral sheet [Fig. 14.3(a)]. The oxygens at the tip of the tetrahedrons replace the OH^- of the octahedral sheet. In this position these oxygens complete their electron shell by accepting one from the aluminum atom in the octahedral sheet. One sheet of kaolinite is about 7.2 Å thick. The illite mineral consists of an octahedral sheet between two silica sheets [Fig. 14.3(b)]. Some of the silicon atoms in the tetrahedron are replaced by aluminum. This is called *isomorphous substitution*. Potassium ions are held between the sheets. In weathered illites these may be partially replaced by other ions, such as calcium or magnesium. The thickness of an illite sheet is about 10 Å. The montmorillonite mineral has the same basic structure as the illite except for the fact that no potassium exists between sheets [Fig. 14.3(c)]. In addition, iron and magnesium often replace aluminum in the octahedral sheet. The spacing between unit sheets of montmorillonite is variable and may increase from 9.6 Å to almost complete separation of the sheets when large amounts of water are adsorbed to the surfaces.

We can see how the differences in the crystal structure of the minerals lead to very different properties. In kaolinite the oxygen atoms in the basal plane of the octahedral sheet are shared with the silica sheet. The bonding between

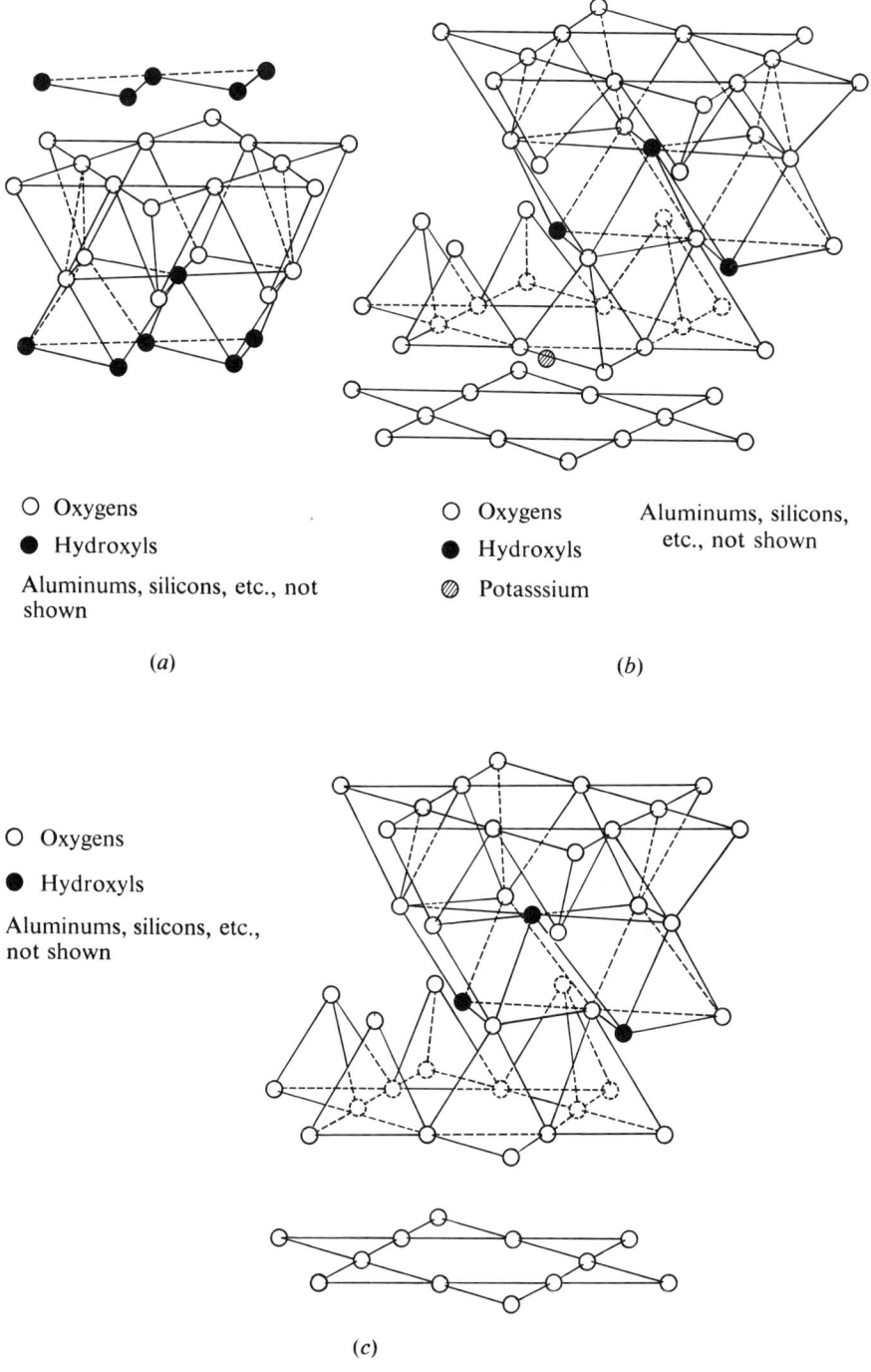

FIGURE 14.3 *Structure of clay mineral groups:* (*a*) *kaolinite;* (*b*) *illite;* (*c*) *montmorillonite.* [After Grim (1953).]

the octahedral sheet and the silica sheet is therefore of the ionic and covalent types. On the other hand, when one kaolinite sheet is stacked over another, the hydroxyls of the octahedral are attached to the oxygen of the silica by means of hydrogen bonds. Such bonds are weaker than primary bonds and the two kaolinite sheets may become separated.

In the montmorillonite structure, the octahedral sheet is bonded to silica sheets on both sides. When two montmorillonite sheets are stacked on top of each other, the oxygens of each sheet are opposite to each other and only secondary bonds are present. These are very weak and the sheets are easily separated. The space in between would be occupied by water molecules and cations. Experiments have shown [Bradley and Grim (1948)] that at high water content montmorillonite sheets may become completely separate.

14.4 Repulsive potential

The clay minerals carry a net negative charge. This phenomenon has been attributed to the following reasons. In the ideal crystal, the positive and negative charges are balanced. However, at the edges of the sheets, the continuity of the structure is broken, resulting in unbalanced charges. Usually these broken bonds produce a net negative charge for the clay particle, but along the broken edges local concentration of positive charges often occurs. A second cause is isomorphous substitution of the silicon in the tetrahedron sheet by aluminum or other ions of lower valence. Magnesium and iron may replace aluminum in the octahedral sheet. A third cause is the dissociation of the hydrogen of hydroxyls exposed around the broken edges if the clay exists in an alkaline solution.

The negatively charged clay sheets create an electric field around them. The negative charges of the clay are balanced by cations such as Na^+ and Ca^{2+}. These are held (adsorbed) to the clay by electrostatic attraction. If the clay particle is surrounded by water, the cations would possess considerable mobility at large distances from the clay but would be tightly held at close range. The negatively charged clay plate and the positively charged cation cloud is called the *diffuse double layer* or simply the *double layer*.

REPULSIVE POTENTIAL. The field intensity and the cation concentration may be described by the Guoy-Chapman equation. In Fig. 14.4(a) the clay is represented by a sheet of infinite extent, and the negative charges are assumed to be uniformly distributed over this surface. One considers a small element of volume dV which contains within it a number of positive charges. If the positive charges have a valence v and concentration n, then the charge density ρ is equal to $v\epsilon n$, where ϵ is an electrostatic unit of charge. Since 4π lines of force per unit area emanate from each point charge, the change in the field intensity F in the distance dx is given by

$$\frac{dF}{dx} = -\frac{4\pi\rho}{\lambda} = -\frac{4\pi v\epsilon n}{\lambda} \tag{14.1}$$

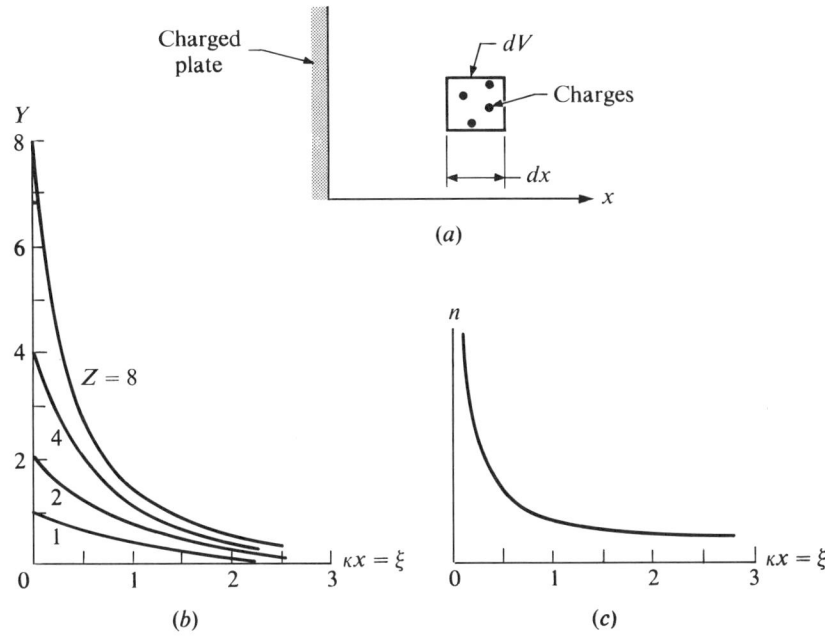

FIGURE 14.4 *Potential and charge distribution in the double layer.* [After Verwey and Overbeek (1948).]

in which λ is the dielectric constant of the medium. Equation (14.1) may also be written

$$\frac{d^2\Psi}{dx^2} = -\frac{4\pi v \epsilon n}{\lambda} \tag{14.2}$$

in which Ψ denotes the electric potential. The relative concentration n/n_0 of charges corresponding to energy levels E and E_0 respectively, are given by the Boltzmann equation:

$$\frac{n}{n_0} = e^{(E_0 - E)/kT}$$

In this expression, T is the absolute temperature, and k the Boltzmann constant. If we let n_0 be the concentration of the cations at zero potential (which occurs at infinite distance from the clay), then E_0 is 0 and Boltzmann's equation gives the ratio of n to n_0 as

$$n = n_0 e^{-E/kT} = n_0 e^{-v\epsilon\Psi/kT} \tag{14.3}$$

Substituting this in Eq. (14.2), one gets the differential equation

$$\frac{d^2\Psi}{dx^2} = -\frac{4\pi}{\lambda} v \epsilon n_0 e^{-v\epsilon\Psi/kT} \tag{14.4}$$

The solution of this equation for a single particle, which represents the case of a clay particle in a thin suspension, in which the neighboring particles are far away and exert no influence, may be obtained with the following boundary conditions. The potential at the surface is Ψ_0, while at very large distance, it is 0; therefore, when

$$x = 0, \quad \Psi = \Psi_0; \qquad x = \infty, \quad \Psi = 0$$

The solution is

$$e^{Y/2} = \frac{e^{Z/2} + 1 + (e^{Z/2} - 1)e^{-\zeta}}{e^{Z/2} + 1 - (e^{Z/2} - 1)e^{-\zeta}} \tag{14.5}$$

in which

$$Y = v\epsilon\Psi/kT \qquad \zeta = \kappa x$$

$$Z = v\epsilon\Psi_0/kT \qquad \kappa = \frac{8\pi n\epsilon^2 v^2}{\lambda kT}$$

For large values of x, Eq. (14.5) may be approximated by

$$Y = 4e^{-\zeta}$$

or

$$\Psi = \frac{4kT}{v\epsilon} e^{-\kappa x} \tag{14.6}$$

Hence at some distance from the clay surface, the potential decays exponentially with x. The solution for Ψ is plotted graphically in Fig. 14.4(b), and Fig. 14.4(c) shows the relationship between the cation concentration n and x.

In the case of a clay aggregate where the clay particles are close together, one may solve Eq. (14.4) for two parallel plates at distance $2d$ apart [Fig. 14.5(a)]. The boundary conditions are

$$\text{when } x = 0, \quad \Psi = \Psi_0; \qquad x = d, \quad \Psi = \Psi_d, \quad \frac{d\Psi}{dx} = 0$$

The solution is obtained numerically. The results are presented in Fig. 14.5(b), which shows the potential Ψ_d midway between two parallel plates as a function of the distance d.

REPULSIVE PRESSURE. From these results we can find the repulsive pressure between two plates [Fig. 14.6(a)]. At any point between the plates, the force exerted by the field on the charges is $\rho(d\Psi/dx)$. This quantity is a function of x, since ρ and $d\Psi/dx$ are both functions of x. In order to maintain equilibrium with $\rho(d\Psi/dx)$, which varies with x, this force must be balanced by a change in the water pressure p. Hence, we have

$$\frac{dp}{dx} + \rho \frac{d\Psi}{dx} = 0 \tag{14.7}$$

Substituting in Eqs. (14.1) and (14.4) we have

$$\frac{dp}{dx} - \frac{\lambda}{4\pi}\left(\frac{d^2\Psi}{dx^2}\frac{d\Psi}{dx}\right) = 0$$

from which we obtain

$$p - \frac{\lambda}{8\pi}\left(\frac{d\Psi}{dx}\right)^2 = \text{constant}$$

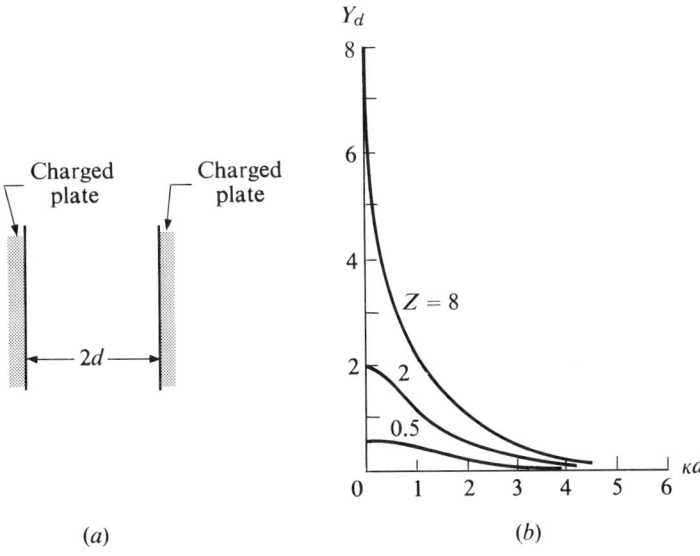

FIGURE 14.5 *Potential midway between two charged plates.* [After Verwey and Overbeek (1948).]

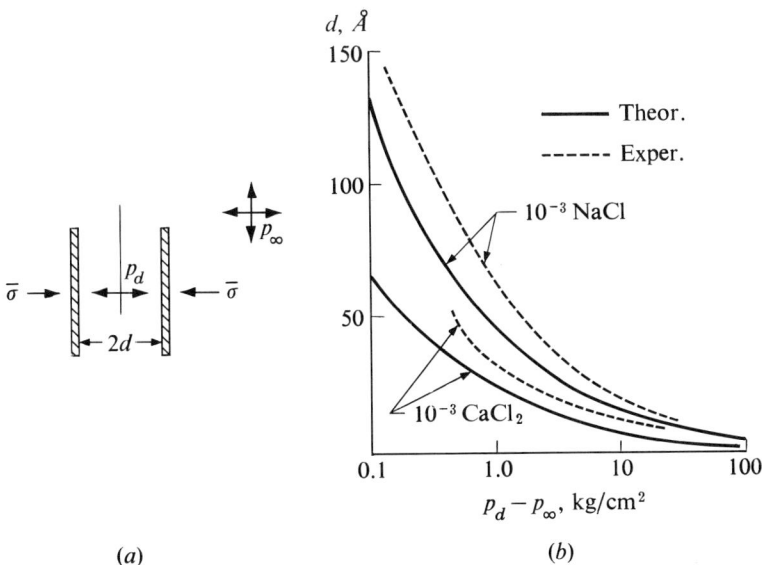

FIGURE 14.6 *Repulsive pressure midway between two clay sheets.* [After Bolt (1956).]

We note that midway between the two plates, $x = d$, $d\Psi/dx = 0$. Hence

$$p - \frac{\lambda}{8\pi}\left(\frac{d\Psi}{dx}\right)^2 - p_d = 0 \qquad (14.8)$$

in which p_d denotes the pressure at $x = d$. If p_∞ is the hydrostatic pressure in the solution at infinite distance away, then the difference between p_d and p_∞ is the net pressure keeping the two plates apart. This pressure tends to push the plates farther apart unless an external stress of equal magnitude is applied to keep the two plates in this particular position.* The term $p_d - p_\infty$ is evaluated by integration of Eq. (14.7) between the limits of $\Psi = \Psi_d$ and $\Psi = 0$. Therefore,

$$p_d - p_\infty = \int_{\Psi=0}^{\Psi=\Psi_d} dp = -\int_0^{\Psi_d} \rho\, d\Psi = 2nkT\left(\cosh\frac{ve\psi_d}{kT} - 1\right) \qquad (14.9)$$

Equation (14.9) is plotted graphically in Fig. 14.6(b). It demonstrates the relationship between the repulsive pressure $(p_d - p_\infty)$ and the distance d between the two plates for n_0 equal to 10^{-3}. It shows the two curves obtained for NaCl ($v = 1$) and CaCl$_2$ ($v = 2$). A complete treatise may be found in Verwey and Overbeek (1948).

There are a number of important objections to the application of the double-layer theory to clays. For example, the energy of the water molecules adsorbed to the clay and to the cations (hydrated water) is not considered. Other points were discussed by Low (1959a). Also, the Guoy-Chapman equation can be solved for a constant charge on the clay surface instead of a constant surface potential. This may be a better approximation to the behavior of the clay. The effect of this and other factors was evaluated by Bolt (1955). Nevertheless, the theory is useful as a concept that enables us to outline the various factors affecting the forces between clay particles. It cannot be relied upon to provide accurate quantitative results on soil properties.

The several factors that affect the repulsive potential are illustrated schematically in Fig. 14.7. The two factors considered are the cation con-

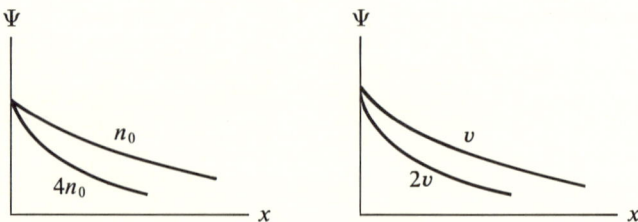

FIGURE 14.7 *Effect of ion concentration and valence on the repulsive potential.* [After Lambe (1958a).]

---

* This external stress is discussed in Sec. 14.11.

centration in the porewater n_0, and the valence of the cation v. If the potential is the same at the surface of the clay particle, we see that increasing the concentration or the valence of the cation reduces the potential.

14.5 Attractive potential

The attractive forces between clay particles have been attributed largely to van der Waals' forces [Verwey and Overbeek (1948), Lambe (1953)]. The attraction between two induced dipoles may be described as follows. A dipole (Fig. 14.8) has a dipole moment equal to ϵl, in which ϵ is the positive or negative charge and l is the distance between the centers of the positive and negative charges. The dipole produces an electric field at point P at a distance x equal to

$$F = \frac{\epsilon l}{\lambda x^3} \qquad (14.10)$$

in which λ is the dielectric constant.

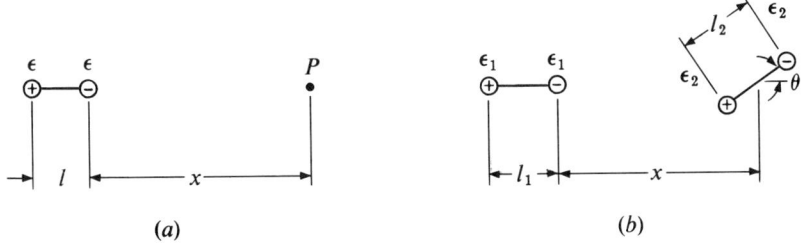

FIGURE 14.8 *Interaction between dipoles.*

We now consider the field from the first dipole (dipole moment $\epsilon_1 l_1$) acting on a second dipole (dipole moment $\epsilon_2 l_2$) located at distance x from the first [Fig. 14.8(b)]. The interaction energy between the two dipoles is the work required to bring the second dipole to its present position. This may be written as

$$E = -\epsilon_2 l_2 F \cos \theta = -\frac{(\epsilon_2 l_2)(\epsilon_2 l_2)}{\lambda x^3} \cos \theta^* \qquad (14.11)$$

The orientation of the particular dipole θ is, however, dependent on the thermal motion of the molecules. The relative number n of dipoles oriented at a given angle θ out of a total of N dipoles is given by Boltzmann's law. Since we are

---

* Consider the dipole being brought to point P with its axis perpendicular to the direction of the field. The work required to do this is zero, since the positive and negative charges are equal. The dipole is then rotated to the position shown in Fig. 14.8. The work done to move the positive charge and negative charge through a distance $\frac{1}{2} \cos \theta$ is equal to $-\frac{1}{2} \epsilon l F \cos \theta$. Hence the total work is equal to $W = -\epsilon l F \cos \theta$.

concerned with material behavior on the macro scale, we must take into account the average condition of many molecules. Taking this average, the energy becomes

$$E = \frac{C}{3} \frac{(\epsilon_1 l_1)^2 (\epsilon_2 l_2)^2}{\lambda k T x^6} \qquad (14.12)$$

in which C is a constant.* This equation shows that the attractive potential is inversely proportional to the sixth power of distance. The attractive force is therefore inversely proportional to the seventh power of distance.

Equation (14.12) is obtained from electrostatic considerations. Recent studies [see Dzyaloshinskii et al. (1961)] taking into consideration the interaction between molecules emitting electromagnetic radiation show that for large distances the attractive force varies as the inverse of the eighth power of the distance. For small distances the relationship reduces down to proportionality with the inverse of the seventh power of distance. For a summary of this problem and the experimental results see Derjaguin (1960).

14.6 Flocculation and dispersion

The behavior of clay particles in a suspension can be interpreted by a qualitative evaluation of the attractive and repulsive potentials. In Fig. 14.9 are shown net repulsive and attractive potentials plotted as functions of the distance d between two clay particles. Since the repulsive potential decays exponentially with x whereas the attractive potential is inversely proportional to x^7, the net potential may be represented by the curves in Fig. 14.9 depending on the relative values of the two potentials. Curve a shows increasing repulsion with decreasing distance. Hence in the fluid suspension the clay particles are kept apart and the particles remain in suspension. This is the *dispersed state*. In the dispersed state the clay particles remain in suspension until they settle to the bottom under their own weight. Curve c shows increasing attraction with decreasing distance. Hence, when two particles encounter each other during random motion they are attracted together. This process continues, and large aggregations of particles are formed which settle out rapidly because of their large weight. This process is called *flocculation*. Curve b represents an intermediate position.

Since the repulsive potential is seriously affected by such factors as cation concentration and valence and temperature, it means that changes in these factors also change the quality of the net-potential curves. Thus a type-b curve may be changed to a type-c curve if the ion concentration is increased. The observed phenomenon that sediments often flocculate in marine water is attributed to the relatively high ion concentration in salt water. Another example is the dispersion of a clay when the calcium cations (valence 2) are replaced by sodium cations (valence 1).

---

* For a derivation of this equation see Rice and Teller (1949).

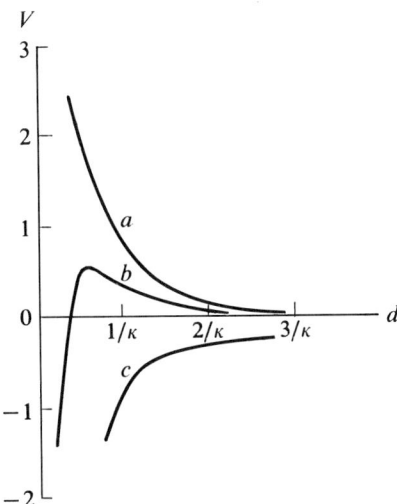

FIGURE 14.9 *Net potential curves.*

14.7 Clay structure

If the net attractive potential is uniformly distributed over the clay particle, the two particles come together parallel to each other, as this is the position of minimum energy. However, as described in Sec. 14.3, there often exist local concentrations of positive charges along edges while the overall charge of the particle is negative. The charge distribution is unsymmetric and the clay behaves as a polar particle. Hence, when the repulsive potential is small, these edges or corners are attracted to the negatively charged face of the clay (see, for example, Schofield and Samson, 1954). Thus we may have a corner-to-plate or edge-to-plate contact. Such a clay is said to have a *flocculent or cardhouse structure*. Clays with the particles lying more or less parallel to each other are said to have a dispersed structure. Figure 14.10(a) and (b) illustrate the dispersed and flocculent structures, respectively. The structure of clays can be observed through the petrographic microscope and the electron microscope [Morgenstern and Tchalenko (1967b), Barden (1972)], and these studies have verified the existence of the dispersed and flocculent structures. Many common clays have structures that lie between these extremes. Figure 14.10(d) is an electron micrograph of a kaolinite clay. It shows particles grouped in clusters in roughly parallel positions. The orientation of the clusters is random.

Clay deposits formed through the flocculation of the particles would be expected to possess a flocculent structure. Such is the case, for example, with many marine and estuarine deposits. If a deposit is formed by the gradual sedimentation of dispersed particles, a structure intermediate between flocculent and dispersed may result. Many fresh-water sediments belong in this

FIGURE 14.10 *Clay structure: (a) dispersed structure; (b) flocculent structure; (c) remolded structure; (d) electron micrograph of a kaolinite clay.* [(d) Courtesy of C. A. Moore.]

category [Mitchell (1956)]. It can be visualized from Fig. 14.10 that the flocculent structure is accompanied by a relatively high void ratio (or water content when saturated). Skempton and Northey (1952) have shown that when consolidated under the same pressure, the Horton clay, which was flocculated in a marine environment, had a much higher liquidity index than the Gosport clay, which was sedimented in fresh water.

If the clay is consolidated one dimensionally, the flocculent structure is gradually broken down by the increasing pressure. The clay particles tend to occupy more-or-less parallel positions. Their orientation would be perpendicular to the direction of the force. Thus consolidation under high pressures usually produces a parallel orientation of the clay particles similar to that in a dispersed structure. Remolding also breaks down the flocculent structure. Owing to the randomness of the remolding action, dispersed structure may exist only in localized pockets [Fig. 14.10(c)]. The clay particles have parallel orientation within each pocket. Shear displacement would tend to produce parallel orientation of particles in the direction of shear.

14.8 Adsorbed water

The plastic property of clays is largely the result of the interaction of the electric field around the clay with the polar water molecule. Figure 14.11(a) illustrates the water molecule as made up of one oxygen and two hydrogen atoms. It can be seen that the negative and positive charges are balanced but that their distribution is not uniform. In other words, the molecule is a dipole with the positive charges concentrated at one end and the negative charges concentrated at the other end.

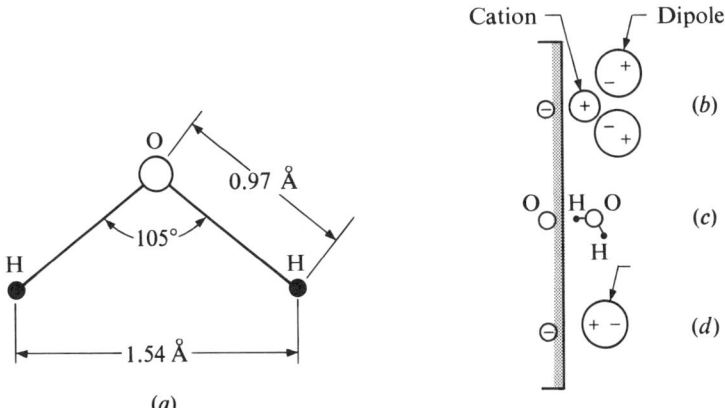

FIGURE 14.11 *Water adsorption.* [After Low and Lovell (1959).]

The water molecules that are held to the clay are called *adsorbed water*. Low and Lovell (1959) suggested several possible mechanisms by which water may be adsorbed. At low water content the cations closest to the clay are held tightly to the negative charge sites on the clay by electrostatic attraction. The water molecules are held to the cations as hydrated water [Fig. 14.11(b)]. Second, a water molecule that comes close to an oxygen atom in the clay lattice may be adsorbed by means of hydrogen bonding. The hydrogen atom may be shared between the oxygen atom in the clay and the oxygen atom in the water molecule [Fig. 14.11(c)]. Finally, the electric field of the clay may interact with the dipolar water molecule. The dipole is attracted to the charged plate and oriented with the positive pole toward the clay [Fig. 14.11(d)].

The exact structure of the adsorbed water has not been definitely established. Based on hydrogen bonding between clay and water, Hendricks and Jefferson (1938) and later Barshad (1949) suggested definite crystalline structures for the adsorbed water. The presence of cations complicates the problem in that it tends to disrupt the water structure. However, many indirect experiments suggest that the adsorbed water closest to the clay may have some sort of structural configuration and that it has an energy state well below that of ordinary water. This water has been found to have a greater viscosity and smaller mobility [see Rosenquist (1959), Low (1959b), Pickett and Lemcoe (1959)] as well as a different density [De Wit and Arens (1950), Anderson and Low (1958)] than that of ordinary water. The degree of fixity of the water molecules decreases with increasing distance from the surface of the clay. The first molecular layer is held very tightly to the clay, whereas several molecular layers away, the water may be quite mobile.

The foregoing studies do not prove the hypothesis of an adsorbed water structure. It can only be said that the results of the studies are compatible with the hypothesis. It must be noted, however, that other hypotheses on the nature of adsorbed water may be equally agreeable with the experimental data. For example, Martin (1960) in a review of the experimental data suggested that the adsorbed water may be a "two dimensional" liquid. In this model the water molecules are held to the clay by bonds. However they may still possess considerable freedom of movement on the clay surface.

A number of interesting soil properties that are time-dependent may be related to the properties of the adsorbed water. The gain of soil strength with time at a constant water content, called *thixotropy*, has been observed in a number of clay soils [Moretto (1948), Skempton and Northey (1952), Mitchell (1960b)]. Presumably this factor may also be responsible for differences between properties of laboratory-consolidated soils and those of soils consolidated in the field over a period of years. These gradual changes in soil property are considered by some [Leonards (1962), for example] to be the result of the slow development of the adsorbed water structure with time. Secondary consolidation is also related to the adsorbed water layer. Leonards and Girault (1961) measured the secondary consolidation in a clay after the water was replaced by a nonpolar liquid (CCl_4). The secondary consolidation is considerably less than that in the undisturbed clay. This suggests that the adsorbed water may exert an important influence on the viscoelastic behavior of the soil-particle contacts.

14.9 Ion exchange

It is shown in Sec. 14.3 that the negative charge of the clay particles is balanced by the cloud of positive ions that collects near the surface of the clay. Some kinds of ions are more strongly attracted to the clay than others. With most clays it has been found that the affinity of the ions to the clay shows the following order:

$$\underset{\text{Trivalent ions}}{Al^{3+}} > \underset{\text{Bivalent ions}}{Ca^{2+} > Mg^{2+}} > \underset{\text{Monovalent ions}}{K^+ > Na^+ > Li^+}$$

In other words, sodium ions may be replaced by calcium ions. This phenomenon is called *ion exchange*. It must be noted that the series above is not unique but may change with the type of clay mineral, the cations involved, and other experimental conditions. Ion exchange may be accomplished in the laboratory by removal of the adsorbed cations through chemical reaction and replacement with a different cation. For example, the calcium may be removed by the reaction

$$Ca\ clay + HCL \rightarrow H\ clay + CaCl_2$$

The $CaCl_2$ is removed by washing with distilled water. A sodium clay is then obtained by titration with NaOH:

$$H\ clay + NaOH \rightarrow Na\ clay + H_2O$$

Ion exchange may also be accomplished by mass action. To obtain a sodium clay the material is made into a suspension with a high concentration of sodium ions.

Ion-exchange capacity is the amount of ions, usually expressed in milliequivalents per gram of clay, that is exchangeable. The larger the surface area and the larger the negative charge of the clay, the larger also is the exchange capacity. Typical values of ion-exchange capacities for the three common clay minerals are given in Table 14.1.

TABLE 14.1 *Values of Ion-exchange Capacities*

Clay mineral	Exchange capacity, meq/100 g
Kaolinite	3-15
Illite	10-40
Montmorillonite	80-150

Ion exchange often results in profound effects on the physical properties of the clay. As can be seen in Fig. 14.4 and Eq. (14.5), the electric potential and field, as well as the repulsive pressure between two clay plates, are dependent upon the valence of the ion. Specific examples are given in the succeeding sections.

14.10 Compression and swelling of clay

The physical-chemical properties of the clay minerals are of great value in understanding the mechanical behavior of clay on the macro scale. A number of interesting investigations have illuminated the compression and swelling of clays.

Equation (14.9) gives the repulsion between two parallel clay plates. Let us ignore the attractive potential for the time being. Since p_∞ is the pressure in the water at infinite distance, where the field intensity is 0, it is the same as the hydrostatic pressure. Hence $p_d - p_\infty$ is the repulsive pressure that is in excess of the hydrostatic pressure. The clay plates therefore tend to be pushed apart by this repulsive pressure. If the plates are to be prevented from moving farther and farther apart, a pressure $p_d - p_\infty$ in excess of the hydrostatic pressure must be applied to the two particles. To illustrate this point we consider an idealized clay mass composed of particles in parallel. A stress σ is applied in a direction perpendicular to the particle orientation. The difference between σ and the hydrostatic pressure is $\bar{\sigma}$. From Eq. (2.28), we have

$$\bar{\sigma} = \sigma - u$$

If we pass a section through the midpoint between two plates [Fig. 14.6(a)], then the effective stress $\bar{\sigma}$ must be resisted by the repulsive pressure or

$$\bar{\sigma} = p_d - p_\infty$$

Hence, the effective stress is equal to the repulsive pressure between two parallel plates. If the effective stress $\bar{\sigma}$ is increased, the plates are forced closer together. The distance d is reduced and $p_d - p_\infty$ increases [see Eq. (14.9) or Fig. 14.6] until it is equal to the new value of $\bar{\sigma}$. In the same way, if $\bar{\sigma}$ is reduced, the plates are pushed apart by the repulsive pressure. The distance d increases until $p_d - p_\infty$ is reduced to the new value of $\bar{\sigma}$. The above process is thus completely reversible.

We now consider a consolidation test in which the applied stress is steadily increased. The increasing stress compresses the clay. If the clay particles are perfectly parallel to each other, the compression results in pushing the clay particles closer and closer together. As the stress is gradually reduced, the clay swells or rebounds as the particles move apart. We see that the process is the same as that described in the preceding paragraph. It follows then the stress-compression relationship (i.e., the pressure vs. void ratio curve in a consolidation test) is the same as the relationship between the repulsive pressure and distance d described by Eq. (14.9) and Fig. 14.6. It also means that the compression and rebound curves should be identical.

We may expect the consolidation and rebound curves to agree best with the repulsive pressure vs. distance relationship if a clay has a dispersed structure and is under conditions where the repulsive potential exercises a predominant influence, that is, the attractive potential is small by comparison. Experiments with fine-grained, purified sodium montmorillonite have shown that a qualitative agreement does exist between the measured pressure vs. void ratio curve and that predicted by the double-layer theory [Bolt (1956)]. The comparisons are reproduced in Fig. 14.6. The experimental curves represent the rebound and reloading of the specimens. The virgin consolidation curve would be somewhat different because initially the clay particles may not be oriented parallel to each other. However, the virgin consolidation improves the particle orientation and during subsequent rebound and reloading, the particle orientation undergoes relatively little change. Then the volume changes reflect primarily the changes in the distance between nearly parallel clay particles.

The agreement between the measured pressure vs. void ratio curve and that calculated from the double-layer theory becomes less satisfactory with calcium and magnesium clays and with clays of larger particle sizes [see Mitchell (1960a), Olson and Mitronovas (1960), for example]. With divalent cations, the repulsive potential is reduced and the effect of the attractive potential may be appreciable. Also the departure from the dispersed structure becomes more important. Instead of occupying nearly parallel positions, many clay particles may be in contact with each other. Then mechanical effects such as aggregation of particles and bending of clay plates may contribute substantially to the volume change behavior.

In a natural soil, there is usually a mixture of clay minerals and cations. Presence of silt and sand particles would interfere with the parallel orientation of the clay particles. Hence, their behavior may be considerably different from the predictions of the double-layer theory. For a thorough review of the research on compression and structure of clays up to 1962 see Meade (1964).

The repulsive pressure given in Eq. (14.9) is also useful in predicting the effect on consolidation and swelling properties of clays when one adsorbed cation is replaced by another through ion exchange. For example, if a divalent cation such as calcium replaces the monovalent sodium cation initially adsorbed to the clay, the repulsive pressure is reduced substantially, and a new repulsion vs. distance relationship results. In this case swelling and compression are smaller for a given stress change. Experiments with swelling clay [for example, Ladd (1959)] have given results that are in good qualitative agreement with theoretical predictions. Such results have great significance, as they provide a rational basis for the use of chemical additives to alter the mechanical properties of clays. They also make it possible to understand the changes in properties of clay deposits due to natural causes such as ion exchange brought about by percolating groundwaters with high ion concentration. Such problems are particularly prominent with clays of high exchange capacity, such as the expansive clays in the western United States.

14.11 Shear strength of clay

When shear failure occurs along one or more slip planes, the original soil structure is disrupted along these planes and the shear displacement produces a nearly parallel orientation of the clay particles along the slip surface. To destroy the initial structure, work must be done against the forces that are in equilibrium prior to shear. The equilibrium position is one of minimum net potential. The work required to displace a particle from this position depends on the potential field at the contact point. The attractive potential is attributed to van der Waals' bonding. If the particles are cemented together at the contacts, the cementing bonds would have to be overcome. If the shear displacement breaks apart two kaolinite particles stacked together, the H bonding between them must be overcome. In the case of illite particles it would be the K bonding. Lambe (1960) suggested that these are some of the forces that may contribute to cohesion.

At failure, the relative displacement along slip planes involves the slipping of one clay plate over another. The stresses necessary to produce this type of displacement are probably influenced by the viscosity of the adsorbed water between the plates and are cohesive in nature. Also, there is particle rotation owing to interference by other particles, as the alignment along the slip plane cannot be perfect. The internal friction would consist of the resistance developed as the particles rotate and slide over each other, as well as the friction between the particles at the contact points (see Sec. 8.12).

When we consider the internal friction, which is the stress-dependent part of the shear strength, we must reconsider the simple definition of the effective stress given in Sec. 2.7. The simple definition is that the effective stress $\bar{\sigma}$ is the force normal to a section transmitted across the soil contact points per unit area of soil. The preceding section shows that if the clay particles are parallel, the effective stress is simply the net repulsion between the particles, there being no contact between the particles. Figure 14.12 shows three possible kinds of "contacts." Type (a) is the case of two parallel plates just described. Type (b) contains both physical contact and repulsion, and the effective stress is the sum of the two. If two particles are perpendicular to each other, as shown in (c), the repulsion is very small and the effective stress is almost equal to the contact force. Thus, even though the measured effective stress may be the same in all three cases, the shearing resistance developed is very different in nature. Hence the value of $\bar{\phi}_e$ may be expected to vary over a wide range depending on the kind of contact available under the given $\bar{\sigma}$, which, in turn, is dependent on the soil structure and the magnitude of the attraction and repulsion between particles.

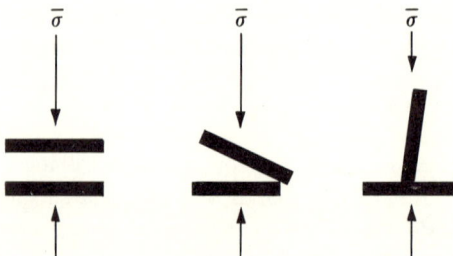

FIGURE 14.12 *Three possible types of contacts between clay particles.*

For detailed examples of the effect of structure on strength properties, the reader may refer to the studies by Lambe (1958b), Seed and Chan (1959), Olson (1962), and Kenney (1967).

14.12 Composition and properties of natural soils

As pointed out in Sec. 14.1, the fine fractions of soils usually contain a variety of clay minerals as well as nonclay minerals, the commonest ones being quartz, feldspar, calcite, and mica. Occasionally a monoclay soil may be encountered. Such a soil has only one clay mineral in the fine fraction. One may well expect the clay-mineral constituents of a soil to exert a profound effect on soil behavior. However, it must be recognized that even with monoclay soils, the extrapolation from clay-mineral behavior to soil behavior involves uncertainties. Extensive data on the composition and properties of natural soils have been collected by Lambe and Martin (1953, 1954, 1955, 1956, 1957). They showed that the clay minerals in natural soils are sometimes poorly crystallized, unlike the standard minerals whose properties are well established. Second, in a mixture of minerals, there may be chemical reactions between the minerals that alter the properties of the constituents. Organic matter and iron oxide are found to affect soil behavior in a manner far out of proportion to their percentage in the clay composition. In addition, mixtures of clay minerals may exist as simple mixtures or the minerals may be interstratified. Finally, the clay minerals may occur as aggregates that are made up of a large number of clay particles. In this case the aggregates may be larger than clay size (2 microns) and may behave more like silt particles. Figure 14.1(b) shows kaolinite particles stacked together to form aggregates that are considerably larger than 2 microns.

In spite of these reservations, knowledge of the clay-mineral composition provides a valuable basis for understanding the soil behavior. To illustrate some of the relationships between clay composition and clay properties, it is instructive to examine the data of Lambe and Martin, some of which is reproduced in Table 14.2.

The Atterberg limits reflect the clay-mineral constituents in the fine fraction. The data show that, in general, soils containing large percentages of montmorillonite have higher limits than those that contain illite and kaolinite. It is also important to note the differences between soils 1, 2, and 3. Soils 1 and 2 contain 70 percent montmorillonite, but the results of mechanical analysis show only 24 and 20 percent finer than 2 microns. This means that many of the montmorillonite particles are aggregated. Thus less surface area is available for the adsorption of water and the Atterberg limits are low for a montmorillonite soil. Compare this with soil 3, which contains 45 percent montmorillonite, but most of it finer than 2 microns. Its Atterberg limits are higher than those of soils 1 and 2, even though it contains less montmorillonite.

Montmorillonites, because of their large negative charge and osmotic pressure, are capable of large volume expansions (swelling) when water is

TABLE 14.2 Composition and Properties of Some Natural Soils[a]

No.	Soil	Part. size <0.07 mm, %	<2µ, %	Composition, %	Atterberg limits, %	Engr. prop.
1	Brown clay, Bombay, India	79	24	Montmor. 70	LL = 69; PL = 30	Natural γ_d = 87 pcf
2	Yellow clay, Kingston, Jamaica	81	20	Montmor. 70 Illite 5	LL = 99; PL = 28	
3	Grey clay, Mobile, Ala.	95	55	Montmor. 45 Illite 25	LL = 106; PL = 40; w = 80	γ_d = 107 pcf w_{opt} = 18%
4	Tan silty clay, Wellfleet, Mass.	41	13	Illite 20 Kaolinite 15	LL = 29; PL = 18	γ_d = 87 pcf w_{opt} = 29%
5	Red clay, South Piedmont	17	3	Halloysite 35 Mica 45	LL = 49; PL = 40	c_u = 910 psf p_c = 2.9 kg/cm^2 k_x/k_y = 4.0
6	Dark grey clay, St. Lawrence Valley	100	78	[Illite 40 Chlorite 10][b] Feldspar 20	LL = 55.2; PL = 22.1; w = 53.6	γ_d = 107 pcf w_{opt} = 18%
7	Brown silt, Fairbanks, Alaska	90	5	Quartz 35 [Vermiculite 27 Illite 20]	LL = 32; PL = 26	c_u = 800 psf at w_{opt}
8	Grey clay, Chicago, Ill.	100	64	[Illite 55 Chlorite 25] Quartz 10 Carbonates 10	LL = 57.9; PL = 20.6; w = 39.7	c_u = 970 psf p_c = 3.8 kg/cm^2 k_x/k_y = 2.7
9	Sandy clay, Vereeniging, South Africa	66	45	[Montmor. 45 Illite 10] Quartz 25	LL = 55; PL = 25; SL = 13	Vol. changes cause severe problems
10	Organic silt, Freedom, Me.	91 (Inorg. portion)	16	Organ. matter 60 Quartz 25 Feldspar 15	LL = 537; PL = 273; w = 680 After drying, LL = 228; PL = 148	

[a] Data from Lambe and Martin (1954, 1955, 1956, 1957).
[b] [ ] indicates interstratified.

available. Most natural soils containing appreciable amounts of montmorillonite are also susceptible to swelling. Soil 9 is known locally as an "expansive clay," and volume changes have caused many serious problems to structures located on this soil. Experience with expansive soils in the western United States shows that the Atterberg limits and the percentage finer than 1 micron are valuable indicators of the expansive properties of clay [Holtz and Gibbs (1954)].

The mineral halloysite has the shape of a rolled-up sheet. Soil 5 consists of 35 percent halloysite, but only 3 percent of the soil is finer than 2 microns. Because of its shape and the aggregation, the soil has a very low dry density. It is frequently difficult to compact clays containing halloysite to high dry densities. Nevertheless, this does not necessarily imply a low strength. The clay used for the construction of Sasuma Dam in Africa contains a large percentage of halloysite in large aggregated clusters. It has a true angle of internal friction (ϕ_e) of 36 deg, which is very large for a clay soil [Terzaghi (1958)].

The influence of organic matter on the Atterberg limits is illustrated by soil 10. Drying destroys a considerable portion of the organic matter and drastically alters the Atterberg limits.

Since most natural soil deposits are consolidated under some overburden pressure, they would show preferred particle orientation in a horizontal direction (unless the preconsolidation pressure is very small or the clay has a highly flocculent structure). This preferred orientation is reflected in the ratio of the horizontal to vertical permeability k_x/k_y of soils 6 and 8.

The values of the undrained shear strength c_u in Table 14.2 should not be taken as representative of the soil composition. The shear strength is so sensitive to the water content and void ratio that it can vary over a wide range. More important is the relationship between the cohesive and frictional com-

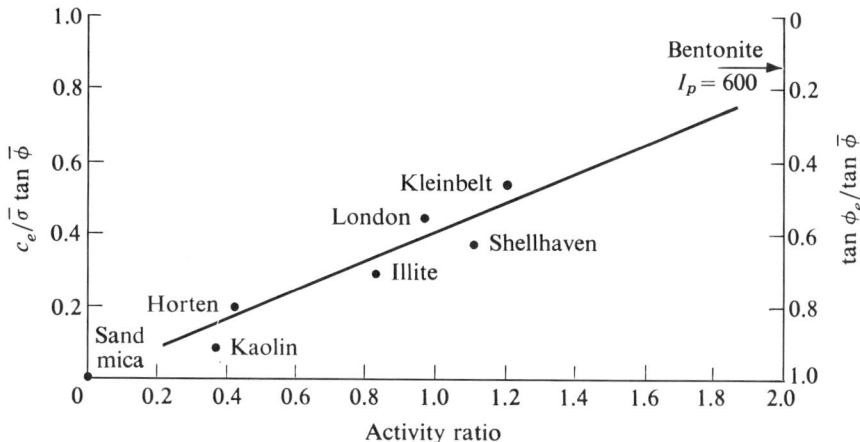

FIGURE 14.13 *Relationship between activity and the cohesive and frictional components of shear strength.* [After Skempton (1953).]

ponents of the shear strength and the mineral composition. For a normally consolidated clay the shear strength is (Sec. 8.8)

$$s = \bar{\sigma} \tan \bar{\phi}$$

The cohesive and frictional components of the shear strength can be expressed as $c_e/\bar{\sigma} \tan \bar{\phi}$ and $\tan \phi_e/\tan \bar{\phi}$, respectively, where c_e and ϕ_e are Hvorslev's true cohesion and true angle of internal friction. Skempton (1953a) has shown that these components bear a close relationship to the activity ratio, and this is shown in Fig. 14.13. The activity ratio (Sec. 1.8) is a fairly good index of the magnitude of the surface forces. Figure 14.13 shows that for a nonclay soil like sand mica the true cohesion is 0. The true cohesion increases as the activity ratio increases. With bentonite (montmorillonite) the shear strength is mostly cohesion. Similar data obtained by Gibson (1953) show that the true angle of internal friction decreases as the plasticity index I_p increases (Fig. 14.14).

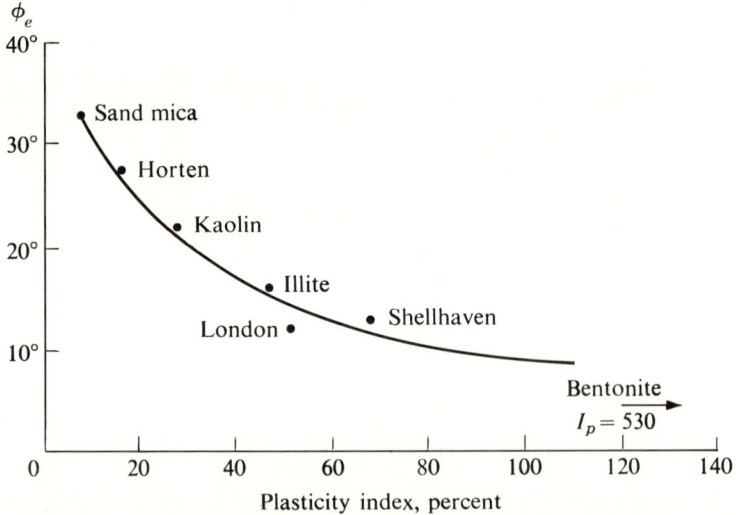

FIGURE 14.14 *Relationship between ϕ_e and plasticity index.* [After Gibson (1953).]

Bibliography

ACUM, W. E. A. and L. FOX (1951), "Computation of Load Stresses in a Three Layer Elastic System," *Geotechnique*, 2, p. 293.
AIR FORCE SPECIAL WEAPONS CENTER (1962), "Air Force Design Manual, Principles and Practices for Design of Hardened Structures," *Tech. Docum. Rept. No. AFSWC—TDR—62-138*, Kirtland Air Force Base, N. M.
AITCHISON, G. D. (1960), "Relationships of Moisture Stress and Effective Stress Functions in Unsaturated Soils," *Pore Pressure and Suction in Soils*. London: Butterworth & Co. Ltd.
ALDRICH, H. P. (1952), "Importance of the Net Load to the Settlement of Buildings in Boston," *Contrib. Soil Mech. 1953*, Boston Soc. Civil Engrs., p. 370.
ANDERSON, D. M. and P. F. LOW (1958), "The Density of Water Adsorbed by Lithium, Sodium, and Potassium Bentonite," *Proc. Soil Sci. Soc. Am.*, 22, p. 99.
ANDRESEN, A. and L. Bjerrum (1967), "Slides in Subaqueous Slopes in Loose Sand and Silt," *Marine Geotechnique*, ed. A. F. Richards, University of Illinois Press, p. 221 (also *Norweg. Geotech. Inst. Pub.* No. 81).
BARDEN, L. (1972), "The Influence of Structure on Deformation and Failure in Clay Soil," *Geotechnique*, 22, p. 159.
BARNES, H. E. (1952), "Soil Investigation Employing a New Method of Layer Value Determination," *Highway Res. Bd. Bull.* 65.
BARRON, R. A. (1948), "Consolidation of Fine-Grained Soils by Drain Wells," *Trans. Am. Soc. Civil Engrs.*, 113, p. 718.
BARSHAD, I. (1949), "The Nature of Lattice Expansion and Its Relation to Hydration in Montmorillonite and Vermiculite," *Am. Mineralogist*, 34, p. 675.
BEREZANTZEV, V. G., V. S. KHRISTOFOROV, and V. N. GOLUBKOV (1961), "Load Bearing Capacity and Deformation of Piled Foundations," *Proc. 5th Intern. Conf. Soil Mech. Found. Eng.*, 2, p. 11.
BIOT, M. A. (1941), "General Theory of Three-Dimensional Consolidation," *J. Appl. Phys.* 12, p. 155.
BISHOP, A. W. (1954), "Correspondence," *Geotechnique*, 4, p. 43.
BISHOP, A. W. (1955), "Use of Slip Circle for Stability Analysis," *Geotechnique*, 5, p. 7.
BISHOP, A. W. (1957), "Some Factors Controlling the Pore Pressures Set Up During the Construction of Earth Dams," *Proc. 4th Intern. Conf. Soil Mech. Found. Eng.*, 2, p. 294.

BISHOP, A. W. (1958), "Test Requirements for Measuring the Coefficient of Earth Pressure at Rest," *Proc. Conf. on Earth Pressure Problems, Brussels*, 1, p. 2.

BISHOP, A. W. (1966), "The Strength of Soils as Engineering Materials," *Geotechnique*, 16, p. 89.

BISHOP, A. W., I. ALPAN, G. E. BLIGHT, and I. B. DONALD (1960a), "Factors Controlling the Strength of Partially Saturated Cohesive Soils," *Proc. Am. Soc. Civil Engrs. Res. Conf. on Shear Strength of Cohesive Soils*, p. 503.

BISHOP, A. W. and L. BJERRUM (1960), "The Relevance of the Triaxial Test to the Solution of Stability Problems," *Proc. Am. Soc. Civil Engrs. Res. Conf. on Shear Strength of Cohesive Soils*, p. 437.

BISHOP, A. W. and D. J. HENDEL (1962), *The Measurement of Soil Properties in the Triaxial Test*. London: E. J. Arnold & Son Ltd.

BISHOP, A. W., M. F. KENNARD, and A. D. M. PENMAN (1960b), "Pore-pressure Observations at Selset Dam," *Pore Pressure and Suction in Soils*. London: Butterworth & Co. Ltd.

BJERRUM, L. (1954), "Theoretical and Experimental Investigations on the Shear Strength of Soils," *Norweg. Geotech. Inst. Publ. 5*, Oslo.

BJERRUM, L. (1963), "Allowable Settlement of Structures," *Proc. European Conf. on Soil Mechanics and Found. Eng.*, Wiesbaden, 2, p. 135.

BJERRUM, L. (1967), "Progressive Failure in Slopes of Over-consolidated Plastic Clay and Clay Shales," *J. Soil Mech. and Found. Eng. Div., Am. Soc. Civil Engrs.*, 93, SM5, p. 1.

BJERRUM, L. (1972), "Embankments on Soft Ground," Performance of Earth and Earth Supported Structures, *Am. Soc. Civil Engrs.*, 2, p. 1.

BJERRUM, L., C. J. F. CLAUSEN, and J. M. DUNCAN (1972), "Earth Pressures on Flexible Structures—A State of the Art Report," *Proc. 5th European Conf. on Soil Mech. and Found. Eng.*, Madrid, 2, p. 167.

BJERRUM, L. and A. EGGESTAD (1963), "Interpretation of Loading Test on Sand," *Proc. European Conf. on Soil Mech. and Found. Engr.*, Wiesbaden, 1, p. 199.

BJERRUM, L., W. JÖNSON, and C. OSTENFELD (1957), "Settlement of a Bridge Abutment on Friction Piles," *Proc. 4th Intern. Conf. Soil Mech. Found. Eng.*, 2, p. 14.

BJERRUM, L. and R. KIRKEDAM (1958), "Some Notes on Earth Pressures in Stiff Fissured Clay," *Proc. Conf. on Earth Pressure Problems, Brussels*, 1, p. 15.

BJERRUM, L. and B. KJAERNSLI (1957), "Analysis of the Stability of Some Norwegian Clay Slopes," *Geotechnique*, 7, p. 1.

BJERRUM, L., S. KRINGSTAD, and O. KUMMENEJE (1961), "The Shear Strength of a Fine Sand," *Proc. 5th Intern. Conf. Soil Mech. Found. Eng.*, 1, p. 29.

BJERRUM, L. and A. LANDVA (1966), "Direct Simple Shear Tests on a Norwegian Quick Clay," *Geotechnique*, 16, p. 1.

BJERRUM, L. and A. ØVERLAND (1957), "Foundation Failure of an Oil Tank in Fredrikstad, Norway," *Proc. 4th Intern. Conf. Soil Mech. Found. Eng.*, 1, p. 287.

BLIGHT, G. E. (1963), "The Utilization of Soil Suction in the Design of Earth Embankments," *Proc. 3rd Reg. Conf. for Africa on Soil Mech. and Found. Eng.*, 1, p. 141.

BLIGHT, G. E. (1967), "Effective Stress Evaluation for Unsaturated Soils," *J. Soil Mech. and Found. Div., Am. Soc. Civil Engrs.*, 93, SM2, p. 125.

BOLT, G. H. (1955), "Analysis of Validity of Gouy-Chapman Theory of the Electric Double Layer," *J. Colloid Sci.*, 10, p. 206.

BOLT, G. H. (1956), "Physical Chemical Analysis of Compressibility of Pure Clay," *Geotechnique*, 6, p. 86.

BOUSSINESQ, J. (1885), *Application des potentials a l'etude de l'equelibre et de movement des solids elastique*. Paris: Gauthier-Villars.

BOZOZUK, M. (1963), "The Modulus of Elasticity of Leda Clay from Field Measurements," *Canad. Geotech. J.*, 1, p. 43.

BOZOZUK, M. and G. A. LEONARDS (1972), "The Gloucester Test Fill," Performance of Earth and Earth Supported Structures, *Am. Soc. Civil Engrs.*, 1, p. 299.
BRADLEY, W. F. and R. E. GRIM (1948), "Colloidal Properties of Layer Silicates," *J. Phys. Colloidal Chem.*, 52, p. 1404.
BUISMAN, A. S. K. (1936), "Results of Long Duration Settlement Tests," *Proc. 1st Intern. Conf. Soil Mech. Found. Eng.*, 1, p. 103.
BURMISTER, D. M. (1943), "Theory of Stresses and Displacements in Layered Systems and Applications to the Design of Airport Runways," *Proc. Highway Res. Bd.*, 23, p. 126.
BURMISTER, D. M. (1948), "The Importance and Practical Use of Relative Density in Soil Mechanics," *Proc. Am. Soc. Testing Mat.*, 48, p. 1249.
CARMEN, P. C. (1956), *Flow of Gases through Porous Media.* New York: Academic Press.
CASAGRANDE, A. (1932), "Research on the Atterberg Limits of Soils," *Public Roads*, 13, No. 8, p. 121.
CASAGRANDE, A. (1936), "The Determination of the Pre-Consolidation Load and Its Practical Significance," *Proc. 1st Intern. Conf. Soil Mech. Found. Eng.*, 3, p. 60.
CASAGRANDE, A. (1937), "Seepage through Dams," *Contrib. Soil Mech. 1940*, Boston Soc. Civil Engrs., p. 295.
CASAGRANDE, A. (1948), "Classification and Identification of Soils," *Trans. Am. Soc. Civil Engrs.*, 113, p. 901.
CASAGRANDE, A. (1949), "Soil Mechanics in the Design and Construction of Logan Airport," *Contrib. Soil Mech. 1943*, Boston Soc. Civil Engrs., p. 176.
CASAGRANDE, A. (1950), "Notes on the Design of Earth Dams," *J. Boston Soc. Civil Engrs.*, 37, p. 405.
CASAGRANDE, A. (1965), "Role of the 'Calculated Risk' in Earthwork and Foundation Engineering," *J. Soil Mech. and Found. Div., Am. Soc. Civil Engrs.*, 91, SM4, p. 1.
CASAGRANDE, A. and R. E. FADUM (1944), "Application of Soil Mechanics in Designing Building Foundations," *Trans. Am. Soc. Civil Engrs.*, 109, p. 383.
CASAGRANDE, A. and S. D. WILSON (1951), "Effect of Rate of Loading on the Strength of Clays and Shales at Constant Water Content," *Geotechnique*, 2, p. 251.
CASAGRANDE, L. (1932), "Naeherungsmethoden zur Bestimmung von Art und Menge der Sickerung durch geschuettete Daemme," Thesis, Technische Hochschule, Vienna.
CHAMECKI, S. (1956), "Structural Rigidity in Calculating Settlements," *J. Soil Mech. and Found. Div., Am. Soc. Civil Engrs.*, 82, SM1.
CHEN, L. S. (1948), "An Investigation of Stress-Strain and Strength Characteristics of Cohesionless Soils," *Proc. 2nd Intern. Conf. Soil Mech. Found. Eng.*, 5, p. 35.
CHRISTIAN, J. T., J. W. BOEHMER, and P. P. Martin (1972), "Consolidation of a Layer under a Strip Load," *J. Soil Mech. and Found. Div., Am. Soc. Civil Engrs.*, 98, SM7, p. 693.
CHURCHILL, R. V. (1941), *Fourier Series and Boundary Value Problems.* New York: McGraw-Hill Book Co., Inc.
CLEVENGER, W. A. (1956), "Experiences with Loess as Foundation Material," *J. Soil Mech. and Found. Div., Am. Soc. Civil Engrs.*, 82, SM3.
COULOMB, C. A. (1776), "Essais sur une application des regles des maximis et minimis à quelques problems de statique relatifs à l'architecture," *Mem. Acad. Roy. Pres. Divers*, Sav. 5, 7, Paris.
CRAWFORD, C. B. (1959), "Influence of Rate of Strain on Effective Stresses in Sensitive Clay," *Am. Soc. Testing Mat. Spec. Tech. Publ.* 254, p. 36.
CRAWFORD, C. B. and K. N. BURN (1962), "Settlement Studies on Mt. Sinai Hospital Toronto," *The Eng. J.*, 45, No. 12, p. 31.
CRAWFORD, C. B. and J. G. Sutherland (1971), "The Empress Hotel, Victoria,

British Columbia, Sixty-Five Years of Foundation Settlements," *Can. Geot. J.* 8, p. 527.
CRONEY, D., J. D. COLEMAN, and W. P. M. BLACK (1958), "Movement and Distribution of Water in Soil," *Highway Res. Bd. Spec. Rept. 40*, p. 226.
CUMMINGS, A. E., G. O. KIRKHOFF, and R. B. PECK (1950), "Effect of Driving Piles into Soft Clay," *Trans. Am. Soc. Civil Engrs.*, 115, p. 275.
D'APPOLONIA, D. J., E. D'APPOLONIA, R. F. BRISSETTE (1968), "Settlement of Spread Footings on Sand," *J. Soil Mech. and Found. Div., Am. Soc. Civil Engrs.*, 94, No. SM3, p. 735.
DAVIS, W. C. and W. G. JONES (1954), "Flexible Pavement Design by Group Index Method," *Highway Res. Bd. Res. Rept. 16B*,
DE JONG, J. and N. R. MORGENSTERN (1971), "The Influence of Structural Rigidity on The Foundations of CN Tower, Edmonton," *Canad. Geot. J.*, 8, p. 527.
DE JOSSELYN DE JONG, G. (1958), "Longitudinal and Transverse Diffusion in Granular Deposits," *Trans. Am. Geophys. Union*, 39, p. 67.
DE JOSSELYN DE JONG, G. (1959), *Statics and Kinematics in the Failable Zone of a Granular Material*. Delft: Waltman.
DERJAGUIN, B. V. (1960), "The Force between Molecules," *Sci. Am.*, 203, No. 1, p. 47.
DESAI, C. S. (1974), "Numerical Design-Analysis for Piles in Sands," *J. Geotech. Div., Am. Soc. Civil Engrs.*, 100, GT6, p. 613.
DESAI, C. S. and J. F. ABEL (1972), *Introduction to the Finite Element Method*. New York: Van Nostrand Reinhold Co.
DE WIT, C. T. and P. L. ARENS (1950), "Moisture Content and Density of Some Clay Minerals and Some Remarks on the Hydration Pattern of Clay," *Trans. 4th Intern. Cong. Soil Sci.*, 2, p. 59.
DOMASCHUK, L. and N. H. WADE (1969), "A Study of Bulk and Shear Moduli of Sand," *J. Soil Mech. and Found. Div., Am. Soc. Civil Engrs.*, 95, SM2, p. 561.
DRUCKER, D. C., R. E. GIBSON, and D. J. HENKEL (1955), "Soil Mechanics and Work Hardening Theories of Plasticity," *Proc. Am. Soc. Civil Engrs.*, 81, No. 798.
DRUCKER, D. C. and W. PRAGER (1952), "Soil Mechanics and Plastic Analysis or Limit Design," *Quart. App. Math.*, 10, p. 157.
DRUCKER, D. C., W. PRAGER, and H. J. GREENBERG (1951), "Extended Limit Design Theorems for Continuous Media," *Quart. Appl. Math.*, 9, p. 381.
DUNCAN, J. M. and C. Y. CHANG (1970), "Non-linear Analysis of Stress and Strain in Soils," *J. Soil Mech. and Found. Div., Am. Soc. Civil Engrs.*, 96, SM5, p. 1629.
DURGUNOGLU, H. T. and J. K. MITCHELL (1973), "Static Penetration Resistance of Soils," Report for NASA Headquarters, Univ. of Calif., Berkeley.
DZYALOSHINSKII, I. E., E. M. LIFSHITZ, and L. P. PITAEVSKII (1961), "General Theory of Van der Waal's Forces," *Soviet Phys.—Usp.*, Sept.–Oct. 1961, p. 153 (orig. *Uspekhi Fizicheskikh Nauk*, 73, Nos. 3–4).
EGGESTAD, A. (1963), "Deformation Measurements below a Model Footing on the Surface of Dry Sand," *Proc. European Conf. on Soil Mech. and Found. Eng.*, Wiesbaden, 1, p. 233.
ELIAS, V. and H. STORCH (1970), "Control and Performance during Construction of a Highway Embankment on Weak Soils," *Highway Res. Rec.*, No. 323, p. 60.
ESMIOL, E. E. (1957), "Seepage through Foundations Containing Discontinuities," *J. Soil Mech. and Found. Div., Am. Soc. Civil Engrs.*, 83, SM1.
EWING, W. M., W. S. JARDETSKY, and F. PRESS (1957), *Elastic Waves in Layered Media*. New York: McGraw-Hill Book Co., Inc.
FELLENIUS, W. (1936), "Calculation of the Stability of Earth Dams," *Trans. 2nd Cong. on Large Dams*, 4, p. 445.
FENNEMAN, N. M. (1931), *Physiography of Western United States*. New York: McGraw-Hill Book Co., Inc.

FENNEMAN, N. M. (1938), *Physiography of Eastern United States.* New York: McGraw-Hill Book Co., Inc.

FIELDS, K. E. and W. L. WELLS (1944), "Pendleton Levee Failure," *Trans. Am. Soc. Civil Engrs.*, 109, p. 1400.

FINN, W. D. L. (1963), "Boundary Value Problems of Soil Mechanics," *J. Soil Mech. and Found. Div., Am. Soc. Civil Engrs.*, 89, SM5, p. 39.

FISK, H. N. (1952), "Mississippi Valley Geology and Its Relation to River Regime," *Trans. Am. Soc. Civil Engrs.*, 117, p. 667.

FLINT, R. F. (1957), *Glacial and Pleistocene Geology.* New York: John Wiley & Sons, Inc.

FOX, F. N. (1948), "The Mean Elastic Settlement of a Uniformly Loaded Area at a Depth below the Ground Surface," *Proc. 2nd Intern. Conf. Soil Mech. Found. Eng.*, 1, p. 129.

GIBBS, H. J. and W. G. HOLTZ (1957), "Research on Determining the Density of Sands by Spoon Penetration Test," *Proc. 4th Intern. Conf. Soil Mech. Found. Eng.*, 1, p. 35.

GIBSON, R. E. (1953), "Experimental Determination of the True Cohesion and True Angle of Internal Friction in Clays," *Proc. 3rd Intern. Conf. Soil Mech. Found. Eng.*, 1, p. 126.

GIBSON, R. E. and P. LUMB (1953), "Numerical Solution of Some Problems in the Consolidation of Clay," *Proc. Inst. Civil Eng.*, 2, pt. 1, p. 182.

GILBOY, G. (1934), "Mechanics of Hydraulic-fill Dams," *Contrib. Soil Mech. 1940*, Boston Soc. Civil Engrs., p. 127.

GRIM, R. E. (1948), "Some Fundamental Factors Influencing the Properties of Soil Materials," *Proc. 2nd Intern. Conf. Soil Mech. Found. Eng.*, 3, p. 8.

GRIM, R. E. (1953), *Clay Mineralogy.* New York: McGraw-Hill Book Co., Inc.

GRIM, R. E. (1962), *Applied Clay Mineralogy.* New York: McGraw-Hill Book Co., Inc.

HAMILTON, E. L. (1964), "Consolidation Characteristics and Related Properties of Sediments from Experimental Mohole," *J. Geophys. Res.*, 69, p. 4257.

HANSBO, S. (1960), "Consolidation of Clay with Special Reference to the Influence of Vertical Sand Drains," *Proc. 18 Swed. Geotech. Inst.*, Stockholm.

HANSEN, J. B. (1953), *Earth Pressure Calculations.* Danish Tech. Press, Inst. Dan. Civil Eng., Copenhagen.

HANSEN, J. B. (1961), "A General Formula for Bearing Capacity," *Bull. No. 11, Danish Geotech. Inst.*, Copenhagen.

HANSEN, J. B. (1962), "Relationships between Stability Analysis with Total and Effective Stresses," *Bull. 15, Danish Geotech. Inst.*, Copenhagen.

HARR, M. E. (1962), *Groundwater and Seepage.* New York: McGraw-Hill Book Co., Inc.

HAYTHORNTHWAITE, R. M. (1960), "Stress and Strain in Soils," *Plasticity*, p. 185. London: Pergamon Press Ltd.

HAYTHORNTHWAITE, R. M. (1960), "Mechanics of the Triaxial Test for Soils," *J. Soil Mech. and Found. Div. Am. Soc. Civil Engrs.*, 86, SM5, p. 35.

HAYTHORNTHWAITE, R. M. (1961), "Methods of Plasticity in Land Locomotion Studies," *Proc. 1st Intern. Conf. Mech. Soil-Vehicle Systems*, Turin.

HAZEN, A. (1911), "Discussion of Dams on Sand Foundations," *Trans. Am. Soc. Civil Engrs.*, 73, p. 199.

HENDRICKS, S. B. and M. E. JEFFERSON (1938), "Structure of Kaolin and Talc-Pyrophyllite Hydrates and Their Bearing on Water Sorption of Clays," *Am. Mineralogist*, 23, p. 863.

HENKEL, D. J. (1956), "Effect of Overconsolidation on the Behavior of Clays During Shear," *Geotechnique*, 6, p. 139.

HENKEL, D. J. (1957), "Investigations of Two Long-Term Failures in London Clay

Slopes at Wood Green and Northolt," *Proc. 4th Intern. Conf. Soil Mech. Found. Eng.*, 2, p. 315.

HENKEL, D. J. (1959), "The Relationship between the Strength, Porewater Pressure, and Volume-change Characteristics of Saturated Clays," *Geotechnique*, 9, p. 119.

HENKEL, D. J. (1960), "The Shear Strength of Saturated Remolded Clays," *Proc. Am. Soc. Civil Engrs. Res. Conf. on Shear Strength of Cohesive Soils*, p. 169.

HILDEBRAND, F. B. (1962), *Advanced Calculus for Applications.* Englewood, N.J.: Prentice-Hall, Inc.

HILF, J. W. (1948), "Estimating Construction Pore Pressures in Rolled Earth Dams," *Proc. 2nd Intern. Conf. Soil Mech. Found. Eng.*, 3, p. 234.

HILF, J. W. (1956), "An Investigation of Pore-water Pressure in Compacted Cohesive Soils," *Bur. Reclamation Tech. Memo 654*, U.S. Dept. of Interior, Denver.

HILF, J. W. (1957), "Compacting Earth Dams with Heavy Tamping Rollers," *J. Soil Mech. and Found. Div., Am. Soc. Civil Engrs.*, 83, SM2.

HILL, R. (1950), *The Mathematical Theory of Plasticity.* London: Oxford-Clarendon Press.

HOGENTOGLER, C. A. (1936), "Essentials of Soil Compaction," *Proc. Highway Res. Bd.*, 16, p. 309.

HOGENTOGLER, C. A. (1940), "Discussion on Flexible Surfaces," *Proc. Highway Res. Bd.*, 20, p. 329.

HOLTZ, W. G. and H. J. GIBBS (1954), "Engineering Properties of Expansive Clays," *J. Soil Mech. and Found. Div., Am. Soc. Civil Engrs.*, 80, No. 516.

HOLTZ, W. G. and H. J. GIBBS (1956), "Triaxial Shear Tests on Pervious Gravelly Soils," *J. Soil Mech. and Found. Div., Am. Soc. Civil Engrs.*, 82, SM1.

HOUSEL, W. S. (1943), "Earth Pressure on Tunnels," *Trans. Am. Soc. Civil Engrs.*, 108, p. 1037.

HU, G. (1965), "Bearing Capacity of Foundations with Overburden Shear," *Sols-Soils*, No. 13, p. 11.

HUTCHINSON, J. N. (1961), "A Landslide on a Thin Layer of Quick Clay at Furre, Norway," *Geotechnique*, 11, p. 69.

HVORSLEV, M. J. (1937), "Uber die Festigheitseigenschaften Gestorter Bindinger Boden," *Ingeniorvidenskabelige Skrifter*, No. 45, Danmarks Naturvidenskabelige Samfund, Køhenhavn.

HVORSLEV, M. J. (1949), *Subsurface Exploration and Sampling of Soils for Civil Engineering Purposes.* Waterways Expt. Sta., U.S. Corps of Engrs., Vicksburg, Miss.

IRELAND, H. O. (1954), "Stability Analysis of the Congress Street Open Cut," *Geotechnique*, 4, p. 163.

JAMES, R. G. and P. L. BRANSBY (1970), "Experimental and Theoretical Investigation of a Passive Earth Pressure Problem," *Geotechnique*, 20, p. 17.

JAMES, R. G. and J. A. LORD (1972), "An Experimental and Theoretical Study of an Active Earth Pressure Problem Relevant to Braced Cuts in Sand," *Proc. 5th European Conf. on Soil Mech. and Found. Eng., Lisbon*, 1, p. 29.

JANBU, N. (1955), "Application of Composite Slip Surface for Stability Analysis," *Proc. European Conf. on Stability of Earth Slopes, Stockholm*, 3, p. 43.

JANBU, N., L. BJERRUM, and B. KJAERNSLI (1956), "Veiledning ved losning av fundamenteringsoppgaver," *Norweg. Geotech. Inst. Publ. 16*, Oslo.

JENNINGS, J. E. (1961), "A Comparison Between Laboratory Prediction and Field Observation of Heave of Buildings on Desiccated Soils," *Proc. 5th Intern. Conf. on Soil Mech. and Found. Eng.*, 1, p. 689.

JOHNSON, A. W. and J. R. SALLBERG (1960), "Factors That Influence Field Compaction of Soils," *Highway Res. Bd. Bull. 272.*

JOINT ASCE–USCOLD Comm. on Current U.S. Practice in the Design and Construction of Arch Dams, Embankment Dams, Concrete Gravity Dams (1967), "Design Criteria for Large Dams," *Am. Soc. Civil Engrs.*

Jones, A. (1961), "Tables of Stresses in Three-layer Elastic Systems," *Highway Res. Bd. Bull. 342.*

Jurgenson, L. (1934), "The Application of Theories of Elasticity and Plasticity to Foundation Problems," *Contrib. Soil Mech. 1940,* Boston Soc. Civil Engrs., p. 184.

Karpoff, K. P. (1954), "Pavlovsky's Theory for Phreatic Line and Slope Stability," *Proc. Am. Soc. Civil Engrs.,* 80, No. 386.

Kenney, T. C. (1967), "The Influence of Mineral Composition on the Residual Strength of Natural Soils," *Proc. Geotechnical Conf., Oslo,* 1, p. 123.

Kjaernsli, B. (1958), "Test Results, Oslo Subway," *Proc. Conf. on Earth Pressure Problems, Brussels,* 2, p. 108.

Ko, H. Y. and R. F. Scott (1967), "A New Soil Testing Apparatus," *Geotechnique,* 17, p. 40.

Kolb, C. R. and W. G. Shockley (1957), "Mississippi Valley Geology, Its Engineering Significance," *J. Soil Mech. and Found. Div., Am. Soc. Civil Engrs.,* 83, SM3.

Kötter, F. (1903), "Die Bestimmung des Druckes an Gekrümmten Gleitflächen," *Ber. Akad. der Wiss.,* Berlin.

Krumbein, W. C. and F. J. Pettijohn (1938), *Manual of Sedimentary Petrography.* New York: Appleton-Century.

Ladd, C. C. (1959), "Mechanism of Swelling by Compacted Clay," *Highway Res. Bd. Bull. 245.*

Ladd, C. C. (1964), Stress-Strain Modulus of Clay in Undrained Shear, *J. Soil Mech. and Found. Div. Am. Soc. Civil Engrs.,* 90, SM5, p. 103.

Lambe, T. W. (1953), "The Structure of Inorganic Soils," *Proc. Am. Soc. Civil Engrs.,* 79, No. 315.

Lambe, T. W. (1958a), "Engineering Behavior of Compacted Clays," *J. Soil Mech. and Found. Div., Am. Soc. Civil Engrs.,* 84, SM2, p. 1655.

Lambe, T. W. (1958b), "Structure of Compacted Clay," *J. Soil Mech. and Found. Div., Am. Soc. Civil Engrs.,* 84, SM2.

Lambe, T. W. (1960), "A Mechanistic Picture of Shear Strength in Clay," *Proc. Am. Soc. Civil Engrs. Res. Conf. on Shear Strength of Cohesive Soils,* p. 555.

Lambe, T. W. (1962), "Pore Pressures in a Foundation Clay," *J. Soil Mech. and Found. Div., Am. Soc. Civil Engrs.,* 88, SM2, p. 19.

Lambe, T. W. (1963), "An Earth Dam for the Storage of Fuel Oil," *Proc. 2nd Pan-Am. Conf. Soil Mech. Found. Eng.,* 2, p. 257.

Lambe, T. W. (1964), "Methods of Estimating Settlement," *J. Soil Mech. and Found. Div., Am. Soc. Civil Engrs.,* 90, SM5, p. 43.

Lambe, T. W. and R. T. Martin (1953, 1954, 1955, 1956, 1957), "Composition and Engineering Properties of Soil," *Proc. Highway Res. Bd.,* 32, p. 576; 33, p. 515; 34, p. 566; 35, p. 661; 36, p. 693.

Lane, K. S. (1957), "Effect of Lining Stiffness on Tunnel Loading," *Proc. 4th Intern. Conf. Soil Mech. Found. Eng.,* 2, p. 223.

Leonards, G. A. (ed.), (1962), "The Engineering Properties of Soils," *Foundation Engineering,* p. 66. New York: McGraw-Hill Book Co., Inc.

Leonards, G. A. and P. Girault (1961), "A Study of the One Dimensional Consolidation Test," *Proc. 5th Intern. Conf. Soil Mech. Found. Eng.,* 1, p. 213.

Leonards, G. A. and J. Narain (1963), "Flexibility of Clay and Cracking of Earth Dams," *J. Soil Mech. and Found. Div., Am. Soc. Civil Engrs.,* 89, SM2, p. 47.

Leonards, G. A. and B. K. Ramiah (1959), "Time Effects in the Consolidation of Clays," *Am. Soc. Testing Mat. Spec. Tech. Publ. 254,* p. 116.

Li, C. Y. (1959), "Construction Pore Pressures in an Earth Dam," *J. Soil Mech. and Found. Div., Am. Soc. Civil Engrs.,* 85, SM5, p. 43.

Lo, K. Y. (1962), "Secondary Compression of Clays," *J. Soil Mech. and Found. Div., Am. Soc. Civil Engrs.,* 87, SM4, p. 61.

Lo, K. Y. (1970), "The Operational Strength of Fissured Clays," *Geotechnique*, 20, p. 57.
Lobeck, A. K. (1939), *Geomorphology*. New York: McGraw-Hill Book Co., Inc.
Low, P. F. (1959a), "Discussion on Physico-Chemical Properties of Soils: Ion Exchange Phenomena," *J. Soil Mech. and Found. Div., Am. Soc. Civil Engrs.*, 85, SM2.
Low, P. F. (1959b), "Viscosity of Water in Clay Systems," *Proc. 8th Natl. Conf. Clays and Clay Minerals*, p. 170. London: Pergamon Press Ltd.
Low, P. F. and C. W. Lovell (1959), "The Factor of Moisture in Frost Action," *Highway Res. Bd. Bull. 225*.
Mansur, C. I. (1957), "Laboratory and in-situ Permeability of Sand," *J. Soil Mech. and Found. Div., Am. Soc. Civil Engrs.*, 83, SM1.
Mansur, C. I. and R. I. Kaufman (1956), "Control of Underseepage, Mississippi River Levees," *J. Soil Mech. and Found. Div., Am Soc. Civil Engrs.*, 82, SM1.
Mansur, C. I. and A. H. Hunter (1970), "Pile Tests. Arkansas River Project," *J. Soil Mech. and Found. Div., Am. Soc. Civil Engrs.*, 96, SM5, p. 1545.
Marsland, A. (1961), "A Study of a Breach in an Earthen Embankment Caused by Uplift Pressures," *Proc. 5th Intern. Conf. Soil Mech. Found. Eng.*, 2, p. 663.
Martin, R. T. (1960), "Adsorbed Water on Clay: A Review," *Proc. 9th Natl. Conf. Clays and Clay Minerals*, p. 28. London: Pergamon Press.
McClelland, B., J. A. Focht, and W. J. Emrich (1969), "Problems in Design and Installation of Offshore Piles," *J. Soil Mech. and Found. Div., Am. Soc. Civil Engrs.*, 95, SM6, p. 1491.
McLaughlin, W. W. and O. L. Stokstad (1946), "Design of Flexible Surfaces in Michigan," *Proc. Highway Res. Bd.*, 25, p. 39.
McNamee, J. and R. E. Gibson (1960), "Plane Strain and Axially Symmetric Problems of the Consolidation of a Semi-infinite Clay Stratum," *Quart. J. Mech. Appl. Math.*, 13. p. 210.
Meade, R. H. (1964), "Removal of Water and Rearrangement of Particles During the Compaction of Clayey Sediments—Review," *U.S. Geol. Survey, Prof. Paper 497B*.
Meyerhoff, G. G. (1951), "Ultimate Bearing Capacity of Foundations," *Geotechnique*, 2, p. 301.
Meyerhoff, G. G. (1953), "The Bearing Capacity of Foundations under Eccentric and Inclined Loads," *Proc. 3rd Intern. Conf. Soil Mech. Found. Eng.*, 1, p. 440.
Meyerhoff, G. G. (1956), "Penetration Tests and Bearing Capacity of Cohesionless Soils," *J. Soil Mech. and Found. Div., Am. Soc. Civil Engrs.*, 82, SM1, p. 5.
Michaels, A. S. and C. S. Lin (1954), "The Permeability of Kaolinite," *Ind. Eng. Chem.*, 46, p. 1239.
Mindlin, R. D. (1936), "Pressure Distribution on Retaining Walls," *Proc. 1st Intern. Conf. Soil Mech. Found. Eng.*, 3, p. 155.
Mitchell, J. K. (1956), "Fabric of Natural Clays and Its Relation to Engineering Properties," *Proc. Highway Res. Bd.*, 35, p. 693.
Mitchell, J. K. (1960a), "The Application of Colloidal Theory to the Compressibility of Clays," *Interparticle Forces in Clay—Water—Electrolyte Systems*, ed. R. H. G. Parry, p. 292. Commonwealth Sci. and Indus. Res. Org., Melbourne.
Mitchell, J. K. (1960b), "Fundamental Aspects of Thixotropy in Soils," *J. Soil Mech. and Found. Div., Am. Soc. Civil Engrs.*, 86, SM3, p. 19.
Mitchell, J. K. and W. S. Gardner (1971), "Analysis of Load Bearing Fills over Soft Subsoils," *J. Soil Mech. and Found. Div., Am. Soc. Civil Engrs.*, 97, SM11, p. 1549.
Moretto, O. (1948), "Effect of Natural Hardening on the Unconfined Compression Strength of Remolded Clays," *Proc. 2nd Intern. Conf. Soil Mech. Found. Eng.*, 1, p. 137.

MORGENSTERN, N. R. and V. E. PRICE (1965), "Analysis of the Stability of General Slip Surfaces," *Geotechnique*, 15, p. 79.
MORGENSTERN, N. R. and J. S. TCHALENKO (1967a), "Microscopic Structures in Kaolin Subjected to Direct Shear," *Geotechnique*, 17, p. 309.
MORGENSTERN, N. R. and J. S. TCHALENKO (1967b), "The Optical Determination of Preferred Orientation in Clays and Its Application to the Study of Microstructure in Consolidated Kaolin," *Proc. Royal Soc. (London), Ser. A.*, 300, p. 218.
MUSKAT, M. (1937), *Flow of Homogeneous Fluids through Porous Media.* Ann Arbor, Mich.: Edwards Brothers, Inc.
NADAI, A. L. (1963), *Theory of Flow and Fracture of Solids*, Vol. 2, McGraw-Hill Book Co., Inc.
NEWMARK, N. M. (1935), "Simplified Computation of Vertical Pressures in Elastic Foundations," *Univ. of Illinois Eng. Expt. Sta. Circ. 24*.
NEWMARK, N. M. (1940), "Stress Distribution in Soils," *Proc. Purdue Conf. on Soil Mech. and Its Applications*, Purdue University, p. 295.
NEWMARK, N. M. (1942), "Influence Charts for Computation of Stresses in Elastic Foundations," *Univ. Illinois Eng. Expt. Sta. Bull. 338*.
OLSEN, H. W. (1960), "Hydraulic Flow through Saturated Clays," *Proc. 9th Natl. Conf. Clays and Clay Minerals*, p. 131. London: Pergamon Press Ltd.
OLSON, R. E. (1962), "The Shear Strength Properties of Calcium Illite," *Geotechnique*, 12, p. 23.
OLSON, R. E. (1963), "Effective Stress Theory of Soil Compaction," *J. Soil Mech. and Found. Div., Am Soc. Civil Engrs.*, 89, SM2, p. 27.
OLSON, R. E. and F. M. MITRONOVAS (1960), "Shear Strength and Consolidation Characteristics of Calcium and Magnesium Illite," *Proc. 9th Natl. Conf. Clays and Clay Minerals*, p. 185. London: Pergamon Press Ltd.
PALMER, L. A. and E. S. BARBER (1940), "Soil Displacement under Circular Loaded Areas," *Proc. Highway Res. Bd.*, 20, p. 279.
PECK, R. B. (1943), "Earth Pressure Measurements in Open Cuts, Chicago Subway," *Trans. Am. Soc. Civil Engrs.*, 108, p. 1008.
PECK, R. B. (1953), "Subsoil Exploration—Denver Coliseum," *Proc. Am. Soc. Civil Engrs.*, 79, No. 326.
PECK, R. B. (1969a), "Advantages and Limitations of the Observational Method in Applied Soil Mechanics," *Geotechnique*, 19, p. 171.
PECK, R. B. (1969b), "Deep Excavations and Tunneling in Soft Ground," *7th Intern. Conf. on Soil Mechanics and Found. Eng.*, State of Art Vol., p. 225.
PECK, R. B. and F. G. BRYANT (1953), "The Bearing Capacity Failure of the Transcona Elevator," *Geotechnique*, 3, p. 201.
PECK, R. B., W. E. HANSON, and T. H. THORBURN (1974), *Foundation Engineering*. New York: John Wiley & Sons, Inc.
PECK, R. B. and W. V. KAUN (1948), "Description of a Flow Slide in Loose Sand," *Proc. 2nd Intern. Conf. Soil Mech. Found. Eng.*, 2, p. 31.
PECK, R. B. and M. E. UYANIK (1955), "Observed and Computed Settlements of Structures in Chicago," *Univ. Illinois Eng. Expt. Sta. Bull. 429*.
PERLOFF, W. H. and I. E. POMBO (1969), "End Restrain Effects in the Triaxial Test," *Proc. 7th Intern. Conf. on Soil Mech. and Found. Eng.*, 1, p. 327.
PETERSON, R. and N. PETERS (1963), "Heave of Spillway Structures on Clay Shales," *Canad. Geotech. J.*, 1, p. 5.
PICKETT, A. G. and M. M. LEMCOE (1959), "An Investigation of the Shear Strength of the Clay-Water System by Radio-frequency Spectroscopy," *J. Geophys. Res.*, 64, p. 1579.
POLUBARINOVA-KOCHINA, P. YA. (1962), *Theory of Ground Water Movement*, trans. by J. M. R. deWiest. Princeton, N.J.: Princeton University Press.
POOROOSHASB, H. B., I. HOLUBEC, and A. N. SHERBOURNE (1966, 1967), "Yielding

and Flow of Sand in Triaxial Compression," *Canad. Geot. J.*, 3, p. 179 and 4, p. 376.
PRANDTL, L. (1920), "Uber die Harte Plastisher Korper, Nachrichten Kon. Gesell. der Wissenschaften," *Math. Phys. Klasse*, p. 74.
PROCTOR, R. R. (1933), "Design and Construction of Rolled Earth Dams," *Eng. News Record*, 8, No. 10.
RAYMOND, G. P. (1972), "The Kars (Ontario) Embankment Foundation, Performance of Earth and Earth Supported Structures," *Am. Soc. Civil Engrs.*, 1, p. 319.
REINUS, E. (1955), "The Stability of the Slopes of Earth Dams," *Geotechnique*, 5, p. 181.
RENDULIC, L. (1937), "Ein Grundgesetz der Tonmechanik und sein Experimentaller Beweis," *Der Bauingenieur*, 18, p. 459.
RICE, F. O. and E. TELLER (1949), *The Structure of Matter*. New York: John Wiley & Sons, Inc.
RICHART, F. E. (1957), "A Review of Theories for Sand Drains," *J. Soil Mech. and Found. Div., Am. Soc. Civil Engrs.*, 83, SM3.
RICO, A., G. MORENO, and G. GARCIA (1969), "Test Embankments on Texcoco Lake," *Proc. 7th Intern. Conf. on Soil Mech. and Found. Eng.*, 2, p. 669.
ROSCOE, K. H. (1953), "An Apparatus for the Application of Simple Shear to Soil Samples," *Proc. 4th Intern. Conf. Soil Mech. and Found. Eng.*, 1, p. 186.
ROSCOE, K. H. (1970), *The Influence of Strain in Soil Mechanics, Geotechnique*, 20, p. 129.
ROSCOE, K. H., R. H. BASSETT, and E. R. L. COLE (1967), "Principal Axes Observed During Simple Shear of a Sand," *Proc. Geotechnical Conf., Oslo*, 1, p. 231.
ROSCOE, K. H., A. N. SCHOFIELD, and A. THURAIRAJAH (1963a), "Yielding of Clays in States Wetter than Critical," *Geotechnique*, 13, p. 211.
ROSCOE, K. H., A. N. SCHOFIELD, and A. THURAIRAJAH (1963b), "An Evaluation of Test Data for Selecting a Yield Criterion for Soils, Laboratory Shear Testing of Soils," *Am. Soc. Testing Mat., Spec. Tech. Pub.*, No. 361.
ROSCOE, K. H., A. N. SCHOFIELD, and C. P. WROTH (1958), "On the Yielding of Soils," *Geotechnique*, 8, p. 22.
ROSENQUIST, I. T. (1959), "Physico Chemical Properties of Soils: Soil Water Systems," *J. Soil Mech. and Found. Div., Am. Soc. Civil Engrs.*, 85, SM2.
ROWE, P. W. (1962), "The Stress-dilatancy Relation for Static Equilibrium of an Assembly of Particles in Contact," *Proc. Roy. Soc. (London)*, Ser. A269, p. 500.
SCHEIDEGGER, A. E. (1957), *The Physics of Flow through Porous Media*. New York: The Macmillan Company.
SCHIFFMAN, R. L., C. C. LADD, and A. T. F. CHEN (1964), "The Secondary Consolidation of Clay," *Symp. on Rheology and Soil Mech., Intern. Union of Theor. and Appl. Mech., Grenoble*.
SCHIFFMAN, R. L., A. T. F. CHEN, and J. C. JORDAN (1967), "An Analysis of Consolidation Theories," *J. Soil Mech. and Found. Div., Am. Soc. Civil Engrs.*, 95 SM1, p. 285.
SCHLEICHER, F. (1926), "Zur Theorie des Baugrundes," *Der Bauingenieur*, 7, p. 931, 949.
SCHMERTMANN, J. H. (1953), "Undisturbed Consolidation Behavior of Clay," *Trans. Am. Soc. Civil Engrs.*, 120, p. 1201.
SCHMERTMANN, J. H. (1970), "Static Cone to Compute Static Settlement over Sand," *J. Soil Mech. and Found. Div., Am. Soc. Civil Engrs.*, 96, SM3, p. 1011.
SCHOFIELD, R. K. and H. SAMSON (1954), "Flocculation of Kaolinite Due to the Attraction of Oppositely Charged Faces," *Disc. Faraday Soc.*, 18, p. 135.
SCOTT, R. F. (1963), *Principles of Soil Mechanics*. Reading, Mass.: Addison-Wesley Publishing Co., Inc.

SEED, H. B. and C. K. CHAN (1959), "Structure and Strength Characteristics of Compacted Clays," *J. Soil Mech. and Found. Div., Am. Soc. Civil Engrs.*, 85, No. 5.
SEED, H. B. and L. B. REESE (1955), "Action of Soft Clay Along Friction Piles," *J. Soil Mech. and Found. Div., Am. Soc. Civil Engrs.*, 81, No. 842.
SEED, H. B., R. J. WOODWARD, Jr., and R. LUNDGREN (1964), "Clay Mineralogical Aspects of the Atterberg Limits," *J. Soil Mech. and Found. Div., Am. Soc. Civil Engrs.*, 90, SM4, p. 107.
SEROTA, S. and R. A. J. JENNINGS (1959), "The Elastic Heave of the Bottom of Excavations," *Geotechnique*, 9, p. 62.
SEVALDSON, R. A. (1956), "The Slide in Lodalen," *Geotechnique*, 6, p. 167.
SHARPE, C. F. S. (1938), *Landslides and Related Phenomena*. New York: Columbia University Press.
SHEA, P. H. and H. E. WHITSETT (1958), "Predicting Seepage Under Dams on Multilayered Foundations," *J. Soil Mech. and Found. Div., Am. Soc. Civil Engrs.*, 84, SM3.
SHIELD, R. T. (1953), "Mixed Boundary Value Problems in Soil Mechanics," *Quart. Appl. Math.*, 11, p. 61.
SHIELD, R. T. (1955), "On Coulomb's Law of Failure in Soils," *J. Mech. Phys. Solids*, 4, p. 10.
SKEMPTON, A. W. (1942), "An Investigation of the Bearing Capacity of a Soft Clay," *J. Inst. Civil Engrs.*, 18, p. 307.
SKEMPTON, A. W. (1948a), "The Geotechnical Properties of a Deep Stratum of Post-glacial Clay at Gosport," *Proc. 2nd Intern. Conf. Soil Mech. Found. Eng.*, 1, p. 145.
SKEMPTON, A. W. (1948b), "The Rate of Softening of Stiff Fissured Clays with Special Reference to London Clay," *Proc. 2nd Intern. Conf. Soil Mech. and Found. Eng.*, 2, p. 50.
SKEMPTON, A. W. (1951), "The Bearing Capacity of Clays," *Proc. British Bldg. Res. Cong.*, 1, p. 180.
SKEMPTON, A. W. (1953), "The Colloidal Activity of Clays," *Proc. 3rd Intern. Conf. Soil Mech. Found. Eng.*, 1, p. 57.
SKEMPTON, A. W. (1954), "The Pore Pressure Coefficients A and B," *Geotechnique*, 4, p. 148.
SKEMPTON, A. W. (1959), "Cast In-Situ Bored Piles in Clay," *Geotechnique*, 4, p. 153.
SKEMPTON, A. W. (1964), "Long Term Stability of Clay Slopes," *Geotechnique*, 14, p. 77.
SKEMPTON, A. W. and A. W. BISHOP (1955), "The Gain in Stability Due to Porepressure Dissipation in a Soft Clay Foundation," *Trans. 5th Cong. on Large Dams*, 1, p. 613.
SKEMPTON, A. W. and L. BJERRUM (1957), "A Contribution to Settlement Analysis of Foundations on Clay," *Geotechnique*, 7, p. 168.
SKEMPTON, A. W. and H. Q. GOLDER (1948), "Practical Examples of $\phi = 0$ Analysis of Slopes in Clays," *Proc. 2nd Intern. Conf. Soil Mech. Found. Eng.*, 2, p. 63.
SKEMPTON, A. W. and J. HUTCHINSON (1969), "Stability of Natural and Embankment Foundations," *Proc. 7th Intern. Conf. on Soil Mech.*, State of Art Vol., p. 291.
SKEMPTON, A. W. and D. H. MACDONALD (1956), "The Allowable Settlement of Buildings," *Proc. Inst. Civil Engrs.*, 5, No. 3, p. 727.
SKEMPTON, A. W. and R. D. NORTHEY (1952), "The Sensitivity of Clays," *Geotechnique*, 3, p. 30.
SKEMPTON, A. W. and D. J. PETLEY (1967), "The Strength Along Structural Discontinuities in Stiff Clays," *Proc. Geotechnical Conf., Oslo*, 2, p. 29.
SKEMPTON, A. W. and W. H. WARD (1952), "Investigation Concerning a Deep Cofferdam in the Thames Estuary Clay at Shellhaven," *Geotechnique*, 3, p. 119.

SOKOLOVSKI, V. V. (1956), *Statics of Soil Media*, trans. by D. H. Jones and A. N. Schofield. London: Butterworth & Co. Ltd.
SOUTHWELL, R. V. (1946), *Relaxation Methods in Theoretical Physics*. London: Oxford University Press.
SOWERS, G. F. (1962), "Shallow Foundations," *Foundation Engineering*, ed. G. A. Leonards, p. 525. New York: McGraw-Hill Book Co., Inc.
STATE HIGHWAY COMMISSION OF KANSAS (1947), "Design of Flexible Pavement Using the Triaxial Method," *Highway Res. Bd. Bull.*, 8.
STREETER, V. L. and E. B. WYLIE (1975), *Fluid Mechanics*. New York: McGraw-Hill Book Co.
SU, H. H. and R. H. PRYSOCK (1972), "Settlement Analysis of Two Highway Embankments," Performance of Earth and Earth-Supported Structures, *Am. Soc. Civil Engrs.*, 1, p. 465.
SYLWESTROWICZ, W. (1953), "Experimental Investigation of the Behavior of Soil Under a Punch or Footing," *J. Mech. Phys. Solids*, 1, p. 258.
TAN, T. K. (1958), "Secondary Time Effects and Consolidation of Clays," *Scientia Sinica*, 7, No. 11, p. 1060.
TAYLOR, D. W. (1937), "Stability of Slopes," *Contrib. Soil Mech. 1940*, Boston Soc. Civil Engrs., p. 337.
TAYLOR, D. W. (1942), "Research on Consolidation of Clays," *Dept. of Civil and Sanitary Engr.*, Serial No. 92, Mass. Inst. Tech.
TAYLOR, D. W. (1948), *Fundamentals of Soil Mechanics*. New York: John Wiley & Sons, Inc.
TERZAGHI, K. (1925a), "Principles of Soil Mechanics. III. Determination of Permeability of Clay," *Eng. News Record*, 95, p. 832.
TERZAGHI, K. (1925b), *Erdbaumechanik auf Bodenphysikalischer Grundlage*. Vienna: Deuticke.
TERZAGHI, K. (1929), "Effect of Minor Geologic Details on the Safety of Dams," *Am. Inst. Mining Met. Engrs. Tech. Publ. 215*, p. 31.
TERZAGHI, K. (1934), "Large Retaining Wall Tests. I. Pressure of Dry Sands," *Eng. News Record*, 112, p. 136.
TERZAGHI, K. (1936a), "Distribution of Lateral Pressure on Sand on the Timbering of Cuts," *Proc. 1st Intern. Conf. Soil Mech. Found. Eng.*, 1, p. 211.
TERZAGHI, K. (1936b), "A Fundamental Fallacy in Earth Pressure Calculations," *Contrib. Soil Mech. 1940*, Boston Soc. Civil Engrs., p. 277.
TERZAGHI, K. (1936c), "The Shearing Resistance of Saturated Soils and the Angle between the Planes of Shear," *Proc. 1st Intern. Conf. Soil Mech. Found. Eng.*, 1, p. 54.
TERZAGHI, K. (1936d), "Stress Distribution in Dry and Saturated Sand Above a Yielding Trap-Door," *Proc. 1st Intern. Conf. Soil Mech. Found. Eng.*, 1, p. 307.
TERZAGHI, K. (1937), "Settlement of Structures in Europe and Methods of Observations," *Trans. Am. Soc. Civil Engrs.*, 103, p. 1432.
TERZAGHI, K. (1941), "General Wedge Theory of Earth Pressures," *Trans. Am. Soc. Civil Engrs.*, 106, p. 68.
TERZAGHI, K. (1943a), "Liner-plate Tunnels on the Chicago Subway," *Trans. Am. Soc. Civil Engrs.*, 108, p. 970.
TERZAGHI, K. (1943b), *Theoretical Soil Mechanics*. New York: John Wiley & Sons, Inc.
TERZAGHI, K. (1954), "Anchored Bulkheads," *Trans. Am. Soc. Civil Engrs.*, 119, p. 1243.
TERZAGHI, K. (1957), "Varieties of Submarine Slope Failures," *Norweg. Geotech. Inst. Publ. 25*, Oslo.
TERZAGHI, K. (1958), "Design and Performance of the Susumua Dam," *Proc. Inst. Civil Engrs.*, 9, p. 369.

TERZAGHI, K. (1960), *From Theory to Practice in Soil Mechanics*, ed. by L. Bjerrum, A. Casagrande, R. B. Peck, and A. W. Skempton. New York: John Wiley & Sons, Inc.
TERZAGHI, K. and T. M. LEPS (1958), "Design and Performance of Vermillion Dam, California," *J. Soil Mech. and Found. Div., Am. Soc. Civil Engrs.*, 84, SM3.
TERZAGHI, K. and R. B. PECK (1967), *Soil Mechanics in Engineering Practice*, 2nd ed. New York: John Wiley & Sons, Inc.
TERZAGHI, K. and R. B. PECK (1957), "Stabilization of an Ore Pile by Drainage," *J. Soil Mech. and Found. Div., Am Soc. Civil Engrs.*, 83, SM1.
TIMOSHENKO, S. P. and J. N. GOODIER (1951), *Theory of Elasticity*, 2nd ed. New York: McGraw-Hill Book Co., Inc.
TSCHEBOTARIOFF, G. P. (1962), "Retaining Structures," *Foundation Engineering*, ed. by G. A. Leonards, p. 438. New York: McGraw-Hill Book Co., Inc.
VERDEYEN, J. and V. ROISIN (1961), "Relationships between External Concentrated Load Effects and the Characteristics of Flexible Bulkheads," *Proc. 5th Intern. Conf. Soil Mech. Found. Eng.*, 2, p. 501.
VERWEY, E. J. W. and J. TH. G. OVERBEEK (1948), *Theory of Stability of Lyophobic Colloids*. Amsterdam: Elsevier Publishing Co.
VESIC, A. S. (1967), "Ultimate Loads and Settlements of Deep Foundations in Sand," *Bearing Capacity and Settlement of Foundations*, Duke Univ., p. 53.
VESIC, A. S. (1973), "Analysis of Ultimate Loads of Shallow Foundations," *J. Soil Mech. and Found. Div., Am. Soc. Civil Engrs.*, 99, SM1, p. 45.
VESIC, A. S. (1963), "Bearing Capacity of Deep Foundations in Sand," *Highway Res. Rec. 39*.
WARD, W. H. (1952), "Physics of Deglaciation in Central Baffin Island," *J. Glaciology*, 2, p. 9.
WARD, W. H. and T. K. CHAPLIN (1957), "Existing Stresses in Several Old London Underground Tunnels," *Proc. 4th Intern. Conf. Soil Mech. Found. Eng.*, 2, p. 256.
WARD, W. H., A. D. M. PENMAN, and R. E. GIBSON (1955), "Stability of a Bank on a Thin Peat Layer," *Geotechnique*, 5, p. 154.
WARD, W. H., S. G. SAMUELS, and M. E. BUTLER (1959), "Further Studies of the Properties of London Clay," *Geotechnique*, 14, p. 33.
WATERWAYS EXPERIMENT STATION (1953), "Unified Soil Classification System," *U.S. Corps of Engrs. Tech. Memo 3-357*, Vicksburg, Miss.
WENZEL, L. K. (1942), "Methods for Determining Permeability of Water-bearing Materials, *U.S. Geol. Survey Water Supply Paper 887*.
WHITMAN, R. V. (1959), "Notes on the Basic Principles of Shear Strength," *Mass. Inst. Tech. Soil Eng. Div. Publ. 102*.
WHITMAN, R. V. and K. A. HEALY (1962), "Shear Strength of Sands During Rapid Loading," *J. Soil Mech. and Found. Div. Am. Soc. Civil Engrs.*, 88, SM2.
WOODS, K. B., R. D. MILES, and C. W. LOVELL, JR. (1962), "Origin Formation, and Distribution of Soils in North America," *Foundation Engineering*, ed. by G. A. Leonards, p. 1. New York: McGraw-Hill Book Co., Inc.
WORLEY, H. E. (1943), "Triaxial Testing Methods Usable in Flexible Pavement Design," *Proc. Highway Res. Bd.*, 23, p. 109.
WROTH, C. P. and B. SIMPSON (1972), "An Induced Failure at a Trial Embankment. Part II Finite Element Computations." Performance of Earth and Earth-Supported Structures, *Am. Soc. Civil Engrs.*, 1, p. 65.
WU, T. H. (1957), "Relative Density and Shear Strength of Sands," *J. Soil Mech. and Found. Div., Am. Soc. Civil Engrs.*, 83, SM1.
WU, T. H. (1958), "Geotechnical Properties of Glacial Lake Clays," *J. Soil Mech. and Found. Div., Am. Soc. Civil Engrs.*, 84, SM3.
WU, T. H. and S. BERMAN (1953), "Earth Pressure Measurement in Open Cut, Contract D8, Chicago Subway," *Geotechnique*, 3, p. 248.

ZANGAR, C. N. (1953), "Theory and Problems of Water Percolation," *Bur. Reclamation Eng. Monograph 8*, U.S. Dept. of Interior, Denver.

ZEEVAERT, L. (1957), "Consolidation of Mexico City Volcanic Clay," *ASTM Spec. Tech. Publ. 232*, p. 18.

ZIENKIEWICZ, O. C. (1967), *The Finite Element Method in Structural and Continuum Mechanics*. New York: McGraw-Hill Publishing Co.

Index

Index

A

Activity ratio, 19, 415
Adhesion, 238, 287–288
Aeolian soils, 369, 379–381
Alluvial soils, 79, 369–374
Anisotropic materials, flow through, 64–65
Aquifer, 50, 52, 60, 84
Atterberg limits, 17–27, 386, 413–415
Atomic bonds, 393, 395
Attractive potential, 403–404, 410, 412
Auger borings, 385

B

Back-swamp deposits, 372, 374
Beam subjected to deflection, 150–153
Bearing capacity, 170, 227–247, 258, 270, 271, 272, 275, 311, 322, 323–326, 336, 338, 339–340, 358–365
Bearing capacity factors, 236, 243–244, 272
Boiling, 44, 45, 46, 75–77, 82
Boltzmann equation, 399
Borings, 384–385, 389–390
Boussinesq equations, 131, 138, 173
Braced excavations, 293–299, 306–309, 322, 328
Braided-stream deposits, 369–372, 377
Brucite, 395

C

Capillary action, 30, 31–33, 208
Capillary rise, 32, 33, 45
Cations, 398, 400, 402, 407–408, 411
Cementation, 114, 380
Channel fillings, 374
Characteristics, method of, 230, 360
Chlorite, 396
Circular arc method, 247–254, 258, 271, 327
Clay, 2–3, 20, 22, 25, 26, 97, 104, 372, 375, 378–379, 382–383, 393–416
Clay blanket, 82
Clay minerals, 7, 17, 18, 19, 21, 393–398, 409, 413
Clays:
  compacted, 330–332
  sensitive, 8, 21, 113
  stiff-fissured, 323, 324
  swelling, 382–383, 411

Coefficient of permeability (*see* Permeability)
Cohesion, 188, 214, 412
Cohesion, true, 216–217, 416
Compaction, 13–17, 28, 208, 330, 344, 415
  AASHO, 15
  California Bearing Ratio, 15
  kneading, 15
  Proctor, 15, 16
Compatibility, 134, 135
Complex potential, 56
Complex velocity, 57
Compressibility, 95, 97, 99, 100, 107–110, 124, 125, 164, 181, 379, 381, 390
Compression index, 10
Cone penetration test, 162, 387–388
Confining pressure, 161, 163, 198, 199, 382
Consolidated-drained test (*see* Drained test)
Consolidated-undrained test, 194, 196, 198, 204, 210, 212–213, 218, 219, 334
Consolidation, 107, 160, 163, 164, 184, 314, 316, 320, 325–326, 352–358, 365–366, 381
  coefficient of, 117, 124, 353–354
  degree of, 115–116, 119–120, 123, 128
  field, 111, 114
  percentage (*see* Consolidation, degree of)
  rate of, 114, 366
  theory of, 116–120, 178
Consolidation test, 99, 107–110, 128, 164–165, 167, 410
Contact surface, 41
Continuity equation, 52–54, 116, 348–352
Coulomb's theory, 282–289, 292, 294, 297, 303, 308
Covalent bonds, 395, 396
Cracking, 286–288, 341
Cracking of a dam, 153
Creep, 92–93
Critical gradient, 76–77

Critical height of a cut, 286–287
Critical void ratio, 198
Culmann's method, 284, 289–291

D

Dams, 58–59, 76–77, 81–82 (*see also* Earth dams)
Darcy's law, 34–35, 48, 49, 50, 52, 77–78, 349
Deformation conditions for earth pressures, 277–280, 293–295, 303–307
Density, 9, 10, 13–17, 370
Density, dry, 10, 11, 13–14, 15, 16, 28, 415
Density index (*see* Relative density)
Density of solid particles, 10
Dipoles, 403, 407–408
Dispersion, 404
Double layer, 398–399, 402, 410–411
Drained condition, 94, 102, 160–161, 190, 195, 258, 292–293, 322, 325, 327
Drained state (*see* Drained condition)
Drained test, 189–190, 194, 196, 198, 200, 205, 213
Drains, 70, 72, 82, 343
Drawdown, 312–314, 332–333, 344
Dupuit's assumption, 72, 73

E

Earth dams, 69, 72–74, 79–80, 209, 312, 330–334, 338
Earth pressure, 156, 157, 276–310, 322–323, 328
  active, 224–227, 272, 274, 277–286, 292–293, 295–296
  coefficient of, 224, 278, 284, 306
  distribution, 224, 278, 285–286, 288, 291–292, 294–298, 303, 306–308, 309
  field measurement of, 297–299
  passive, 224–227, 234, 274, 277–286, 295–296

Earth pressure (*cont.*)
 at rest, 99, 224, 277–278, 299–300, 331
 due to surcharge load, 289–292
 on tunnel linings, 303–305
Eccentric loads on foundations, 237–238
Effective diameter, 6, 39
Effective stress, 40–44, 46, 96, 115, 123, 157, 181, 189–190, 196–204, 208–209, 213, 410, 412
Elasticity, 91–92, 93, 130–159, 291
Electric potential, 399–403, 409
Electrical resistivity, 389
Embankments, 72–74, 315–317, 330, 344
Energy, interaction, 403–404
Energy dissipation, 259–263, 270
Equilibrium, 132–133, 134, 221, 229, 249–250, 254–255, 257, 259, 288, 300, 400, 411
Equipotential lines, 55, 56, 57, 59–61, 64, 70, 76–77, 84, 350
Eskers, 378
Excavation, 220, 273, 293–294, 299, 305–307, 317–318, 327
Expansibility, 97

F

Failure, 187–189, 207–208, 211, 221–222, 249–250, 259–260, 271, 318, 320, 321–323, 337–338
Failure by heave, 44, 76
Failure envelope, 188–189, 191–193, 196–206, 208, 214, 216
Failure plane (*see* Failure surface)
Failure surface, 99, 192–193, 204–208, 218, 222–223, 227, 228–231, 232–236, 238, 243, 245, 246–250, 252, 254, 256, 262, 267–268, 280–286, 288–291, 294–295, 307, 309, 328, 343, 359, 361, 364, 411–412 (*see also* Slip line)
Failure theories, 187–189, 211

Field, electric, 398, 403, 410
Finite element method, 346
Finite differences, 346–348, 352, 357, 360–361
Flocculation, 404–405
Flow:
 between parallel plates, 35, 36
 rate of, 34, 49, 50, 52–53, 56, 63, 116, 348, 353
 through tubes, 36
Flow lines, 56, 59, 60, 61, 63, 65, 66, 68, 70, 71, 74, 80, 84, 350
Flow net, 59–61, 64, 65, 66, 67, 68, 70, 71, 72, 73, 80, 82, 258, 292, 321
Flow slides, 8, 330
Fluvial soils (*see* Alluvial soils)
Footings, model, 161, 228
Foundations, 324–325, 334–341, 344, 390–391
 on clay, 174–179, 315–317, 323–326
 on cohesionless soils, 160–162, 179–182, 245, 258, 328–330,
 deep, 181, 237–246, 334–337, 339–340
 footing, 168, 184, 227–237, 344, 391
 mat, 171, 183, 339
 pier, 334–337
 pile, 275, 334–337
 raft (*see* Foundations, mat)
 shallow, 235–237, 245, 246
 strip (*see* Load, strip)
Fourier integral, 154
Fourier series, 118, 150–151
Friction:
 between soil and foundation, 238, 336–337
 between soil and wall, 281–282, 284–285

G

General wedge theory, 294–295, 307
Geophysical methods, 388–389
Glacial-fluvial soils, 377–378
Glacial lake deposits, 378–379, 390

Glacial soils, 369, 374–379
Gravel, 2, 22, 43, 78, 372, 377–378, 390
Groundwater, 61–62, 105, 181, 320–321, 411
Group index, 22
Grouser plate, 260–263
Guoy-Chapman equation, 398, 402

H

Halloysite, 396, 415
Head, 31, 45, 55, 59, 63, 64, 68–70, 350
Head, pressure, 30, 45
Hooke's law, 91–92, 131–132, 134
Hydraulic gradient, 34, 43, 48, 59, 63, 76, 77, 78, 80
Hydraulic radius, 37, 39
Hydrogen bonds, 395, 407–408, 412
Hydrometer analysis, 4–6

I

Illite, 19, 20, 396–397, 412–414
Index properties, 1–21, 27, 28, 383, 413–414
Influence charts, 139–143, 167
Influence number, 140, 148, 161
Internal friction, 188, 214–215, 387, 412
Internal friction, true, 216–217, 415–416
Invariants:
 strain, 90
 stress, 90
Ion exchange, 408–409, 411
Ionic bonds, 395, 396
Isochrones, 119, 120, 123
Isomorphous substitution, 396, 398

K

Kames, 378

Kaolinite, 19, 20, 393–394, 396–398, 412–414
Kinematically admissible state, 259–262, 270
Kötter's equations, 231

L

Laminar flow, 35
Layered medium, stresses in, 145–146
Limestone, 383–384
Liquefaction, 330
Liquid limit, 17–19, 22, 25
Liquidity index, 20, 406
Load, strip, 142–145, 227–237, 267–270, 272, 360–365
Load tests, 168, 169–170, 184, 328, 390
Loading conditions, 311–314, 322, 325, 346
Loading, rate of, 216–217, 321, 325
Loess, 380–381
Logarithmic spiral, 228, 232, 234–235, 238, 240, 247, 262, 286, 307
Lower bound solutions, 258–263, 274

M

Marine deposits, 405
Marl, 369
Meander belt deposits, 372–374
Meniscus, 31, 32, 33, 42, 208
Method of slices, 254–258, 273
Modulus, 168, 177, 180
 bulk, 92, 93, 161, 164
 constrained, 99, 108
 shear, 92, 94, 161
Modulus of elasticity, 10, 91–92, 102, 104, 149, 163, 389
Mohr's circle, 88, 90, 100, 102, 103, 104, 188, 192–193, 197, 203–204, 205, 216, 226–227, 228–229, 238–240, 264, 265, 274, 281, 361

Mohr-Coulomb theory of failure, 187–189, 212, 222, 263, 271
Mohr's envelope (*see* Failure envelope)
Moisture-density curve, 13–16
Moisture content (*see* Water content)
Montmorillonite, 19, 20, 382, 396–398, 410, 413–416
Moraines, 375–377

N

Natural levees, 372, 374
Natural slopes, stability of, 318, 328, 341–343
Nonhomogeneous materials, flow through, 65–68, 78, 350–352
Normally consolidated clay, 111, 166, 167, 199-200, 213, 216, 322–323, 374, 378–379
Numerical solutions, 346–366

O

Octahedral sheet, 395–398
Organic soils, 26, 413–415
Outwash deposits, 371–372, 375–377, 380
Overconsolidated clay (*see* Preconsolidated clay)
Overconsolidation ratio, 196

P

Particle shape, 6–7, 198
Particle size, 2, 3, 39, 369, 375
Particle-size distribution, 3–6, 13, 27, 48, 198, 216, 375, 386
Pavement, 145, 147, 157
Peat, 369
Pedological classification, 26
Penetration resistance, 387–388, 391
Permeability, 34, 39–40, 46, 48–52, 78–79, 80, 124, 350–351, 353, 415

Permeability tests, 48–50, 78
Phreatic surface (*see* Seepage, line of)
Piezometer, 30, 80, 98, 158, 343
Piezometric level (*see* Piezometric surface)
Piezometric surface, 50, 59, 60, 61, 63, 64, 78, 98, 105, 158
Piston sampler, 387
Plane strain, 86, 87, 89, 134, 135, 136, 155, 212
Plane stress, 86, 134–135
Plastic flow, 94, 130, 187, 263–264, 267, 271, 299
Plastic limit, 10, 17, 19
Plastic potential, 264
Plasticity, 93, 131, 213, 221, 259–260, 263, 270, 271, 276, 294, 358
  chart, 20, 22
  index, 18, 19, 20, 22, 416
Point-bar deposits, 372, 374
Poiseuille's law, 36
Poisson's ratio, 91–92, 102, 161, 163, 389
Poreair pressure, 34, 41–42, 208, 210
Porewater pressure, 30–31, 33–34, 40–44, 45–46, 94–98, 102, 104–107, 115, 117, 119–120, 123, 128, 159, 185, 191, 193–203, 207, 208–210, 217, 220, 257–258, 292–293, 315–321, 325–327, 330–336, 342–343, 345, 352–358, 365–366
Porewater pressure coefficient (*see* Porewater pressure parameter)
Porewater pressure, excess hydrostatic, 94–98, 117, 158, 194, 200, 220, 316–321, 325–327, 330–336, 357–358
Porewater pressure parameter, 94–98, 102, 105, 158–159, 171, 195–196, 202, 317, 330, 331, 344
Porosity, 9, 11
Porous medium 35, 37
Potential function, 55
Preconsolidated clay, 111, 114, 166, 176, 196, 199–202, 208, 216,

Index

Preconsolidated clay (*cont.*)
    271, 305, 322–323, 326, 375
Preconsolidation pressure, 111, 112,
    113–114, 127
Pressure, artesian, 46
Pressure vs. void ratio relationship,
    110–114, 167, 216, 381, 410
Pressure on yielding base, 153–156,
    299–303
Primary consolidation, 121
Principal axes, 85, 90, 281
Principal plane, 85, 87, 103, 223,
    281, 300
Principal stress, 85, 87, 90, 93, 103,
    141, 142, 144, 157, 158,
    165, 167, 172, 188–189,
    192, 205–208, 211–212,
    223–224, 300, 317, 359,
    361
Pumping test, 50–51, 78, 390

Q

Quicksand, 44

R

Rankine's theory, 223–227, 280–281,
    285, 286, 294–295, 297, 306–308
Relative density, 17, 180–181, 380,
    387, 390–391
Remolded soil, 113, 336
Representative samples. 385–386, 389
Repulsive potential, 398–401, 405
Repulsive pressure, 400–403, 404, 409–
    411
Residual soils, 369, 381–382
Retaining walls, 280–281, 308, 309,
    328
Reynolds number, 78
Rock, decomposed, 382–384
Roundness, 6

S

Safety, factor of, 77, 82, 245, 251–256,
    272, 273, 314, 318, 320–323,
    328, 332–333, 337–338, 340,
    342–343, 390
Sampling, soil, 385–387, 390
Sand, 2–3, 8, 22–24, 78–79, 80, 97,
    105, 180, 370–372, 377–380,
    390
Sand drain, 119
Saturation, degree of, 9, 11, 12, 208–
    210
Secondary consolidation, 121, 124,
    179, 408
Sediments, 369–370, 372–378,
    404
Seepage, 43–44, 46, 48–84, 257–258,
    292–293, 320–321, 332, 334,
    348–352, 366
Seepage, line of, 67–69, 70, 72–74, 82
Seepage force, 61, 74–76, 78, 132
Sensitivity, 8
Settlement, 127, 128, 129, 147–150,
    150–152, 157, 160–186, 338,
    340–341, 344, 365, 381
    allowable, 173, 182, 338
    due to concentrated load, 147
    consolidation, 160, 163–167, 170–
        172, 173, 174, 177, 183
    differential, 173, 174, 183, 340–341
    of earth dam, 150–151
    field observations of, 174–182
    immediate, 160, 168, 170, 172,
        174, 179, 183
    of loaded area, 147–150
    rate of, 178–179
    of rigid foundation, 148
    tolerable (*see* Settlement, allowable)
Shale, 382–383
Shape factor, 37, 39, 236–237, 243
Shear, simple, 99
Shear failure, 94
Shear strength, 93, 94, 187–220, 230,
    254, 256, 258, 260, 292, 297,
    314–317, 320–332, 334, 336–
    337, 344, 374, 387–388, 411,
    413, 415–416
    of clays, 190, 199–212, 316–328,
        331, 333, 336–337
    of cohesionless soils, 190, 197–199,
        212, 214–215, 217–218, 329–330

Shear strength (*cont.*)
    drained, 189-190, 192-193, 203-205, 218, 297, 317, 327
    residual, 216, 328-330
    undrained, 10, 190-192, 203-205, 207, 210, 218, 297, 314, 316-317, 320-328, 330, 334, 337, 344, 415
Shear test, direct, 99-100, 189-192, 198, 218
Sheepsfoot rollers, 16
Sheet pile, 82, 293, 308-309, 348-350
Sheeting (*see* sheet pile)
Shrinkage limit, 17, 21, 27-28
Sieve analysis, 3-4
Sign convention, 135
Silica sheet, 395-398
Silt, 2-3, 8, 20, 22, 25, 26, 78, 80, 372, 377-378, 380
Size effect, 167-170
Slickenside, 216, 328
Slip line, 230-231, 264-267, 272, 300, 359-360 (*see also* Failure surface)
Slip surface (*see* Failure surface)
Slopes on clay, 317-323, 326-328
Soil classification, 1, 21-27, 385
Soil classification system:
    AASHO, 21-22, 26-27
    Unified, 22-26, 27
Soil horizons, 26
Soil map, 367-368
Soil pressure:
    allowable (*see* Soil pressure, design)
    design, 182, 184, 185, 245-246, 272, 337
Soundings, 162, 387-392
Specific gravity, 10, 11, 12
Sphericity, 6
Split-spoon sampler, 385-387, 390
Stability number, 251-253, 273, 327
Stability of slopes, 247-258, 273-275, 312-323, 326-327, 341-343, 344
Standard penetration test, 162, 180-182, 387-388, 391
Statically admissible state, 259-262
Stokes' law, 4
Strain distribution, 161-162
Strain:
    octahedral normal, 90
    octahedral shear, 90
    shear, 88, 89, 92, 93, 99, 104
    tensile, 152-153
    volumetric, 90, 92, 93, 104, 108, 111, 164
Strain rates, 260, 263-267, 271
Strains at a point, 88-90
Stream function, 55
Stress distribution, 142-145, 156, 173
Stress function, 135, 136, 150-152, 154-155
Stress history, 110
Stress:
    normal, 85, 87, 90, 144, 188, 192, 218, 231
    octahedral normal, 90, 92
    octahedral shear, 90, 92
    shear, 87, 88, 90, 94, 103, 141, 142, 144, 188-189, 192, 218, 302-303
    total, 41-42, 43, 196-197, 200, 202-203, 317, 327, 331, 334, 343, 344
Stress path, 161, 163, 165, 167, 177-178, 185, 189, 191, 192, 196, 200-202, 213, 325-327
Stress relaxation, 92-93, 305
Stress-strain curve (*see* Stress-strain relations)
Stress-strain relations, 90-94, 98-102, 125, 131-132, 187, 195, 221, 264, 380, 383
Stresses:
    at a point, 85-88
    due to concentrated load, 136-138, 291
    due to loaded area, 138-139
Structure:
    dispersed, 8, 405-407
    flocculent, 8, 405-406, 415
    single grained, 8
    soil, 7-8, 112, 381, 405-407, 412-413
Subgrade, 22, 28, 145, 147, 157
Subsoil exploration, 365, 384-392
Surface tension, 31, 32, 42
Swedish circle, 247
Swelling, 382-383, 409-411, 413-415

T

Taylor's series, 347
Thin-walled sampler, 386–387
Thixotropy, 408
Till, 375–377
Time-consolidation curves, 120, 126
Time factor, 119
Tortuosity, 39
Transfer conditions, 65–68
Triaxial test, 100–102, 104, 163, 185, 186, 192–196, 331
Tunnels, 303–305

U

Unconfined compression test, 204, 210, 323
Unconsolidated undrained test (see Undrained test)
Undisturbed samples, 162, 179, 385–387, 389
Undrained condition, 94, 95, 97, 102, 160–161, 163, 190, 195, 203, 207, 316, 321, 324–325, 327
Undrained state (see Undrained condition)
Undrained test, 193–194, 196–197, 204, 208, 219–220, 314
Uniformity coefficient, 6
Unit weight (see Density)
Uplift pressure, 87
Upper bound solutions, 258–263, 270–271, 274

V

Valley trains, 378
Van der Waal's forces, 395, 403, 412
Vane shear test, 323
Varved clays, 378
Velocity field, 263–270, 281
Velocity of displacement 260, 263–271
Velocity of flow, 34–39, 43, 56, 57, 59, 61–64, 82, 128, 350–351
Vemiculite, 396
Vertical cut, stability of, 206–208, 286–287
Viscoelasticity, 92–93, 125–126
Viscosity, 38, 92, 125, 408, 412
Void ratio, 9, 10, 11, 12, 212
Volume change, 107–114, 266, 268

W

Wash borings, 385
Water, adsorbed, 17, 107–408, 412
Water content, 9, 10, 381, 386–387
Water content, optimum, 13, 14
Water table, 30, 33, 42, 43, 379
Wave velocity, 389
Weathering, 381–382, 396
Weight-volume relationships, 9–13
Wells, 50, 60, 84, 390
Wind-blown sand, 379–380
Work, 259–263, 270, 403
Work hardening, 213–214

Y

Yield function, 264
Yield surface, 211–212
Yielding (see Plastic flow)
Young's modulus (see Modulus of elasticity)